Remote Sensing of Ocean and Coastal Environments

Remote Sensing of Ocean and Coastal Environments

Edited by

Meenu Rani

Department of Geography, Kumaun University,
Nainital, Uttarakhand, India

Kaliraj Seenipandi

Central Geomatics Laboratory (CGL),
National Center for Earth Science Studies (NCESS),
Thiruvananthapuram, Kerala, India

Sufia Rehman

Department of Geography, Faculty of Natural Sciences,
Jamia Millia Islamia, New Delhi, Delhi, India

Pavan Kumar

College of Horticulture and Forestry,
Rani Lakshmi Bai Central Agricultural University,
Jhansi, Uttar Pradesh, India

Haroon Sajjad

Department of Geography, Faculty of Natural Sciences,
Jamia Millia Islamia, New Delhi, Delhi, India

ELSEVIER

Elsevier
Radarweg 29, PO Box 211, 1000 AE Amsterdam, Netherlands
The Boulevard, Langford Lane, Kidlington, Oxford OX5 1GB, United Kingdom
50 Hampshire Street, 5th Floor, Cambridge, MA 02139, United States

Library of Congress Cataloging-in-Publication Data
A catalog record for this book is available from the Library of Congress

British Library Cataloguing-in-Publication Data
A catalogue record for this book is available from the British Library

ISBN: 978-0-12-819604-5

For information on all Elsevier publications visit our website at https://www.elsevier.com/books-and-journals

Publisher: Candice Janco
Acquisitions Editor: Louisa Munro
Editorial Project Manager: Sara Valentino
Production Project Manager: Bharatwaj Varatharajan
Cover Designer: Mark Rogers

Idea by Prem Pandey and Design by Mr. Kumar Krishna Mohan
Sr. Graphic Designer, Shiv Nadar University

Typeset by TNQ Technologies

Contents

Contributors

S.K. Aditya National Centre for Earth Science Studies (NCESS), Thiruvananthapuram, Kerala, India

R. Aneesh Kumar Environmental Technology Division, Council of Scientific & Industrial Research National Institute for Interdisciplinary Science and Technology (CSIR-NIIST), Thiruvananthapuram, Kerala, India

K. Anoop Krishnan National Centre for Earth Science Studies (NCESS), Thiruvananthapuram, Kerala, India

J. Ansari Environmental Technology Division, Council of Scientific & Industrial Research National Institute for Interdisciplinary Science and Technology (CSIR-NIIST), Thiruvananthapuram, Kerala, India; Academy of Scientific and Innovative Research (AcSIR), Ghaziabad, Uttar Pradesh, India

Trisha Chakraborty Department of Geography, Jadavpur University, Kolkata, West Bengal, India

N. Chandrasekar Centre for GeoTechnology, Manonmaniam Sundaranar University & Francis Xavier Engineering College, Tirunelveli, Tamil Nadu, India

B.S. Chaudhary Department of Geophysics, Kurukshetra University, Kurukshetra, Haryana, India

Debajit Datta Department of Geography, Jadavpur University, Kolkata, West Bengal, India

Prashant Ghadei Department of Geography, Central University of Tamil Nadu, Thiruvarur, Tamil Nadu, India

Kalita Himangshu Haryana Space Applications Centre (HARSAC), (Department of Science & Technology, Haryana) CCS HAU Campus, HISAR, Haryana, India

Daji Huang State Key Laboratory of Satellite Ocean Environment Dynamics, Second Institute of Oceanography, Ministry of Natural Resources, Hangzhou, China

K. Ibrahim-Bathis Department of Engineering, Politeknik Negeri Pontianak, Pontianak, West Kalimantan, Indonesia

Jeenu Jose National Centre for Earth Science Studies (NCESS), Thiruvananthapuram, Kerala, India

V. Kashyap Environmental Technology Division, Council of Scientific & Industrial Research National Institute for Interdisciplinary Science and Technology (CSIR-NIIST), Thiruvananthapuram, Kerala, India; Academy of Scientific and Innovative Research (AcSIR), Ghaziabad, Uttar Pradesh, India

A. Krishnakumar National Centre for Earth Science Studies (NCESS), Thiruvananthapuram, Kerala, India

Pavan Kumar College of Horticulture and Forestry, Rani Lakshmi Bai Central Agricultural University, Jhansi, Uttar Pradesh, India

Md. Masroor Department of Geography, Jamia Millia Islamia, New Delhi, Delhi, India

Shafique Matin Environment Protection Agency (EPA), Wexford, Ireland

M.A. Mohammed-Aslam Department of Geology, School of Earth Sciences, Central University of Karnataka, Gulbarga, Karnataka, India

Mrinmoyee Naskar Department of Geography, Jadavpur University, Kolkata, West Bengal, India; Department of Geography, Baruipur College, Baruipur, West Bengal, India

Sohini Neogy Department of Geography, Jadavpur University, Kolkata, West Bengal, India

M.K. Rafeeque National Centre for Earth Science Studies, Thiruvananthapuram, Kerala, India

K.K. Ramachandran National Centre for Earth Science Studies (NCESS), Thiruvananthapuram, Kerala, India

M. Rameshan National Centre for Earth Science Studies, Thiruvananthapuram, Kerala, India

Meenu Rani Department of Geography, Kumaun University, Nainital, Uttarakhand, India

Sufia Rehman Department of Geography, Faculty of Natural Sciences, Jamia Millia Islamia, New Delhi, Delhi, India

R.G. Rejith Minerals Section, Materials Science and Technology Division, National Institute for Interdisciplinary Science and Technology (CSIR-NIIST), Council of Scientific & Industrial Research, Thiruvananthapuram, Kerala, India; Academy of Scientific and Innovative Research (AcSIR), Ghaziabad, Uttar Pradesh, India

R.A. Renjith Minerals Section, Materials Science and Technology Division, National Institute for Interdisciplinary Science and Technology (CSIR-NIIST), Council of Scientific & Industrial Research, Thiruvananthapuram, Kerala, India

Asit Kumar Roy Department of Geography, Jadavpur University, Kolkata, West Bengal, India

P.M. Saharuba Environmental Technology Division, Council of Scientific & Industrial Research National Institute for Interdisciplinary Science and Technology (CSIR-NIIST), Thiruvananthapuram, Kerala, India

Haroon Sajjad Department of Geography, Faculty of Natural Sciences, Jamia Millia Islamia, New Delhi, Delhi, India

Praveen Sathee Sankar Department of Physics, St. Thomas College, Kozhencherry, Kerala, India

Sakhre Saurabh Environmental Technology Division, Council of Scientific & Industrial Research National Institute for Interdisciplinary Science and Technology (CSIR-NIIST), Thiruvananthapuram, Kerala, India; Academy of Scientific and Innovative Research (AcSIR), Ghaziabad, Uttar Pradesh, India

Kaliraj Seenipandi National Centre for Earth Science Studies (NCESS), Thiruvananthapuram, Kerala, India

Sulochana Shekhar Department of Geography, Central University of Tamil Nadu, Thiruvarur, Tamil Nadu, India

K. Shravanraj Minerals Section, Materials Science and Technology Division, National Institute for Interdisciplinary Science and Technology (CSIR-NIIST), Council of Scientific & Industrial Research, Thiruvananthapuram, Kerala, India

Ram Kumar Singh Department of Natural Resources, TERI School of Advanced Studies, New Delhi, India

M.K. Sreeraj National Centre for Earth Science Studies, Thiruvananthapuram, Kerala, India

M. Sundararajan Minerals Section, Materials Science and Technology Division, National Institute for Interdisciplinary Science and Technology (CSIR-NIIST), Council of Scientific & Industrial Research, Thiruvananthapuram, Kerala, India; Academy of Scientific and Innovative Research (AcSIR), Ghaziabad, Uttar Pradesh, India

S. Venkatesan Department of Geology, National College, Trichy, Tamil Nadu, India

M. Vrinda Department of Geology, Govt. College, Kasaragod, Kerala, India

Wenbin Yin State Key Laboratory of Satellite Ocean Environment Dynamics, Second Institute of Oceanography, Ministry of Natural Resources, Hangzhou, China

Biographies

Dr. Meenu Rani received her MTech degree in remote sensing from the Birla Institute of Technology, Ranchi, India. She is currently affiliated with the Department of Geography, Kumaun University, Nainital, Uttarakhand, India. She has worked on remote-sensing applications as a junior research fellow at the Haryana Space Applications Centre, as a research associate on the Indian Council of Agricultural Research, and at the G.B. Pant National Institute of Himalayan Environment and Sustainable Development. Dr. Rani has authored and coauthored several peer-reviewed scientific research papers and presented work at many national and international conferences, including in the United States, Italy, and China. She has been awarded various fellowships from the International Association for Ecology, Future Earth Coasts, and the Scientific Committee on Antarctic Research Scientific Research Programme. She was awarded an Early Career Scientist achievement in 2017 from Columbia University, New York, New York, USA.

Dr. Kaliraj Seenipandi is a scientist at the Central Geomatics Laboratory, National Centre for Earth Science Studies, Thiruvananthapuram. He received his MSc in remote sensing and geoinformation technology with a first from Madurai Kamaraj University, Madurai, and his MTech in geomatics from the Indian Institute of Surveying and Mapping, Survey of India, Hyderabad. He was awarded the DST-INSPIRE Fellowship (both JRF and SRF) for his PhD in remote sensing–geotechnology from the Centre for GeoTechnology, Manonmaniam Sundaranar University, Tirunelveli, Tamil Nadu, India. He has specialized in remote sensing, geoinformatics, and GIS modeling of earth and environmental processes. He has published over 30 research papers in the earth and environment fields and more than 25 proceedings in national and international conferences. He was awarded the Young Scientist of the Year 2016 award by the International Foundation for Environment and Ecology, Kolkata, in association with the Confederation of Indian Universities, New Delhi, and the Green Technologist of the Year 2017 award by the Scientific and Environmental Research Institute, New Delhi, in association with the Indian Institute of Ecology and Environment, New Delhi. His research interests are in the fields of remote sensing, geoinformatics, GIS modeling, earth and environmental dynamics, coastal vulnerability assessment, and natural resource monitoring and management.

Ms. Sufia Rehman is a doctoral candidate in the Department of Geography, Jamia Millia Islamia, New Delhi, India. She has completed her bachelor's in geography and subsequently obtained her master's degree in geography from Jamia Millia Islamia. She is the recipient of a Gold Medal in Master of Arts. She specializes in remote sensing and GIS and hydrological studies. Her areas of interest include coastal ecosystem conservation and management, climate change, and disaster management. She has made a remarkable contribution to water-related research in areas such as coastal landscape vulnerability and flood vulnerability. She has presented her research in national and international conferences. She has many research papers in journals of international repute and book chapters to her credit. Ms. Rehman has been awarded many scholarships from various agencies.

Dr. Pavan Kumar is a Faculty Member at the College of Horticulture and Forestry, Rani Lakshmi Bai Central Agricultural University, Jhansi, U.P., India. He obtained his PhD from the Faculty of Natural Sciences, Jamia Millia Islamia, New Delhi. He obtained his BSc (botany) and MSc (environmental science) from Banaras Hindu University, Varanasi, India, and subsequently obtained a master's degree in remote sensing (MTech) from the Birla Institute of Technology, Mesra Ranchi, India. His current research interests include climate change and coastal studies. He is the recipient of an Innovation China National Academy Award for Remote Sensing. Dr. Kumar has published 50 research papers in international journals and authored a number of books. He has visited countries such as the United States, France, the Netherlands, Italy, China, Indonesia, Brazil, and Malaysia for various academic and scientific assignments, workshops, and conferences. Dr. Kumar is a member of the International Association for Vegetation Science (France) and the Institution of Geospatial and Remote Sensing Malaysia.

Haroon Sajjad is Professor in the Department of Geography, Faculty of Natural Sciences, Jamia Millia Islamia, New Delhi, India. He obtained his B.Sc, M.Sc, M.Phil and Ph.D degrees all from Aligarh Muslim University, Aligarh, India. His present research interests include environmental management, sustainable development, watershed management and applications of remote sensing and GIS. He has four books to his credit. He has published more than hundred research papers in journals of repute. Prof. Sajjad has presented fifty research papers at national and international conferences including at Sapienza University of Rome, Italy, University of British Columbia, Canada, University of Western Cape, Bellville, South Africa, and University of Brighton, U.K. He has delivered invited talks at various universities. Ten doctoral degrees to the scholars have been awarded under his supervision. He has chaired academic sessions at various conferences. He is the reviewer of many scientific research journals and member of scientific bodies.

Foreword

It has been 60 years since the first low-earth orbital weather satellite, TIROS-1, was launched. To perform remote sensing, the satellite used just a simple TV camera pointed down from space to observe the earth. Despite the rudimentary data obtained from the satellite's slow-scan camera, the field of meteorology was revolutionized by the use of remote sensing data.

In the following years, remote sensing has increased in its breadth of applications, with satellites now being used to guide decisions and inform stakeholders in subjects that range from coastal ecosystem monitoring to hazard mitigation and land use and management. The data provided by remote-sensing platforms will continue to serve a critical role in guiding scientists, consultants, engineers, environmental managers, and policy makers as we navigate the challenges presented by the changing climate.

Remote Sensing of Ocean and Coastal Environments provides a robust foundation for all who strive to understand the theory and processes behind remote sensing, and familiarizes the reader with the current state-of-the-art methodologies in the application of remote sensing to ecological, economic, and risk management problems. This work was made possible, in part, by the strong support of CLIVAR (Climate and Ocean: Variability, Predictability, and Change), one of the four core projects of the World Climate Research Programme (WCRP). Through CLIVAR's commitment to knowledge transfer, education, capacity building, and outreach—notably through the establishment of international conferences and workshops, bringing together senior scientists and early career researchers—the organization has facilitated collaboration and positively influenced many careers, including my own.

Understanding the humble origins of our field, and looking forward to a bright future guided by technological innovations, we scientists in the remote sensing community take up the mantle of responsibility handed to us by past generations. I believe that it is the duty of all of those who work in the earth sciences to understand how to maximize the tools at their disposal for the well-being of our planet. And there are few tools in our scientific arsenal as powerful as the fleets of watchful eyes in the sky.

Dr. Noel C. Baker
ALTIUS and PICASSO satellite missions, Royal Belgian Institute for Space Aeronomy (BIRA-IASB), Brussels, Belgium

1

Remote sensing of Ocean and Coastal Environment — Overview

Meenu Rani[1], Md. Masroor[2], Pavan Kumar[3]

[1]DEPARTMENT OF GEOGRAPHY, KUMAUN UNIVERSITY, NAINITAL, UTTARAKHAND, INDIA;
[2]DEPARTMENT OF GEOGRAPHY, JAMIA MILLIA ISLAMIA, NEW DELHI, DELHI, INDIA;
[3]COLLEGE OF HORTICULTURE AND FORESTRY, RANI LAKSHMI BAI CENTRAL AGRICULTURAL UNIVERSITY, JHANSI, UTTAR PRADESH, INDIA

Chapter outline

1. Introduction

Satellite imagery has greatly contributed to the mapping of coastal ecosystems and provided estimates of land coverage and alteration in the coastal ecosystem (Tripathy et al., 2018; Burke et al., 2001; Hochberg et al., 2003). Advances in the design of sensors and data analysis techniques make remote sensing systems more practical and desirable for use in coastal ecosystem management, such as estuaries, wetlands, and coral reefs (Held et al., 2003; Conchedda et al., 2008; Lyons et al., 2012; Rani et al., 2018). Multispectral and hyperspectral sensors are used to monitor coastal land cover, coastal water dissolved substances, and biotic/abiotic suspended particle concentrations. Coastal ecosystems are highly complex in terms of geography. To validate and accurately

Remote Sensing of Ocean and Coastal Environments. https://doi.org/10.1016/B978-0-12-819604-5.00001-9

measure the remotely sensed information, an effective field data collection sampling technique is required using ships, navigation marks, and field instruments. Ocean remote sensing is mainly concerned with collecting and interpreting information from a remote point of view on coast, sea, land, and atmosphere. The remote sensing platforms can range from towers above Earth's surface to aircraft at low and medium altitudes and satellites in space, depending on the requirements of resolution and cost limitations (Klemas, 2015; Colomina and de la Tecnologia, 2008; Watts et al., 2012). Previously, satellites provided wide spatial coverage, consistent revisits, and multispectral images for coastal environment transformation analyses, but lacked high spatial resolution required for different uses. High-resolution satellite images are now available, but the high resolution and regular, flexible overflights provided by aerial sensors are more appropriate for a variety of applications, such as coastline delineation, land use–land cover change analyses, wetland mapping, and oil slick tracking (Klemas et al., 1993; Klemas, 2015). Using large-scale mapping and simulation of applications, satellite images can be combined with global positioning system (GPS) location and used as layers in geographic information systems (GIS). Unmanned aerial platforms are cheaper in comparison to manned aircraft platforms. GPS-controlled unmanned aerial vehicles (UAVs) are capable of obtaining very high-resolution images of particular landscape features with revisit times set by the pilot. In coastal environmental studies, UAVs like drones, balloons, blimps, and quadcopters are now being used effectively (Eisenbeiss and Sauerbier, 2011). Table 1.1 summarizes the specifications of all known past, current, and future satellites that have been widely used in marine applications.

Coastal regions will be affected by changing atmospheric and ocean temperatures, weather patterns, sea-level rise, ocean chemistry, and rising demands from a rapidly increasing global population over the next few decades. Without appropriate resource management plans, these changes can lead to increased threats to ecosystem services, human health, and land and economic prosperity. The aim of this chapter is to familiarize the reader with different techniques, advantages, and challenges related to coastal remote sensing applications.

2. Satellite remote sensing

Satellite observations of thermal infrared ocean currents help scientists and other users obtain real time data of current in the oceans (Schwab and Bedford, 1994; Robinson, 2004). High-resolution satellite measurements of sea surface temperature (SST) are suitable for analyzing western boundary currents such as the Kuroshio and the Gulf Stream with broad temporal and spatial displacements. Global change, accurate, precise, and long-term observations of SST are also relevant for large-scale studies (Kwon et al., 2010; Lee and Cornillon, 1995; Schmitz and Holland, 1986). Fish and wildlife communities used SST data to monitor marine habitats in many parts of the world.

Table 1.1 Summary of satellite systems, programs, and payload products commonly used in marine remote sensing.

Satellite system	Launch	Expected (EOL)	Satellite program	Agencies	Orbit	Altitude (km)	Satellite status	Payload	Last update
ADEOS	1996-08-17	1997-06-30	Advanced earth observing satellite (original name: Midori)	JAXA	SunSync	797	Inactive	AVNIR ILAS-I IMG NSCAT OCTS POLDER RIS TOMS	2015-07-27 19:41:38
ADEOS-2	2002-12-14	2003-10-25	Advanced earth observing satellite (original name: Midori)	JAXA	SunSync	812	Inactive	AMSR DCS (ADEOS) GLI ILAS-II POLDER SeaWinds	2015-07-27 19:42:57
COMPIRA	≥2020	≥2025	Coastal and ocean measurement mission with precise and innovative radar altimeter	JAXA	DRIFT	TBD	Planned	Altimeter (COMPIRA) SHIOSAI	2019-12-30 09:09:57
COMS	2010-06-26	≥2020	Communication, oceanography, and meteorology satellite	KMA KARI ME MLTM	GEO	35,786	Operational	GOCI MI	2019-12-28 15:48:45
ERS-1	1991-07-17	2000-03-10	European remote-sensing satellite	ESA	SunSync	785	Inactive	AMI-SAR AMI-SCAT ATSR LRR (ESA) PRARE RA	2019-10-30 02:17:06
ERS-2	1995-04-21	2011-07-06	European remote-sensing satellite	ESA	SunSync	785	Inactive	AMI-SAR AMI-SCAT ATSR-2 GOME LRR (ESA) MWR (Envisat) PRARE RA	2019-10-30 02:22:16

Continued

Table 1.1 Summary of satellite systems, programs, and payload products commonly used in marine remote sensing.—cont'd

Satellite system	Launch	Expected (EOL)	Satellite program	Agencies	Orbit	Altitude (km)	Satellite status	Payload	Last update
GEO-KOMPSAT-2A	2018-12-04	≥2029	Communication, oceanography and meteorology satellite	KMA KARI ME MLTM	GEO	35,786	Operational	AMI KSEM/CM KSEM/MG KSEM/PD	2019-08-02 13:32:54
GEO-KOMPSAT-2B	≥2020	≥2031	Communication, oceanography, and meteorology satellite	KMA KARI ME MLTM	GEO	35,786	Planned	GEMS GOCI-II	2019-12-30 08:53:18
IRS-P3	1996-03-21	2004-10-15	Indian remote-sensing satellite	ISRO	SunSync	817	Inactive	IXAE MOS WiFS (IRS-P3)	2015-07-28 12:39:15
NOAA-6	1979-06-27	1987-03-31	National Oceanic and Atmospheric administration—4th generation	NOAA NASA	SunSync	840	Inactive	AVHRR Argos HIRS/2 MSU SEM/MEPED SEM/TED SSU	2015-07-28 20:38:17
NOAA-7	1981-06-23	1986-06-07	National Oceanic and Atmospheric Administration—4th generation	NOAA NASA	SunSync	860	Inactive	AVHRR/2 Argos HIRS/2 MSU SEM/MEPED SEM/TED SSU	2015-07-28 20:39:21
RASAT	2011-08-17	≥2020	Remote sensing satellite	TÜBITAK-UZAY	SunSync	690	Unclear	OIS	2019-12-29 00:32:09
SWOT	≥2021	≥2024	Surface water and ocean topography	NASA CNES	DRIFT	891	Planned	Altimeter KaRIN MW radiometer	2019-10-30 19:59:13

TIROS-N	1978-10-13	1981-02-27	National Oceanic and Atmospheric Administration—4th generation	NOAA NASA	SunSync	850	Inactive	AVHRR Argos HIRS/2 MSU SEM/MEPED SEM/TED SSU	2015-07-29 13:09:18
Beijing-1	2005-10-27	2010	Beijing	NRSCC	SunSync	699	Presumably inactive	CMT SLIM6	2019-12-27 20:34:43
CFOSAT	2018-10-29	≥2022	Chinese-French oceanography satellite	CNSA CNES	SunSync	500	Operational	SCAT (CFOSAT) SWIM	2019-10-30 18:50:25
CFOSAT follow-on	≥2022	≥2027	Chinese-French oceanography satellite	CNSA CNES	SunSync	500	Considered	SCAT (CFOSAT) SWIM	2019-10-30 18:54:20
MOS-1	1987-02-19	1995-11-29	Marine observatory satellite	JAXA	SunSync	908	Inactive	MESSR MSR VTIR	2015-07-28 19:53:25
MOS-1B	1990-02-07	1996-04-25	Marine observatory satellite	JAXA	SunSync	908	Inactive	MESSR MSR VTIR	2013-02-22 17:12:36
OceanSat-1 (IRS-P4)	1999-05-26	2010-08-08	Satellite for the ocean	ISRO	SunSync	723	Inactive	MSMR OCM (OceanSat-1)	2019-10-29 15:53:58
OceanSat-2	2009-09-23	≥2020	Satellite for the ocean	ISRO	SunSync	730	Operational	OCM (OceanSat-2) OSCAT ROSA (OceanSat)	2020-01-03 17:26:50
OceanSat-3	≥2020	≥2025	Satellite for the ocean	ISRO	SunSync	723	Planned	A-DCS OCM (OceanSat-3) OSCAT SSTM	2019-12-30 08:59:26
OceanSat-3A	≥2020	≥2025	Satellite for the ocean	ISRO	SunSync	723	Planned	OCM (OceanSat-3) OSCAT SSTM	2019-12-30 09:27:05
ScatSat-1	2016-09-26	≥2021	Satellite for the ocean	ISRO	SunSync	732	Operational	OSCAT	2020-01-03 17:28:26
SMOS	2009-11-02	≥2020	Soil moisture and ocean salinity	ESA CDTI CNES	SunSync	755	Operational	GPS (ESA) MIRAS STA	2019-12-29 15:43:13

Thermal infrared remote sensing was first used by oceanography and meteorological societies to gain widespread acceptance (Dzwonkowski et al., 2014; Stammer et al., 2002). Thermal infrared sensors have been mounted on fully operational meteorological satellites for over 40 years to provide cloud-high temperature images; when no clouds occur, they observe SST variation. Thermal infrared instruments used to derive SST include Advanced Very High-Resolution Radiometer (AVHRR) of National Oceanic and Atmospheric Administration (NOAA) Polar-orbiting Environmental Operational Satellites, moderate resolution imaging spectro-radiometer aboard National Aeronautics and Space Administration (NASA) Earth Observing System Terra and Aqua satellites, the geostationary operational environmental satellite imagery and long-range scanning radiometer on the European Remote Sensing Satellite (ERS-2) (Gentemann et al., 2003; Purkis and Klemas, 2011; Samberg, 2007; Cracknell and Hayes, 2007; Klemas, 2011). Table 1.2 shows the oceanographic satellite database and which frequencies are used for transmitting data for microwave active or passive remote sensing.

2.1 Oil spill

Large scale oil spills could destroy wetland, marine life, and estuarine animal ecosystems. In order to mitigate the damage caused by a spill and enhance prevention and clean-up efforts, shipping carriers, oil companies, and other accountable authorities must immediately obtain information on the source of the leak, size, and nature of the spill; speed and direction of oil movement; current, wave, and wind information to predict future oil flow and dispersion (Klemas, 2010; Nwilo and Badejo, 2006; Kennish, 2002). Most large oil spills in the oceans were caused due to tanker landings, crashes, and break-ups resulting in a high proportion of oil floating over the ocean surface and endangering the aquatic and coastal ecosystem. Many times, remote sensors observed data to track and forecast oil direction and possible movement (Blumer et al., 1971). These observed data helped direct recovery and preventive measures, including protective booms and shipments of skimming vessels. Users of remote sensing oil spill tracking data include the Coast Guard, oil companies, environmental agencies, and shipping/fishing/insurance and defense agencies (Klemas, 2010). The key operating data requirements for oil spill incidents include regular site images to track the spill dynamics. Sensors of satellites and aircrafts meet these requirements and provide multitemporal images on different resolutions at high frequency intervals to monitor oil spills. These sensors also provide key inputs for modeling of drift prediction and controlling activities. Most of these sensors use electromagnetic waves. Oil fluoresces in the ultraviolet field seem to have substantially higher reflectance. However, ultraviolet light is easily absorbed in the atmosphere and can only be used on low-altitude aircraft to avoid this scattering. Sun glint, hydrothermal materials, and wind slicks can also confuse ultraviolet sensors. To reduce this uncertainty, sensors are used in conjunction with other thermal infrared sensors and radar. The development of low-cost digital cameras

Table 1.2 All frequencies used for transmitting data to and from oceanography satellites or for microwave active or passive remote sensing.

Satellite	Space agency	Launch	Expected (EOL)	Service	Direction	Polarization	Comment
Envisat	ESA	2002-03-01	2012-04-08	RA-2	Active	–	S-band radar altimeter
ERS-1	ESA	1991-07-17	2000-03-10	AMI-SCAT	Active	–	C-band scatterometer
ERS-2	ESA	1995-04-21	2011-07-06	AMI-SCAT	Active	–	C-band scatterometer
GOES-10	NOAA	1997-04-25	2006-12-01	SDL	S–E	Linear	Raw data
GOES-11	NOAA	2000-05-03	2011-12-05	PDR/GVAR	E–S	Linear	Processed images/soundings
GOES-12	NOAA	2001-07-23	2010-05-10	PDR/GVAR	E–S	Linear	Processed images/soundings
GOES-13	NOAA	2006-05-24	≥2020	PDR	E–S	Linear	Processed images/soundings
GOES-14	NOAA	2009-06-27	≥2020	PDR	E–S	Linear	Processed images/soundings
GOES-15	NOAA	2010-03-04	≥2020	PDR	E–S	Linear	Processed images/soundings
GOES-16	NOAA	2016-11-19	≥2027	DCPR	E–S	Linear	DCP reports
GOES-17	NOAA	2018-03-01	≥2029	GRB	E–S	RHCP and LHCP	Processed images/soundings
GOES-8	NOAA	1994-04-13	2004-05-05	PDR/GVAR	S–E	Linear	Processed images/soundings
GOES-9	NOAA	1995-05-23	2003-05-22	PDR/GVAR	S–E	Linear	Processed images/soundings
GOES-T	NOAA	≥2020	≥2031	DCPC	E–S	RHCP	DCP command
JASON-1	NASA	2001-12-07	2013-07-01	DORIS	E–S	RHCP	Precise positioning by DORIS
JASON-2	NASA	2008-06-20	2019-10-01	TM and raw data	S–E	RHCP and LHCP	Telemetry and altimeter real-time data
JASON-3	NASA	2016-01-17	≥2021	TM and raw data	S–E	RHCP and LHCP	Telemetry and altimeter real-time data
KOMPSAT-5	KARI	2013-08-22	≥2020	CORI	Active	–	X-band SAR
Meteosat-10	EUMETSAT	2012-07-05	≥2024	HRIT	E–S	RHCP	High resolution data
Meteosat-11	EUMETSAT	2015-07-15	≥2024	HRIT	S–E	Linear	High resolution data (not in use)
Meteosat-4	EUMETSAT	1989-03-06	1995-11-08	HRI	S–E	Linear	High resolution data
Meteosat-5	EUMETSAT	1991-03-02	1998-06-01	HRI	S–E	Linear	High resolution data
Meteosat-6	EUMETSAT	1993-11-20	2007-04-27	HRI	E–S	RHCP	High resolution data
Meteosat-7	EUMETSAT	1997-09-02	2006-12-05	HRI	S–E	Linear	High resolution data

Continued

Table 1.2 All frequencies used for transmitting data to and from oceanography satellites or for microwave active or passive remote sensing.—cont'd

Satellite	Space agency	Launch	Expected (EOL)	Service	Direction	Polarization	Comment
Meteosat-8	EUMETSAT	2002-08-28	2016-07-04	HRIT	S–E	Linear	High resolution data (not in use)
Meteosat-9	EUMETSAT	2005-12-21	≥2024	HRIT	S–E	Linear	High resolution data (not in use)
NOAA-15	NOAA	1998-05-13	≥2020	HRPT	S–E	RHCP	Full information data
NOAA-16	NOAA	2000-09-21	2014-06-09	HRPT	S–E	LHCP	Full information data
NOAA-18	NOAA	2005-05-20	≥2020	Command	E–S	RHCP	Commands
NOAA-19	NOAA	2009-02-06	≥2020	HRPT	S–E	RHCP	Full information data
NOAA-20	NOAA	2017-11-18	≥2024	TT & C	S–E	RHCP	Telemetry and commands
OceanSat-1 (IRS-P4)	ISRO	1999-05-26	2010-08-08	MSMR	Passive	H andV	Window channel
OceanSat-2	ISRO	2009-09-23	≥2020	OSCAT	Active	–	Ku-band scatterometer
OceanSat-3	ISRO	≥2020	≥2025	OSCAT	Active	–	Ku-band scatterometer
OceanSat-3A	ISRO	≥2020	≥2025	OSCAT	Active	–	Ku-band scatterometer
TRMM	NASA	1997-11-27	2015-04-08	PR	Active	–	Ku-band precipitation radar

on aircraft and multispectral sensors on satellites are widely used and their visible wavelengths also have a reasonable atmospheric transmission window. Oil has higher reflectivity than water in the visible region, and can be observed even more easily by using horizontally spaced filters.

2.2 Current velocity extraction and mapping

Feature-tracking sequential satellite images are used to track the displacement of specified ocean features such as areas of different water temperatures, surface slicks, and chlorophyll plumes over time between consecutive images to quantify surface flow. AVHRR thermal infrared images, Synthetic Aperture Radar (SAR) radar images, and sea-viewing large field-of-view (SeaWiFS) ocean-color images were all used for current velocity (Kuo and Yan, 1994; Liu et al., 2006). Satellite feature-tracking techniques were used to measure water currents in areas like the Gulf Stream, the Gulf of Mexico, Ireland's west coast, California Current, Kuroshio Current, and New Zealand's coastline.

A major drawback of this technique is that the cloud cover also obscures the surface features of visible and thermal infrared oceans (Breaker et al., 1994; Romeiser and Runge, 2007). A more contemporary technique, using interferometric SAR (InSAR) tracking, allows improved spatial resolution and efficiency to receive data from satellites anywhere in the world. This enables the imaging of line-of-sight surface velocity fields with SARs spatial resolution of the order of meters for satellites within a wide range of tens to hundreds of kilometers (Monaldo et al., 2003; Tarquis Alfonso et al., 2014).

2.3 Sea surface temperature

For a wide range of oceanography research, accurate large-scale, long-term SST measurements are significant. Satellite-derived high-resolution SST observations are suitable for monitoring boundary currents such as the Canary and the Gulf Stream, which exhibit displacements on a large scale. Long-term series of reliable, global SSTs are required to assess the health of coral reefs, supporting a wide range of marine diversity. The thermal infrared red radiance measured over all the oceans varies mainly with SST, which makes it easy to accurately determine the SST if certain atmospheric corrections are included. After the launch of AVHRR on NOAA- 7, infrared satellite SST observations have continued for almost 3 decades and have contributed to global climate studies, weather forecasting, physical oceanography research, and regional support of ship routing and fishing (Nagaraja Rao et al., 1989; Llewellyn-Jones et al., 1984; Riegl and Purkis, 2012). Spatiotemporal studies include variation in SST pattern related to interannual climate phenomena such as El Niño and La Niña cycles in the equatorial region of the Pacific and Atlantic Oceans (Espinoza Villar et al., 2009; Liu et al., 2005). In coastal upwelling studies, where increasing cold water carries nutrients to the sea surface, encouraging phytoplankton and zooplankton to attract and grow large concentrations of fish is another important application of SST sensing (Bell et al., 2004; Tozzi et al., 2004; Yan, 1993).

2.4 Sea surface salinity

Sea surface salinity (SSS) is crucial to estimate global water balance and evaporation rates, and to understand currents. Furthermore, low-salinity water is indicative sources of fresh water, such as rivers and streams that are feeding the ocean. Such rivers often carry natural and anthropogenic contamination from the land to the sea and can directly impact marine ecosystems with a higher level of salinity (Schroeder et al., 2012; Burrage et al., 2008). Airborne microwave radiometers are capable of measuring sea surface salinity and are widely used for various applications. The power receiving radiometer antennae in microwave radiometry is proportional to emissivity of microwave and ocean surface temperature. Salinity is measured as parts per 1000 (ppt) and average salinity of the ocean is 35 ppt (Droppleman et al., 1970). This means that the dissolved salt occurs at a concentration of 35 ppt or 3.5% with the remaining 96.5% water molecules. The salinity of the sea surface was the most important oceanic parameter that was not measured from satellites. However, newly advanced instruments designed to help SSS from satellites are available now. For example, a fixed two-dimensional interferometric antenna is used by the European Soil Moisture and Ocean Salinity satellite. The satellite sensor can detect salinity with high accuracy up to 50 km spatial resolution (Bell et al., 2004; Glenn et al., 2004; Yan, 1993).

2.5 Coastal currents

Ocean flow is determined by various physical factors such as wind friction, tides, and ocean density while atmospheric circulation, including winds, are caused by convection due to latitudinal temperature variation and the influence of Earth's Coriolis effect (Catry et al., 2009). Oceanic currents are considered as water movement from one place to the next and measured in knots or meters per second (Brill et al., 1993). Air movement primarily induces surface currents as the wind passes over the water. Currents have speed of three–four knots in the influence of generating wind speed because the main ocean currents cover a large distance. The Coriolis movement deflects the currents to turn right in the northern hemisphere, forcing them to pass in circular and gyres patterns in the clockwise direction while counterclockwise in the south. Major ocean currents of world includes the Humboldt, Kuroshio, Oyashio, Alaska, and California currents in the Pacific Ocean; Brazil, Benguela, Canary, Gulf Stream, and Labrador currents in the Atlantic Ocean; and Agulhas, Muzambique (Cox, 1975; Carton et al., 2000; Lebreton et al., 2012), and Somali currents in the Indian Ocean (Shankar et al., 2002; McCreary et al., 1993; Wyrtki, 1973). The most significant offshore and ocean currents are wind generated, wave-driven, tidal currents and buoyant channel plumes. These local currents may be short-term (daily) or long-term (seasonal) according to duration. Difference in ocean density depends on temperature and salinity. Warm water is less dense and steps up to the surface whereas colder, salt-laden water goes down (Vachon et al., 1995). Deep ocean currents cannot be detected using airborne or satellite remote sensors and are considered a limitation of remote sensing in the studies of coastal and ocean remote sensing (Tralli et al., 2005).

3. Geographical information system

Although digital elevation model (DEM) visualizations are helpful for data interpretation and analysis, a key issue remains how to provide complete DEM to the coastal zone user in a portable digital form (useable in GIS) that maximizes available data resolutions (Yin et al., 2012; Marfai and King, 2008; Mills et al., 2005). This is particularly important as recent data are usually available in the form of high spatial resolution than the previous data (NOAA and US Geological Survey) used to create the basic DEM. However, these new datasets must not be forced to be gridded down to lower resolutions just to fit in with the DEM. The vertical accuracy of the DEM varies spatially due to the wide range of temporal datasets and data collection techniques used in the acquisition of source data (Parker, 2002). Merged uniformly spaced grid cell models are useful to handle such complexity.

4. Fundamental techniques

Oceanographers and coastal experts recognize two fundamental techniques for coastal and offshore recent observations, Eulerian and Lagrangian methods. Eulerian methods calculate the rate of water flow in the ocean past a point while the Lagrangian method measures the shifting of water parcels in the ocean by tracking the location of chemical tracers or subsurface drifters (Morang and Gorman, 2005).

4.1 Eulerian technique

The Eulerian technique generally includes current meters on buoy moorings set to the ocean bottom and record currents on different depths at a particular location. Arrays of these moorings with current meters on different depths are used in marine water for a day to a month to track currents at particular locations such as harbor channel entrances and tidal inlets (Gorman et al., 1998; Morang et al., 1997a,b). The impeller current meters directed by vane are used to track ocean currents, calculating existing wind speed in the current. Thus, vane rotation rate is related to the ocean current speed. Current speed and direction calculations are stored in the memory of a computer chip (Webb, 1996; Bleck et al., 1995). A sound pulse can be used to recover the current meter that triggers an acoustic connection that removes from the anchor the cable and instrument kit.

4.2 Lagrangian technique

Lagrangian methods include releasing ocean drifters, which are record acoustically, visually, or through radio waves. The drifters are designed to move with the flowing water, helping researchers determine the current velocity and direction. Usually, Lagrangian current measuring methods are used in assessment of water movement, emissions control, ice flow modeling, and current global ocean research. Ocean drifters are primarily designed to detect water current flow at different depths (Mendoza and Mancho, 2010; Lehahn et al., 2007; Molcard et al., 2005). A common configuration of such Lagrangian drifters involves a float or a float connected to a current drogue by a cable. The drogue fix for a particular depth acts as an underwater sail as the ocean current pushes it. The drogue surface ensures that instead of being moved by the wind, drifters track the movement of water. The surface float supplies buoyancy, and contains recorded digital and satellite data. Satellites determine the location of drifter from the transmitting signal and send data to ground stations where the position of drift is fixed. Surface float communications with GPS satellites can also acoustically track drifting of subsurface buoys to determine exact positions.

References

Bell, M.J., Martin, M.J., Nichols, N.K., 2004. Assimilation of data into an ocean model with systematic errors near the equator. Quart. J. R. Meteorol. Soc. J. Atmos. Sci. Appl. Meteorol. & Phys. Oceanogr. 130 (598), 873–893.

Bleck, R., Dean, S., O'Keefe, M., Sawdey, A., 1995. A comparison of data-parallel and message-passing versions of the miami isopycnic coordinate ocean model (MICOM). Parallel Comput. 21 (10), 1695–1720.

Blumer, M., Sanders, H.L., Grassle, J.F., Hampson, G.R., 1971. An ocean of oil: a small oil spill. Environment 13 (2), 2–12.

Breaker, L.C., Krasnopolsky, V.M., Rao, D.B., Yan, X.H., 1994. The feasibility of estimating ocean surface currents on an operational basis using satellite feature tracking methods. Bull. Am. Meteorol. Soc. 75 (11), 2085–2096.

Brill, R.W., Holts, D.B., Chang, R.K.C., Sullivan, S., Dewar, H., Carey, F.G., 1993. Vertical and horizontal movements of striped marlin (*Tetrapturus audax*) near the Hawaiian Islands, determined by ultrasonic telemetry, with simultaneous measurement of oceanic currents. Mar. Biol. 117 (4), 567–574.

Burke, L., Kura, Y., Kassem, K., Revenga, C., Spalding, M., McAllister, D., Caddy, J., 2001. Coastal Ecosystems. World Resources Institute, Washington, DC.

Burrage, D., Wesson, J., Miller, J., 2008. Deriving sea surface salinity and density variations from satellite and aircraft microwave radiometer measurements: application to coastal plumes using STARRS. IEEE Trans. Geosci. Rem. Sens. 46 (3), 765–785.

Carton, J.A., Chepurin, G., Cao, X., 2000. A simple ocean data assimilation analysis of the global upper ocean 1950–95. Part II: Results. J. Phys. Oceanogr. 30 (2), 311–326.

Catry, T., Ramos, J.A., Jaquemet, S., Faulquier, L., Berlincourt, M., Hauselmann, A., et al., 2009. Comparative foraging ecology of a tropical seabird community of the Seychelles, western Indian Ocean. Mar. Ecol. Prog. Ser. 374, 259–272.

Colomina, I., de la Tecnologia, P.M., July 2008. Towards a new paradigm for high-resolution low-cost photogrammetryand remote sensing. In: The International Archives of the Photogrammetry, Remote Sensing and Spatial Information Sciences, ISPRS Congress, Beijing, China, XXXVII. Part B, vol. 1, pp. 1201–1206.

Conchedda, G., Durieux, L., Mayaux, P., 2008. An object-based method for mapping and change analysis in mangrove ecosystems. ISPRS J. Photogramm. Remote Sens. 63 (5), 578–589.

Cox, M.D., 1975. A baroclinic numerical model of the world ocean: preliminary results. Numer. Model. Ocean Circ. 107–120.

Cracknell, A.P., Hayes, L., 2007, Second edn. Introduction to remote sensing. Taylor and Fancis Group, Boca Raton, FL, ISBN 13:9780850664096.

Droppleman, J.D., Mennella, R.A., Evans, D.E., 1970. An airborne measurement of the salinity variations of the Mississippi River outflow. J. Geophys. Res. 75 (30), 5909–5913.

Dzwonkowski, B., Park, K., Lee, J., Webb, B.M., Valle-Levinson, A., 2014. Spatial variability of flow over a river-influenced inner shelf in coastal Alabama during spring. Continent. Shelf Res. 74, 25–34.

Eisenbeiss, H., Sauerbier, M., 2011. Investigation of UAV systems and flight modes for photogrammetric applications. Photogramm. Rec. 26 (136), 400–421.

Espinoza Villar, J.C., Ronchail, J., Guyot, J.L., Cochonneau, G., Naziano, F., Lavado, W., et al., 2009. Spatio-temporal rainfall variability in the Amazon basin countries (Brazil, Peru, Bolivia, Colombia, and Ecuador). Int. J. Climatol. J. R. Meteorol. Soc. 29 (11), 1574–1594.

Gentemann, C.L., Donlon, C.J., Stuart-Menteth, A., Wentz, F.J., 2003. Diurnal signals in satellite sea surface temperature measurements. Geophys. Res. Lett. 30 (3).

Glenn, S., Schofield, O., Dickey, T.D., Chant, R., Kohut, J., Barrier, H., et al., 2004. The expanding role of ocean color and optics in the changing field of operational oceanography. Oceanography 17 (2), 86–95.

Gorman, L.T., Morang, A., Larson, R.L., 1998. Monitoring the coastal environment; Part IV: mapping, shoreline change, and bathymetric analysis. J. Coast Res. 14 (1), 61–92.

Held, A., Ticehurst, C., Lymburner, L., Williams, N., 2003. High resolution mapping of tropical mangrove ecosystems using hyperspectral and radar remote sensing. Int. J. Rem. Sens. 24 (13), 2739–2759.

Hochberg, E.J., Andréfouët, S., Tyler, M.R., 2003. Sea surface correction of high spatial resolution Ikonos images to improve bottom mapping in near-shore environments. IEEE Trans. Geosci. & Remote Sens. 41 (7), 1724–1729.

Kennish, M.J., 2002. Environmental threats and environmental future of estuaries. Environ. Conserv. 29 (1), 78–107.

Klemas, V., 2010. Tracking oil slicks and predicting their trajectories using remote sensors and models: case studies of the Sea Princess and Deepwater Horizon oil spills. J. Coast Res. 789–797.

Klemas, V., 2011. Beach profiling and LIDAR bathymetry: an overview with case studies. J. Coast Res. 27 (6), 1019–1028.

Klemas, V.V., 2015. Coastal and environmental remote sensing from unmanned aerial vehicles: an overview. J. Coast Res. 31 (5), 1260–1267.

Klemas, V.V., Dobson, J.E., Ferguson, R.L., Haddad, K.D., 1993. A coastal land cover classification system for the NOAA Coastwatch Change Analysis Project. J. Coast Res. 862–872.

Kuo, N.J., Yan, X.H., 1994. Using the shape-matching method to compute sea-surface velocities from AVHRR satellite images. IEEE Trans. Geosci. Rem. Sens. 32 (3), 724–728.

Kwon, Y.O., Alexander, M.A., Bond, N.A., Frankignoul, C., Nakamura, H., Qiu, B., Thompson, L.A., 2010. Role of the Gulf Stream and Kuroshio–Oyashio systems in large-scale atmosphere–ocean interaction: a review. J. Clim. 23 (12), 3249–3281.

Lebreton, L.M., Greer, S.D., Borrero, J.C., 2012. Numerical modelling of floating debris in the world's oceans. Mar. Pollut. Bull. 64 (3), 653–661.

Lee, T., Cornillon, P., 1995. Temporal variation of meandering intensity and domain-wide lateral oscillations of the gulf stream. J. Geophys. Res. Oceans 100 (C7), 13603–13613.

Lehahn, Y., d'Ovidio, F., Lévy, M., Heifetz, E., 2007. Stirring of the northeast Atlantic spring bloom: a Lagrangian analysis based on multisatellite data. J. Geophys. Res. Oceans 112 (C8).

Liu, J.P., Li, A.C., Xu, K.H., Velozzi, D.M., Yang, Z.S., Milliman, J.D., DeMaster, D.J., 2006. Sedimentary features of the Yangtze river-derived along-shelf clinoform deposit in the East China sea. Continent. Shelf Res. 26 (17–18), 2141–2156.

Liu, Z., Vavrus, S., He, F., Wen, N., Zhong, Y., 2005. Rethinking tropical ocean response to global warming: the enhanced equatorial warming. J. Clim. 18 (22), 4684–4700.

Llewellyn-Jones, D.T., Minnett, P.J., Saunders, R.W., Zavody, A.M., 1984. Satellite multichannel infrared measurements of sea surface temperature of the NE Atlantic Ocean using AVHRR/2. Quart. J. R. Meteorol. Soc. 110 (465), 613–631.

Lyons, M.B., Phinn, S.R., Roelfsema, C.M., 2012. Long term land cover and seagrass mapping using Landsat and object-based image analysis from 1972 to 2010 in the coastal environment of South East Queensland, Australia. ISPRS J. Photogramm. Remote Sens. 71, 34–46.

Marfai, M.A., King, L., 2008. Potential vulnerability implications of coastal inundation due to sea level rise for the coastal zone of Semarang city, Indonesia. Environ. Geol. 54 (6), 1235–1245.

McCreary Jr., J.P., Kundu, P.K., Molinari, R.L., 1993. A numerical investigation of dynamics, thermodynamics and mixed-layer processes in the Indian Ocean. Prog. Oceanogr. 31 (3), 181–244.

Mendoza, C., Mancho, A.M., 2010. Hidden geometry of ocean flows. Phys. Rev. Lett. 105 (3), 038501.

Mills, J.P., Buckley, S.J., Mitchell, H.L., Clarke, P.J., Edwards, S.J., 2005. A geomatics data integration technique for coastal change monitoring. Earth Surf. Process. Landf. J. Br. Geomorphol. Res. Gr. 30 (6), 651–664.

Molcard, A., Griffa, A., Özgökmen, T.M., 2005. Lagrangian data assimilation in multilayer primitive equation ocean models. J. Atmos. Ocean. Technol. 22 (1), 70–83.

Monaldo, F., Kerbaol, V., Clemente-Colón, P., Furevik, B., Horstmann, J., Johannessen, J., et al., 2003. The SAR Measurements of Ocean Surface Winds: A White Paper for the 2nd Workshop on Coastal and Marine Applications of SAR, Longyearbyen, Spitsbergen, Norway, 8–12 September 2003. ESA SP, p. 565.

Morang, A., Gorman, L.T., 2005. Monitoring coastal geomorphology. Encycl. Coast. Sci. 663–674.

Morang, A., Larson, R.L., Gorman, L.T., 1997a. Monitoring the coastal environment; Part III: geophysical and research methods. J. Coast Res. 13 (4), 1964–1985.

Morang, A., Larson, R.L., Gorman, L.T., 1997b. Monitoring the coastal environment; Part I: waves and currents. J. Coast Res. 13 (1), 111–133.

Nagaraja Rao, C.R., Stowe, L.L., McClain, E.P., 1989. Remote sensing of aerosols over the oceans using AVHRR data theory, practice and applications. Int. J. Rem. Sens. 10 (4–5), 743–749.

Nwilo, P.C., Badejo, O.T., 2006. Impacts and Management of Oil Spill Pollution along the Nigerian Coastal Areas.

Parker, B., 2002. The integration of bathymetry, topography and shoreline and the vertical datum transformations behind it. Int. Hydrogr. Rev. 3 (3), 14–26.

Purkis, S.J., Klemas, V.V., 2011. Remote Sensing and Global Environmental Change. John Wiley & Sons.

Rani, M., Rehman, S., Sajjad, H., Chaudhary, B.S., Sharma, J., Bhardwaj, S., Kumar, P., 2018. Assessing coastal landscape vulnerability using geospatial techniques along Vizianagaram–Srikakulam coast of Andhra Pradesh, India. Nat. Hazards 94 (2), 711–725.

Riegl, B.M., Purkis, S.J., 2012. Coral reefs of the Gulf: adaptation to climatic extremes in the world's hottest sea. In: Coral Reefs of the Gulf. Springer, Dordrecht, pp. 1–4.

Robinson, I.S., 2004. Measuring the Oceans from Space: The Principles and Methods of Satellite Oceanography. Springer Science & Business Media.

Romeiser, R., Runge, H., 2007. Detailed analysis of ocean current measuring capabilities of TerraSAR-X in several possible along-track InSAR modes on the basis of numerical simulations. IEEE Trans. Geosci. Rem. Sens. 45, 21–35.

Samberg, A., 2007. The state-of-the-art of airborne laser systems for oil mapping. Can. J. Rem. Sens. 33 (3), 143–149.

Schmitz Jr., W.J., Holland, W.R., 1986. Observed and modeled mesoscale variability near the Gulf stream and Kuroshio extension. J. Geophys. Res. Oceans 91 (C8), 9624–9638.

Schroeder, T., Devlin, M.J., Brando, V.E., Dekker, A.G., Brodie, J.E., Clementson, L.A., McKinna, L., 2012. Inter-annual variability of wet season freshwater plume extent into the Great Barrier Reef lagoon based on satellite coastal ocean colour observations. Mar. Pollut. Bull. 65 (4–9), 210–223.

Schwab, D.J., Bedford, K.W., 1994. Initial implementation of the great lakes forecasting system: a real-time system for predicting lake circulation and thermal structure. Water Qual. Res. J. 29 (2–3), 203–220.

Shankar, D., Vinayachandran, P.N., Unnikrishnan, A.S., 2002. The monsoon currents in the north Indian Ocean. Prog. Oceanogr. 52 (1), 63–120.

Stammer, D., Wunsch, C., Giering, R., Eckert, C., Heimbach, P., Marotzke, J., et al., 2002. Global ocean circulation during 1992–1997, estimated from ocean observations and a general circulation model. J. Geophys. Res. Oceans 107 (C9), 1–1.

Tarquis Alfonso, A.M., Platonov, A., Matulka, A.M., Grau, J., Sekula, E., Diez, M., Redondo Apraiz, J.M., 2014. Application of multifractal analysis to the study of SAR features and oil spills on the ocean surface. Nonlinear Process Geophys. 21 (2), 439–450.

Tozzi, S., Schofield, O., Falkowski, P., 2004. Historical climate change and ocean turbulence as selective agents for two key phytoplankton functional groups. Mar. Ecol. Prog. Ser. 274, 123–132.

Tralli, D.M., Blom, R.G., Zlotnicki, V., Donnellan, A., Evans, D.L., 2005. Satellite remote sensing of earthquake, volcano, flood, landslide and coastal inundation hazards. ISPRS J. Photogramm. Remote Sens. 59 (4), 185–198.

Tripathy, B.R., Seenipandi, K., Sajjad, H., Joshi, P.K., Chaudhary, B.S., Kumar, P., 2018. Monitoring of seasonal variability and movement of suspended sediment concentrations along the Thiruvananthapuram coast, southern India, using the Landsat OLI sensor. Ocean Sci. 14 (5), 1085–1092.

Vachon, P.W., Krogstad, H.E., Ikeda, M., Dobson, F.W., 1995. Synthetic aperture radar imagery of ocean waves. Oceanogr. Appl. Remote Sens. 237–255.

Watts, A.C., Ambrosia, V.G., Hinkley, E.A., 2012. Unmanned aircraft systems in remote sensing and scientific research: classification and considerations of use. Rem. Sens. 4 (6), 1671–1692.

Webb, D.J., 1996. An ocean model code for array processor computers. Comput. Geosci. 22 (5), 569–578.

Wyrtki, K., 1973. Physical oceanography of the Indian ocean. In: The Biology of the Indian Ocean. Springer, Berlin, Heidelberg, pp. 18–36.

Yan, C.Y., 1993. A plate reconstruction of the south west Pacific 0–100 Ma. In: Proceedings of Ocean Drilling Program, Scientific Results, vol. 130, pp. 697–709.

Yin, J., Yin, Z., Wang, J., Xu, S., 2012. National assessment of coastal vulnerability to sea-level rise for the Chinese coast. J. Coast Conserv. 16 (1), 123–133.

2

Ocean and coastal remote sensing: platforms, sensors, instruments, data products, tools, and techniques

Haroon Sajjad, Sufia Rehman
DEPARTMENT OF GEOGRAPHY, FACULTY OF NATURAL SCIENCES, JAMIA MILLIA ISLAMIA, NEW DELHI, DELHI, INDIA

Chapter outline

1. Introduction

Ocean as a major constituent of Earth's surface plays a key role in the hydrological cycle and provides essential marine resources. It supports biodiversity, functioning of various ecological phenomena, and acts as a source of recreation to human beings (Board, 1993). These are vast, hardly accessible, and less understood due to lack of sufficient data and their dynamic nature. Ocean and coastal studies assume significance to reveal their various dynamics, understanding the regional characteristics, and analyzing the inherent vulnerabilities at various scales (Bastos et al., 2016). The Intergovernmental Panel on Climate Change in its fifth assessment report highlighted that ocean absorbs nearly 93% of the energy due to extra greenhouse effect, which may be experienced at a depth of 1000 m (Stocker, 2014). Land resources have always played a vital role in providing food resources to human beings, but the growing trade-offs between anthropogenic activities

Remote Sensing of Ocean and Coastal Environments. https://doi.org/10.1016/B978-0-12-819604-5.00002-0

and environmental implications have resulted in severe degradation of this resource (Fischer et al., 2012). The critical food demand has led to the hunt for other substitutes. As compared to land, oceans are a reliable source of food due to their vastness and enduring resource potential. In spite of various discoveries of hydrothermal vents, the physiological traits and biochemistry of oceans have been less explored due to insufficient techniques. Human interference has also increased with time, which has threatened coastal ecosystems and fragile organisms. Thus monitoring and assessment of oceans are important in order to maintain their potential for biological and ecosystem functioning.

Recent development in technology, data, and theories have broadened the spheres of geospatial techniques and allowed exploration of hidden dimensions of the coastal environment (Al-Tahir et al., 2006). Remote sensing datasets can be effective in identifying the critical physical properties of coastal ecosystems such as the health of aquatic vegetation, surface water temperature, types of aquatic organisms, and so on. Thus remote sensing plays a vital role in quantifying the functioning of aquatic ecosystems (Richardson and Ledrew, 2006).

2. Platforms and sensors

Remote sensing data products (satellite and airborne) are proved to be advantageous in analyzing the changes occurring on the coastal ecosystems at different temporal and spatial scales and have minimized the dependency on field investigations (Klemas, 2010). Recent development in technology, data, and theories have broadened the sphere of geospatial techniques and allowed exploration of the various dimensions of the coastal environment (Yang, 2009). Remote sensing is being carried out with the help of instruments known as sensors, placed in satellite orbit or having a suborbital platform. There are two types of sensors, active and passive. Active sensors illuminate themselves to collect the information of the features while passive sensors acquire the reflected energy of the object provided by the sun (Klemas, 2009b).

Remote sensing datasets have been used by scholars for more than a century. The first ground information through remote sensing was obtained from an airborne platform in 1858 introduced by Frenchman Gaspard Felix Tournachon. Development of airborne platforms were increased during World War I and World War II. Earlier developments in remote sensing were witnessed during wars and reconnaissance applications. Later, they were extensively utilized for monitoring and analyzing Earth's natural resources (Jensen, 2009). Optical remote sensors are critical for coastal assessment. These sensors use the visible and infrared wavelengths of electromagnetic spectrum for acquiring information based on the concentration of the components. Optical sensors are categorized on the basis of spectral band properties as panchromatic, multispectral, and hyperspectral. Hyperspectral images are effective in providing detailed information about the targeted objects and have been found effective in coastal management (Teodoro, 2016). Remote sensing provides real-time monitoring of coastal environments due to its continuous synoptic coverage. Recently, these datasets have been used for various applications in providing a basis for validating the data for numerical simulation, boundary conditions, in situ sediment concentration measurement, and assisting in three-dimensional coastal modeling (Jiang et al., 2016).

3. Data products

Remote sensing datasets are providing impactful details about the coastal regimes and assist in examining the changes driven by anthropogenic and natural phenomena. A wide range of remote sensing data products have been used by scholars for coastal dynamic modeling (Van Zuidam et al., 1998), salt marsh vegetation (Belluco et al., 2006), analyzing the impervious surface characteristics (Chormanski et al., 2008), coastal mapping (Elaksher, 2008), water quality assessment (Bierman et al., 2011), coastal erosion risk assessment (Bio et al., 2015), coastal environment monitoring (Jiang et al., 2016), and coastal zone management (Colak et al., 2019). Literature review of previous coastal studies has revealed that a variety of datasets have been utilized by scholars for analyzing the changes occurring along riverine areas. These datasets include aerial photographs, digital elevation models, hyperspectral and multispectral data, SeaWiFS, Light Detection and Ranging, and Landsat products (Belluco et al., 2006; Barale et al., 2008; Bio et al., 2015; Anderson et al., 2016; Colak et al., 2019). Data products of Modis, SeaWiFS, Landsat, and Sentinel 2 play an important role in providing detailed information of the coastal areas. These satellites provide multispectral, multiresolution, temporal, and multifrequency datasets, allowing close monitoring and examination of changes with bands specifically devoted to ocean exploration (Fig. 2.1). Availability of these datasets has widened the applicability of remote sensing in ocean and coastal policy implementation (Levy et al., 2018). Landsat 4–5, 7 ETM+, and 8 OLI products have largely been used to produce the bathymetry map of the ocean floor. Remote sensing datasets have provided wider applicability in coastal hazard monitoring and assessment (Rajeesh and Dwarakish, 2015). Airborne radar data with high resolution has been found useful in analyzing local flood inundation assessment. The recent development of high-resolution microwave datasets and advance land imager have also increasingly been utilized for turbidity and inundation assessment (Rehman et al., 2019). Integration of synthetic aperture radar data with GIS can effectively be utilized for flood and landslide assessments (Rahman and Thakur, 2018).

Sea surface temperature is also important in affecting the dynamics and biological productivity. Advanced very high resolution radiometer (AVHRR) and National Oceanic and Atmospheric Administration Optimum Interpolation (OI) sea surface temperature (SST) sensors have been found capable for large-scale surface temperature assessment (Goela et al., 2016; Klein et al., 2019). Advancement in satellite datasets are provide potential for providing accurate and timely information that could be used for strategic purposes (Rajeesh and Dwarakish, 2015). It has also been identified that Landsat 8 and

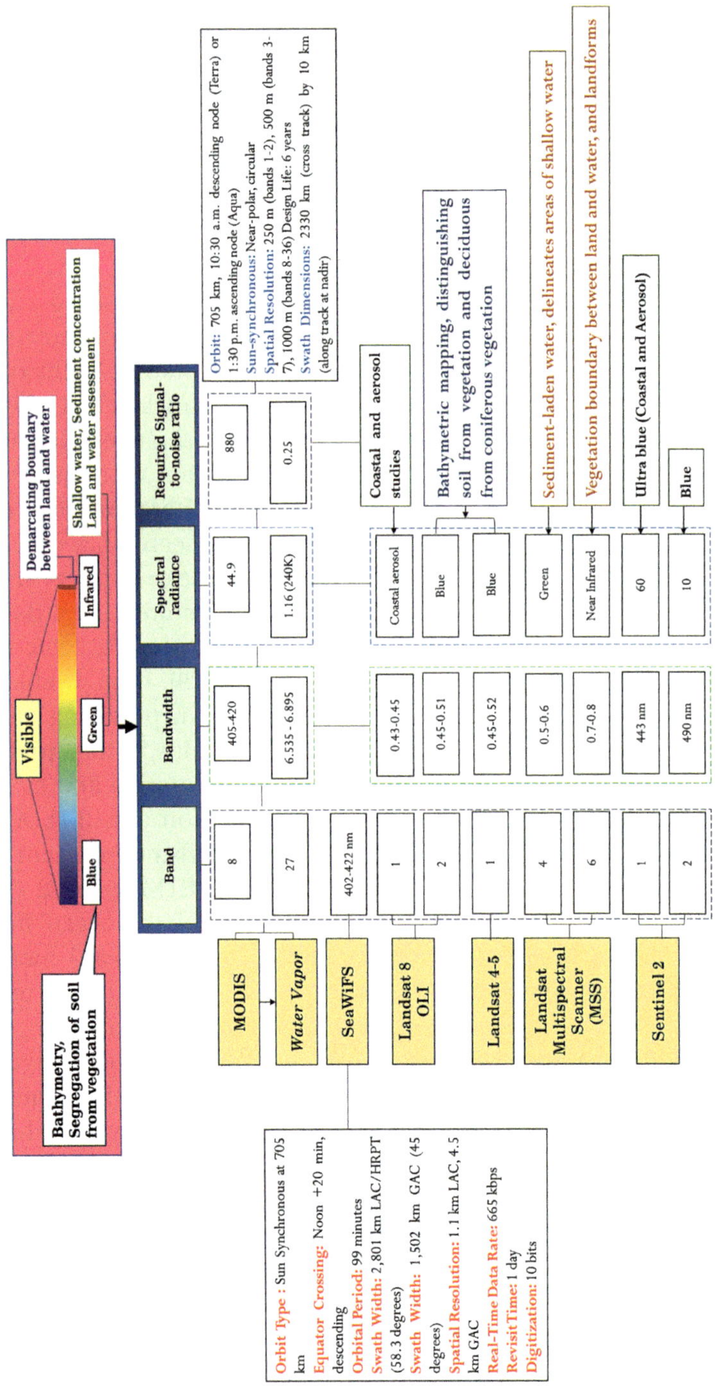

FIGURE 2.1 Important remote sensing systems and their respective bands used for coastal assessment.

Sentinel 2 have assisted in coastal monitoring of ocean floor features observation, sediment concentration, and phytoplankton. Landsat archives have provided huge datasets since 1982 that can be used to analyze multiyear changes occurring on the ocean floor (Lyons et al., 2011). It was designed for land surface applications but also utilized for suspended sediment assessment. However, these assessments also have several challenges. Groom et al. (2019) emphasized that inadequacy of high-resolution ocean color data must be considered by space agencies. Infrared sensors such as spinning enhanced visible and infrared imager and AVHRR are useful in obtaining ocean surface temperature up to 10 μm while an advanced microwave sensor is capable of obtaining ocean temperature up to 1 mm (Storto and Oddo, 2019). Spaceborne global navigation satellite systems-reflectometry is the latest advanced remote sensing technique assist in detecting geophysical parameters of the ocean floor (Dong and Jin, 2019). Sediment dynamics affect the morphological traits of the oceans involving deltas, lagoons, and changes in the shoreline. These observations require effective satellite observations and appropriate empirical modeling. Remote sensing color data is capable of capturing various suspended phenomena, erosion, and deposition along coastal areas (Benincasa et al., 2019).

4. Tools and techniques

Coastal regions are fragile and dynamic zones due to their varied geomorphological conditions, and their complex physiological characteristics require scientific knowledge for effectual management practices (Petihakis et al., 2018). A holistic approach is required to understand the biological and physiochemical marine processes in and around coastal areas (Cocquempot et al., 2019). Various scholars have utilized a number of methods to analyze coastal regions as the morphology of beaches and shoreline dynamics change (Ojeda and Guillén, 2008). Earlier, expensive and time-consuming classical surveying equipment was used in ocean studies, which lacked large scale spatial assessment (Muehe, 2011). During the 1930 and 1940s, active remote sensing was used for military purposes before the beginning of World War II. Environmental signals at that time were considered noise, which later was used in remote sensing for detecting ground features (National Academies of Sciences, Engineering, and Medicine, 2015).

Advancement in computer technology and remote sensing data products have greatly contributed to the coastal studies and provided timely and real-time monitoring of coastal dynamics at spatial scales. The need for analyzing coastal areas quantitatively have shifted its dependence from qualitative to quantitative approaches (Fig. 2.2). These approaches included large datasets and a huge number of methods ranging from simple analytical approaches to more advanced statistical techniques for predicting shoreline changes (Le Cozannet et al., 2014). The hydrodynamic model is an example of an integrated model based on various datasets such as river flow, waves, surges, pattern of currents, and so on, which can be acquired from remote sensing datasets (radars and scatterometers). Thermal and microwave wavelengths of the spectrum are now being

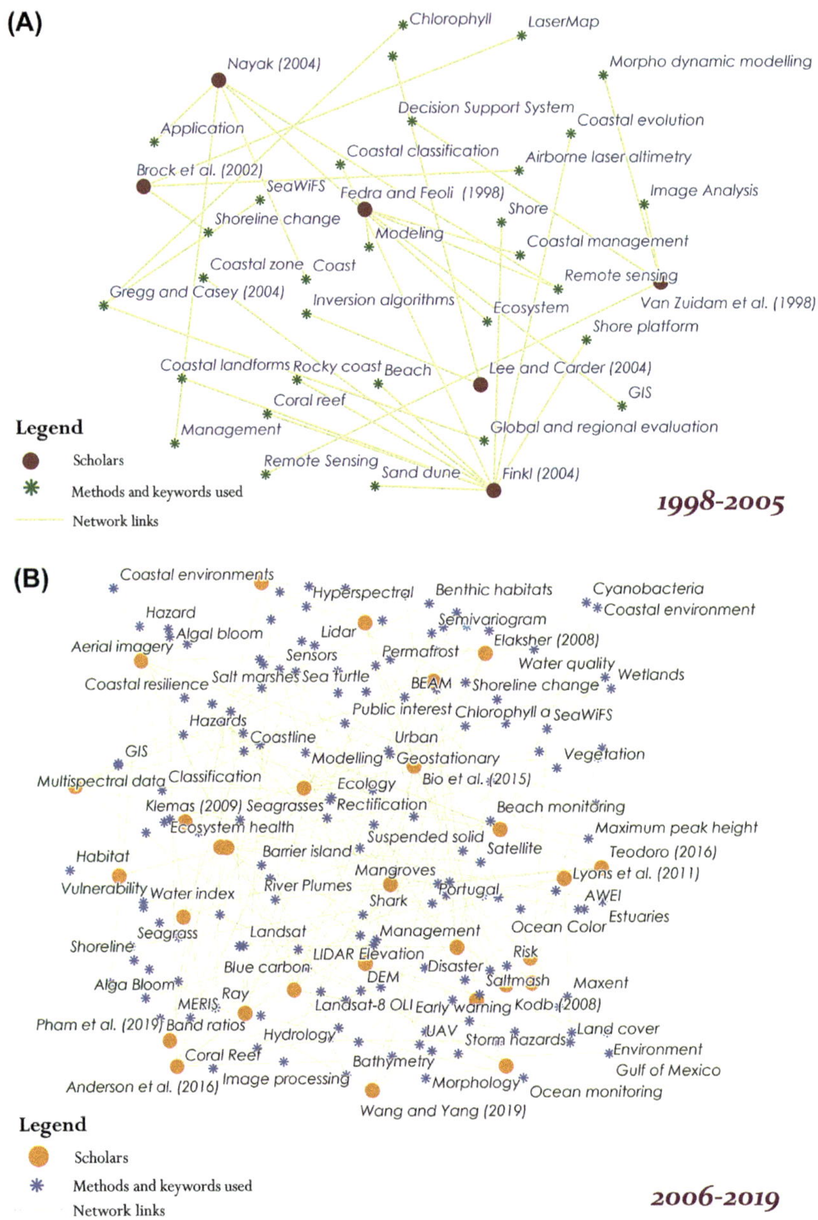

FIGURE 2.2 Network analysis of various authors and methods; (A) 1998–2005 and (B) 2006–2019.

utilized for examining sea surface temperature and salinity (Klemas, 2009a). Conventional methods are constant and lack in providing sufficient information about coastal fragility. Remote sensing, which is temporally consistent and provides instantaneous coverage, has many merits over conventional techniques, and thus offers a range of information about coastal and inland ocean traits (Matthews, 2011). Availability of various terrestrial and oceanographic remote sensing products ranging from global to regional level have largely been incorporated in coastal decision-making processes (Dean and Populus, 2013). Past, present, and future observations are required for analyzing the coastal dynamics for effective monitoring and management of coastal areas. Remote sensing data renders mapping of physicochemical (temperature, water transparency, salinity, and pollutants), hydromorphological (discharge, freshwater, shoreline evolution, currents), and biological conditions of the ocean (Doxaran et al., 2019).

Fig. 2.2 shows the network analysis between authors and respective methods used for examining the ocean and coastal areas. Risk assessment, biogeochemical models, inversion algorithm, distributed runoff modeling, inversion algorithm, GIS modeling, histogram thresholding, cluster analysis, discriminant analysis, factor analysis, principal components analysis, semivariogram, geographically weighted regression, support vector machines, water index, automatic water extraction index (AWEI), decision support system, image analysis, and morphodynamic modeling were identified as prominent models in coastal studies (Fedra and Feoli, 1998; Van Zuidam et al., 1998; Sanchez-Hernandez et al., 2007; Chormanski et al., 2008; Bierman et al., 2011; Ramesh et al., 2015; Bio et al., 2015; Jiang et al., 2016; Colak et al., 2019; Wang and Yang, 2019).

4.1 Prominent indices used in coastal studies

Integration of geodatabase with GIS significantly contributed to coastal studies (McLaughlin et al., 2002). High-resolution satellite data have magnificently assisted in analyzing shoreline dynamics and comprehensive coastal assessment. Coastal studies as discussed earlier were more oriented toward analyzing the physiogeological characteristics of the landscape to identify the magnitude of the hazard event (Klein and Nicholls, 1999). Sensitive areas can also be distinguished through examination of geological and geomorphic characteristics of the landscape as slope, bathymetry, sea level rise, and so on (Small and Nicholls, 2003). Indices are used to analyze the significant perturbations in coastal areas due to climate change–induced sea-level rise, erosion, storm surge, and spill impact. These indices are mainly aimed at segregating the areas into different categories having similar characteristics. These classifications help in management and policy implication for the exposed areas. These indices have been used for various purposes ranging from vulnerability to ecological sensitivity assessment. Borja et al. (2000) utilized a marine biotic index (BI) to examine the ecological stability around the European coast. They emphasized that BI may help in consolidating the long-term environmental sustainability by identifying the changes in water quality and sea bottom characteristics induced by anthropogenic activities.

A coastal vulnerability database helps to discern the areas by integrating the past and current state of the coast and assist in effective policy formulation. Vittal Hegde and Radhakrishnan Reju (2007) combined socioeconomic and physical factors to formulate an effective coastal vulnerability index (CVI) for the eastern coast of Mangalore. They emphasized that computation of CVI largely depends on the type and quality of data. Xu (2006) used a modified normalized difference water index (MNDWI) for enhanced water assessment and noise-free delineation of open water. Environmental noise hinders in identifying open water and to overcome this problem Acharya et al. (2018) combined the normalized difference water index (NDWI), normalized difference vegetation index, MNDWI, and AWEI for easy assessment of the surface water. They further suggested that use of multiscenes with machine learning models may also be effective in surface water assessment. Kovacs et al. (2004) used high-resolution satellite data and leaf area index for monitoring mangrove conditions. Cecchi et al. (2014) developed a new biotic index on the basis of coral ligenous macroalgal assemblages. Simulation models, statistical techniques, and indices have been widely used to analyze the trophic level of the waters. Karydis (2009) highlighted that quantification of eutrophication is not easy due to difficulty in discrimination of its affecting factor. He suggested that ecological indicators, diversity indices, and multimetric index are effective in examining the eutrophication of ocean waters. Izaguirre et al. (2011) examined the variability in storm surge height using the satellite data. Thus remote sensing data has largely assisted in idealizing the sensitivity and vulnerability of coastal areas.

4.2 Challenges in coastal remote sensing

Satellite remote sensing has contributed to in situ measurements temporally and spatially to a great extent. Operational oceanography and simulation models have widely assisted in monitoring of disaster management and coastal conservation globally. Though remote sensing has helped in the analysis of various characteristics of coastal areas, it also faces several challenges that need to be addressed. One of the biggest challenges in coastal assessment is the wide variation in sea surface dynamics and meteorological conditions at various scales. Various high-resolution satellites are available for mapping and analyzing coastal dynamics; however, their use at regional and local levels is limited due to their high cost and inadequate archive (Nascimento et al., 2013). It is also difficult to discern the merits and demerits of coastal anomalies and observation at various levels. Though a large archive of various satellite data is available, the exploration of various phenomena such as sea waves, SST, and surface winds are still limited (Chapron et al., 2008).

These challenges can be effectively minimized using effective modeling and appropriate data. Efforts should be made to provide processed data in each format for various end users, which will enhance the application for coastal assessment and analysis of each phenomenon. Earth observation may be combined with operational sensors data for temporal and synoptic coastal assessment. Full utilization of satellite products and observations together may be used for societal benefits. Furthermore, improving fundamental knowledge through development of appropriate empirical and theoretical models will help in coastal assessment at various scales. Altimeter measurement integrated with radiometer and scatterometers will also provide valuable insights about the sea structure and associated phenomena. Availability of baseline data will impart the understanding of inherent coastal characteristics. Machine learning algorithms with satellite observation have widened the scope of effectual monitoring of the coastal areas.

References

Acharya, T., Subedi, A., Lee, D., 2018. Evaluation of water indices for surface water extraction in a Landsat 8 scene of Nepal. Sensors 18 (8), 2580. https://doi.org/10.3390/s18082580.

Al-Tahir, A., Baban, S.M.J., Ramlal, B., 2006. Utilizing emerging geo-imaging technologies for the management of tropical coastal environments. West Indian J. Eng. 29 (1), 11–22.

Anderson, C., Carter, G., Funderburk, W., 2016. The use of aerial RGB imagery and LIDAR in comparing ecological habitats and geomorphic features on a natural versus man-made barrier island. Remote Sens. 8 (7), 602.

Barale, V., Jaquet, J.M., Ndiaye, M., 2008. Algal blooming patterns and anomalies in the Mediterranean Sea as derived from the SeaWiFS data set (1998–2003). Remote Sens. Environ. 112 (8), 3300–3313.

Bastos, L., Bio, A., Iglesias, I., 2016. The importance of marine observatories and of RAIA in particular. Front. Mar. Sci. 3, 140.

Belluco, E., Camuffo, M., Ferrari, S., Modenese, L., Silvestri, S., Marani, A., Marani, M., 2006. Mapping salt-marsh vegetation by multispectral and hyperspectral remote sensing. Remote Sens. Environ. 105 (1), 54–67.

Benincasa, M., Falcini, F., Adduce, C., Sannino, G., Santoleri, R., 2019. Synergy of satellite remote sensing and numerical ocean modelling for coastal geomorphology diagnosis. Remote Sens. 11 (22), 2636.

Bierman, P., Lewis, M., Ostendorf, B., Tanner, J., 2011. A review of methods for analysing spatial and temporal patterns in coastal water quality. Ecol. Indicat. 11 (1), 103–114.

Bio, A., Bastos, L., Granja, H.M., Pinho, J.L., Gonçalves, J.A., Henriques, R.F., et al., 2015. Methods for coastal monitoring and erosion risk assessment: two Portuguese case studies. Revista de Gestao Costeira Integrada 15 (47), 63-63.

Board, O.S., National Research Council, 1993. Oceanography in the Next Decade: Building New Partnerships. National Academies Press.

Borja, A., Franco, J., Pérez, V., 2000. A marine biotic index to establish the ecological quality of soft-bottom benthos within European estuarine and coastal environments. Mar. Pollut. Bull. 40 (12), 1100–1114.

Cecchi, E., Gennaro, P., Piazzi, L., Ricevuto, E., Serena, F., 2014. Development of a new biotic index for ecological status assessment of Italian coastal waters based on coralligenous macroalgal assemblages. Eur. J. Phycol. 49 (3), 298–312.

Chapron, B., Garello, R., Weissman, D.E., September 2008. Ocean remote sensing: challenges for the future. In: OCEANS 2008. IEEE, pp. 1–7.

Chormanski, J., Van de Voorde, T., De Roeck, T., Batelaan, O., Canters, F., 2008. Improving distributed runoff prediction in urbanized catchments with remote sensing-based estimates of impervious surface cover. Sensors 8 (2), 910–932.

Cocquempot, L.M., Delacourt, C., Paillet, J., Riou, P., Lamarche, S., Ballu, V., et al., 2019. Coastal ocean and nearshore observation: a French case study. Front. Mar. Sci. 6, 324.

Colak, T.I., Senel, G., Goksel, C., 2019. Coastline zone extraction using Landsat-8 oli imagery, case study: Bodrum Peninsula, Turkey. Int. Arch. Photogramm. Remote Sens. & Spat. Inf. Sci. 42 (4/W12).

Dean, A.M., Populus, J., 2013. In: Meaden, G.J., Aguilar-Manjarrez, J. (Eds.), Remote Sensing and GIS Integration, pp. 147–189.

Dong, Z., Jin, S., 2019. Evaluation of spaceborne GNSS-R retrieved ocean surface wind speed with multiple datasets. Remote Sens. 11 (23), 2747.

Doxaran, B., Dogliotti, M., Senechal, 2019. Editorial for the special issue "remote sensing in coastal zone monitoring and management—how can remote sensing challenge the broad spectrum of temporal and spatial scales in coastal zone dynamic? Remote Sens. 11 (9), 1028. https://doi.org/10.3390/rs11091028.

Elaksher, A.F., 2008. Fusion of hyperspectral images and lidar-based dems for coastal mapping. Optic Laser. Eng. 46 (7), 493–498.

Fedra, K., Feoli, E., 1998. GIS technology and spatial analysis in coastal zone management. EEZ Technol. 3, 171–179.

Fischer, G., Hizsnyik, E., Prieler, S., van Velthuizen, H., Wiberg, D., 2012. Scarcity and Abundance of Land Resources: Competing Uses and the Shrinking Land Resource Base. SOLAW Background Thematic Report - TR02.

Goela, P.C., Cordeiro, C., Danchenko, S., Icely, J., Cristina, S., Newton, A., 2016. Time series analysis of data for sea surface temperature and upwelling components from the southwest coast of Portugal. J. Mar. Syst. 163, 12–22.

Groom, S.B., Sathyendranath, S., Ban, Y., Bernard, S., Brewin, B., Brotas, V., et al., 2019. Satellite ocean colour: current status and future perspective. Front. Mar. Sci. 6, 485.

Izaguirre, C., Méndez, F.J., Menéndez, M., Losada, I.J., 2011. Global extreme wave height variability based on satellite data. Geophys. Res. Lett. 38 (10).

Jensen, J.R., 2009. Remote Sensing of the Environment: An Earth Resource Perspective. Pearson Education India, 2/e.

Jiang, D., Hao, M., Fu, J., 2016. Monitoring the coastal environment using remote sensing and GIS techniques. Appl. Stud. Coast. & Mar. Environ.. https://doi.org/10.5772/62242.

Karydis, M., University of the Aegean, Lesvos I. (Greece). Department of Environmental Studies, December 2009. Eutrophication assessment of coastal waters based on indicators: a literature review. In: Proceedings of the International Conference on Environmental Science and Technology, vol. 1. University of the Aegean, Chania (Greece).

Klein, K.P., Lantuit, H., Heim, B., Fell, F., Doxaran, D., Irrgang, A.M., 2019. Long-term high-resolution sediment and sea surface temperature spatial patterns in arctic nearshore waters retrieved using 30-year Landsat archive imagery. Remote Sens. 11 (23), 2791. https://doi.org/10.3390/rs11232791.

Klein, R.J., Nicholls, R.J., 1999. Assessment of Coastal Vulnerability to Climate Change. Ambio, pp. 182–187.

Klemas, V., 2010. Remote sensing techniques for studying coastal ecosystems: an overview. J. Coast Res. 27 (1), 2–17.

Klemas, V.V., 2009a. Remote sensing of coastal resources and environment. Aplinkos tyrimai, inžinerija ir vadyba Environ. Res. Eng. & Manag. 48 (No. 2), 11–18. Kaunas, Technologija. ISSN 1392-1679.

Klemas, V.V., 2009b. Sensors and techniques for observing coastal ecosystems. In: Remote Sensing and Geospatial Technologies for Coastal Ecosystem Assessment and Management. Springer, Berlin, Heidelberg, pp. 17–44.

Kovacs, J.M., Flores-Verdugo, F., Wang, J., Aspden, L.P., 2004. Estimating leaf area index of a degraded mangrove forest using high spatial resolution satellite data. Aquat. Bot. 80 (1), 13–22.

Le Cozannet, G., Garcin, M., Yates, M., Idier, D., Meyssignac, B., 2014. Approaches to evaluate the recent impacts of sea-level rise on shoreline changes. Earth Sci. Rev. 138, 47–60. https://doi.org/10.1016/j.earscirev.2014.08.005.

Levy, G., Vignudelli, S., Gower, J., 2018. Enabling earth observations in support of global, coastal, ocean, and climate change research and monitoring. Int. J. Remote Sens. 39 (13), 4287–4292. https://doi.org/10.1080/01431161.2018.1464101.

Lyons, M., Phinn, S., Roelfsema, C., 2011. Long Term Land Cover and Seagrass Mapping Using Landsat Sensors from 1972–2010 in the Coastal Environment of South East Queensland. Australia.

Matthews, M.W., 2011. A current review of empirical procedures of remote sensing in inland and near-coastal transitional waters. Int. J. Remote Sens. 32 (21), 6855–6899.

McLaughlin, S., McKenna, J., Cooper, J.A.G., 2002. Socio-economic data in coastal vulnerability indices: constraints and opportunities. J. Coast Res. 36, 487–497. https://doi.org/10.2112/1551-5036-36.spl.487.

Muehe, D., 2011. Erosão Costeira - Tendência ou Eventos Extremos? O Litoral entre Rio de Janeiro e Cabo Frio, Brasil/Coastal Erosion. Revista da Gestão Costeira Integrada 11 (3), 315–325. https://doi.org/10.5894/rgci28.

Nascimento, W.R., Souza-Filho, P.W.M., Proisy, C., Lucas, R.M., Rosenqvist, A., 2013. Mapping changes in the largest continuous Amazonian mangrove belt using object-based classification of multisensor satellite imagery. Estuar. Coast Shelf Sci. 117, 83–93. https://doi.org/10.1016/j.ecss.2012.10.005.

National Academies of Sciences, Engineering, and Medicine, 2015. A Strategy for Active Remote Sensing amid Increased Demand for Radio Spectrum. National Academies Press.

Ojeda, E., Guillén, J., 2008. Shoreline dynamics and beach rotation of artificial embayed beaches. Mar. Geol. 253 (1–2), 51–62.

Petihakis, G., Perivoliotis, L., Korres, G., Ballas, D., Frangoulis, C., Pagonis, P., et al., 2018. An integrated open-coastal biogeochemistry, ecosystem and biodiversity observatory of the eastern Mediterranean–the Cretan Sea component of the POSEIDON system. Ocean Sci. 14 (5), 1223–1245.

Rahman, M.R., Thakur, P.K., 2018. Detecting, mapping and analysing of flood water propagation using synthetic aperture radar (SAR) satellite data and GIS: a case study from the Kendrapara District of Orissa State of India. Egypt. J. Remote Sens. & Space Sci. 21, S37–S41.

Rajeesh, R., Dwarakish, G.S., 2015. Satellite oceanography– A review. Aquat. Procedia 4, 165–172. https://doi.org/10.1016/j.aqpro.2015.02.023.

Ramesh, R., Chen, Z., Cummins, V., Day, J., D'Elia, C., Dennison, B.,., Kremer, H., 2015. Land–ocean interactions in the coastal zone: past, present & future. Anthropocene 12, 85–98.

Rehman, S., Sahana, M., Hong, H., Sajjad, H., Ahmed, B.B., 2019. A systematic review on approaches and methods used for flood vulnerability assessment: framework for future research. Nat. Hazards 1–24.

Richardson, L.L., Ledrew, E.F., 2006. Remote sensing and the science, monitoring, and management of aquatic coastal ecosystems. Remote Sens. Aquat. Coast. Ecosyst. Processes 1. https://doi.org/10.1007/1-4020-3968-9_1.

Sanchez-Hernandez, C., Boyd, D.S., Foody, G.M., 2007. Mapping specific habitats from remotely sensed imagery: support vector machine and support vector data description based classification of coastal saltmarsh habitats. Ecol. Inf. 2 (2), 83–88.

Small, C., Nicholls, R.J., 2003. A global analysis of human settlement in coastal zones. J. Coast Res. 584–599.

Stocker, T. (Ed.), 2014. Climate Change 2013: The Physical Science Basis: Working Group I Contribution to the Fifth Assessment Report of the Intergovernmental Panel on Climate Change. Cambridge University Press.

Storto, A., Oddo, P., 2019. Optimal Assimilation of daytime SST retrievals from SEVIRI in a regional ocean prediction system. Remote Sens. 11 (23), 2776. https://doi.org/10.3390/rs11232776.

Teodoro, A.C., 2016. Optical satellite remote sensing of the coastal zone environment — an overview. Environ. Appl. Remote Sens. 165–196. https://doi.org/10.5772/61974.

Van Zuidam, R.A., Farifteh, J., Eleveld, M.A., Tao, C., 1998. Developments in remote sensing, dynamic modelling and GIS applications for integrated coastal zone management. J. Coast Conserv. 4 (2), 191–202.

Vittal Hegde, A., Radhakrishnan Reju, V., 2007. Development of coastal vulnerability index for Mangalore coast, India. J. Coast Res. 1106–1111.

Wang, X., Yang, W., 2019. Water quality monitoring and evaluation using remote-sensing techniques in China: a systematic review. Ecosys. Health Sustain. 5 (1), 47–56.

Xu, H., 2006. Modification of normalised difference water index (NDWI) to enhance open water features in remotely sensed imagery. Int. J. Remote Sens. 27 (14), 3025–3033. https://doi.org/10.1080/01431160600589179.

Yang, X., 2009. Remote sensing, geospatial technologies and coastal ecosystems. In: Remote Sensing and Geospatial Technologies for Coastal Ecosystem Assessment and Management. Springer, Berlin, Heidelberg, pp. 1–14.

Further reading

Ahmad, H., 2019. Applications of remote sensing in oceanographic research. Int. J. Oceanogr. & Aquac. 1–9. https://doi.org/10.23880/ijoac-16000159.

Alesheikh, A.A., Ghorbanali, A., Nouri, N., 2007. Coastline change detection using remote sensing. Int. J. Environ. Sci. Technol. 4 (1), 61–66.

Brock, J.C., Purkis, S.J., 2009. The emerging role of lidar remote sensing in coastal research and resource management. J. Coast Res. 1–5.

Brock, J.C., Wright, C.W., Sallenger, A.H., Krabill, W.B., Swift, R.N., 2002. Basis and methods of NASA airborne topographic mapper lidar surveys for coastal studies. J. Coast Res. 1–13.

Dunkin, L., Reif, M., Altman, S., Swannack, T., 2016. A spatially explicit, multi-criteria decision support model for loggerhead sea turtle nesting habitat suitability: a remote sensing-based approach. Remote Sens. 8 (7), 573.

Finkl, C.W., 2004. Coastal classification: systematic approaches to consider in the development of a comprehensive scheme. J. Coast Res. 166–213.

Gregg, W.W., Casey, N.W., 2004. Global and regional evaluation of the SeaWiFS chlorophyll data set. Remote Sens. Environ. 93 (4), 463–479.

Kääb, A., 2008. Remote sensing of permafrost-related problems and hazards. Permafr. Periglac. Process. 19 (2), 107–136.

Lee, Z., Carder, K.L., 2004. Absorption spectrum of phytoplankton pigments derived from hyperspectral remote-sensing reflectance. Remote Sens. Environ. 89 (3), 361–368.

Mancini, F., Dubbini, M., Gattelli, M., Stecchi, F., Fabbri, S., Gabbianelli, G., 2013. Using unmanned aerial vehicles (UAV) for high-resolution reconstruction of topography: the structure from motion approach on coastal environments. Remote Sens. 5 (12), 6880–6898.

Matthews, M.W., Odermatt, D., 2015. Improved algorithm for routine monitoring of cyanobacteria and eutrophication in inland and near-coastal waters. Remote Sens. Environ. 156, 374–382.

Melesse, A.M., Weng, Q., Thenkabail, P.S., Senay, G.B., 2007. Remote sensing sensors and applications in environmental resources mapping and modelling. Sensors 7 (12), 3209–3241.

Mishra, D., Gould, R., 2016. Preface: remote sensing in coastal environments. Remote Sens. 8, 665.

Nayak, S., 2004. Role of remote sensing to integrated coastal zone management. In: XXth Congress of the International Society for Photogrammetry and Remote Sensing (Istanbul, Turkey), Commission, vol. 7, p. 18.

Pham, T.D., Xia, J., Ha, N.T., Bui, D.T., Le, N.N., Tekeuchi, W., 2019. A review of remote sensing approaches for monitoring blue carbon ecosystems: mangroves, seagrassesand salt marshes during 2010–2018. Sensors 19 (8), 1933.

Saatchi, S., Buermann, W., Ter Steege, H., Mori, S., Smith, T.B., 2008. Modeling distribution of Amazonian tree species and diversity using remote sensing measurements. Remote Sens. Environ. 112 (5), 2000–2017.

Toure, S., Diop, O., Kpalma, K., Maiga, A.S., 2019. Shoreline detection using optical remote sensing: a review. ISPRS Int. J. Geo-Inf. 8 (2), 75.

Williamson, M.J., Tebbs, E.J., Dawson, T.P., Jacoby, D.M., 2019. Satellite remote sensing in shark and ray ecology, conservation and management. Front. in Mar. Sci. 6, 135.

3

Ocean remote sensing for seasonal predictability of phytoplankton (chl-a) biomass in the Southern Indian coastal water region using Landsat 8 OLI images

Kaliraj Seenipandi[1], K.K. Ramachandran[1], Prashant Ghadei[2], Sulochana Shekhar[2]

[1]NATIONAL CENTRE FOR EARTH SCIENCE STUDIES (NCESS), THIRUVANANTHAPURAM, KERALA, INDIA; [2]DEPARTMENT OF GEOGRAPHY, CENTRAL UNIVERSITY OF TAMIL NADU, THIRUVARUR, TAMIL NADU, INDIA

Chapter outline

Remote Sensing of Ocean and Coastal Environments. https://doi.org/10.1016/B978-0-12-819604-5.00003-2

1. Introduction

Phytoplankton biomass is an indicator of the wealthiest aquatic environment; it is a primary food source for many aquatic living organisms and fish communities (Chauhan et al., 2002). Within almost all types of phytoplankton species, chlorophyll-a (chl-a) occurs as photosynthetically active pigments that influence the amount of backscattering from water and other components (Balzano et al., 2015). The phytoplankton communities in coastal waters are diverse in nature and include approximately tens of thousands of phytoplankton species (Solanki et al., 2001). Seawater at each 100 mL contains thousands of flora and fauna, especially phytoplankton species within every 100 m^2 of the area (Desortiva, 1981; Bosart and Sprigg, 1998). In a coastal water column, phytoplankton productivity is highly sensitive to physical and chemical properties, and depends on environmental and climatic factors that are noteworthy shifting blooms or phenology of marine species, such as changing of salinity, temperature, and nutrients resulting from natural and anthropogenic impacts (Beaver and Crisman, 1991; Dey and Singh, 2003).

In the Southern Indian coastal water, phytoplankton biomass is characterized by seasonal hydrodynamic processes like salinity, temperature, river discharge, and inflow of nutrients via littoral current that affects diversity of phytoplankton, zooplankton, sea grasses, coral reefs, and so on. In the coastal water column, phytoplankton (chl-a) concentration faces seasonal variability due to changes in sea surface temperature (SST) and salinity. The dispersal of organic and inorganic materials (i.e., phytoplankton and suspended sediments) can distinctly extract from water based on the amount of spectral reflectance from the water column (Sarangi, 2011). The combination of suitable bands is used to estimate the phytoplankton biomass based on differentiating spectral reflectance of water and chl-a components (Chauhan et al., 2002; Sarangi et al., 2008). In the aquatic environment, phytoplankton biomass (chl-a) and spatiotemporal variability is often too difficult to characterize using in situ survey and conventional methods. Remote sensing techniques provide a near-real-time synoptic overview of the large areas commonly used for monitoring phytoplankton biomass variability in coastal water (Reinart and Kutser, 2006; Lee and Carder, 2002).

Remote sensing provides an effective platform for coastal and ocean dynamic studies based on multispectral images with a variety of spectral, spatial, and radiometric resolutions, as evidenced by much research worldwide. Mapping of phytoplankton using spectral reflectance properties of the image provides insight for understanding the phenology of marine ecosystems in coastal water (Nagamani et al., 2011). Multispectral images of the different satellites including Landsat TM, ETM+, and Operational Land Imager (OLI); Sentinel 2 MSI, Oceansat, and MODIS, are used for estimating phytoplankton (chl-a) biomass in the large area of aquatic environment (Nagamani et al., 2008; Barale, 2010; Palmer et al., 2015). The pigments (chl-a) of phytoplankton reflect unique amounts of spectral radiance from the water column that is relatively straightforward to retrieve phytoplankton biomass from remotely sensed images (Sarangi, 2011; Chen and Quan, 2013). Satellite images acquired on a multitemporal scale are used for measuring spatiotemporal variability of phytoplankton biomass based on discriminating spectral properties of the coastal water environment (Behrenfeld and Boss, 2006; Geider et al., 1997; Ahn et al., 2008). Mapping the phytoplankton (chl-a) biomass using multispectral images provide insights of spatiotemporal dynamics, bloom initiation, and peak and declining stages (Siegel et al., 2013). Measurement of phytoplankton (chl-a) in coastal water using multispectral images of Landsat TM and ETM+, and OLI provides positive correlation compared to in situ measurement (Vaillancourt et al., 2004). Many researchers have used Landsat TM and ETM+, and OLI images for measurement of phytoplankton (chl-a) biomass using multivariate algorithms based on different band combination analysis (Burrage and Wesson, 2008; Wallhead et al., 2014; Kaliraj et al., 2019a,b). The empirical algorithm is used for estimation of phytoplankton (chl-a) biomass using multispectral images of Landsat, Oceansat, SeaWiFS, MODIS, Sentinel, and others based on specific spectral signature that is discriminating phytoplankton from water properties. Mapping of phytoplankton (chl-a) biomass and its spatiotemporal variability provides primary information for understanding growth and distribution of flora and fauna species in coastal and marine ecosystems.

2. Study area profile

The Southern Indian coastal water experiences dynamic changes in phytoplankton (chl-a) biomass, which plays a significant role in growth and productivity of coastal and marine ecosystems. The study area covers the seawater column with the areal extends of 8440.22 km^2 along the shoreline length of 410.72 km between the Keelakarai and Thiruvananthapuram coastal stretch. The geographical extend of the study area lies between the latitude 7°38′ N and 9°20′ N and the longitude 76°50′ E and 78°53′ E. Fig. 3.1 shows the study area location and chl-a observed stations. The water column constitutes mixing of seawater from the tropical India Ocean with the flows of the two seas, Arabian Sea and Bay of Bengal. The coastal landforms consist of laterite uplands, estuaries, sand dunes, and outlets of the nonperennial river. The coastal water receives river discharges along with runoff matter from different drainage networks like Thamirabarani, Valliyar, Pazhayar, and Vaippar during northeast and southwest monsoons (Kaliraj and Chandrasekar, 2012; Kaliraj et al., 2015, 2016). The water column had a shallow depth (1–10 m) between nearshore and surf zones, and gradually increased to 20–30 m toward a seaward distance of 5–6 km (Kaliraj et al., 2013, 2014a,b, 2017). The water column is seasonally varying in hydrodynamic-like waves, wind and currents, and so on. The average value of wave energy is about 0.5–8.5 kJ/km^2, and the mean current velocity is measured as 0.14 ms^{-1}. The study area prevails subtropical climatic conditions with temperature ranging from 23.78 to 33.95°C and annual average rainfall between 826 and 1456 mm. Monitoring and assessment of phytoplankton (chl-a) and its seasonal variability using multitemporal Landsat 8 OLI images is vital for understanding ocean and coastal hydrodynamic characteristics as well as an indicator to map fish-catching potential in the southern Indian coastal water region.

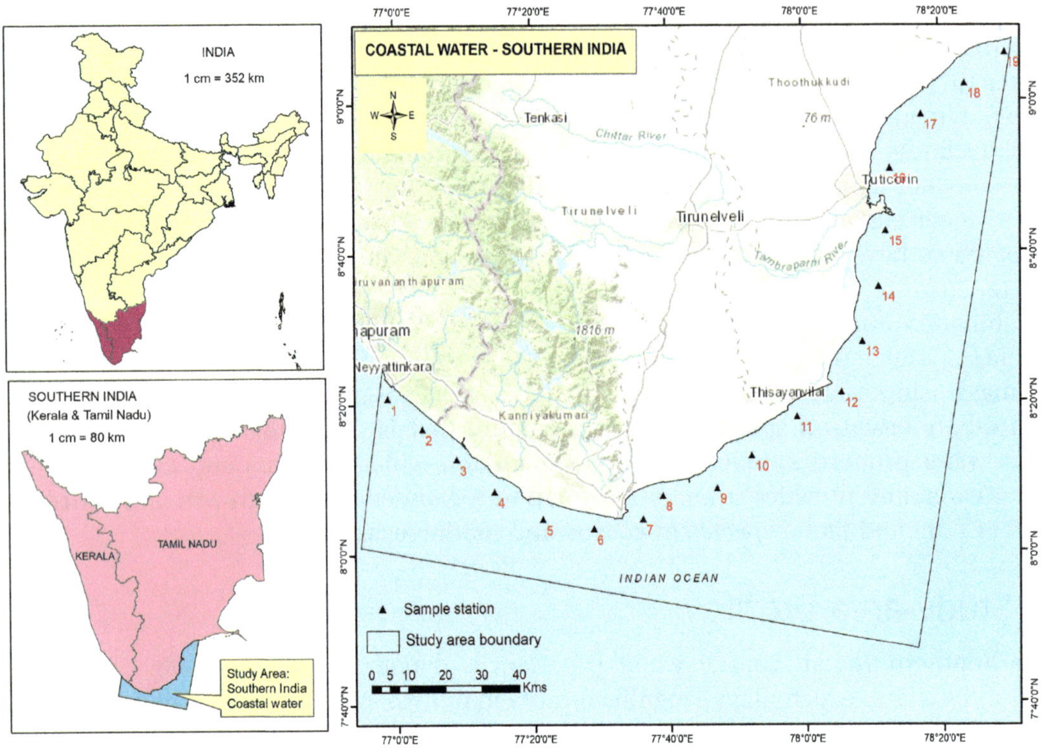

FIGURE 3.1 Geographical location of the study area.

3. Materials and methods

Multitemporal Landsat 8 OLI images are used for estimating phytoplankton (chl-a) and its seasonal variability in the Southern Indian coastal water column for the three seasons (pre-, post-, and during monsoons) during the period of 2017 to 2018. Landsat 8 OLI images (level-1 GeoTIFF data product) are collected from the USGS-Earth Explorer online portal. Landsat 8 satellite has two sensors, OLI and thermal infrared sensor. Landsat 8 OLI images have improved radiometric precision over a 12-bit dynamic range that improves overall signal-to-noise ratio and enables radiometric characterization for exploring land and water conditions (LSDS-1574, 2019). The OLI sensor collects image data for nine spectral bands (visible, NIR, SWIR) with a 30 m spatial resolution except the Panchromatic band (15 m). The two new bands in addition to the legacy Landsat bands (1–5, 7, and Pan)—(1) Coastal/Aerosol band (band 1; wavelength = 0.435–0.451 μm), principally used for ocean color observations, is similar to ALI's band 1; and (2) Cirrus band (band 9; wavelength = 1.36–1.38 μm) aids in the detection of cirrus clouds. Landsat 8 OLI image with multispectral bands (spatial resolution = 30 m) covers wide areas of Earth's landscape while providing sufficient resolution to distinguish land and water features within a 16-day repetitive cycle. Landsat 8 OLI images are analyzed using the empirical algorithm developed by O'Reilly et al. (2000) for mapping phytoplankton (chl-a) with a comprehensive and accurate pattern of spatial and seasonal variability in coastal water regions.

Fig. 3.2 shows the flowchart of systematic methodology for estimating the phytoplankton (chl-a) biomass and its seasonal variability using Landsat 8 OLI images. The systematic analysis of phytoplankton (chl-a) biomass is performed using the following steps: (1) preprocessing of Landsat 8 OLI images for atmospheric and radiometric correction using FLAASH module in ENVI software, (2) masking of land and clouds from layer-stacked OLI images using ERDAS imagine software, (3) demarcating the water column using an isodata algorithm of unsupervised classification technique, and (4) estimating the phytoplankton biomass (chl-a) via single fourth-order polynomial equation developed by O'Reilly et al. (2000) using the Raster Calculator module in ArcGIS software. The Landsat 8 OLI (Level 1T—GeoTiFF) images are calibrated via top-of-atmosphere and dark-object-subtraction techniques for atmospheric and radiometric correction using ENVI 5.3 software. The calibrated OLI image shows spectral reflectance as a reflected value from the objects. Second, a masking technique is applied on an OLI image—multispectral bands for removing land, clouds, and intertidal zones and distinguish the boundary between water and nonwater regions. Finally, the OLI images (FCC format) are analyzed for estimating the chl-a at pixel scale of 30 × 30 m using the single fourth-order polynomial equation that is expressed as

$$log10\,(Chl - a) = a_0 + a_1X + a_2X^2 + a_3X^3 + a_4X^4 \tag{3.1}$$

where $X = log_{10}[R_{rs}(\lambda_b)/R_{rs}(\lambda_g)]$

In Eq. (3.1), chl-a refers phytoplankton biomass (chlorophyll-a concentration) in the water column, a_0–a_4 are the polynomial fit coefficients of chl-a in different wavelengths (λ) derived from in situ sampling, and λ_b and λg refer to blue and green bands of a Landsat 8 OLI image. In this analysis, Eq. (3.1) is substituted with the coefficient values as follows:

$$Chl - a\left(mg/m^3\right) = \left[log10^{0.2511-2.0853\,X+1.5035\,X2-3.1747\,X3+0.3383\,X4}\right] \tag{3.2}$$

in which $X = log_{10}\left(R_{rs} * Band2 / R_{rs} * Band3\right)$

where *Chl-a* refers to chl-a concentration (mg/m^3), *Band2* and *Band3* refer to the blue band (0.45–0.51 μm) and the green band (0.53–0.59 μm) of the Landsat 8 OLI image, *X* refers to the logarithmic value of the numerical value used in the equation, which refers to chl-a coefficient values derived from in situ chl-a measurements. In the combination of selected bands of the OLI image, chl-a reflects a unique spectral signature that differs from water leaving radiance and shows a high degree of consistency for chl-a estimation.

4. Results and discussion

Phytoplankton biomass and seasonal variability is estimated using multitemporal Landsat 8 OLI images. Monitoring and measurement of phytoplankton is the primary information for researchers and stakeholders in the field of coastal and marine studies.

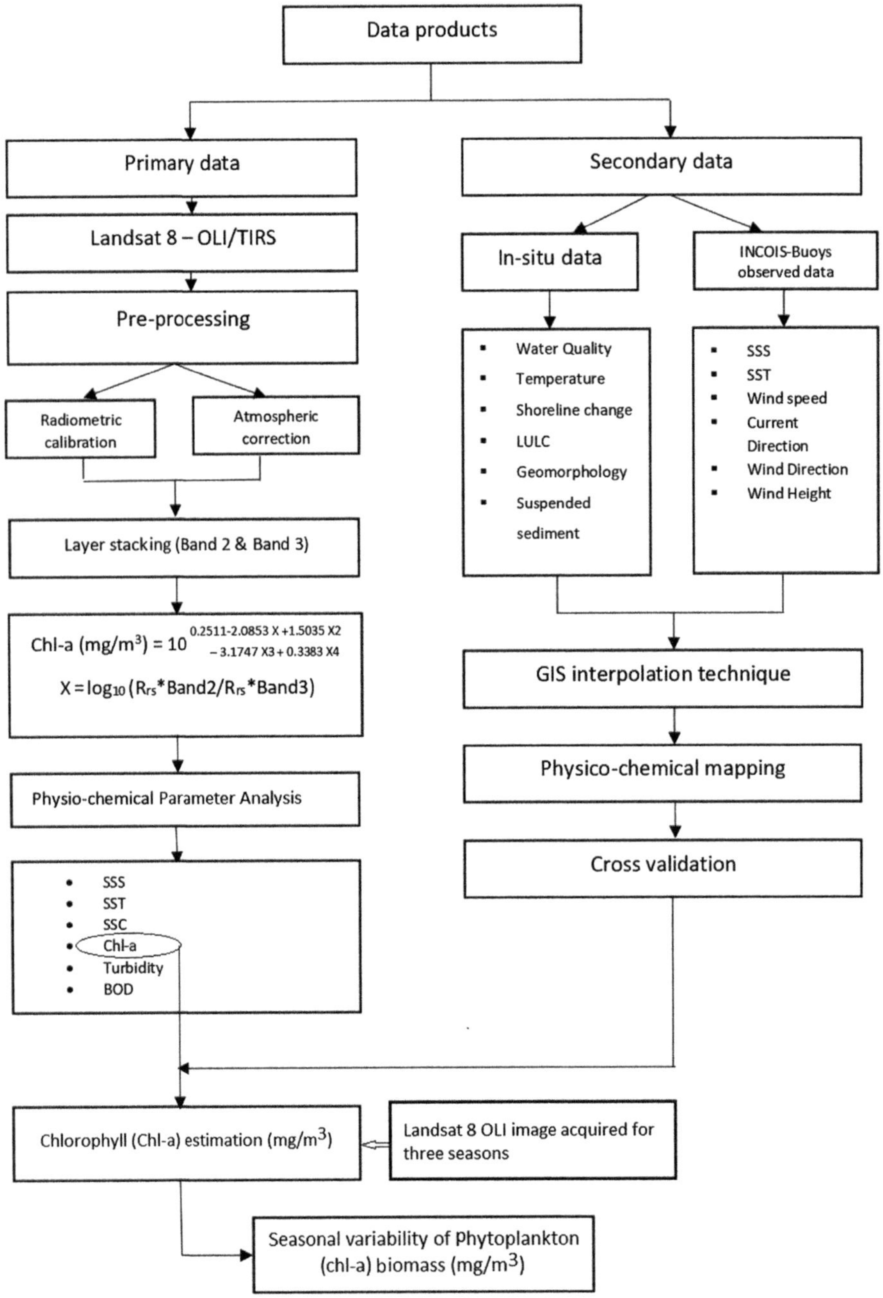

FIGURE 3.2 Methodology flowchart for estimating chl-a seasonal variability.

4.1 Spectral reflectance of Landsat 8 OLI for chl-a extraction

Measuring the phytoplankton (chl-a) biomass from Landsat 8 OLI images uses the empirical algorithm based on distinct spectral properties of clear water and chl-a reflectance. The phytoplankton (chl-a) concentration in coastal water determines the amount of reflectance at different wavelengths, which is quite different from water leaving radiance. In situ measurement of chl-a concentration in coastal water shows tolerable accuracy using a field spectro-radiometer operating at a wavelength of 443, 490, 560, and 670 nm. Hence, the Landsat 8 OLI image is widely used to measure chl-a concentration based on spectral reflectance (Rrs) and water leaving radiance (Lw) using suitable algorithm validation. The Landsat 8 OLI image contains blue and green bands (wavelength (λ) 0.45–0.59 μm) able to clearly differentiate the chl-a reflectance from coastal water and other suspended matters. Spectral in-water measurements of water and chl-a have been propagated for calculating polynomial fit coefficient values, as estimated from spectral profiles using linear regression analysis with R^2 value > 0.75. Phytoplankton (chl-a) occurring in coastal water reflects more radiance in visible bands (<85% at wavelength 0.4–0.6 μm) than that of water leaving reflectance, and it is reversely proportional in the wavelength of NIR bands. Mapping of chl-a in coastal water using a Landsat 8 OLI image shows the actual distribution trend, indicating that the single fourth-order polynomial model after being calibrated for atmospheric and radiometric correction and spectral smoothing. The chl-a is measured with reliable accuracy from the wavelength of the blue band (0.45–0.51 μm) and green band (0.53–0.59 μm) at the reflectance rate > 75%, in which the spectral reflectance is properly responsive to detect chl-a concentration from those band combinations of OLI images that tend to estimate the phytoplankton (chl-a) biomass in coastal water.

4.2 Estimation of phytoplankton (chl-a) biomass

Phytoplankton (chl-a) biomass and seasonal variability in a water column of the Southern Indian coastal and offshore region is estimated using multitemporal Landsat 8 OLI images acquired during 2017–2018. The result reveals that the higher concentration of phytoplankton (chl-a) biomass is estimated in premonsoon with ranges from 0.88 to 3.26 mg/m^3, and it is significantly decreased to 0.67–2.62 mg/m^3 in monsoon and 0.72–3.01 mg/m^3 in postmonsoon. The phytoplankton (chl-a) biomass changes seasonally from place to place based on dynamics of SST, salinity and littoral currents, precipitation and river discharges, and so on. However, the coastal water column within proximity of shallow depth (<10 m depth) of a nearshore area with the distance of 1 km from the shoreline found a higher concentration of phytoplankton (chl-a) biomass at 0.95–3.26 mg/m^3 during all three seasons. During premonsoon (September 2018), the spatial variability of chl-a is estimated at higher range of 0.63–3.25 mg/m^3 due to prevailing suitable SST and salinity along with availability of nutrients through eastward flowing littoral currents from the tropical Indian Ocean and the Arabian Sea. The lower concentration of chl-a is estimated in the majority of areas at the range of 0.56–3 mg/m^3, during postmonsoon (March 2018), while the wide-spreading of lower chl-a is due to lowering river discharge, and the prevailing higher SST and lower salinity is from the Bay of Bengal (BoB) west-flowing currents. Noticeably, in the water column of nearshore, the chl-a concentration is spatially found at higher values of 1.02–3 mg/m^3 and 0.51–2.62 mg/m^3 during postmonsoon and monsoon, respectively. Meanwhile, the chl-a concentration is found at a lower rate of 0.95–2.62 mg/m^3 in offshore wide-spreading water columns during monsoon. It is noted that the spatial variability of chl-a shows higher concentration in nearshore water (shallow depth region) and lower value in offshore water (deep depth region) due to influences of SST, waves, currents, winds, and river discharges.

4.3 Phytoplankton (chl-a) biomass variability in premonsoon (September 2018)

During premonsoon, the phytoplankton (chl-a) biomass spatially varies in the range of 0.63–3.25 mg/m^3. Fig. 3.3 shows the spatial variability of phytoplankton (chl-a) biomass

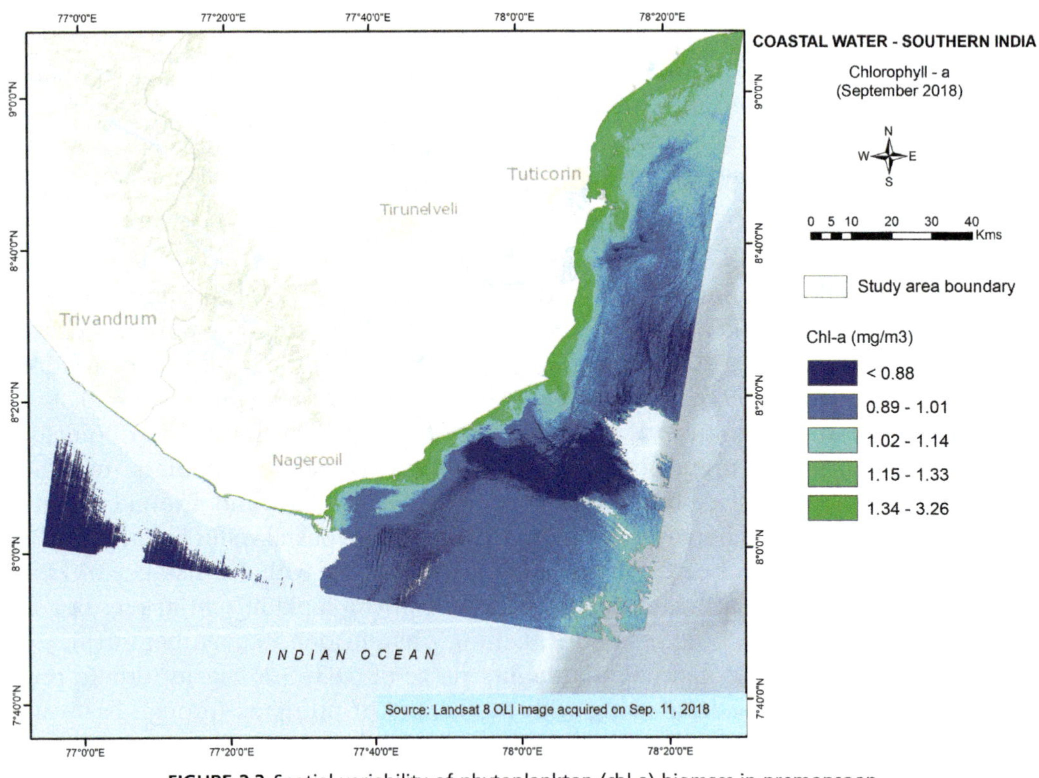

FIGURE 3.3 Spatial variability of phytoplankton (chl-a) biomass in premonsoon.

in premonsoon. Within the coastal water column, the chl-a concentration is found at a higher value of 1.15–3.26 mg/m^3 among all three seasons due to availability of nutrients transported through east-flowing littoral currents. Meanwhile, the offshore (deep sea) wide-spreading water column is estimated as lower phytoplankton (chl-a) biomass with the range of 0.88–1.14 mg/m^3 due to the dynamics of SST and salinity from place to place and at vertical range of the water column that highly influences phytoplankton growth and productivity.

4.4 Phytoplankton (chl-a) biomass variability in monsoon (December 2017)

During monsoon (December 2017), the lower phytoplankton (chl-a) biomass found within coastal water at the rate of 0.51–2.62 mg/m^3 compared to other seasons, wherein, the average value of phytoplankton (chl-a) biomass is estimated as lower rate of 0.749 mg/m^3 than that of pre-monsoon and post monsoon. Within the proximity of coastal water (shallow depth of nearshore), the phytoplankton (chl-a) biomass is estimated as higher value at the range of 0.95–2.62 mg/m^3 than the wide-spreading water column of the offshore region. Fig. 3.4 shows the spatial variability of

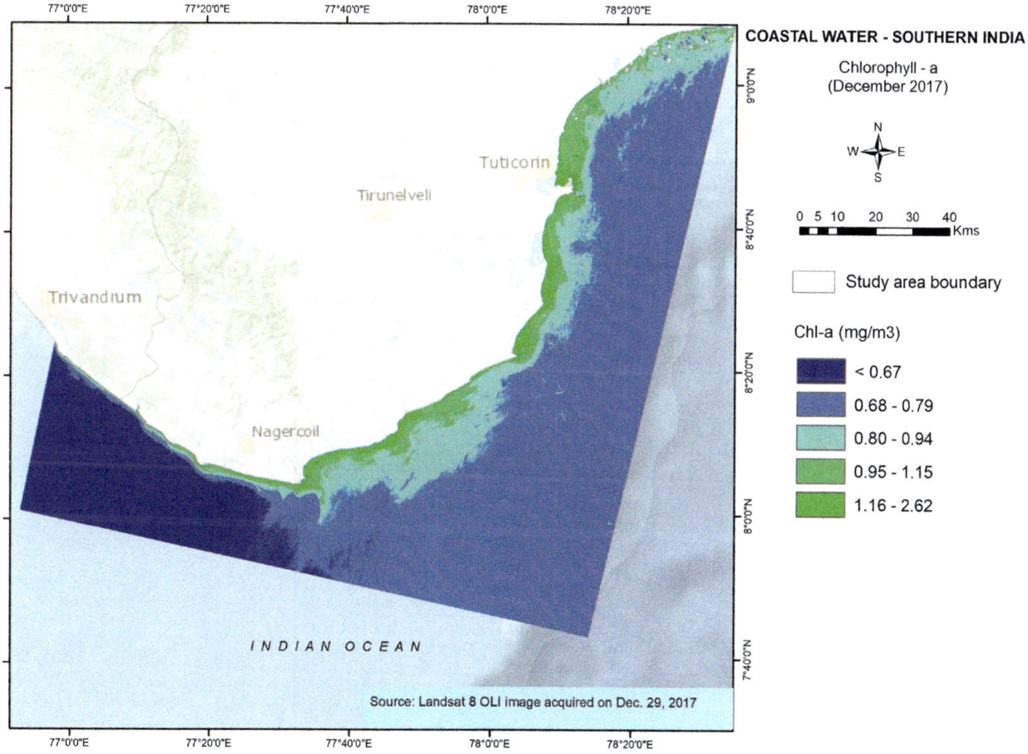

FIGURE 3.4 Spatial variability of phytoplankton (chl-a) biomass in monsoon.

phytoplankton (chl-a) biomass in monsoon period. It is noted that the chl-a concentration found higher value especially in eastern part of the coastal water than offshore (deep-sea water), due to impacts of northeast monsoon that affects SST and salinity characteristics. However, a few spots of coastal water is found at lower phytoplankton (chl-a) biomass due to mixing of freshwater from terrestrial runoff that causing lower salinity and nutrients.

4.5 Phytoplankton (chl-a) biomass variability in postmonsoon (March 2018)

During postmonsoon, the phytoplankton (chl-a) biomass is found at the higher rate in water columns of nearshore (1.04–3.01 mg/m^3) and offshore (0.72–1.03 mg/m^3) compared to the monsoon period. Fig. 3.5 shows the spatial variability of phytoplankton (chl-a) biomass in postmonsoon. The average rate of phytoplankton (chl-a) biomass is estimated as 0.82 mg/m^3 in a wide-spreading water column, whereas the phytoplankton (chl-a) biomass spatial variability is recorded as lower concentration (0.72–0.86 mg/m^3) in a majority of offshore regions due to lower seawater salinity while prevailing higher SST. However, a few pockets of coastal water are estimated as higher concentration of phytoplankton (chl-a) biomass at the range of 1.04–3.01 mg/m^3 due to mixing of nutrients from littoral currents.

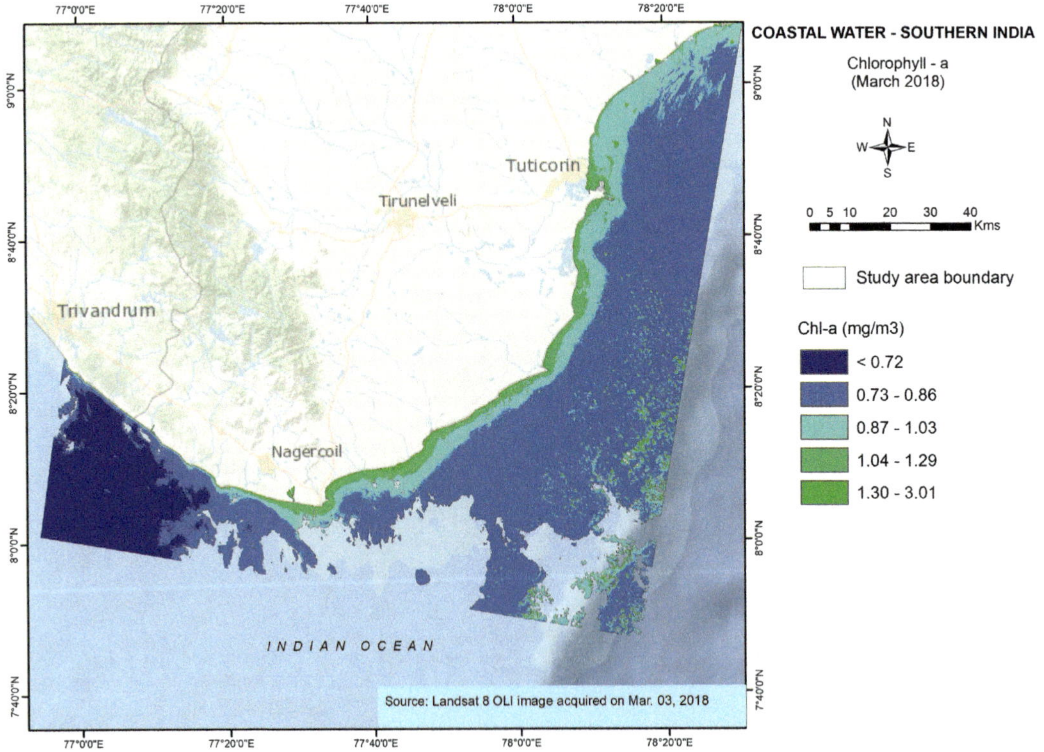

FIGURE 3.5 Spatial variability of phytoplankton (chl-a) biomass in postmonsoon.

4.6 Seasonal variability of phytoplankton (chl-a) biomass during 2017–2018

In the Southern Indian coastal water, the seasonal variability of phytoplankton (chl-a) biomass is characterized using Landsat 8 OLI images. Fig. 3.6 shows the seasonal variability of phytoplankton (chl-a) biomass for the three seasons during 2017–18. The phytoplankton (chl-a) biomass variability for the 19 selected stations (Fig. 3.1) of the Southern Indian coastal water is shown in Table 3.1. The mean of phytoplankton (chl-a) biomass is noted as 0.92 mg/m^3, whereas the overall maximum chl-a concentration is estimated as 1.37 mg/m^3 in station 12, and the overall minimum chl-a concentration is measured at 0.61 mg/m^3 in station 4. In the offshore wide-spreading water column (deep sea), the phytoplankton (chl-a) biomass shows at a lower rate as 0.67–1.14 mg/m^3 during all three seasons, whereas the chl-a concentration was found at a relatively constant rate during monsoon and postmonsoon periods in stations 11–19.

The standard deviation (SD) value for phytoplankton (chl-a) biomass is estimated at 0.03–0.15 mg/m^3, which indicates little spatiotemporal variability during all three seasons, but it changes in the offshore water column (deep-sea) during different seasons. Seasonal variability of phytoplankton (chl-a) biomass shows the higher deviation as 0.15–0.26 mg/m^3 in a few packets of coastal water in stations 1, 11, 12, 18, and 19. The phytoplankton (chl-a) biomass in water columns of the offshore region is measured at a lower rate mainly due to the dynamics of SST and salinity and the lack of nutrient availability. The growth and productivity of phytoplankton (chl-a) in an offshore water column is highly sensitive to physical and chemical properties of seawater that are frequently influencing algal blooms or phenology of marine species by changing salinity, temperature, and nutrients. The chl-a concentration has lower values in the western part of coastal water than the eastern side. Similarly, some packets of coastal water have noticed a lower range of chl-a concentration due to dynamic changes of salinity and SST

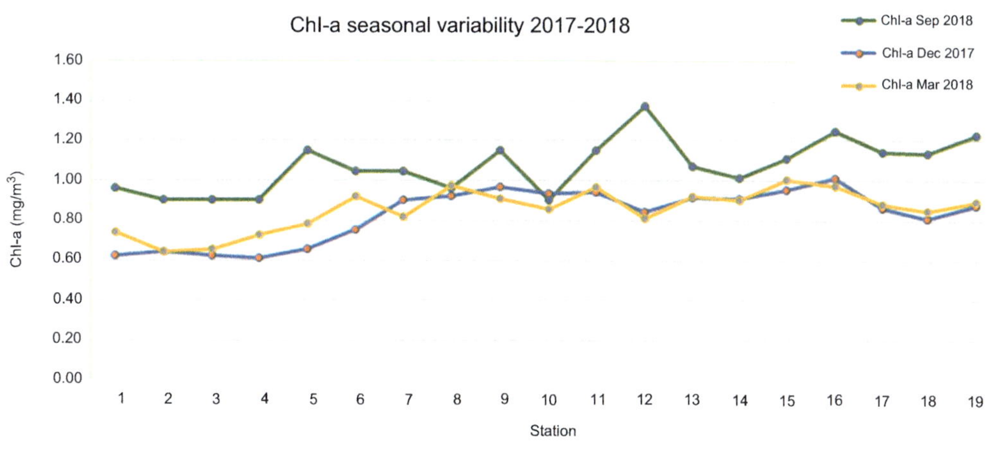

FIGURE 3.6 Seasonal variability of phytoplankton (chl-a) biomass during 2017–2018.

Table 3.1 Chl-a seasonal variability in selected stations along the Southern Indian Coast (Chl-a mean refers average of 3 seasons, SD refers Standard Deviation)

	Chl-a seasonal variability (mg/m^3)						
Stations	**Sep 2018 (pre-monsoon)**	**Dec 2017 (monsoon)**	**Mar 2018 (post monsoon)**	**Chl-a Maximum (mg/m^3)**	**Chl-a Minimum (mg/m^3)**	**Chl-a Mean (mg/m^3)**	**SD**
1	0.96	0.62	0.74	0.96	0.62	0.78	0.17
2	0.90	0.64	0.65	0.90	0.64	0.73	0.15
3	0.90	0.62	0.66	0.90	0.62	0.73	0.15
4	0.90	0.61	0.73	0.90	0.61	0.75	0.15
5	1.15	0.65	0.79	1.15	0.65	0.86	0.26
6	1.05	0.76	0.93	1.05	0.76	0.91	0.15
7	1.05	0.90	0.82	1.05	0.82	0.92	0.11
8	0.96	0.93	0.97	0.97	0.93	0.95	0.03
9	1.15	0.97	0.91	1.15	0.91	1.01	0.12
10	0.90	0.94	0.86	0.94	0.86	0.90	0.04
11	1.15	0.94	0.97	1.15	0.94	1.02	0.11
12	1.37	0.85	0.81	1.37	0.81	1.01	0.31
13	1.08	0.91	0.92	1.08	0.91	0.97	0.09
14	1.02	0.91	0.90	1.02	0.90	0.94	0.06
15	1.11	0.95	1.01	1.11	0.95	1.02	0.08
16	1.25	1.01	0.97	1.25	0.97	1.08	0.15
17	1.14	0.87	0.88	1.14	0.87	0.96	0.15
18	1.14	0.81	0.85	1.14	0.81	0.93	0.18
19	1.23	0.88	0.90	1.23	0.88	1.00	0.20

that influence growth and productivity of phytoplankton and their assemblages. Seasonal variability of phytoplankton (chl-a) biomass shows a higher rate in pre-monsoon; this rate decreases toward postmonsoon and monsoon periods up to values of 1.37–0.90 mg/m^3, respectively. However, in the offshore water column, the phytoplankton (chl-a) biomass is measured at lower rate at a value of $<$0.90 mg/m^3 and noted as a relatively homogenous distribution during all three seasons. The seasonal variability of SST and salinity determines spatiotemporal distribution of phytoplankton (chl-a) biomass and their recurrence strength in the offshore region, but the growth and productivity of phytoplankton (chl-a) biomass in the coastal water column depends on littoral currents and river discharges.

Phytoplankton (chl-a) biomass seasonal variability in water columns of nearshore and offshore varies from place to place depending on changing hydrodynamic factors. Meanwhile, some site-specific packets of water columns found significant amounts of phytoplankton (chl-a) biomass due to additional nutrients from littoral sediments and river discharges. In the Southern Indian coastal water region, the phytoplankton (chl-a) biomass varies at spatiotemporal scale, depending on seasonally influenced hydrodynamic factors.

5. Conclusions

Mapping of phytoplankton (chl-a) biomass and its seasonal variability using Landsat 8 OLI images provides vital information for stakeholders of fisheries and oceanographic studies. It is noted that the band combination with the wavelength of 0.45–1.0 μm of OLI image perfectly reflects the spatiotemporal variability of phytoplankton (chl-a) biomass in a water column of coastal and offshore regions. In the Southern Indian coastal water column, the spatiotemporal variability of phytoplankton (chl-a) biomass shows a higher rate in coastal water columns during all three seasons due to supplying enough nutrients from littoral currents and river discharges. In the offshore, the wide-spreading water column of deep sea found a lower rate of phytoplankton (chl-a) biomass due to dynamic changing SST and salinity, and it is relatively constant with a very little spatiotemporal variability during monsoon. It is noted that the seasonal variability of phytoplankton (chl-a) biomass is due mainly to changing of hydrodynamic factors like SST, salinity, precipitation, littoral currents, and river discharges. Mapping and monitoring of phytoplankton (chl-a) biomass and its seasonal variability is being a primary source for finding the potential fishing zone and helpful for preparing sustainable management plans for coastal and marine ecosystems for long-term scale.

Acknowledgments

The corresponding author is thankful to the Director, National Center for Earth Science Studies (NCESS), Ministry of Earth Sciences, Government of India, for providing continuing support and encouragement. The authors extend their gratitude to IMD, SOI, NIOT, INCOIS, NOAA, NASA, USGS–Earth Explorer, and PSMSL–UK for providing necessary data sources.

References

Ahn, Y.H., Shanmugam, P., Moon, J.E., Ryu, J.H., 2008. Satellite remote sensing of a low salinity waters plume in the East China Sea. Ann. Geophys. 26, 2019–2035.

Balzano, S., Ellis, A.V., Lan, C.L., Leterme, S.C., 2015. Seasonal changes in phytoplankton on the Northeastern shelf of Kangaroo Island (South Australia) in 2012 and 2013. Oceanologia 57, 251–262.

Barale, V., 2010. Oceanography From Space: Revisited. Springer, Ispra, pp. 237–238.

Beaver, J.R., Crisman, T.L., 1991. Temporal variability in algal biomass and primary productivity in Florida lakes relative to longitudinal gradients, organic colour and trophic states. Hydrobiologia 224, 89–97.

Behrenfeld, M.J., Boss, E., 2006. Beam attenuation and phytoplankton (chl-a) concentration as alternative optical indices of phytoplankton biomass. J. Mar. Res. 64, 431–451.

Bosart, L.F., Sprigg, W.A., 1998. The Meteorological Buoy and Coastal Marine Automated Network for the United States. National Academies Press, Washington, DC.

Burrage, D., Wesson, J., 2008. Deriving sea surface salinity and density variations from satellite and aircraft microwave radiometer measurements: application to coastal plumes using STARRS. IEEE Trans. Geosci. Remote Sens. 46 (3), 765–785.

Chauhan, P., Mohan, M., Sarngi, R.K., Kumari, B., Nayak, S., Matondkar, S.G.P., 2002. Surface phytoplankton (chl-a) a estimation in the Arabian Sea using IRS-P4 Ocean Colour Monitor (OCM) satellite data. Int. J. Remote Sens. 23 (8), 1663–1676.

Chen, H., Quan, W., 2013. An improved algorithm for retrieving phytoplankton (chl-a)-a from the Yellow River Estuary using MODIS imagery. Environ. Monit. Assess. 185, 2243–2255.

Dey, S., Singh, R.P., 2003. Comparison of phytoplankton (chl-a) distributions in the Northeastern Arabian Sea and Southern Bay of Bengal using IRS P4 ocean color monitor data. Remote Sens. Environ. 85, 424–428.

Desortiva, B., 1981. Relationship between phytoplankton (chl-a) concentration and phytoplankton biomass in several reservoirs in Czechoslovakia. Int. Revue Ges. Hydrobiol. 66, 153–169.

Geider, R., MacIntyre, H., Kana, T., 1997. Dynamic model of phytoplankton growth and acclimation: responses of the balanced growth rate and the phytoplankton (chl-a) a: carbon ratio to light, nutrient-limitation and temperature. Mar. Ecol. Prog. Ser. 148, 187–200.

Kaliraj, S., Chandrasekar, N., Magesh, N.S., 2014a. Multispectral image analysis of suspended sediment concentration along the Southern coast of Kanyakumari, Tamil Nadu, India. J. Coast. Sci. 1 (1), 63–71. ISSN: 2348 – 6740.

Kaliraj, S., 2016a. Geospatial Analysis of Coastal Geomorphological Vulnerability on South West Coast of Kanyakumari Using Remote Sensing Technology. Unpublished Ph.D. Thesis. ManonmaniamSundaranar University, Tirunelveli, pp. 83–110.

Kaliraj, S., Chandrasekar, N., 2012. Geo-processing model on coastal vulnerability index to explore risk zone along the South West coast of Tamilnadu, India. Int. J. Earth Sci. Eng. 5 (5), 1138–1147.

Kaliraj, S., Chandrasekar, N., Magesh, N.S., 2014b. Impacts of wave energy and littoral currents on shoreline erosion/accretion along the South-West coast of Kanyakumari, Tamil nadu using DSAS and geospatial Technology. Environ. Earth Sci. https://doi.org/10.1007/s12665-013-2845-6.

Kaliraj, S., Chandrasekar, N., Magesh, N.S., 2013. Evaluation of coastal erosion and accretion processes along the SouthWest coast of Kanyakumari, Tamil Nadu using geospatial techniques. Arab. J. Geosci. https://doi.org/10.1007/s12517-013-1216-7.

Kaliraj, S., Chandrasekar, N., Ramachandran, K.K., 2019a. Phytoplankton (chl-a) biomass seasonal variability in the gulf of Mannar, South India: a remote sensing perspective. In: Kumar, P., et al. (Eds.), Applications and Challenges of Geospatial Technology. https://doi.org/10.1007/978-3-319-99882-4_5.

Kaliraj, S., Chandrasekar, N., Ramachandran, K.K., 2016. Mapping of coastal landforms and volumetric change analysis in the South West coast of Kanyakumari, South India using remote sensing and GIS techniques. Egypt. J. Remote Sens. & Space Sci. https://doi.org/10.1016/j.ejrs.2016.12.006.

Kaliraj, S., Chandrasekar, N., Ramachandran, K.K., Selvakumar, S., 2019b. Seawater intrusion vulnerability in the coastal aquifers of Southern India—an appraisal of the GALDIT model, parameters' sensitivity, and hydrochemical indicators. Environ. Sci. Pollut. Res. 26, 9755–9784. https://doi.org/10.1007/s11356-019-04401-0.

Kaliraj, S., Chandrasekar, N., Ramachandran, K.K., Srinivas, Y., Saravanan, S., 2017. Coastal landuse and land cover change and transformations of Kanyakumari coast, India using remote sensing and GIS. Egypt. J. Remote Sens. & Space Sci. https://doi.org/10.1016/j.ejrs.2017.04.003.

Kaliraj, S., Chandrasekar, N., Simon Peter, T., Selvakumar, S., Magesh, N.S., 2015. Mapping of coastal aquifer vulnerable zone in the South West coast of Kanyakumari, South India, using GIS-based DRASTIC model. Environ. Monit. Assess. https://doi.org/10.1007/s10661-014-4073-2.

Lee, Z., Carder, K.L., 2002. Effect of spectral band numbers on the retrieval of water column and band bottom properties from ocean color data. Appl. Opt. 41, 2191–2201.

Nagamani, P.V., Chauhan, P., Dwivedi, R.M., 2008. Development of chlorophyll-a algorithm for ocean colour monitor onboard OCEANSAT-2 satellite. IEEE Geosci. Remote Sens. Lett. 5 (3), 527–531.

Nagamani, P.V., Shikhakolli, R., Chauhan, P., 2011. Phytoplankton variability in the Bay of Bengal during winter monsoon using Oceansat-1Ocean Colour Monitor Data. J. Indian Soc. Remote Sens. 39 (1), 117–126.

O'Reilly, J.E., Maritorena, S., O'Brien, M.C., Siegel, D.A., Toole, D., Menzies, et al., 2000. SeaWiFS Postlaunch calibration and validation Analyses. In: Hooker, S.B., Firestone, E.R. (Eds.), Part 3. NASA Tech. Memo. 2000-206892, 11. NASA Goddard Space Flight Center, Greenbelt, Maryland, pp. 49–58.

Palmer, S.C.J., Odermatt, D., Hunter, P.D., Brockmann, C., Presing, M., Balzter, H., Toth, V.R., 2015. Satellite remote sensing of phytoplankton phenology in Lake Balaton using 10 years of MERIS observations. Remote Sens. Environ. 158, 441–452.

Reinart, A., Kutser, T., 2006. Comparison of different satellite sensors in detecting cyanobacterial bloom events in the Baltic Sea. Remote Sens. Environ. 102 (1-2), 74–85. https://doi.org/10.1016/j.rse.2006.02.013.

Sarangi, R.K., 2011. Impact of cyclones on the Bay of Bengal phytoplankton (chl-a) variability using remote sensing satellites. Indian J. Geo-Mar. Sci. 40 (6), 794–801.

Sarangi, R.K., Nayak, S., Panigrahy, R.C., 2008. Monthly variability of phytoplankton (chl-a) and associated physical parameters in the Southwest Bay of Bengal waters using remote sensing data. Indian J. Mar. Sci. 37 (3), 256–266.

Siegel, D., et al., 2013. Regional to global assessments of phytoplankton dynamics from the SeaWiFS mission. Remote Sens. Environ. 135, 77–91.

Solanki, H.U., Dwivedi, R.M., Nayak, S.R., Jadeja, J.V., Thakar, D.B., Dave, H.B., Patel, M.I., 2001. Application of Ocean Colour Monitor phytoplankton (chl-a) and AVHRR SST for fishery forecast: preliminary validation results off Gujarat coast, Northwest coast of India. Indian J. Mar. Sci. 30, 132–138.

Vaillancourt, R.D., Brown, C.W., Guillard, R.R.L., Balch, W.M., 2004. Light backscattering properties of marine phytoplankton: relationships to cell size, chemical composition and taxonomy. J. Plankton Res. 26, 191–212.

Wallhead, P.J., Garçon, V.C., Casey, J.R., Lomas, M.W., 2014. Long-term variability of phytoplankton carbon biomass in the Sargasso Sea. Glob. Biogeochem. Cycles 28, 825–841.

Further reading

Afdal, 2014. Chl-aVariation in Cirebon Waters and Its Relationship With Nutrient Concentration JurnalOseanologi dan Limnologi di Indonesia. LIPI, Jakarta, Indonesia.

Buditama, G., Damayanti, A., Pin, G.T., 2017. Identifying distribution of chl-aConcentration using Landsat 8 OLI on marine waters area of cirebon. IOP Conf. Ser. Earth Environ. Sci. 98. https://doi.org/10.1088/1755-1315/98/1/012040, 012040.

Geiger, F., Grossi, D., Trembanis, C., Kohut, T., Oliver, J., 2013. Satellite-derived coastal ocean and estuarine salinity in the Mid-Atlantic. Continent. Shelf Res. 63, S235–S242.

Hanintyo, R., Susilo, E., 2016. Comparison of chl-aMeasurement using multi spatial imagery and numerical model in bali strait. J. Front. Mar. Sci. 4 (Scopus).

Jaelani, L.M., Limehuwey, R., Kurniadin, N., Pamungkas, A., Koenhardono, E.S., Sulisetyono, A., April 2016. Estimation of TSS and chl-a concentration from Landsat 8-OLI: the effect of atmosphere and retrieval algorithm, IPTEK. J. Technol. & Sci. 27 (1).

Klemas, V., 2011. Remote sensing of sea surface salinity: an overview with case studies. J. Coast Res. 27 (5), 830–838. https://doi.org/10.2112/jcoastres-d-11-0006.1.

Nurdin, S., Mustapha, M.A., Lihan, T., Zainuddin, M., 2017. Applicability of remote sensing oceanographic data in the detection of potential fishing grounds of Rastrelligerkanagurta in the archipelagic waters of Spermonde, Indonesia. J. Fish. Res. 196 (Scopus).

Yadav, S., Yamashiki, Y., Susaki, J., Yamashita, Y., Ishikawa, K., 2019. Chlorophyll estimation of lake water and coastal water using Landsat-8 and Sentinel-2a Satellite,The international archives of the photogrammetry. Remote Sens. & Spat. Inf. Sci. XLII-3/W7.

Yoshino Watanabe, F.S., Alcântara, E., Rodrigues, T.W., Imai, N.N., Barbosa, C.C., Rotta, L.H., 2015. Estimation of chl-aConcentration and the trophic state of the Barra Bonita hydroelectric reservoir using OLI/Landsat-8 images. Int. J. Environ. Res. Public Health 2015 12, 10391–10417. https://doi.org/10.3390/ijerph120910391.

4

Ocean remote sensing for spatiotemporal variability of wave energy density and littoral current velocity in the Southern Indian offshore

Kaliraj Seenipandi[1], K.K. Ramachandran[1], Pavan Kumar[2]

[1]NATIONAL CENTRE FOR EARTH SCIENCE STUDIES (NCESS), THIRUVANANTHAPURAM, KERALA, INDIA; [2]COLLEGE OF HORTICULTURE AND FORESTRY, RANI LAKSHMI BAI CENTRAL AGRICULTURAL UNIVERSITY, JHANSI, UTTAR PRADESH, INDIA

Chapter outline

1. Introduction

The ocean environment constitutes a variety of hydrodynamic processes that regulate global climate conditions, hence it is very necessary to study wave energy density and

Remote Sensing of Ocean and Coastal Environments. https://doi.org/10.1016/B978-0-12-819604-5.00004-4

littoral current velocity of the ocean and coastal water regions to understand physical and atmospheric interaction. Surface water of oceans and coastal environments produces a variety of energy resources that depend on the potential impact of waves, wind, temperature, salinity, and currents that are classified as tidal energy, wave energy, and thermal energy. Ocean waves (swell) are formed by transferring energy from the motion of atmospheric wind to the ocean surface and releasing a certain amount of energy to the shoreline, causing erosion and accretion of coastal landforms for long-term scale (Kaliraj et al., 2014). Ocean wave energy results from the interacting process of wind, waves (swell), and currents passing over the surface of the seawater. The wave is as long as propagated by the wind just above the seawater surface; there is a certain amount of energy transfer from the wind to the waves, depending on the characteristics of wave height, seawater density, and acceleration due to gravity. Ocean waves (swell) result from wind and water interacting processes, but wave energy depends on oceanic parameters like wave height, wavelength, wind speed, and water density (Reynolds, 1877; Galvin, 1963; Phillips, 1977; Dyer, 1986; Din et al., 2012). In addition, physical parameters like depth (bathymetry), topography (slope), temperature, salinity, and pressure are involved in determining the dispersal of energy of the waves and littoral current flow. In the ocean or coastal environment, wave energy is transported horizontally along with littoral current velocity parallel to a wave crest and releasing to the shoreline, whereas, according to linear wave theory, the average value of wave energy density per unit area is proportional to gravity acceleration, seawater density, and square of wave height (Cruz, 2008; Folley and Whittaker, 2009; Jayaram et al., 2016). Mapping of spatiotemporal variability of wave energy releasing from ocean and coastal water has significant potential in terms of the development of a renewable energy resource because of negligible environmental impact, production of electricity, and a minimal use of fossil fuels in terms of addressing the climate change issue, and finally generating economically viable energy for developing countries.

In general, oceanographic parameters are measured using in situ survey like buoys and other devices, but they provide point data for specific sites. Even though buoy observations provide information on waves, they are not continuous, and synoptic information is not available for most parts of the Indian coast. Ocean remote sensing provides real-time observations of oceanographic parameters for understanding the spatiotemporal variations of ocean states and provides data sources for ocean forecast modeling. On a global scale, many satellites (namely MODIS-Aqua, Oceansat-OCM, GEOSAT, ERS-1, TOPEX, ERS-2, GFO, JASON-1, ENVISAT, JASON-2, CRYOSAT-2, HY-2A, SARAL, JASON-3, and Sentinel-3A) provide synoptic coverage of the ocean surface for measurement of various parameters like significant wave height, wave direction, currents, sea surface temperature, and sea surface salinity. Since 1985, the launch of GEOSAT, the satellite altimetry data has provided continuous coverage of global observations of ocean and coastal dynamics (Lee et al., 2004; Queffeulou, 2004; Kumar et al., 2015; Sepulveda et al., 2015; Uti et al., 2017; Young and Ribal, 2019).

Satellite altimeter measurement provides data sources for global coverage of wave height and wind speed for studying the ocean state forecasting in long-term trends. European Remote-Sensing Satellite (ERS-1) launched in mid-1991; since then, a total of 13 satellite altimeter missions have been operated for measuring oceanic parameters with the two latest satellites, JASON-3 and SENTINEL-3A, launched in 2016. Altimeter data obtained from Sentinel-3A is used for estimating significant wave height at a regional scale. Ocean remote sensing satellites provide imaging and nonimaging data for a 3- to 10-day repeat cycle, however, CRYOSAT-2 has a 369-day repeat cycle with a semirepeat cycle of 30 days. Currently, the ocean remote sensors have greatly improved as the number of altimeters in operation has increased for density of observations in ocean and coastal environments (Yong et al., 2015). Satellite radar altimetry is one of the most important tools for measuring surface wind waves. The continuous observation of coastal wave information is made possible from the ENVISAT, JASON-2, and SARAL missions. Moreover, satellite altimeter along track data were obtained from the AVISO (JASON-2) PISTACH coastal products and ISRO (SARAL). The altimeter data of ENVISAT is widely used for significant wave heights and wave energy measurements like the WAVEWATCH III global wave model. Many ocean remote sensing products (i.e., ocean color image and satellite altimetry data) are freely available to users for studying coastal ocean environments. These data products made possible estimation of wave energy density and littoral current velocity from coastal and ocean water columns, one of the successful studies established widely for understanding impacts of hydrodynamics, erosion and accretion processes, ocean forecasting, and harvesting ocean wave energy (Cruz, 2008; Kaliraj et al., 2014; Jayaram et al., 2015; Kaliraj, 2016).

Satellite altimetry data have been widely used for enhancing several projects established on monitoring and assessment of coastal and ocean environments; for example, the AltiKa altimeter data used for Coastal, Hydrology and Ice (PEACHI) measurement of hydrodynamics and geophysical parameters (i.e., significant wave height, wind speed, gravity, salinity, seawater density and temperature) based on mathematical algorithms (Wang, 1998; Queffeulou, 2004; Kaliraj et al., 2014). The advancement of data processing applied on ocean remote sensing products like AltiKa altimeter, SARAL, Synthetic Aperture Radar (SAR), Cryosat-2, Sentinel-3A, and Sentinel 3B have produced better results for ocean and coastal parametric analysis with final spatial resolution with good agreement with buoy measurements (Kumar et al., 2015; Sepulveda et al., 2015; Uti et al., 2017; Young and Ribal, 2019).

The Southern Indian offshore has seasonally dominated in wave, wind, and current behaviors because of the impacts of Tropical Indian Ocean, Arabian Sea, and Bay of Bengal, which directly affect beach morphodynamics, sediment transport, and erosion and accretion processes along the coastal stretches. Along the Southern Indian offshore, water depths greater than 300 m are characterized by relatively higher wave energy densities than those of shallow coastal water in the nearshore area, hence determining the stability of coastal landforms in this region. Scientific approaches of ocean remote sensing data are used for assessment of wave energy density and littoral current velocity at local and regional scales. Integrated remote sensing and GIS technology provides an effective platform for mapping and assessment of wave energy density and littoral current velocity at spatiotemporal scale.

2. Study area

The Southern Indian offshore experiences spatiotemporal changes in wave, current, and wind due to impacts of seasonal and monsoonal behaviors that play a significant role in morphodynamics of landforms along coastal stretches (Kaliraj, 2016). Geographically, the study area covers the water column of offshore and nearshore that lies between the latitude 7°38′ N and 9°20′ N and the longitude 76°50′ E and 78°53′ E, and the seawater column extends about 8440.22 km^2, parallel to the Keelakarai–Thiruvananthapuram coastline with a shoreline length of 410.72 km. The offshore constitutes wave and current characteristics from the complex hydrodynamic processes of the Tropical India Ocean associated with the Arabian Sea and Bay of Bengal Sea. Fig. 4.1 shows the geographical location of the study area. Along the coastline, stretches are found with a variety of landforms like sandy beaches, plains, terraces, sand dunes, rocky shores, and estuaries. Some packets of young formation of sandy beaches and beach ridges are found due to the action of the constructive action of waves and littoral currents (Kaliraj et al., 2013, 2016). The water column is getting runoff from river discharges of different drainage networks, including Thamirabarani, Valliyar, Pazhayar, and Vaippar during northeast and southwest monsoons (Kaliraj and Chandrasekar, 2012; Kaliraj et al., 2014). The nearshore has been found with shallow depth of water column (between 1 and 10 m) and extends up to surf zones, and the offshore is gradually increased at a depth of 20–30 m toward seaward from a distance of 5–6 km (Kaliraj et al., 2013, 2014, 2017). In the study area subtropical climatic conditions with temperature ranges from 23.78 to 33.95°C prevail and annual average rainfall is between 826 and 1456 mm. The hydrodynamic behaviors of waves, wind, and currents are seasonally varying due to monsoonal effects, wherein the average value of wave energy is about 0.5–8.5 kJ/km^2 and the mean current velocity is measured as 0.14 ms^{-1}. In the nearshore and offshore region, the transmitting of wave energy toward shoreline is due to continuous surface wave and littoral currents. The releasing of wave energy and littoral current velocity to the shoreline determines the processes of erosion and accretion along various parts of the Southern Indian coastal stretches (Kaliraj et al., 2013). Monitoring the spatiotemporal variability of wave energy density and littoral current using ocean remote sensing-based time-series data provides viable information for understanding ocean stats and hydrodynamic characteristics.

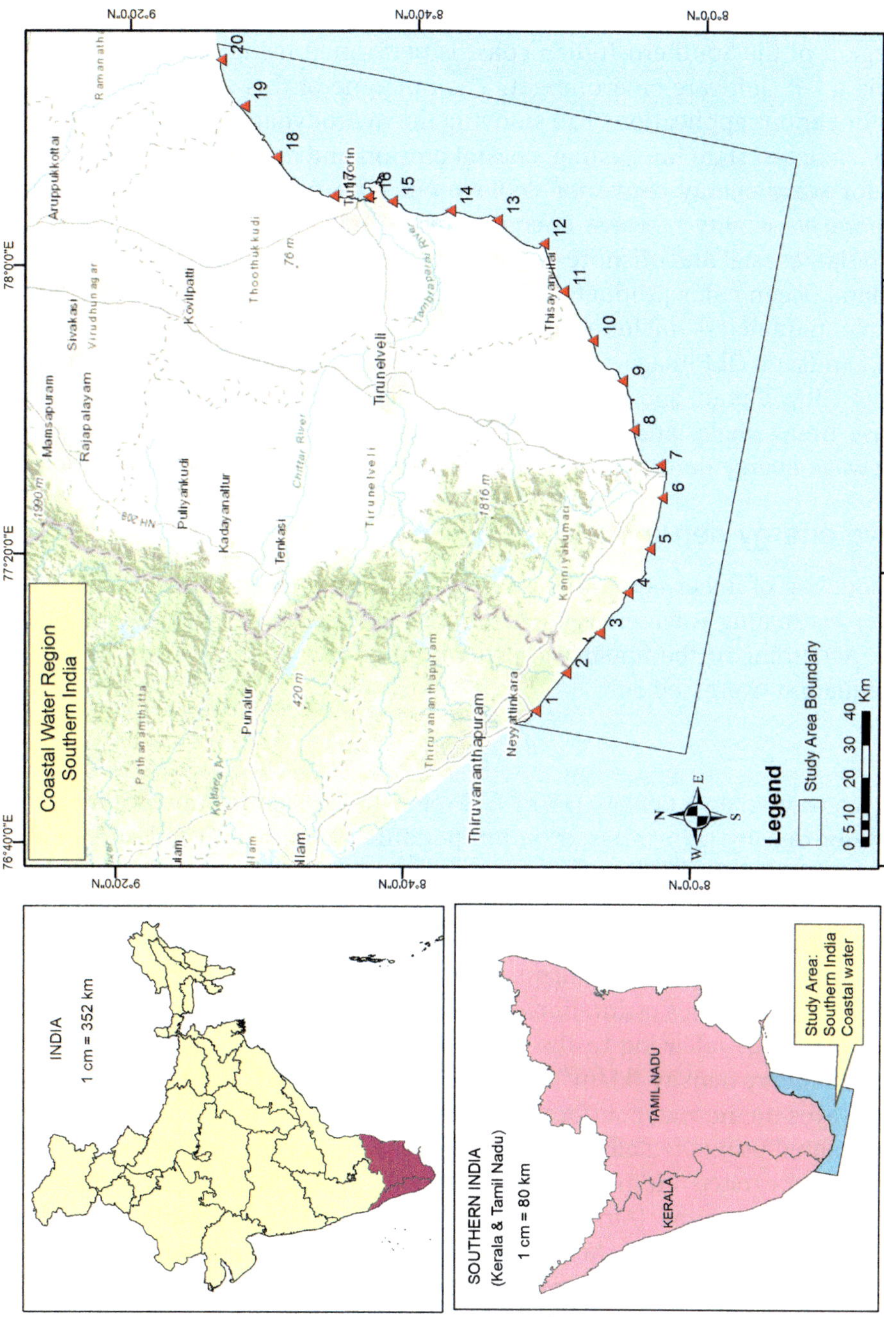

FIGURE 4.1 Geographical location of the study area.

3. Materials and methods

Assessment and mapping of wave energy density and littoral current velocity in the offshore region of the Southern Indian coast is performed using ocean remote sensing data sets via a GIS software environment. The outcome of this study provides vital information for various applications like studying the hydrodynamics of ocean and coastal environments, ocean state forecasting, coastal erosion and accretion processes, and site suitability for wave energy converter deployment. The present study investigates the spatiotemporal variability of wave energy density and littoral current velocity in the Southern Indian coastal and offshore regions using multiple ocean satellite products like altimeter data, ocean color products, and so on for the three seasons during 2017–18. The multiple parameters include significant wave height (SWH), seawater density (subject to Landsat 8 OLI image–derived salinity and temperature), gravity acceleration derived from GRACE map, and bathymetry map–derived measurement of wave breaker height, wave break angle, and offshore slope gradient, which are used as inputs for estimating wave energy density and littoral current velocity using GIS techniques.

3.1 Wave energy density calculation

The GIS modeling of linear equations is executed using a raster layer of the above parameters for estimating wave energy density and littoral current velocity at 10 * 10 m pixel scale. According to the linear wave theory, the wave energy density is calculated using the equation expressed as

$$E = \frac{1}{8}\rho g H^2 \tag{4.1}$$

where E is the wave energy density (kJ/m^2); H refers to the significant wave height (m); ρ is the seawater density (g/m^3) based on temperature, salinity, and pressure at specific locations; and g is the acceleration due to gravity (m/s^2). The energy and power densities of a wave are proportional to the square of the wave amplitude; knowledge of the average wave height is therefore important when considering where to place a wave farm. The mathematical relationship of significant wave height and seawater density along with gravity acceleration involves estimation of wave energy density that is relatively equal to the amount of energy releasing to the shoreline. In the seawater column, the average value of wave energy density (kJ/m^2) per unit area is proportional to seawater density and gravity waves on the water surface to the wave height squared. The significant wave height is estimated using the daily mean significant wave height data based on the level-4 products of integrated altimetry measurements of satellites, namely Sentinel-3A, Sentinel-3B, Jason-3, Saral (AltiKa), Cryosat-2, and CFOSAT available at https://resources.marine.copernicus.eu. Acceleration due to gravity is measured using GRACE maps published by NASA; the Gravity Recovery and Climate Experiment (GRACE) mission consists of Earth gravity datasets of twin satellites that were launched in 2002. GRACE maps show the measurement of Earth's gravity field with a precision greater than any previous instrument. GRACE provides the measurement of the entire gravity field of Earth every 30 days. The time series data of GRACE measurement can be used for assessing the changes in gravity over time, which reveals important details about polar ice sheets, sea level, ocean currents, ocean water cycle, and the earth interior structures. Seawater density (g/m^3) is mathematically estimated at 10 * 10 m pixel of seawater column from the satellite-derived products of temperature, salinity, and pressure using the formula proposed by UNESCO (1981) and is expressed as $\rho = \rho(S,T,0)/p)/1\text{-}K(S,T,p)$, where, K($T$, S, p) are the parameters of temperature, salinity, and pressure. The parameters derived from satellite products are cross-validated using marine-buoy measurements of wave height, seawater density, temperature, and salinity collected from the data archives of the Indian Ocean Forecast System and National Institute of Ocean Technology. The derived products of wave energy density is compared with the results obtained from the WAVEWATCH-III wave model for validation purposes, and it shows excellent agreement with root-mean-square error (RMSE) value of 0.30. Hence, it is agreed that the satellite-derived input parameters like SWH, gravity, and seawater density have produced high accuracy measurement for ocean and coastal studies.

3.2 Littoral current velocity measurement

Measurement of littoral current (also known as longshore current) is performed using the linear multiregression equation proposed by Galvin (1963) and Longuet-Higgins (1970); the equation is represented as

$$V_{LH} = 20.7\, m(gHb)^{\frac{1}{2}} \sin 2\alpha b \tag{4.2}$$

where V_{LH} refers to the littoral current velocity (m/s), the constant value of geometry of the breaking wave is *20.7* (source: SanilKumar et al., 2002), g is the gravity acceleration value (m/s^2) derived from GRACE data sets, m is the slope gradient based on bathymetry chart published by NHO, H_b refers to breaker height (m), and αb refers the breaker angle (degree of angle between wave breaker and shoreline) derived from in situ surveyed data sets. The GIS-based mathematical model is executed with the previous empirical equation using multiple parameters, wherein the strength of littoral current velocity varies approximately proportional to the square root of the wave height and with sin2αb, where the value of αb is the wave incidence angle at breaking or wave breaker. As the position of the breaking line constantly shifts due to the irregularity of natural wave fields and since the distance to the breaker line varies with the wave height, the distribution of the longshore current in the coastal profile will vary accordingly. Multiple thematic layers of input parameters such as current velocity, direction of seasonal movement, bathymetry, and the rip current point are prepared from various datasets and analyzed based on attributes for mapping the spatiotemporal characteristics of littoral current velocity and seasonal movement.

4. Result and discussion

The Southern Indian offshore region found dynamic changes in wave energy density and littoral current velocity due to seasonal and monsoonal impacts. Spatiotemporal variability of these factors have strongly influenced hydrodynamic processes like littoral drift, sediment transport, coastal morphodynamics, and controlling accretion or erosion of the coastal landforms. In the Southern Indian coastal and offshore region, the spatiotemporal variability of wave energy density and littoral current velocity is estimated using multiple ocean satellite products like altimeter data, and ocean color products for the three seasons during 2017–18. In the offshore, intensity of wave energy is determined by the behavior of significant wave height, seawater density, and gravity acceleration subject to their seasonal variability. The result shows the annual average wave energy density within the range of 1.2–10.48 kJ/m^2. Fig. 4.2 shows the annual average (mean) of wave energy density along the Southern Indian coastal and offshore regions. Along the southern coastal stretch, the wave energy density varies from place to place based on interacting forces of waves and currents at the meeting points of coastal configuration structures, and these processes directly affect sediment transport and accretion or erosion of the coastal landforms.

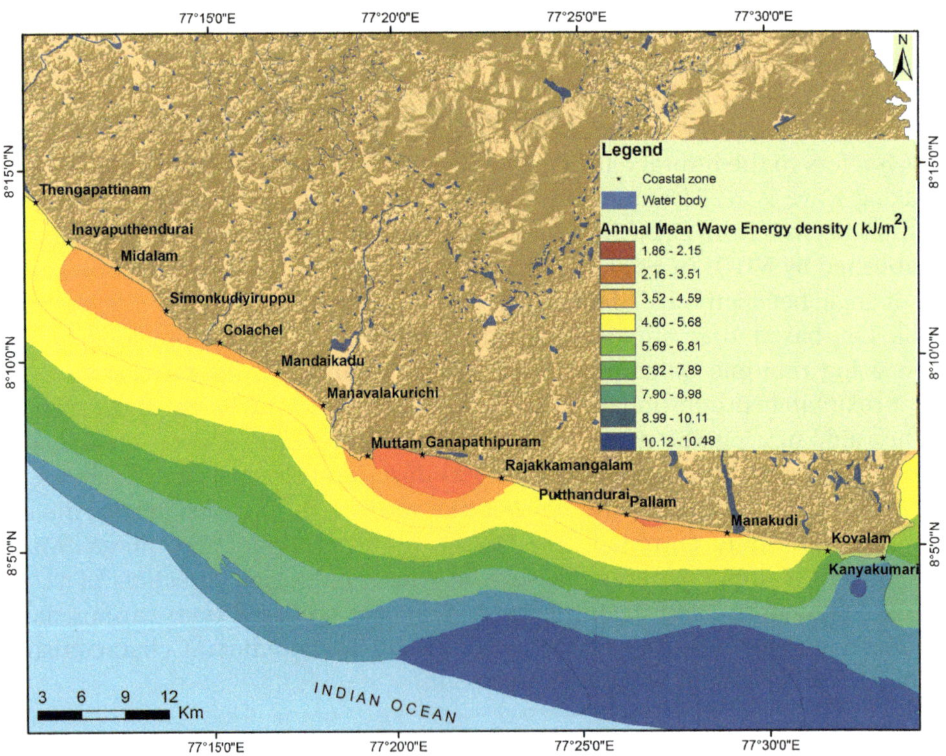

FIGURE 4.2 Annual average of wave energy density in the Southern Indian coastal and offshore region.

4.1 Spatiotemporal variability of wave energy density

Figs. 4.3–4.5 show the spatiotemporal variability of wave energy density in the Southern Indian coastal and offshore water during the three seasons of 2017–18. The Southern Indian offshore region found the highest range of wave energy density during premonsoon (June–September) and it is estimated at 3.46–10.48 kJ/m^2 with a mean value of 6.52 kJ/m^2 due to impacts of southwestern monsoonal wind from Indian Ocean and Arabian Sea. Similarly, the offshore region has as higher rate of wave energy density during monsoon (October–December) at the range of 1.12–9.78 52 kJ/m^2 with a mean value of 6.06 52 kJ/m^2 due to stronger influences of northeastern monsoonal wind velocity from Bay of Bengal Sea. However, the offshore region is recorded as having a lower rate of wave energy density during postmonsoon (March–May), at the range of 1.76–4.09 6.52 kJ/m^2 with a mean value of 3.06 6.52 kJ/m^2, whereas the annual average of wave energy density is estimated at 1.86–5.68 kJ/m^2 in coastal water of the nearshore area and 5.69–10.48 kJ/m^2 in the offshore water column.

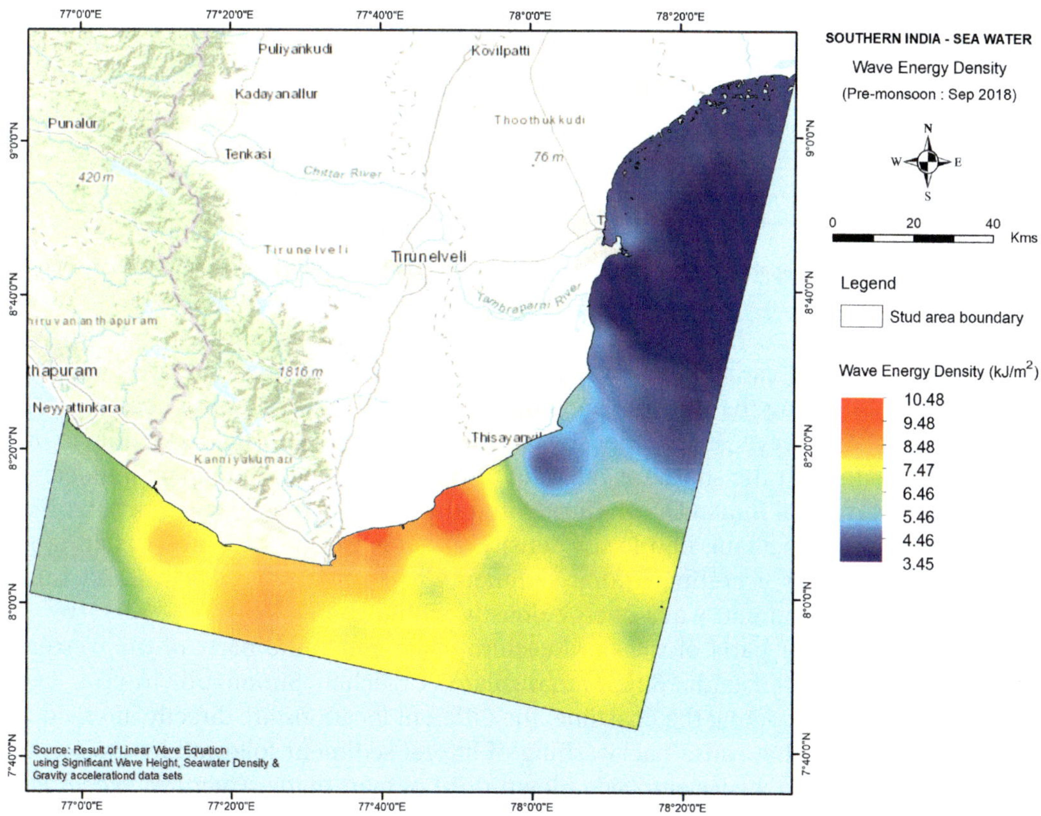

FIGURE 4.3 Spatiotemporal variability of wave energy density in the Southern Indian coastal and offshore during premonsoon (September 2018).

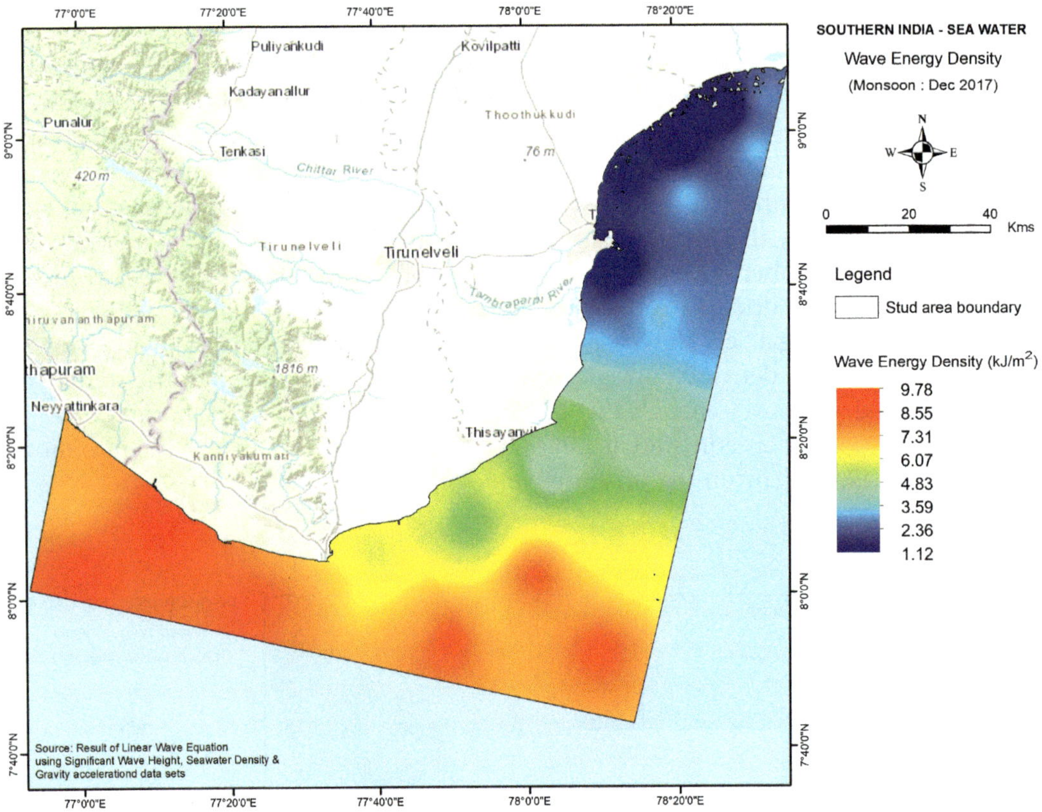

FIGURE 4.4 Spatiotemporal variability of wave energy density in the Southern Indian coastal and offshore during monsoon (October 2017).

The spatiotemporal variability of wave energy density shows the standard deviation value at 0.4–3.24 during the three seasons of 2017–18. The Kanyakumari coastal water column had higher wave energy density at the rate of 6.5–10.48 kJ/m^2 due to its geographical settings at the southern-most tip of Indian subcontinent facing impacts of waves from the Tropical Indian Ocean throughout the year. However, some portions, like Ganapathipuram and Pallam, had lower density of wave energy $<$3.46 kJ/m^2, acting as constructive waves for accreting littoral sediments to form young beaches and sand dunes. Similarly, the annual wave energy density within the range of 2.5–3.5 kJ/m^2 is estimated in the major parts of the northeastern coast and some parts of the western coastal zones near Manakudi, Rajakkamangalam, Colachel, Simonkudiyiruppu, and Midalam coastal zones. Along the coastline, the different locations are directly interacting with energetic waves that cause backwashing of littoral sediment followed by erosion of landforms, especially in the leeward side (down-drift) of man-made structures like jetties, groins, and seawalls, and so forth. However, the adjacent zones of the windward side (up-drift zone) are accreted by littoral sediments that lead to formation of beaches or dunes.

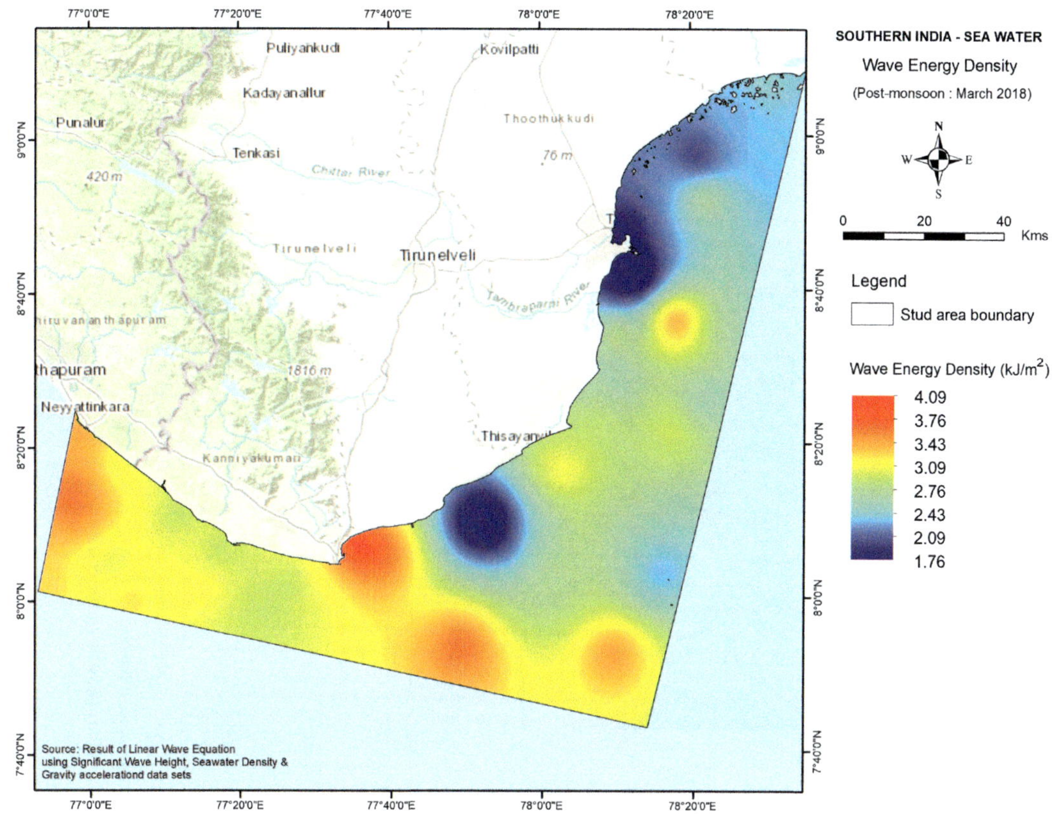

FIGURE 4.5 Spatiotemporal variability of wave energy density in the Southern Indian coastal regions and offshore during postmonsoon (March 2018).

In the western part, the overwashing of sediments by hitting of high energy waves leads to erosion processes in the western part of the coastal stretch. In the midcoastal stretch, a moderate wave energy (2.5–4.5 kJ/m^2) exists between Manavalakurichi and Mandaikadu coastal zones, and it is triggering deposition of littoral sediments with enriched with placer minerals. The higher wave energy releasing into the southerly coastal areas and offshore at a rate of (9.78–10.48 kJ/m^2) are eroding the rocky shore features found in the Kanyakumari–Kovalam coastal stretch.

4.2 Seasonal variability of wave energy density in 2017–2018

In the coastal and offshore water column, the seasonal variability of wave energy shows strong seasonal characteristics due to impacts of significant wave heights, seawater density, and monsoonal wind behaviors between the seasons. Fig. 4.6 shows the seasonal variability of wave energy density in the Southern Indian coastal water during the three seasons of 2017–18. In the Southern Indian coastal water column, the wave energy

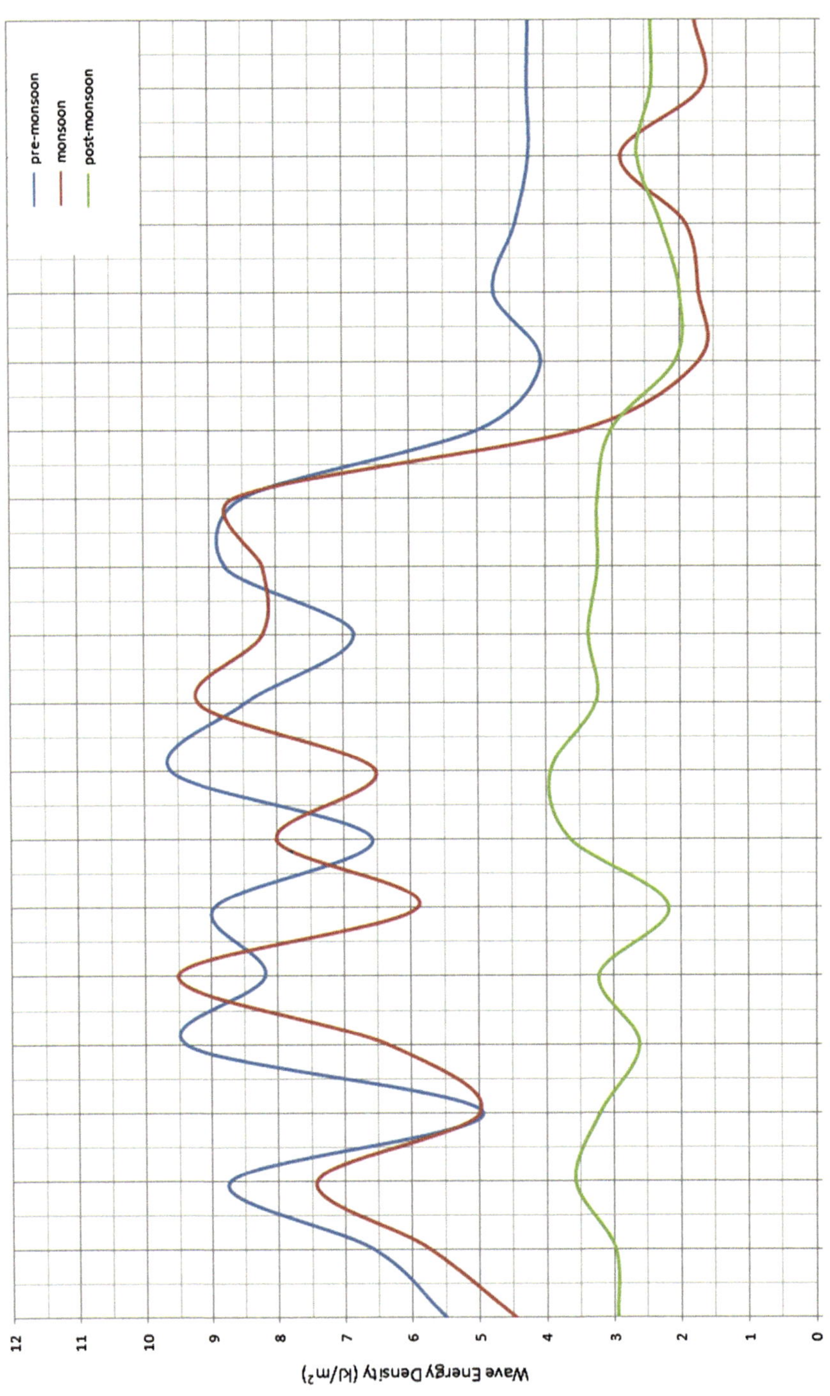

FIGURE 4.6 Seasonal variability of wave energy density in the Southern Indian coastal water during the three seasons of 2017–2018.

Table 4.1 Seasonal variability of wave energy density in 20 stations along the Southern Indian coastline.

Stations	Wave Energy Density (kJ/m²) Sep 2018 (pre-monsoon)	Dec 2017 (monsoon)	Mar 2018 (post-monsoon)	Maximum WE density (kJ/m²)	Minimum WE density (kJ/m²)	Mean WE density (kJ/m²)	SD
1	5.50	4.47	2.95	5.50	2.95	4.30	1.28
2	6.57	5.73	2.99	6.57	2.99	5.10	1.87
3	8.73	7.43	3.57	8.73	3.57	6.58	2.68
4	4.94	4.99	3.20	4.99	3.20	4.38	1.02
5	9.38	6.35	2.62	9.38	2.62	6.12	3.38
6	8.19	9.51	3.22	9.51	3.22	6.97	3.32
7	8.95	5.89	2.18	8.95	2.18	5.67	3.39
8	6.57	8.02	3.64	8.02	3.64	6.08	2.23
9	9.60	6.53	3.92	9.60	3.92	6.68	2.84
10	8.47	9.18	3.25	9.18	3.25	6.96	3.24
11	6.84	8.21	3.36	8.21	3.36	6.14	2.50
12	8.77	8.21	3.22	8.77	3.22	6.73	3.05
13	8.51	8.63	3.22	8.63	3.22	6.79	3.09
14	4.94	3.38	3.01	4.94	3.01	3.78	1.02
15	4.05	1.70	2.05	4.05	1.70	2.60	1.27
16	4.76	1.71	2.00	4.76	1.71	2.82	1.68
17	4.44	1.91	2.26	4.44	1.91	2.87	1.37
18	4.23	2.87	2.62	4.23	2.62	3.24	0.86
19	4.24	1.67	2.41	4.24	1.67	2.77	1.33
20	4.23	1.78	2.42	4.23	1.78	2.81	1.27

density varies in different sectors due to impacts of hydrodynamics seasonal variability like wave heights, seawater density, and flow of water currents. Table 4.1 shows the seasonal variability of wave energy density at the 20 stations located (Fig. 4.1) along the Southern Indian coastline. Form this observation, the wave energy within proximity of coastal water is estimated in the range of 1.67–9.60 kJ/m^2, whereas the higher wave energy (>9.18 kJ/m^2) recorded in some of stations includes 3 and 5–10, while a few stations include 15–20 along the northeastern part having lower energy at the rate of <3.92 kJ/m^2. However, the variability of wave energy density in all these stations shows standard deviation value at 0.4–3.24 during the three seasons of 2017–18. The wave energy density of the offshore water is due to wind direction and speed of the oscillating water column perpendicular to the plane, and hence, the southern offshore experience higher energetic waves throughout the years that are moving toward the Southern Indian coastline. The nearshore water column is recorded with higher wave energy during premonsoon and monsoon periods of 2017–18, but it is at a lower state during postmonsoon. However, within the proximity of the northeastern portion, the coastal water is noticed with lower wave energy due to the presence of the island landmass of Sri Langa and Rameswaram, which is acting like a barrier to reduce waves from open sea surface, whereas the waves reduce energy due to the prevailing lower surface tension effects that cause negligible wavelengths in this region compared to other parts. Moreover, regarding the energetic wave releasing to the mid-shoreline, the significant changes found in the coastal landforms due to backwashing of sediments by higher energetic waves (>6.06 kJ/m^2) during southwest and northeast monsoons and the lower wave energy (<5.05 kJ/m^2) are recorded in summer (March–May) due to diverging the hydrodynamic processes from the open sea surface.

4.3 Littoral current velocity and seasonal movement

In the Southern Indian water column, the movement of littoral current has seasonally coincided with the characteristics of waves, wind, and water currents, whereas the littoral current results from waves breaking at an angle to the shore and releasing wave energy at the coastline. Fig. 4.7 shows the seasonal flow of littoral current along the Southern Indian coastal water. The velocity of littoral current varies in different places

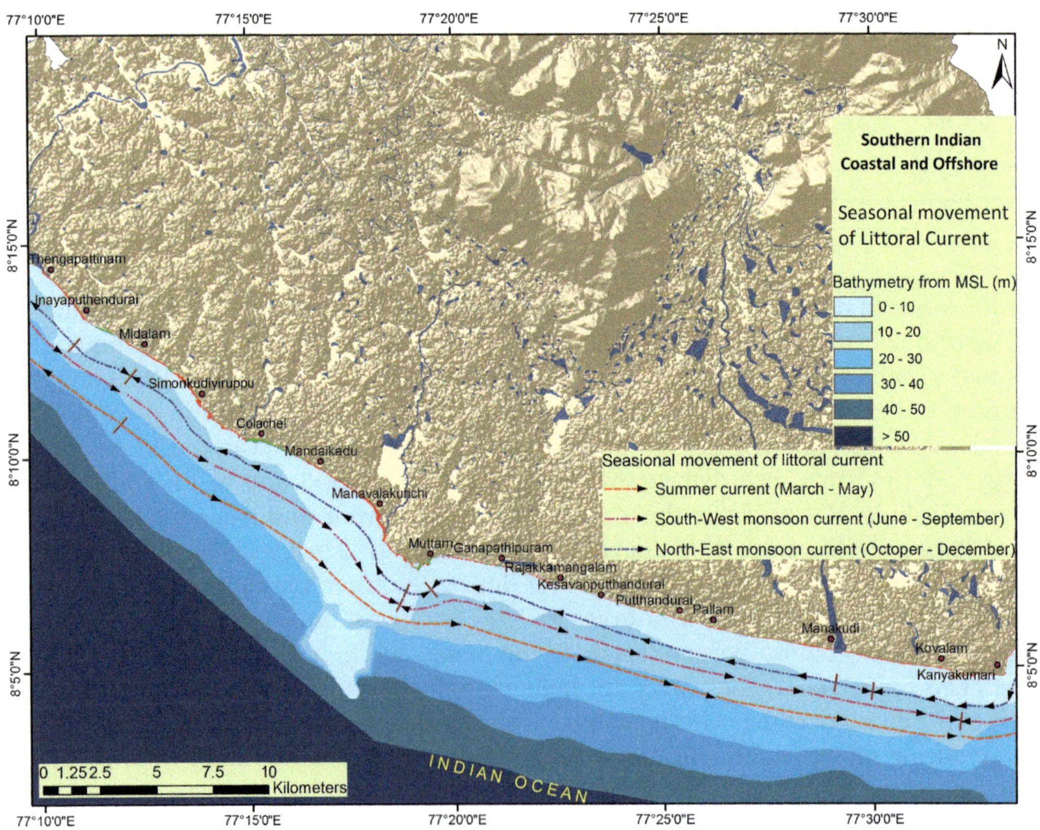

FIGURE 4.7 Seasonal flow of littoral current along the Southern Indian coastal water.

subject to wave angle, offshore slope (bathymetry), and height of wave breaker, and it is estimated in the range of 0.14–0.55 m/s. The littoral current velocity is noted at a higher range (0.32–0.55 m/s) along the southerly coastline, whereas the wave breaks occurring at a steeper offshore slope encounter the eastward flow of littoral currents with higher velocity and are destructive to landforms and manmade structures. However, the north-eastern coastline has recorded a lower current velocity by the action of wider breaking angles at the gentler offshore slope, resulting in northward movement of littoral drift. Many places of nearshore found that higher velocity of littoral current depends on continental slope, shoreline features, and depth of the water, whereas the breaking waves moving parallel to the shoreline conform to the general shape of coastal landforms.

In the Southern Indian coastal water, there are three types of littoral current systems: (1) the westward flowing littoral current with lower velocity during northeast monsoon, and (2) the eastward flowing littoral current in southwest monsoon and summer periods. The littoral currents have seasonally varied in velocity and direction according to waves and monsoonal winds, and it is controlling the littoral drift followed by sediment transport processes. The eastward flowing currents have higher velocity (>0.35 m/s) that carries a larger quantity of littoral sediments from the Arabian Sea and the Tropical Indian Ocean toward the northeastern coastline. However, the lower velocity of westward flowing current leads to accretion processes at various parts of the southern coastline during the northeast monsoon. However, many places face the littoral current with moderate velocity at the range of 0.25–0.35 m/s, and these coastal zones see seasonal changes in landforms. The nearshore within proximity of shallow water shows lower current velocity (0.14–0.22 m/s) during all the three seasons, whereas the coastal zones are accreting with larger littoral sediments. Meanwhile, the littoral current found at high velocity (>0.50 m/s) during summer and southwest monsoon causes severe erosion on the headlands. The circulating littoral current formed as rip-current at a velocity of 0.3–0.5 m/s within proximity of Cape Comorin, Kanyakumari, and Muttam coastal zones induces deformation of landforms in the down drift side. Overall observation reveals that the littoral current flow and velocity controlling littoral drift and sediment transport along the coastline depends on seasonal variability of hydrodynamic characteristics.

5. Conclusions

Monitoring wave energy density and littoral current provides insights for understanding coastal and ocean hydrodynamics and its impact on environmental ecosystems. Ocean remote sensing provides effective platforms for assessment and mapping spatiotemporal variability of wave energy and littoral current on global and regional scales. Satellite oceanographic measurement provides real-time observations of parameters like wave height, wind speed, salinity, temperature, and seawater density for ocean state forecasting. In the Southern Indian offshore and coastal water, the highest range of wave energy density during premonsoon and monsoon than that of summer is due to the strong effect of monsoonal hydrodynamic processes. Meanwhile, the eastward flowing littoral current found at higher velocity causes coastal landform changes in various locations. At coastal zones higher wave energy prevails, but lower littoral current velocity experiences erosion process more than that of other areas due to backwashing of sediments, whereas the coastal zones found the reversing processes are formed as depositional landforms due to swashing of sediments. It is concluded that the ocean remote sensing products can be used for long-term monitoring of wave energy and littoral current and provides vital information for ocean studies and sustainable coastal management.

Acknowledgment

The corresponding author is thankful to the Director, National Center for Earth Science Studies (NCESS), Ministry of Earth Sciences, Government of India, for providing continued support and encouragement. The authors extend their gratitude to INCOIS, NOAA, NASA, USGS–Earth Explorer, and European Union–Copernicus Marine Service Information (CMEMS) for providing necessary data sources.

References

Cruz, J., 2008. Ocean Wave Energy – Current Status and Future Prospects. Springer, ISBN 978-3-540-74894-6, 431pp.

Din, A.H.M., Omar, K.M., Naeijie, M., Ses, S., 2012. Long-term sea level change in the Malaysian seas from multi-mission altimetry data. Int. J. Phys. Sci. 7, 1694–1712.

Dyer, K.R., 1986. Coastal and Estuaries Sediment Dynamics. John Wiley and Sons, London.

Folley, M., Whittaker, T., 2009. Analysis of nearshore wave energy resource. J. Renew. Energy 34 (7), 1709–1715.

Galvin Jr., C.J., 1963. Ph D Thesis. Massachusetts Institute of Technology, Cambridge, Massachusetts.

Jayaram, C., Bansal, S., Krishaveni, A.S., Chacko, N., Chowdry, V.M., Dutta, D., Rao, K.H., Dutt, C.B.S., Sharma, J.R., Dadhwal, V.K., 2016. Evaluation of SARAL/Altika measured significant wave height and wind speed in the Indian Ocean Region. J. Indian Soc. Remote Sens. 44, 225–231. https://doi.org/10.1007/s12524-015-0488-7.

Kaliraj, S., 2016. Geospatial Analysis of Coastal Geomorphological Vulnerability on South West Coast of Kanyakumari Using Remote Sensing Technology. Unpublished Ph.D. Thesis. Manonmaniam Sundaranar University, Tirunelveli, pp. 83–110.

Kaliraj, S., Chandrasekar, N., 2012. Geo-processing model on coastal vulnerability index to explore risk zone along the south west coast of Tamilnadu, India. Int. J. Earth Sci. Eng. 5 (5), 1138–1147.

Kaliraj, S., Chandrasekar, N., Magesh, N.S., 2014. Impacts of wave energy and littoral currents on shoreline erosion/accretion along the South-West coast of Kanyakumari, Tamil Nadu using DSAS and geospatial technology. Environ. Earth Sci. https://doi.org/10.1007/s12665-013-2845-6.

Kaliraj, S., Chandrasekar, N., Magesh, N.S., 2013. Evaluation of coastal erosion and accretion processes along the SouthWest coast of Kanyakumari, Tamil Nadu using geospatial techniques. Arab. J. Geosci. https://doi.org/10.1007/s12517-013-1216-7.

Kaliraj, S., Chandrasekar, N., Ramachandran, K.K., 2016. Mapping of coastal landforms and volumetric change analysis in the South West coast of Kanyakumari, South India using remote sensing and GIS techniques. Egypt. J. Remote Sens. & Space Sci. https://doi.org/10.1016/j.ejrs.2016.12.006.

Kaliraj, S., Chandrasekar, N., Ramachandran, K.K., 2017. Mapping of coastal landforms and volumetric change analysis in the south west coast of Kanyakumari, South India using remote sensing and GIS techniques. Egy. J. Remote Sens. Space Sci 20 (2), 265–282.

Kumar, U.M., Swain, D., Sasamal, S.K., Reddy, N.N., Ramanjappa, T., 2015. Validation of SARAL/Altika significant wave height and wind speed observations over North Indian Ocean. J. Atmos. Sol. Terr. Phys. 135, 174–180.

Lee, H.J., Jo, H.R., Chu, Y.S., Bahk, K.S., 2004. Sediment transport on macrotidal flats in Garolim Bay, west coast of Korea: significance of wind waves and asymmetry of tidal currents. Continent. Shelf Res. 24, 821–832.

Longuet-Higgins, M.S., 1970. Longshore currents generated by obliquely incident sea waves. J. Geophys. Res. 75 (33), 6778–6789. https://doi.org/10.1029/jc075i033p06778.

Phillips, O.M., 1977. The Dynamics of the Upper Ocean, second ed. Cambridge University Press, ISBN 978-0-521-29801-8.

Queffeulou, P., 2004. Long-term validation of wave height measurements from altimeters. Mar. Geodes. 27, 495–510.

Reynolds, O., 1877. On the rate of progression of groups of waves and the rate at which energy is transmitted by waves. Nature 16 (408), 343–344. https://doi.org/10.1038/016341c0.

SanilKumar, V., Anand, N.M., Gowthaman, R., 2002. Variations in nearshore processes along Nagapattinam coast. India. Curr. Sci. 82 (11), 1381–1389. http://drs.nio.org/drs/handle/2264/285.

Sepulveda, H.H., Queffeulou, P., Ardhuin, F., 2015. Assessment of SARAl Altika wave height measurements, using buoy, Jason-2 and Cryosat-2 data comparisons. Mar. Geodes. 38 (sup1), 449–465. https://doi.org/10.1080/01490419.2014.1000470.

UNESCO, 1981. Convention concerning the protection of the world cultural and natural heritage. Report of the World Heritage Committee, (5th Session), Sydney, 26–30 October 1981. https://whc.unesco.org/archive/1981/cc-81-conf003-6_e.pdf.

Uti, M.N., Din, A.H.M., Omar, A.H., 2017. Reliability of wind speed data from satellite altimeter to support wind turbine energy. Int. Archives Photogram. Remote Sens. & Spat. Sci. 42, 215–224.

Wang, Y., 1998. Sea level changes, human impacts and coastal responses in China. J. Coast Res. 14 (1), 31–36.

Yong, W., Jie, Z., Junmin, M., Jing, W., 2015. A wave energy resource assessment in the China's seas based on multi-satellite merged radar altimeter data. Acta Oceanol. Sin. 34, 115–125.

Young, I.R., Ribal, A., 2019. Multi-platform evaluation of global trends in wind speed and wave height. Science 364, 548–552.

5

Ocean remote sensing of seawater salinity and its seasonal variability: a case study of Southern Indian coastal water using Landsat 8 OLI images

Kaliraj Seenipandi[1], K.K. Ramachandran[1], Prashant Ghadei[2], Sulochana Shekhar[2]

[1]NATIONAL CENTRE FOR EARTH SCIENCE STUDIES (NCESS), THIRUVANANTHAPURAM, KERALA, INDIA; [2]DEPARTMENT OF GEOGRAPHY, CENTRAL UNIVERSITY OF TAMIL NADU, THIRUVARUR, TAMIL NADU, INDIA

Chapter outline

1. Introduction

Sea surface salinity (SSS) is dynamically changing physicochemical properties of seawater at a spatiotemporal scale that can influence sea surface temperature (SST), chlorophyll-a (chl-a), and nutrients, especially in coastal water regions (Barre et al., 2008;

 https://doi.org/10.1016/B978-0-12-819604-5.00005-6

Font et al., 2010; Dierssen and Kaylan, 2012; Akhil et al., 2016; Kaliraj et al., 2019a,b). In general, the salinity of seawater is defined as the salt concentration (e.g., sodium and chloride) in seawater. It is measured in practical salinity units (1 PSU = 1 g/kg), which is a unit based on the properties of electrical conductivity of seawater. In 1 Kg of seawater, the mean chemical composition found is chloride (19 g), sodium (11 g), sulfate (3 g), magnesium (1.5 g), calcium (0.35 g), potassium (0.35 g), and other elements in tiny concentrations (Lewis, 1980; Mantyla, 1987; Delcroix and McPhaden, 2002; Ahn et al., 2008; Fournier et al., 2015). The average salinity of the global ocean is 35.5 PSU, and varies from less than 15 PSU at the mouth of the rivers (estuaries) to more than 40 PSU in the Dead Sea (Lewis, 1980; UNESCO, 1981; Mantyla, 1987). Monitoring and assessment of seawater properties including physical, chemical, and biological matter using in situ sampling offers results with high accuracy but not easy for preparing a simultaneous database for spatiotemporal variability (Gabarro et al., 2004; Boutin and Martin, 2006; Brown et al., 2008; Geiger et al., 2013; Abe and Ebuchi, 2014; Bhaskar and Chiranjivi, 2015; Dinnat et al., 2019). Remote sensing technology has become a useful platform to make it possible for assessing water properties of rivers, lakes, reservoirs, coastal, and ocean in an efficient manner (Ellison et al., 1998; Klemas, 2011; Lagerloef, 2014; Durack, 2015; Eladawy et al., 2017).

Applications of satellite remote sensing provide an effective platform for assessing physicochemical properties of coastal and ocean water and climatic aquatic ecology studies (Kaliraj et al., 2014a,b; Lagerloef, 2014). Landsat 8 Operational Land Imager (OLI) images provide evidence for measuring and monitoring SSS for long-term changes in coastal and ocean productivity, habitats, and biological processes (Kaliraj et al., 2019a,b; LSDS-1574, 2019). Mapping SSS seasonal variability is a vital source for finding frequency and intensity of coral bleaching, which is becoming evidence of climate change vulnerability (Marghany et al., 2010; Mishra et al., 2015; Lin et al., 2019). The global mapping of SSS and SST is very much required to predict future climate conditions, which is possible from optical and microwave sensors' ability to detect changes of physicochemical properties. The measurement of SST depends on emissivity of natural thermal energy from seawater (Barre et al., 2008; Klemas, 2011; Dierssen and Kaylan, 2012; Durack, 2015; Sengupta et al., 2016; Seelanki et al., 2018). Satellite sensors with various capabilities are used for measuring the amount of reflected radiation (spectral signature) from Earth's terrain and water surface and directly or indirectly is used for identifying major physicochemical and biological properties within or near real time (Ahn et al., 2008; Momin et al., 2016; Xiang et al., 2017; Dinnat et al., 2019). Satellite images acquired multitemporally are used for monitoring and assessment of SSS, SST, and chl-a, which play a major role in oceanic and climate studies (Burrage and Wesson, 2008; Sharma et al., 2010; Kaliraj et al., 2014a,b; Durack, 2015; Rajabi et al., 2017). Multitemporal remotely sensed images represent synaptic coverage of water bodies that typically are used for assessing the spatiotemporal status of water quality on a regional and global level (Boutin and Martin, 2006; Dierssen and Kaylan, 2012). Satellite remote sensing offers an effective platform to estimate seawater salinity on a regional and global scale at the frequency of satellite overpass, dramatically enhancing our monitoring capabilities relative to in situ observation networks.

Global measurement of SSS at synoptic scale can be derived from different satellites like SMOS, SMAP, MODIS, Oceansat, and Aquarius, at exactitude of 0.2–0.4 PSU and highly correlated with in situ measurement like ARGO floats, RAMA, and OMNI buoys (Delcroix and McPhaden, 2002; Brown et al., 2008; Barre et al., 2008; Font et al., 2010; Marghany et al., 2010; Abe and Ebuchi, 2014; Bhaskar and Jayaram, 2015; Dinnat et al., 2019). Satellite-derived SST and SSS are used for studying ocean assimilation processes at the exact ocean/seawater surface (depth = 0), but it could not be assimilated using ARGO instruments such as the sensing depth > 5 m (Boutin and Martin, 2006; Bhaskar and Jayaram, 2015). NASA's Aquarius mission (June 2011–August 2015) and the European Space Agency's Soil Moisture and Ocean Salinity (SMOS) mission (November 2009 to the present) are capable of measuring SSS from space across the world's oceans, but the derived products available are in poor spatial resolution (i.e., Aquarius = 150 km with temporal resolution of 7 days, and SMOS = 250 km with an average temporal resolution of 10–30 days) (Abe and Ebuchi, 2014; Seelanki et al., 2018). These satellite missions targeting salinity have focused on ocean rather than coastal applications, and the SSS-derived products are coarse in coastal water and not suitable for studies of coastal and estuarine environments (Ratheesh et al., 2013; Lin et al., 2019). Multispectral images of Landsat and MODIS are used worldwide for estimating SSS, SST, and chl-a using a linear regression algorithm, and these outputs noted a higher correlation (R2 > 0.81) with the in situ samplings validating at the time of the satellite passes (Marghany et al., 2010; Salleh et al., 2013; Xiang et al., 2017; Kaliraj et al., 2019a,b).

Monitoring and measurement of SSS is vital for studying the ocean and climate dynamics (Gabarro et al., 2004; Font et al., 2010) and critical to our understanding of ocean circulation and the global water cycle (Durack, 2015). Historically, the measurements of SSS were relatively sparse, as they were limited to in situ measurements from ships, drifters, and moorings. Since the 21st century, the situation has markedly improved due to the growth of ocean remote sensing satellites. Mapping of SSS in the coastal region using satellite images was successfully performed using Landsat TM data and in situ SSS in the estuarine environment (Dierssen and Kaylan, 2012; LSDS-1574, 2019). Satellite remote sensing offers the potential to estimate salinity across entire water bodies at the frequency of satellite overpass, dramatically enhancing our monitoring capabilities relative to in situ observation networks. SSS dynamic studies are at the heart of several major scientific domains that play a vital role in seawater circulation patterns, influence the spatial distribution of many marine organisms, and affect seawater density in both coastal systems and open oceans.

Multiple satellite images of ocean and coastal environments such as Landsat series, IRS series, SPOT, ESA, MODIS, Oceansat, Sentinel, and SAR microwave are available for insight. MODIS and AVHRR satellite images are often used for retrieving SSS for larger areas but with lower scale; this may perform poorly along the coastline due to mixed pixels at the border of land and water (Klemas, 2011; Momin et al., 2016; Rajabi et al., 2017). The derived SSS products from various satellite images are enabling routine monitoring of SSS on synoptic scales and are used for scientific exploitation like monitoring tropical salinity variability, seawater plumes, climate change, and so on. Many researchers have used MODIS images for retrieving SST and SSS from seawater column at a resolution of 1 km using statistical models and artificial neural network algorithms, but these outputs suffer from data and methodical limitations, especially in a coastal water environment (Ratheesh et al., 2013; Marghany et al., 2010; Salleh et al., 2013; Mishra et al., 2015; Xiang et al., 2017). The measurement of SSS and SST using multispectral images of Landsat satellite series like TM, ETM+, and OLI successfully executed for coastal and ocean water with a tolerable accuracy using mathematical algorithms and involve as primary input for coastal and ocean dynamic studies based on data consistency with higher spatial and temporal resolution (Kaliraj et al., 2019a,b). Hence, the present study deals with multitemporal Landsat 8 OLI images for estimation of seawater salinity and its seasonal variability in the Southern Indian coastal water region.

2. Study area

The Southern Indian coastal water prevails dynamic changes in seawater salinity, which plays a significant role in coastal hydrodynamics and air–sea interaction processes, and is a parameter for stratifying climatic and monsoon structure in this region (Kaliraj, 2016). Geographically the study area lies within latitude 7°38′ N to 9°20′ N and longitude 76°50′ E to 78°53′ E, and the coastal water column extends about 8440.22 km^2, parallel to the Keelakarai–Thiruvananthapuram coastline with the shoreline length of 410.72 km. The water column constitutes mixing of seawater from the tropical India Ocean with the flows of the two seas, the Arabian Sea and Bay of Bengal (BoB) (Fig. 5.1). The coastal landforms consist of laterite uplands, estuaries, sand dunes, and outlets of the non-perennial river. The coastal water receives river discharges along with runoff matter from different drainage networks like Thamirabarani, Valliyar, Pazhayar, and Vaippar during northeast and southwest monsoons (Kaliraj and Chandrasekar, 2012; Kaliraj et al., 2015, 2016). The water column was found with shallow depth (1–10 m) between nearshore and surf zones, and gradually increased to 20–30 m toward seaward distance of 5–6 km (Kaliraj et al., 2013, 2014a,b, 2017). The water column is seasonally varying in hydrodynamics like waves, wind and currents, and so on; the average value of wave energy is about 0.5–8.5 kJ/km^2 and the mean current velocity is measured as 0.14 ms^{-1}. The study area had subtropical climatic conditions with temperature ranges from 23.78 to 33.95°C and annual average rainfall between 826 and 1456 mm. Monitoring and assessment of SSS and its seasonal variability using multitemporal Landsat 8 OLI images is vital for understanding ocean and coastal hydrodynamic characteristics in the study area.

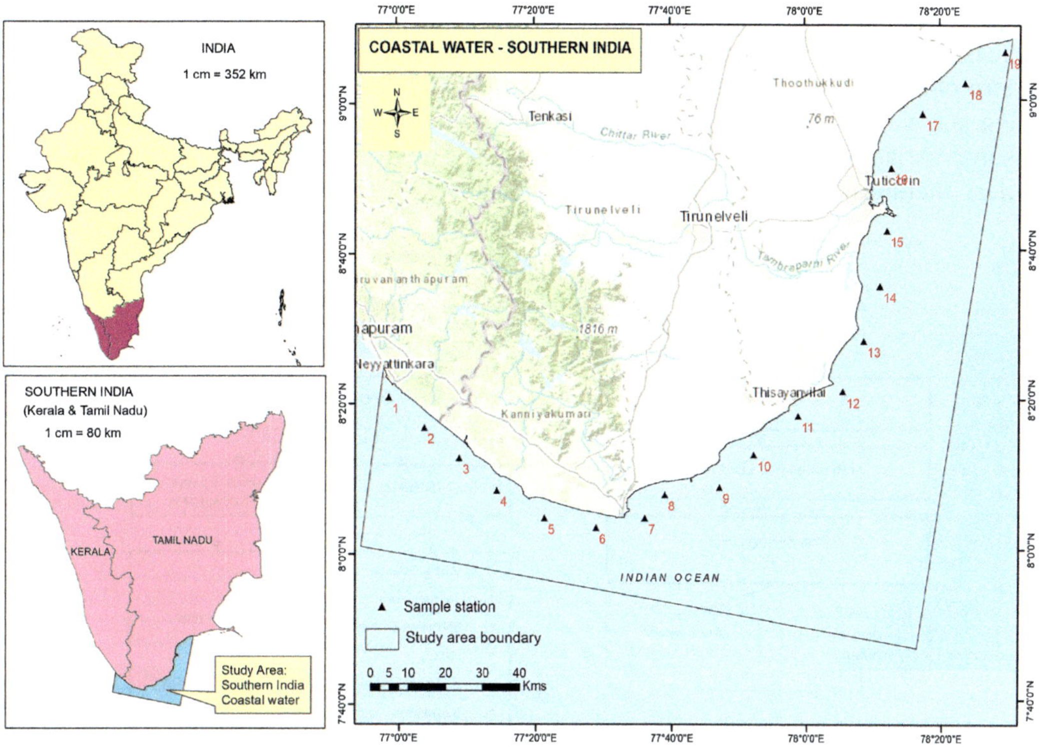

FIGURE 5.1 Geographical location of the study area.

3. Materials and methods

Multitemporal Landsat 8 OLI images are used for estimating SSS and its seasonal variability in the Southern Indian Coastal water column for the three seasons (premonsoon, postmonsoon, and monsoon) during 2017–2018. Landsat 8 OLI images (level-1 GeoTIFF data product) are collected from the USGS–Earth Explorer online portal. Landsat 8 satellite has two sensors: OLI and thermal infrared sensor. Landsat 8 OLI images having improved radiometric precision over a 12-bit dynamic range that improves overall signal-to-noise ratio and enables radiometric characterization for exploring land and water condition (LSDS-1574, 2019). The OLI sensor collects image data for nine spectral bands (visible, near-infrared, short-wave infrared) with a 30 m spatial resolution except Panchromatic band (15 m). The two new bands in addition to the legacy Landsat bands (1–5, 7, and Pan) are coastal/aerosol band (band 1; wavelength = 0.435–0.451 μm), principally used for ocean color observations, which is similar to ALI's band 1; and cirrus band (band 9; wavelength = 1.36–1.38 μm), which aids in the detection of cirrus clouds. Landsat 8 OLI image with multispectral bands (spatial resolution = 30 m) covers wide areas of Earth's landscape while providing sufficient resolution to distinguish land and water features within a 16-day repetitive cycle. Landsat 8 OLI images are analyzed using the least square algorithm for mapping salinity (SSS) with a comprehensive and accurate pattern of spatial and seasonal variability in the coastal water region.

Fig. 5.2 shows the flowchart of systematic methodology for estimating the SSS and seasonal variability using Landsat 8 OLI images. The systematic analysis for SSS estimation includes the following steps: (1) Landsat 8 OLI image preprocessing for

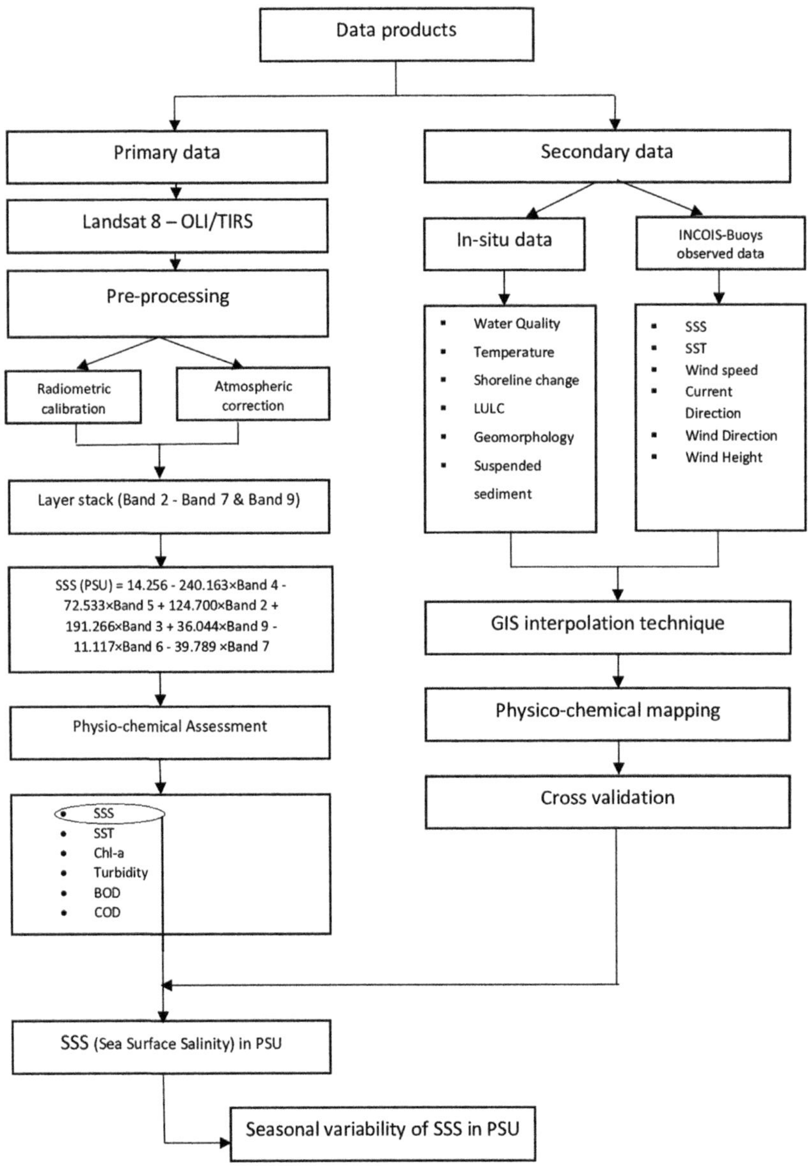

FIGURE 5.2 Methodology flowchart for estimating SSS seasonal variability.

atmospheric and radiometric correction using FLAASH module of ENVI software, (2) layer stacking and land masking for OLI multispectral bands, (3) extraction of land and water column using isodata unsupervised classification technique, and (4) executing the least square algorithm for estimating SSS using ArcGIS–Raster Calculator module. The Landsat 8 OLI (Level 1T–GeoTiFF) images are calibrated via top-of-atmosphere and dark-object-subtraction techniques for atmospheric and radiometric correction using ENVI 5.3 software. The calibrated OLI image shows spectral reflectance as a reflected value from the objects. Second, the masking technique is applied on the OLI image; multispectral bands for removing land, clouds, and intertidal zones and distinguishing the boundary between water and nonwater regions. Finally, the OLI images (FCC format) are analyzed for estimating the SSS at 30 × 30 m pixel scale using the empirical algorithm executed via Map Algebra tool–ArcGIS 10.6 software, expressed as

$$\begin{aligned} SSS(psu) = [&14.256 - 240.163 * Band\,4 - 72.533 * Band\,5 + 124.700 * Band\,2 + 191.266 * Band\,3 \\ &+ 36.044 * Band\,9 - 11.117 * Band\,6 - 39.789 * Band\,7] \end{aligned} \tag{5.1}$$

where *SSS* is SSS (PSU) and *Band 2, 3, … 9* refer to Landsat 8 OLI multispectral bands; the numerical value used in the equation refers to salinity coefficient values derived from in situ salinity measurements.

The Landsat 8 OLI image–derived SSS products are cross-validated using INCOIS–Buoys observed SSS data based on two statistical methods, (1) linear regression analysis (LRA), and (2) root-mean-square error (RMSE) calculation, and the accuracy and correlation between derived and in situ measurement in selected stations within the study area is reported.

4. Results and discussion

The SSS derived from multitemporal Landsat 8 OLI images shows spatial and seasonal variability of salinity along the Southern Indian coastal water, which provides primary information for understanding coastal hydrodynamics and biogeochemical processes. In general, freshwater fluxing from rainfall as well as river discharge affects SSS distribution in ocean and coastal water, respectively (Mishra et al., 2015; Akhil et al., 2016). The results reveal that the SSS is seasonally changing in place to place based on different factors like evaporating seawater, precipitation, and river discharges. During post-monsoon (March–May 2018), the SSS spatial variability is estimated at a higher range of 23.64–36.59 PSU; this is due to widespread potential evaporation whereas lower rainfall ($E > P$) increases salinity concentration in coastal as well as offshore water column. The lower concentration of SSS is estimated in a majority of areas in the range of 20.40–35.18 PSU during premonsoon (September 2018), whereas the wide-spreading of lower salinity is due to eastward flow of lower salinity coastal-water currents from the Arabian Sea, and discharging of river-runoff into the BoB (Sharma et al., 2010; Pant et al., 2015). Noticeably, the SSS concentration is spatially varying at a higher value of 32.40–35.88

PSU in the nearshore water column due to northeastward flowing salty warm-current along the shoreline, whereas the offshore water is found at a lower SSS value of 25.43–30.66 PSU during monsoon (December 2017) due to offshore mixing of precipitation from northeast monsoon. The SSS measurement for three seasons shows higher salinity in nearshore water (shallow depth region) and lower salinity in offshore water (deep depth region) due to influences of SST, waves, currents, winds, and river discharges that mix salinity in coastal water at various ranges depending on spatiotemporal variability of ocean hydrothermodynamics (Vinayachandran and Nanjundiah, 2009; Subrahmanyam et al., 2011; Yuan et al., 2018). The seasonal variability of SSS concentration influences major physicochemical and biological properties of seawater that affects marine and coastal ecosystems due to monsoonal variability of oceanic and coastal processes (Seelanki et al., 2018; Kaliraj et al., 2019a,b).

4.1 SSS spatial variability in monsoon (December 2017)

The spatial variation of SSS is estimated in the range of 25.43–35.88 PSU during monsoon. The coastal water along the nearshore was found with higher SSS concentration up to 32.40–35.88 PSU than that of the offshore region. This is due to circulating cold water current from the Arabian Sea and Tropical Indian Ocean that circulates high salinity parallel to the shoreline during northeast monsoon (Ratheesh et al., 2013; Mishra et al., 2015; Sengupta et al., 2016). Fig. 5.3 shows the SSS spatial variability in coastal

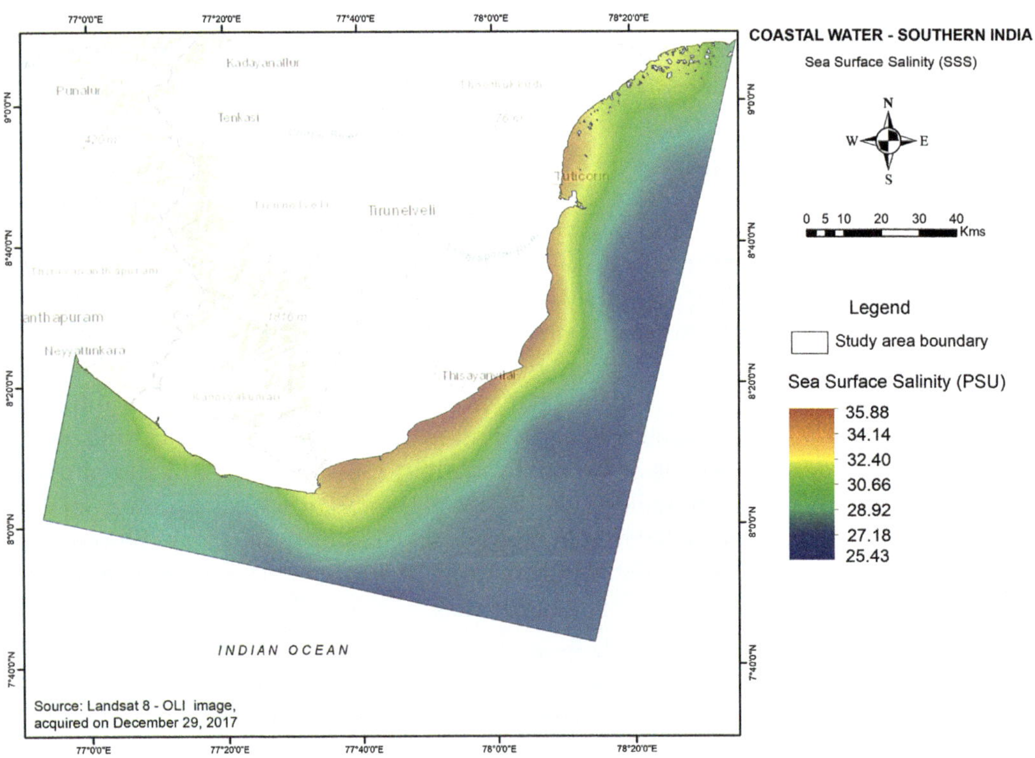

FIGURE 5.3 SSS spatial variability during monsoon (December 2017).

water during monsoon periods. The result reveals that the SSS concentration found higher values in nearshore water than that of offshore (deep-sea water) due to northeast monsoon influences on coastal hydrodynamic processes. It is noted that the coastal water, especially along the eastern shoreline, was found as higher SSS concentration with an average value of 31.07 PSU. However, coastal water in some locations is recorded with lower salinity due to freshwater discharging from river outlets in different locations.

4.2 SSS spatial variability in premonsoon (September 2018)

During premonsoon (September 2018), the coastal water was found as lower SSS concentration (20.40–35.18 PSU) compared to other seasons, wherein the estimated average SSS is lower (25.59 PSU) than that of monsoon and postmonsoon. Fig. 5.4 shows the SSS spatial variability in premonsoon periods. The offshore wide-spreading water column (deep sea) is estimated as lower SSS concentration within a range of 20.40–27.79 PSU. However, in the eastern nearshore, a few pockets of coastal water are noted as higher salinity in the range of 30.25–35.18 PSU. The eastward flowing cold water current transports maximum salinity water (>37 PSU) seen in the Arabian Sea toward BoB and mixes into coastal water along the nearshore (Ratheesh et al., 2013; Kaliraj et al., 2014a,b; Lin et al., 2019). Meanwhile, the offshore water column of the BoB was found with lower salinity (<30 PSU) because of changing the monsoonal physical processes.

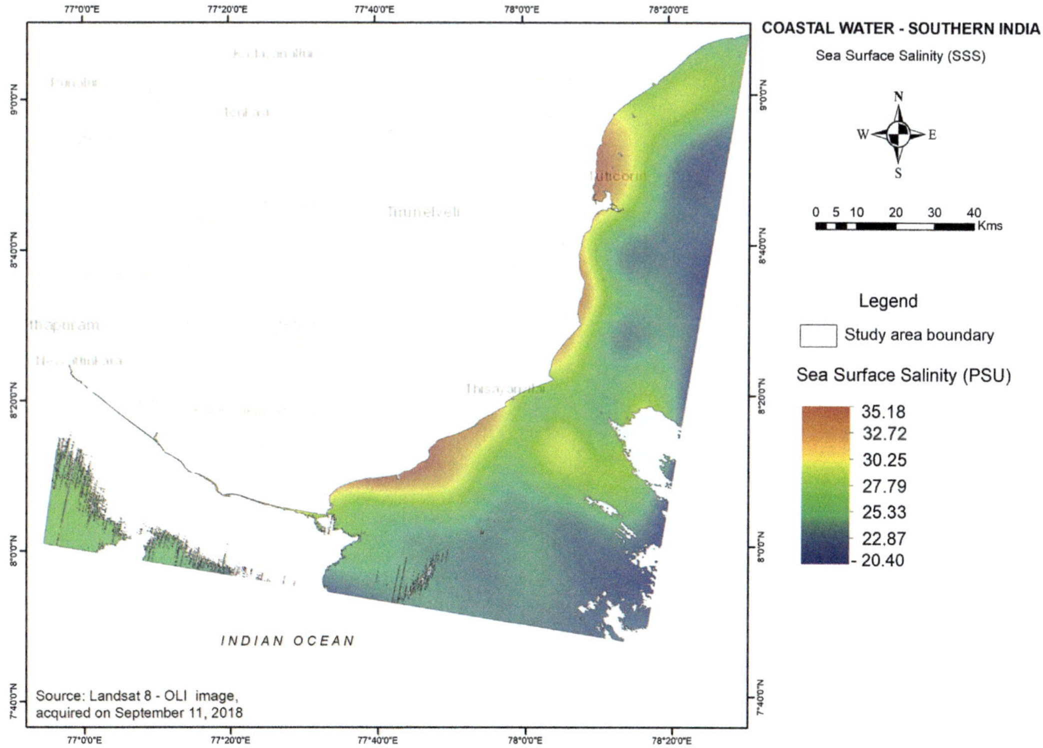

FIGURE 5.4 SSS spatial variability in premonsoon (September 2018).

4.3 SSS spatial variability in postmonsoon (March 2018)

Fig. 5.5 shows the SSS spatial variability during postmonsoon. During postmonsoon (March–May 2018), the average SSS is estimated as 29.18 PSU in the wide-spreading area of coastal water. The SSS spatial variability is recorded as lower concentration (23.64–30.11 PSU) in the majority of offshore water because of higher SST interacting with seawater, triggering flow of warm current from Tropical Indian Ocean (TIO) and BoB into the offshore region (Subrahmanyam et al., 2011; Pant et al., 2015; Yuan et al., 2018). However, a few pockets of coastal water in the nearshore are estimated as higher SSS concentration in the range of 32.28–36.59 PSU due to occurring potential evaporation resulting from higher SST and atmospheric temperature; meanwhile, from the Arabian Sea, eastward flowing seawater is mixing higher salinity into coastal water.

4.4 SSS seasonal variability in 2017–2018

Seasonal variability of SSS in the Southern Indian coastal water is characterized from the SSS measurements using Landsat 8 OLI images for three seasons during 2017–2018. Fig. 5.6 shows seasonal variability of SSS concentration in the Southern Indian coastal water. The SSS measured values at 19 stations (Fig. 5.1 shows geographical locations of

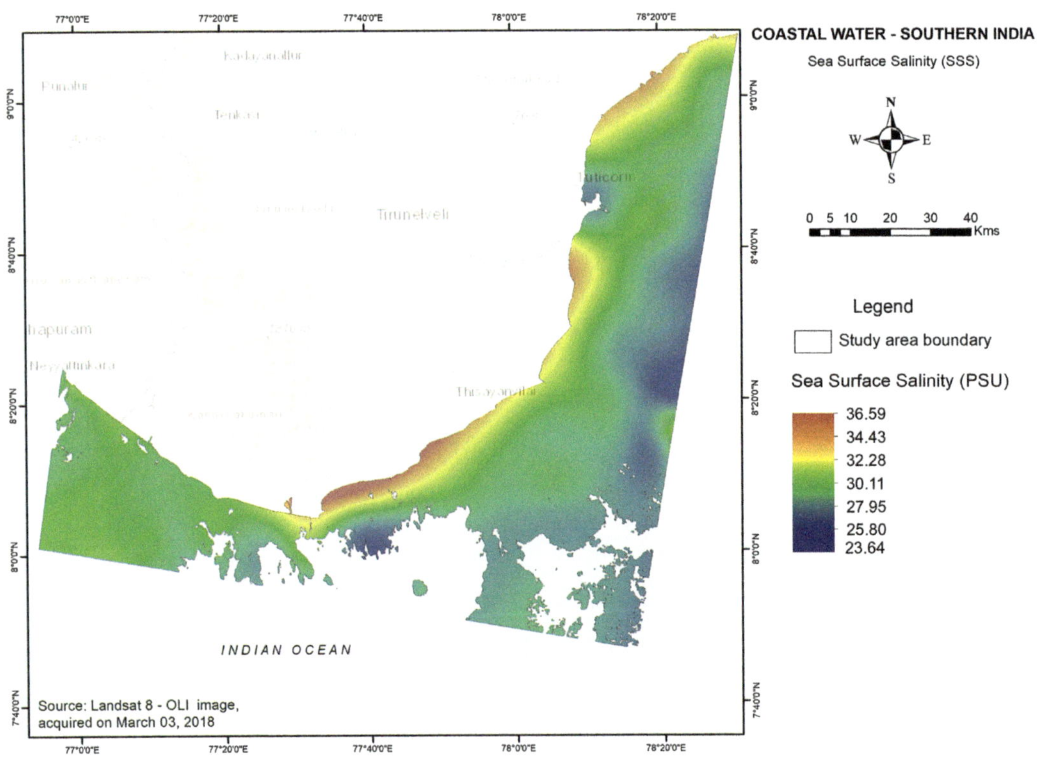

FIGURE 5.5 SSS spatial variability in postmonsoon (March 2018).

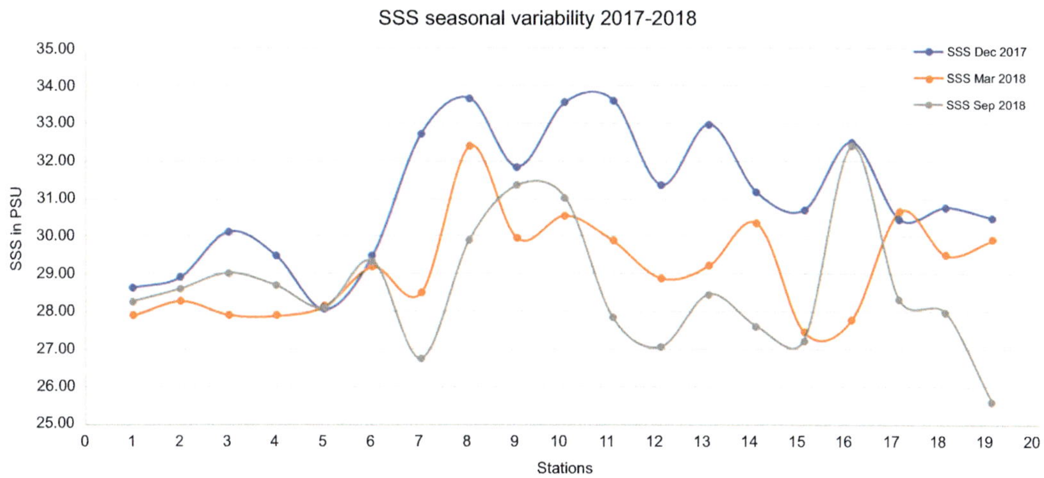

FIGURE 5.6 Seasonal variability of SSS during 2017–2018.

selected stations in the study area) represented obvious spatiotemporal variability of SSS during the season of monsoon (October–December 2017), premonsoon (June–September 2018), and postmonsoon (March–May 2018). Table 5.1 shows SSS seasonal variability for 19 selected stations along the Southern Indian coastal water. The mean of SSS is noted as 29.90 PSU; the overall maximum SSS value is 33.66 PSU in station 11, and the overall minimum SSS value is 25.56 in station 19. In the wide-spreading water column of the offshore (deep sea), the SSS seasonal variability is estimated at 20.40–30.66 PSU and it is relatively constant with very little spatiotemporal variation during pre- and postmonsoons. The SSS seasonal variability shows the standard deviation (SD) value of 0.05–1.94 for the majority of stations, which indicates that SSS concentration in the water column of the nearshore (shallow depth) is relatively constant during all three seasons, but it is changing in offshore water (deep-sea) column during different seasons. SSS seasonal variability shows the higher SD as 2.16–3.07 in a few coastal water packets found over stations 7, 11–13, 16, and 19, whereas SSS concentration is strongly affected by monsoon jets, potential evaporation, surface salinification, and so on. On the other hand, it is due to impacts of hydrodynamic processes like upwelling, deep mixing, and discharging of river runoff. In the offshore water column, the SSS concentration is found as lower values, mainly due to higher SST and warm current flows from BoB and TIO that may cause diluting over wide-spreading areas (Mishra et al., 2015; Momin et al., 2016; Kaliraj et al., 2019a,b). The lower SSS concentration found in offshore water due to precipitation mixes into BoB and the Arabian Sea during northeast and southwest monsoons, respectively. The higher concentration of SSS is estimated due to flowing coastal current from the Arabian Sea and mixes into the southwestern BoB during northeast monsoon. Meanwhile, the transporting warm current over interbasin between the Arabian Sea and BoB during

Table 5.1 SSS seasonal variability in selected stations along the Southern Indian Coast (SSS mean refers average of three seasons; *SD*, standard deviation).

	SSS seasonal variability (PSU)						
Stations	**Dec. 2017 (monsoon)**	**Mar. 2018 (postmonsoon)**	**Sep. 2018 (premonsoon)**	**SSS maximum (PSU)**	**SSS minimum (PSU)**	**SSS mean (PSU)**	**SD**
1	28.64	27.91	28.27	28.64	27.91	28.27	0.36
2	28.92	28.29	28.61	28.92	28.29	28.61	0.31
3	30.11	27.92	29.02	30.11	27.92	29.02	1.09
4	29.47	27.90	28.68	29.47	27.90	28.68	0.78
5	28.06	28.16	28.11	28.16	28.06	28.11	0.05
6	29.48	29.19	29.34	29.48	29.19	29.34	0.14
7	32.72	28.52	26.75	32.72	26.75	29.33	3.07
8	33.66	32.40	29.90	33.66	29.90	31.98	1.91
9	31.84	29.95	31.35	31.84	29.95	31.05	0.98
10	33.55	30.56	31.03	33.55	30.56	31.71	1.61
11	33.60	29.88	27.84	33.60	27.84	30.44	2.92
12	31.36	28.89	27.06	31.36	27.06	29.11	2.16
13	32.97	29.23	28.46	32.97	28.46	30.22	2.41
14	31.17	30.36	27.61	31.17	27.61	29.71	1.86
15	30.71	27.48	27.24	30.71	27.24	28.48	1.94
16	32.50	27.78	32.39	32.50	27.78	30.89	2.69
17	30.46	30.68	28.32	30.68	28.32	29.82	1.31
18	30.77	29.51	27.99	30.77	27.99	29.42	1.39
19	30.49	29.91	25.59	30.49	25.59	28.66	2.67

southwest and northeast monsoon seasons mix lower salinity in the BoB offshore (Bhaskar and Jayaram, 2015; Lin et al., 2019). The SSS concentration has lower values in the western side compared to values of the eastern side of the nearshore.

During 2017–2018, the SSS seasonal variability within coastal water (shallow depth) increases from premonsoon to postmonsoon (summer), up to values of 30. 20–32.55 PSU, reaches maximum value of <32.55 PSU during northeast monsoon (December). However, in offshore water, the SSS concentration is found at a lower value of <30.25 PSU with a relatively homogenous pattern during all three seasons. Hydrodynamic factors changing in different seasons affect SSS concentration in coastal and ocean water column particularly in the BoB, whereas the SSS is influencing at site-specific scale due to other factors like precipitation and freshwater fluxes (Sharma et al., 2010; Momin et al., 2016). In the Southern Indian coastal water, the SSS concentration varies at spatiotemporal scale, depending on seasonally influencing factors (i.e., during monsoon), whereas the other monsoonal factors are controlling SSS concentration such as river discharge, freshwater flux, and precipitation, especially in coastal water column. Meanwhile, offshore water is affected by the amount of evaporation and precipitation.

4.5 SSS validation using linear regression analysis and RMSE estimation

The SSS for the South Indian coastal water column is estimated using Landsat 8 OLI images at pixel scale of 10 * 10 m. These SSS-derived products for three seasons are cross-validated using INCOIS-buoys SSS observed data (in situ) based on two statistical methods, (1) linear regression analysis (coefficient of determination) and (2) RMSE estimation. Fig. 5.7 shows SSS cross-validation using linear regression relationship and RMSE using SSS derived (Landsat 8 OLI) and SSS observed (INCOIS buoys data at station 09 CE) for the period of 2017–2018. The LRA reveals positive correlation with the R^2 value of 0.9618 that indicates good agreement with INCOIS-buoys observed SSS data (in situ measurements) for all three seasons. Table 5.2 shows statistical measurement of linear regression and RMSE for Landsat 8 OLI derived SSS and INCOIS buoys observed SSS data for the period 2017–2018.

The RMSE is estimated for validating the accuracy of SSS derived products using INCOIS-buoys observed SSS data (in situ data) based on the equation

$$\text{RMSE} = \sqrt{\sum_{i=1}^{n} (Xd - Xo)^2 / N} \tag{5.2}$$

where X_d is the SSS derived products, X_o is the INCOIS-buoys observed SSS data, N refers to total number of observed stations. The RMSE estimated value (±0.138 PSU) shows a comprehensive relation between Landsat 8 OLI image derived SSS products and INCOIS buoys observed SSS data. The lower RMSE value indicates the relatively accurate result of SSS derived products using Landsat 8 OLI images. Moreover, the least square algorithm produces higher performance for estimating SSS as compared to a second polynomial order algorithm with minimal residual errors. The relationship between SSS

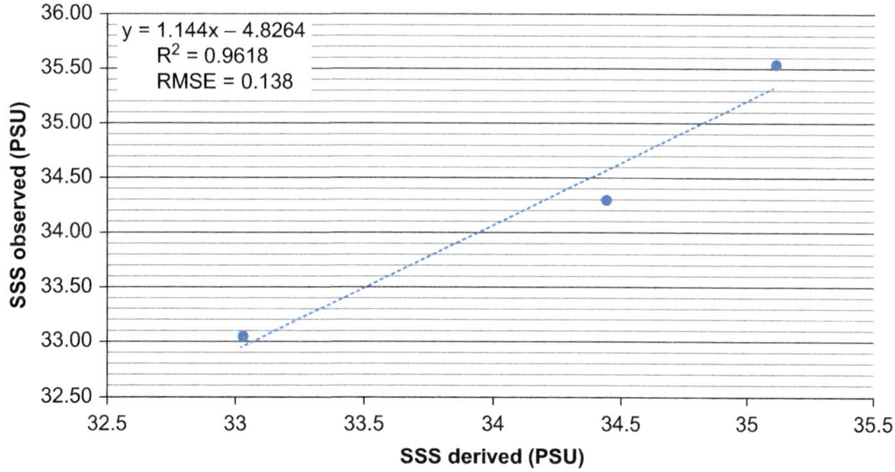

FIGURE 5.7 Linear regression and RMSE estimation using SSS observed (INCOIS buoys data) and SSS derived (Landsat 8 OLI) for the periods 2017–2018; premonsoon, monsoon, and postmonsoon.

Table 5.2 SSS validation for three seasons; premonsoon, monsoon, and postmonsoon using SSS observed (INCOIS buoys data) and SSS derived (Landsat 8 OLI).

INCOIS buoys observed SSS				Landsat 8 OLI derived SSS			
Buoys station	**SSS observed (PSU)**	**Date/time**	**Location**	**Station**	**SSS derived (PSU)**	**Date/time**	**Location**
AD 09	35.52	11-09-2018 05:00:00	73°18′14.5116″E 8°15′3.6792″N	1	35.11	11-09-2018 05:05:29	77°47′19.284″E 8°13′30.622″N
AD 09	33.04	01-01-2018 05:00:00	73°18′14.5116″E 8°15′3.6792″N	2	33.03	29-12-2017 05:06:03	78°5′45.434″E 8°23′54.887″N
AD 09	34.29	03-03-210 05:00:00	73°18′14.5116″E 8°15′3.6792″N	3	34.44	03-03-2108 05:05:34	77°36′22.934″E 8°8′31.625″N

derived products and SSS observed data shows positive correlation as the R^2 value is 0.9618 and RMSE value is ±0.138 PSU; this model can be directly used to retrieve SSS from multispectral bands of Landsat, MODIS, and ASTER satellite data.

5. Conclusions

Satellite remote sensing provides vital sources for estimating major physicochemical properties of coastal water and overcomes measurement difficulties in traditional methods. Estimation of SSS seasonal variability using Landsat 8 OLI images is one of the most important variables for oceanic and climate studies. In the Southern Indian coastal water column, the SSS spatiotemporal variability shows higher concentration in the coastal water column during all three seasons. Whereas in the offshore, the wide-spreading water column of deep sea found lower SSS concentration and is relatively constant with very little spatiotemporal variability during pre- and postmonsoons. It is noted that the SSS seasonal variability is mainly due to changing physical and climatic factors like SST, precipitation, waves, currents, winds, and so on. On the other hand, in the coastal water column, the spatiotemporal variability of SSS concentration depends on river discharges, runoff, and freshwater flux. Mapping and monitoring of SSS seasonal variability shows necessary progress for ocean and climatic studies and is helpful for development of sustainable management plans for environmental ecosystems for the long term.

Acknowledgments

The corresponding author is thankful to the Director, National Center for Earth Science Studies (NCESS), Ministry of Earth Sciences, government of India, for providing continued support and encouragement. The authors extend their gratitude to IMD, SOI, NIOT, INCOIS, NOAA, NASA, USGS–Earth Explorer, and PSMSL-UK for providing necessary data sources.

References

Abe, H., Ebuchi, N., 2014. Evaluation of sea-surface salinity observed by Aquarius. J. Geophys. Res. Oceans 119 (11), 8109–8121. https://doi.org/10.1002/2014JC010094.

Ahn, Y.H., Shanmugam, P., Moon, J.E., Ryu, J.H., 2008. Satellite remote sensing of a low salinity water plume in the East China Sea. Ann. Geophys. 26, 2019–2035.

Akhil, V.P., Lengaigne, M., Vialard, J., Durand, F., Keerthi, M.G., Chaitanyi, A.V.S., Papa, F., Gopalakrishna, V.V., Montegut, C., 2016. A modeling study of processes controlling the Bay of Bengal sea surface salinity interannual variability. J. Geophys. Res. Oceans 121 (12), 8471–8495. https://doi.org/10.1002/2016JC011662.

Barre, H.M.J., Duesmann, B., Kerr, Y.H., 2008. SMOS: the mission and the system. IEEE Trans. Geosci. Remote Sens. 46, 587–593.

Bhaskar, T.V.S.U., Jayaram, C., 2015. Evaluation of Aquarius sea surface salinity with argo sea surface salinity in the tropical Indian ocean. IEEE Geosci. Remote Sens. Lett. 12 (6), 1292–1296. https://doi.org/10.1109/LGRS.2015.2393894.

Boutin, J., Martin, N., 2006. ARGO upper salinity measurements: perspectives for L-band radiometer calibration and retrieved sea surface salinity validation. Geosci. Remote Sens. Lett. IEEE 3, 202–206.

Brown, M.A., Torres, F., Corbella, I., Colliander, A., 2008. SMOS calibration. IEEE Trans. Geosci. Remote Sens. 46, 646–658.

Burrage, D., Wesson, J., 2008. Deriving sea surface salinity and density variations from satellite and aircraft microwave radiometer measurements: application to coastal plumes using STARRS. IEEE Trans. Geosci. Remote Sens. 46 (3), 765–785.

Delcroix, T., McPhaden, M., 2002. Interannual sea surface salinity and temperature changes in the western Pacific warm pool during 1992–2000. J. Geophys. Res. 107 https://doi.org/10.1029/2001JC000862. SRF 3-1–SRF 3-17.

Dierssen, H., Kaylan, R., 2012. Remote sensing of ocean color. In: Earth System Monitoring. Springer Science, New York, NY, USA, pp. 8952–8975.

Dinnat, E.P., Le Vine, D.M., Boutin, J., Meissner, T., Lagerloef, G., 2019. Remote sensing of sea surface salinity: comparison of satellite and in situ observations and impact of retrieval parameters. Rem. Sens. 11, 750. https://doi.org/10.3390/rs11070750.

Durack, P.J., 2015. Ocean salinity and the global water cycle. Oceanography 28, 20–31. https://doi.org/10.5670/oceanog.2015.03.

Eladawy, A., Nadaoka, K., Negm, A., Abdel-Fattah, S., Hanafy, M., Shaltout, M., 2017. Characterization of the northern Red Sea's oceanic features with remote sensing data and outputs from a global circulation model. Oceanologia 59, 213–237.

Ellison, W., Balana, A., Delbos, G., Lamkaouchi, K., Eymard, L., Guillou, C., Prigent, C., 1998. New permittivity measurements of sea water. Radio Sci. 33 (3), 639–648.

Font, J., Camps, A., Borges, A., Martin, N.M., Boutin, J., Reul, N., Kerr, Y.H., Hahne, A., Mecklenburg, S., 2010. SMOS: the challenging sea surface salinity measurement from space. Proc. IEEE 98, 649–665.

Fournier, S., Chapron, B., Salisbury, J., Vandemark, D., Reul, N., 2015. Comparison of space borne measurements of sea surface salinity and colored detrital matter in the Amazon plume. J. Geophys. Res. Oceans 120, 3177–3192. https://doi.org/10.1002/2014JC010109.

Gabarro, C., Vall-llosera, M., Font, J., Camps, A., 2004. Determination of sea surface salinity and wind speed by L-band microwave radiometry from a fixed platform. Int. J. Remote Sens. 25 (1), 111–128.

Geiger, F., Grossi, D., Trembanis, C., Kohut, T., Oliver, J., 2013. Satellite-derived coastal ocean and estuarine salinity in the Mid-Atlantic. Continent. Shelf Res. 63, S235–S242.

Kaliraj, S., 2016. Geospatial Analysis of Coastal Geomorphological Vulnerability on South West Coast of Kanyakumari Using Remote Sensing Technology. Unpublished Ph.D. Thesis. ManonmaniamSundaranar University, Tirunelveli, pp. 83–110.

Kaliraj, S., Chandrasekar, N., 2012. Geo-processing model on coastal vulnerability index to explore risk zone along the south west coast of Tamilnadu, India. Int. J. Earth Sci. Eng. 5 (5), 1138–1147.

Kaliraj, S., Chandrasekar, N., Magesh, N.S., 2013. Evaluation of coastal erosion and accretion processes along the southwest coast of Kanyakumari, Tamil Nadu using geospatial techniques. Arab. J. Geosci. https://doi.org/10.1007/s12517-013-1216-7.

Kaliraj, S., Chandrasekar, N., Magesh, N.S., 2014a. Multispectral image analysis of suspended sediment concentration along the Southern coast of Kanyakumari, Tamil Nadu, India. J. Coast. Sci. 1 (1), 63–71. ISSN: 2348 – 6740.

Kaliraj, S., Chandrasekar, N., Magesh, N.S., 2014b. Impacts of wave energy and littoral currents on shoreline erosion/accretion along the south-west coast of Kanyakumari, Tamil nadu using DSAS and geospatial technology. Environ. Earth Sci. https://doi.org/10.1007/s12665-013-2845-6.

Kaliraj, S., Chandrasekar, N., Simon Peter, T., Selvakumar, S., Magesh, N.S., 2015. Mapping of coastal aquifer vulnerable zone in the south west coast of Kanyakumari, South India, using GIS-based DRASTIC model. Environ. Monit. Assess. https://doi.org/10.1007/s10661-014-4073-2.

Kaliraj, S., Chandrasekar, N., Ramachandran, K.K., 2016. Mapping of coastal landforms and volumetric change analysis in the south west coast of Kanyakumari, South India using remote sensing and GIS techniques. Egypt. J. Remote Sens. & Space Sci. https://doi.org/10.1016/j.ejrs.2016.12.006.

Kaliraj, S., Chandrasekar, N., Ramachandran, K.K., Srinivas, Y., Saravanan, S., 2017. Coastal landuse and land cover change and transformations of Kanyakumari coast, India using remote sensing and GIS. Egypt. J. Remote Sens. & Space Sci. https://doi.org/10.1016/j.ejrs.2017.04.003.

Kaliraj, S., Chandrasekar, N., Ramachandran, K.K., 2019a. Phytoplankton (chl-a) biomass seasonal variability in the Gulf of mannar, south India: a remote sensing perspective. In: Kumar, P., et al. (Eds.), Applications and Challenges of Geospatial Technology. https://doi.org/10.1007/978-3-319-99882-4_5.

Kaliraj, S., Chandrasekar, N., Ramachandran, K.K., Selvakumar, S., 2019b. Seawater intrusion vulnerability in the coastal aquifers of southern India—an appraisal of the GALDIT model, parameters' sensitivity, and hydrochemical indicators. Environ. Sci. Pollut. Res. 26, 9755–9784. https://doi.org/10.1007/s11356-019-04401-0.

Klemas, V., 2011. Remote sensing of sea surface salinity: an overview with case studies. J. Coast Res. 27 (5), 830–838. https://doi.org/10.2112/jcoastres-d-11-00060.1.

Lagerloef, G., 2014. In: Njoku, E. (Ed.), Sea Surface Salinity, Encyclopedia of Remote Sensing. Springer-Verlag, XXV, p. 939.

Lewis, E.L., 1980. The practical salinity scale 1978 and its antecedents. IEEE J. Ocean. Eng. OE 5 (1), 3–8.

Lin, X., Qiu, Y., Cha, J., Guo, X., 2019. Assessment of aquarius sea surface salinity with argo in the Bay of Bengal. Int. J. Remote Sens. https://doi.org/10.1080/01431161.2019.1615654.

LSDS-1574, 2019. Landsat 8 (L8) Data Users Handbook. https://prd-wret.s3-us-west-2.amazonaws.com/assets/palladium/production/atoms/files/LSDS1574_L8_Data_Users_Handbook-v5.0.pdf.

Mantyla, A.W., 1987. Standard seawater comparisons updated. J. Phys. Ocean. 17, 543–548.

Marghany, M., Hashim, M., Cracknell, A.P., 2010. Modelling Sea Surface Salinity from MODIS Satellite Data. Springer-Verlag Berlin Heidelberg, pp. 545–556. ICCSA 2010, Part I, LNCS 6016.

Mishra, A.K., Dwivedi, S., Shrivastava, A., 2015. High resolution simulation of the salinity variability in the Bay of Bengal and Arabian Sea during the years 1998-2014 using an ocean circulation model. Discovery 39 (180), 173–179. ISSN: 2278-5469).

Momin, I.M., Mitra, A.K., Mahapatra, D.K., Ragagopal, E.N., 2016. Review of recent evaluation of sea surface salinity with satellite estimates in the tropical Indian ocean. J. Appl. Remote Sens. 10. https://doi.org/10.1117/1.jrs.10.046008.

Pant, V., Girishkumar, M.S., Bhaskar, T.V.S.U., Ravichandran, M., Papa, F., Thangaprakash, V.P., 2015. Observed interannual variability of near-surface salinity in the Bay of Bengal. J. Geophys. Res. Oceans 120 (5), 3315–3329. https://doi.org/10.1002/2014JC010340.

Rajabi, S., Hasanlou, M., Safari, A.R., 2017. Spatial and temporal analysis of sea surface salinity using satellite imageryin Gulf of Mexico. Int. Arch. Photogramm. Remote Sens. & Spat. Inf. Sci. XLII-4/W4.

Ratheesh, S., Sharma, R., Sikhakolli, R., Kumar, R., Basu, S., 2013. Assessing sea surface salinity derived by aquarius in the Indian ocean. IEEE Geosci. Remote Sens. Lett. 11 (4), 719–722. https://doi.org/10.1109/LGRS.2013.2277391.

Ratheesh, S., Mankad, B., Basu, S., Kumar, R., Sharma, R., 2013. Assessment of satellite-derived sea surface salinity in the Indian Ocean. Geosci. Remote Sens. Lett. IEEE 10, 428–431. https://doi.org/10.1109/LGRS.2012.2207943.

Salleh, D., Shattri, M., Biswajeet, P., Lawal, B., Ahmad, M., 2013. Potential fish habitat mapping using MODIS-derived sea surface salinity, temperature and chlorophyll-a data: south China sea coastal areas, Malaysia. Geocarto Int. 28, 546–560.

Seelanki, V., Sreenivas, P., Prasa, K.V.S.R., 2018. Impact of aquarius sea-surface salinity assimilation in improving the ocean analysis over Indian ocean. Mar. Geodes 41 (2), 144–158. https://doi.org/10.1080/01490419.2017.1422817.

Sengupta, D., Raj, G.N.B., Ravichandran, M., Lekha, J.S., Papa, F., 2016. Near-surface salinity and stratification in the north Bay of Bengal from moored observations. Geophys. Res. Lett. 43 (9), 4448–4456. https://doi.org/10.1002/2016GL068339.

Sharma, R., Agarwal, N., Momin, I.M., Basu, S., Agarwal, V.K., 2010. Simulated sea surface salinity variability in the tropical Indian ocean. J. Clim. 23 (24), 6542–6554. https://doi.org/10.1175/2010JCLI3721.1.

Subrahmanyam, B., Murty, V.S.N., Heffner, D.M., 2011. Sea surface salinity variability in the tropical Indian ocean. Remote Sens. Environ. 115 (3), 944–956. https://doi.org/10.1016/j.rse.2010.12.004.

UNESCO, 1981. The practical salinity scale 1978 and the international equation of state of seawater 1980. Techn. Pap. mar. sci 36, 25.

Vinayachandran, P.N., Nanjundiah, R.S., 2009. Indian Ocean sea surface salinity variations in a coupled model. Clim. Dynam. 33, 245–263. https://doi.org/10.1007/s00382-008-0511-6.

Xiang, Y., Bei, X., Xiangyang, Y., Yebao, L., Buli, W., Cui, B., Xin, L., 2017. Retrieval of remotely sensed sea surface salinity using MODIS data in the Chinese Bohai Sea. Int. J. Remote Sens. 38, 7357–7373.

Yuan, X., Salama, M., Su, Z., 2018. An observational perspective of sea surface salinity in the Southwestern Indian ocean and its role in the south asia summer monsoon. Remote Sens. 10 (12), 1930. https://doi.org/10.3390/rs10121930.

6

Impacts of placer mining on groundwater and air quality in the Chavara coastal stretch of Kerala: remote sensing and GIS-based approach

V. Kashyap[1,2], R. Aneesh Kumar[1], P.M. Saharuba[1], Sakhre Saurabh[1,2], J. Ansari[1,2]

[1]*ENVIRONMENTAL TECHNOLOGY DIVISION, COUNCIL OF SCIENTIFIC & INDUSTRIAL RESEARCH NATIONAL INSTITUTE FOR INTERDISCIPLINARY SCIENCE AND TECHNOLOGY (CSIR-NIIST), THIRUVANANTHAPURAM, KERALA, INDIA;* [2]*ACADEMY OF SCIENTIFIC AND INNOVATIVE RESEARCH (AcSIR), GHAZIABAD, UTTAR PRADESH, INDIA*

Chapter outline

Remote Sensing of Ocean and Coastal Environments. https://doi.org/10.1016/B978-0-12-819604-5.00006-8

1. Introduction

Heavy mineral (HM) sand deposits in Chavara are a conglomeration of minerals such as ilmenite, rutile, leucoxene, monozite, zircon, and silllimanite. Kerala's HM deposit is a world class deposit since the concentration of these deposits are above 60%. The coastal tract where these mineral deposits were heeded was in the Neendakara Kayamkulam grooves. This deposit was commonly called the Chavara deposit, and covers a total of 22 km in length (Department of Mining and Geology [DMG Kerala], n.d.). Kerala Minerals and Metals Ltd. and Indian Rare Earths Ltd. are the two public sector companies engaged in the field of mining activities in Chavara Kerala for the past eight decades (Indian Rare Earths Ltd. [IREL Ltd.], 2017; Kerala Minerals and Metals Ltd. [KMML], 2008a,b). Inland mining is a type of mining method adopted by the companies in the mining area. The current study focuses on the inland mining activities and associated impact of this method on groundwater quality and air quality. For purposes of the study, a mined-out land portion of the Chavara, Panmana, and Vadakkumthala villages and a nonmined land portion of Neendakara and Chavara villages of the Karunagapally taluk of the Kollam district were selected. Selection of mined-out land and nonmined land was done in order to have a comparative study. Inland mining employed by companies is a physical process, and no chemicals are being used for heavies' upgradation on site, hence there is no chance for chemical contamination of water due to mining; this was confirmed by physiochemical analysis of the water samples. For physiochemical analysis of groundwater (GW), nine samples from the mining area and eight samples from the nonmining area were collected and analyzed in the lab. Analysis showed that all the samples from the mining area were affected in terms of salinity, whereas only 25% from the nonmining area had an impact on GW quality. For evaluating the level of saltwater contamination, $Cl^-/(CO_3^{2-}+HCO_3^-)$ ratio has been adopted for the current study (Revelle, 1941; Klassen et al., 2014). The results were plotted in a geographic information system (GIS) environment to understand the spatial variation of the degree of contamination due to inland mining. Through the plotting it has been identified that the well samples collected near mine pits and near Thiruvananthapuram Shornur (TS) canal showed higher contamination levels than other well water samples. To unmask the reason behind the saltwater contamination the mining method adopted by the company has been examined. The study revealed that dredging of TS canal, unscientific exploitation, and simultaneous nonfilling of mines were factors contributing to saltwater contamination.

Considering the air pollution aspect, the current project has little significance since the mining method is a wet process. The only concern is the dust emission caused by the transport of heavies to a mineral separation plant (MSP) as the panchayat roads connecting the MSP and mining site are narrow, and the movement of trucks through these roads create traffic congestion and dust emission.

Mining of the rare earths is needed by the country because end products like ilmenite, rutile, zircon, and sillimanite have significant usage in various industries and have export value (IREL Ltd., 2017; KMML, 2008a,b). The refilling of the tailings reduces the background radiation level of the area (Concertation of Minerals, 2016). The importance of rare earth minerals are shown in Table 6.1 (KMML, 2008'b), which shows the applicability level of rare earths in industries ranging from the paper industry to the nuclear industry. As per the new national mineral policy, the Government of India is encouraging beach sand minerals exploration (National Mineral Policy, 2019).

2. Materials and methods

2.1 Study area

This particular investigation was carried out at two areas, site A and site B. The first area was a mined-out land portion with an area of 1.52 sq.km, which extends between longitudes 76°31′ 30.952″ E to 76°31′ 9.380″ E and latitudes 8°59′ 28.299″ N to 9°0′ 42.771″ and is situated in Chavara, Panmana, and Vadakkumthala villages (Site A). The second area selected was a nonmined land portion with an area of 0.42 sq.km, which extends between longitudes 76°31′ 39.936″ E to 76°31′ 53.716″ and latitudes 8°58′ 24.283″ N to 8°58′ 57.318″ situated in Neendakara and Chavara villages (Site B). The study areas are located at a distance of 14.80 km from Kollam (Fig. 6.1).

Table 6.1 Applications of rare earths minerals.

Rare earths minerals	Applications
Ilmenite	In the manufacturing of synthetic rutile, production of titanium tetrachloride, manufacturing of ferroalloys such as ferrotitanium, manufacturing of welding electrodes, manufacturing of titanium white, a pigment, and other titanium salts
Titanium dioxide pigment (rutile)	Paints, printing inks, plastic, paper, rubber, textiles, and ceramics
Zircon	Ceramics, foundries, refractories, zirconium chemicals, zirconium metals, and nuclear technology
Sillimanite	High temperature refractory and ceramic industry
Monazite	An important ore of thorium, which is used in nuclear power generation

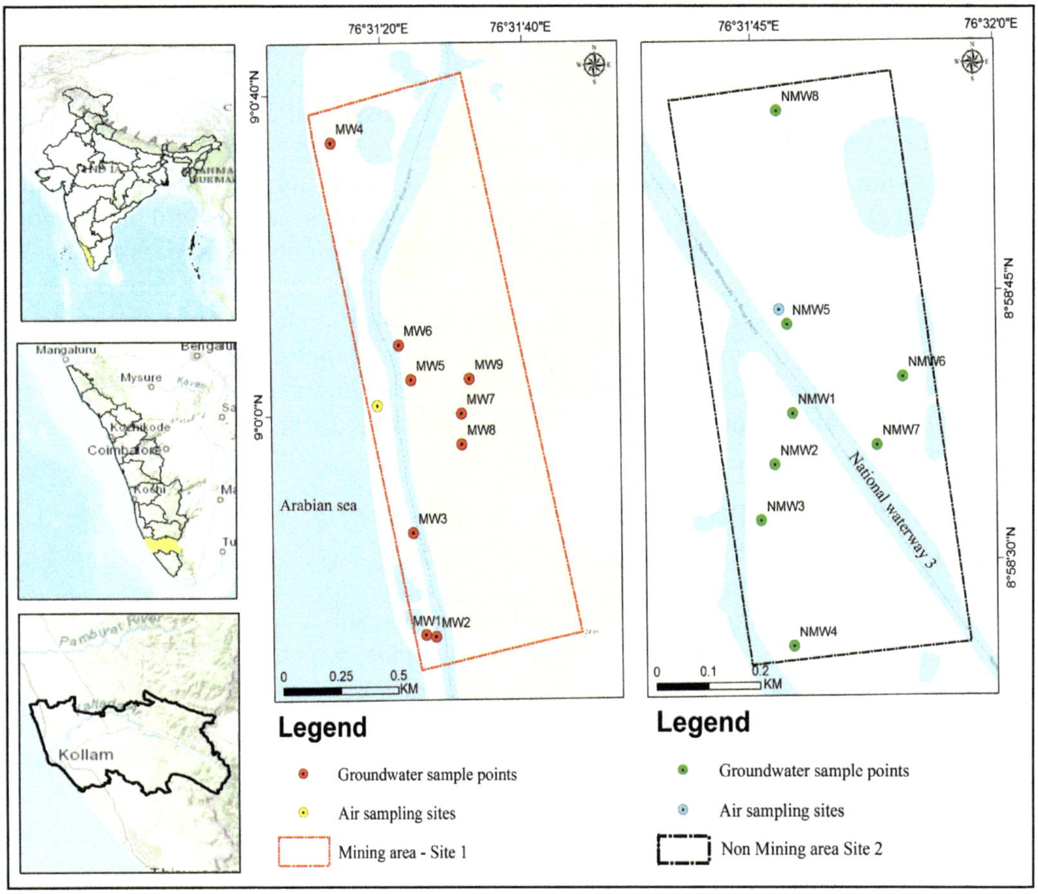

FIGURE 6.1 Study area.

2.2 Sample collection and analysis

2.2.1 Water environment

Samples for the current study are GW samples collected from open dug wells. A total of nine samples were collected from site A during the premonsoon period of year 2018, and eight samples were collected from site B during the premonsoon period of year 2018. Samples were collected after pumping the well and stored in polyethylene cans. Samples were brought to the lab and preserved as per the preservation techniques mentioned in standard protocol for collection and preservation (IS 3025-1, 1987).

For site A, drinking water parameters such as pH, conductivity, total dissolved solids (TDS), alkalinity, ammonia-N, total chloride, calcium, magnesium, sodium, potassium, iron, silica, phosphate, sulfate and nitrite −N were analyzed. All the analysis was done as per Indian Standards IS: 3025 (Table 6.2). For site B, drinking water parameters such as pH, TDS, alkalinity, and chloride were analyzed. All the analysis was done as per Indian Standards IS: 3025.

Table 6.2 Specific tests and method used for analysis of groundwater samples-IS 3025.

Sl. No	Specific tests or types of tests performed	Specification, standard (method), or technique used	Range of testing/limit of detection	Methodology
1	pH	IS 3025 (PART 11)—1983 (reaffirmed 2017)	1.00—14.00	Electrometric method
2	Conductivity	IS3025(PART 14)—1984 (reaffirmed 2013)	10 μS/cm—80 mS/cm	Conductance cell
3	TDS	IS3025(PART 16)—1984 (reaffirmed 2017)	10—500 mg/L	Gravimetric method
4	Alkalinity	IS3025(PART 23)—1986 (reaffirmed 2014)	10—500 mg/L	Indicator method
5	Ammonia-N	IS3025 (PART 34)—1988 (reaffirmed 2014)	1.0—100.0 mg/L	Titrimetric method
6	Total chloride	IS3025 (PART 32)—1988 (reaffirmed 2014)	10—500 mg/L	Potentiometric method
7	Calcium	IS3025 (PART 40)—1991 (reaffirmed 2014)	1.0—400.0 mg/L	EDTA titrimetric method
8	Magnesium	IS3025 (PART 46)—1994 (reaffirmed 2014)	1.0—250.0 mg/L	Volumetric method using EDTA
9	Sodium	IS3025 (PART 45)—1993 (reaffirmed 2014)	1.0—100.0 mg/L	Flame photometry method
10	Potassium	IS3025 (PART 45)—1993 (reaffirmed 2014)	1.0—100.0 mg/L	Flame photometry method
11	Iron	IS3025 (PART 53)—2003 (reaffirmed 2014)	0.050—0.500 mg/L	1,10 phenanthrolin method
12	Silica	IS3025 (PART 35)—1988 (reaffirmed 2014)	0.50—8.00 mg/L	Molidbosilcate method
13	Phosphate	IS3025 (PART 31)—1988 (reaffirmed 2014)	0.10—2.00 mg/L	Vanadomolybdo-Phosphoric acid method
14	Sulfate	IS3025 (PART 24)—1986 (reaffirmed 2014)	5.0—40.0 mg/L	Turbidity method
15	Nitrite-N	IS3025 (PART 34)—1988 (reaffirmed 2014)	0.10—2.00 mg/L	Nesslerization method

2.3 Method of study

The indicator for measuring the saltwater contamination used in the present investigation was $Cl^-/(CO_3^{2-} + HCO_3^-)$ ratio. Based on the range of $Cl^-/(CO_3^{2-} + HCO_3^-)$ ratio, the level of saltwater contamination has been categorized, as mentioned in Table 6.3 (Todd, 1959; Sarada and Bhushanavathi, 2015).

Table 6.3 Indicator for saltwater contamination.

Range of $Cl^-/(CO_3^{2-} + HCO_3^-)$	Saltwater contamination level
Less than 0.5	Normal groundwater (no saltwater contamination)
Between 0.5 and 1.30	Slightly contaminated groundwater
Between 1.30 and 2.80	Moderately contaminated groundwater
Between 2.80 and 6.60	Injuriously contaminated groundwater
Between 6.60 and 15.50	Highly contaminated groundwater(near sea water)
Greater than 200	Sea water

The wells that were having saline water contamination are plotted using ArcMap 10.3. Geostatistics are considered an assembly of numerical techniques that deal with characterization of spatial attributes by making use of primarily random models in a manner that is the same for time series analysis characterizing temporal data (Olea, 1999). In geostatistical methods, the kriging technique is quite popular. This technique is applicable to cases such as determining GW level, estimation of hydrochemical distribution of soil properties, and other estimations (Kuswantoro et al., 2012). The spatial interpolation tool of the ArcMap was used to find the distribution and pattern of the salinity over the study areas. An ordinary kriging method was used, as it is the one that best fits with the interpolation of GW chemical parameters. Kriging is one of the geostatistical methods based on statistical models. The kriging method includes autocorrelation, by which there will be statistical relationships among the measured points. Due to this, these geostatistical techniques have capabilities of both producing a prediction surface and providing some measure of certainty or accuracy predictions (Olea, 1999; Kuswantoro et al., 2012).

2.3.1 The kriging formula

In kriging the prediction of an unmeasured location is derived on the basis of weightage of surrounding measured values.

$$\widehat{Z}(S_0) = \sum_{i=1}^{N} \lambda_i Z(Si)$$

where:
$Z\ (Si)$ = value (measured) located at i-th location
λ_i = measured value at i-th location having an unspecified weight
S_0 = the location where prediction required
N = the number of measured values

The weights in the kriging method are not only based on distance between the measured points and prediction location but also on overall spatial arrangement of measured points. To apply spatial arrangement in weights, the spatial auto relation must be quantified. So in the ordinary kriging method, weight (λ_i) is dependent upon various factors such as the distance to prediction location, fitted model to measured points, and the spatial relationships among the measured values around the prediction location (ESRI, 2016).

2.4 Air environment

Analysis of air quality has been made by setting up two high-volume samplers in the MSP and in site B, during February–April 2018. Monitoring was done twice a week for 24 h/day. The PM_{10} pollutant was analyzed by using APM 460 NL Respirable Dust sampler [16]. PM_{10} particles were collected on QM-A quartz fiber filters (Whatman, UK) using mass-flow-controlled high-volume samplers. The QM-A filters were precombusted in a muffle furnace at a temperature of 550°C for 6 h, in a desiccator for a period of 48 h prior to sample collection. These preparations are necessary for exterminating impurities and moisture from filter papers. After sampling, the filters were kept in desiccator for at least 48 h prior to final weight measurement. This prevented the interruption of absorbed moisture during sampling (Sen et al., 2016). The results obtained will be compared with the permissible standard (Table 6.4) (NAAQS, 2009). The analysis was done as per the *Guidelines for the Measurement of Ambient Air Pollutants*, published by Central Pollution Control Board.

3. Results and discussion

3.1 Water environment

3.1.1 Site A

The analysis results of the samples collected from Site A is mentioned in Table 6.8. The results were compared with IS 10,500:2012 drinking water standard of India. pH were within the limits for all the samples. All samples except MW5 exhibited TDS values less

Table 6.4 National ambient air quality standard for PM_{10}.

Pollutant	Time-weighted average	Concentration in ambient air		Methodology of collection
		Industrial, residential, rural, and other areas	Ecologically sensitive area (notified by central government)	
Particulate matter, PM_{10}, μg/m^3	Annual[a] 24 h[b]	60 100	60 100	Gravimetric method

[a]Annual arithmetic mean of minimum 104 measurements in a year at a particular site taken twice a week, 24 hourly at uniform intervals.

[b]24 hourly or 8 hourly or 1 hourly monitored values, as applicable, shall be complied with 98% of the time in a year. 2% of the time, they may exceed the limits but not on two consecutive days of monitoring.

than 500 mg/L. High calcium content was found in samples MW1 and MW2. All samples contained chloride ions at levels less than 250 mg/L except for MW1, which exhibited a high chloride value of 297.5 mg/L. Iron, phosphate, alkalinity, nitrite, and magnesium were within the permissible limits in all samples.

The analysis showed that in most of the wells, water quality parameters were within the permissible limits in the absence of alternate source as per IS 10500 2012 and no chemical contamination occurs in the mining site. This agrees with the fact that here mining is done through physical process only. The only concerning factor will be saltwater contamination caused by deep dredging.

3.1.2 Evaluation of saltwater contamination in site A samples

Saltwater contamination was evaluated using the ratio as mentioned. Carbonate alkalinity, bicarbonate alkalinity, and total alkalinity along with chloride levels are mentioned in Table 6.5 (Sarada and Bhushanavathi, 2015).

Thus from the evaluation using $Cl^-/(CO_3^{2-} + HCO_3^-)$, from the nine samples collected from the mining site (site A), it can be seen that saltwater contamination exists in all nine well water samples. More than 50% of the samples were slightly contaminated, and the rest were moderately contaminated. From the interpolation it was found that, out of the total area, 19.36% is moderately contaminated and 80.64% of the area is slightly contaminated (Fig. 6.2).

3.1.3 Site B

By the detailed physiochemical analysis of site A water samples, we were able to identify that the concerning factor with regard to mining in this area is saltwater contamination. In order to have a comparison, eight water samples from the nonmining area (site B) were collected. For site B, only alkalinity, chloride, and TDS values were analyzed as no chemical contamination is warranted. The analysis result is shown in Table 6.6.

Compared to drinking water standards, samples NW7 and NW4 showed higher TDS values, and samples NW6, NW4, and NW2 showed higher alkalinity.

Table 6.5 Concentration values of CO_3^{2-}, HCO_3^- and Cl^- in mg/L for site A.

Sample ID	Carbonate alkalinity (CO_3^{2-})	Bicarbonate alkalinity (HCO_3^-)	Total alkalinity	Chloride as Cl^-	$Cl^-/(CO_3^{2-} + HCO_3^-)$
MW1	0	197	197	297.5	1.51
MW2	0	77	77	89	1.16
MW3	0	66	66	107	1.62
MW4	0	28	28	22	0.79
MW5	0	134	134	162	1.21
MW6	0	38	38	71	1.87
MW7	0	70	70	44	0.63
MW8	0	52	52	28	0.54
MW9	0	28	28	20	0.71

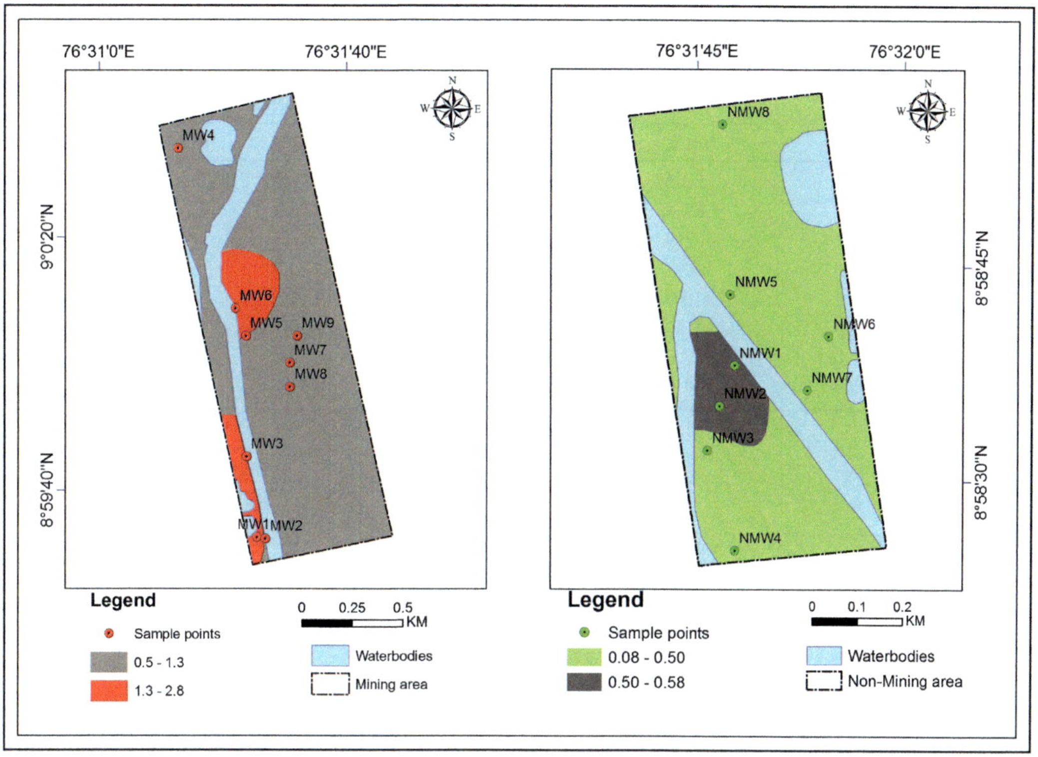

FIGURE 6.2 Distribution of saltwater contamination in study areas.

3.1.4 Evaluation of saltwater contamination in site B samples

Similar to site A samples, site B samples were evaluated using $Cl^-/(CO_3^{2-} + HCO_3^-)$. Carbonate alkalinity, bicarbonate alkalinity, and total alkalinity along with chloride levels are mentioned in Table 6.7.

From the nonmining site it can be clearly seen that only two samples showed saltwater contamination that are slightly contaminated; only 11.88% of the area is affected and the rest is unaffected in terms of saltwater contamination (Fig. 6.2). The saltwater contamination may be due to the proximity of the site with TS canal where dredging activities were being carried out by NWAI.

3.2 Air environment

Three months of monitoring of PM_{10} for 24 h from March 2, 2018 to April 27, 2018 was carried out at mining and nonmining areas (Table 6.9).

Table 6.6 Analysis of water samples from site B (premonsoon 2018).

Sample	pH	TDS	Alkalinity		Chloride
			Carbonate	Bicarbonate	
ID		mg/L	mg/L	mg/L	mg/L
NMW1	6.8	258.3	0	109.2	63
NMW2	6.8	567.4	0	232	119
NMW3	7.1	483.9	0	302	70
NMW4	7.0	560	0	423.2	77
NMW5	7.1	215.9	0	181	14
NMW6	7.2	37.3	0	260.8	52.5
NMW7	7.0	177.6	0	145.8	17.5
NMW8	6.9	311.9	0	187	42

Table 6.7 Concentration values of CO_3^{2-}, HCO_3^- and Cl^- in mg/L for site B.

Sample ID	Carbonate alkalinity (CO_3^{2-})	Bicarbonate Alkalinity (HCO_3^-)	Total alkalinity	Chloride as Cl^-	$Cl^-/(CO_3^{2-} + HCO_3^-)$
NMW1	0	109.2	109.2	63	0.58
NMW2	0	232	232	119	0.51
NMW3	0	302	302	70	0.23
NMW4	0	423.2	423.2	77	0.18
NMW5	0	181	181	14	0.08
NMW6	0	260.8	260.8	52.5	0.20
NMW7	0	145.8	145.8	17.5	0.12
NMW8	0	187	187	42	0.22

As per the CPCB standard for ambient air quality, the PM_{10} standard for 24 h is 100 μg/m^3. Figs. 6.3 and 6.4 show the PM_{10} values along with permissible limit marked. From the graphs it can be seen that none of the PM_{10} values were exceeding the 100 μg/m^3 limit. Thus the mining does not contribute a major impact on air quality of the region although the dust emission during truck traffic can contribute a minor impact on the air quality.

4. Conclusion

The present investigation identified the major impacts caused by unscientific inland mining. The major impact identified due to current methodology of mining was salt-water contamination. The physiochemical analysis of GW samples showed no significant

Table 6.8 Physiochemical analysis of site A (premonsoon 2016).

Sample ID	pH	Conductivity	TDS	Alkalinity		Ammonia-N	Total chloride	Calcium	Magnesium	Sodium	Potassium	Iron	Silica	Phosphate	Sulfate	Nitrite-N
				Carbonate	Bicarbonate											
		μS/cm	mg/L	mg/L	mg/L	mg/L	mg/L	mg/L	mg/L	mg/L	mg/L	mg/L	mg/L	mg/L	mg/L	mg/L
MW1	7.2	831.5	417.3	0	197	BDL	297.5	82.6	22.8	56.24	12.5	0.25	3.2	0.05	110.7	0.002
MW2	8	931	456	0	77	BDL	89	82	15	140	20	0.19	2.7	0.02	150	0.003
MW3	7	759	372	0	66	BDL	107	48	9	137	16	0.28	3.9	0.03	48	0.084
MW4	8	259	127	0	28	BDL	22	12	8	44	21	0.09	1.0	0.03	30	0.001
MW5	7	1428	701	0	134	BDL	162	44	28	267	2	0.07	5.0	0.03	65	0.002
MW6	7	958	273	0	38	BDL	71	36	8	105	4	0.13	2.6	0.02	52	0.003
MW7	7	520	255	0	70	BDL	44	32	13	72	3	0.17	4.3	0.00	57	BDL
MW8	7	537	263	0	52	BDL	28	32	21	43	21	0.13	2.6	BDL	158	0.006
MW9	7	202.7	99	0	28	BDL	20	10	9	25	1	0.09	2.6	0.02	23	BDL

Table 6.9 PM_{10} concentration at MS plant-mining area and nonmining area.

Sl.no	Date	Location	PM_{10} $\mu g/m^3$
1′	3/02/18	MSP/mining area	66
2	12/02/18		43
3	13/02/18		85
4	25/02/18		73
5	26/02/18		56
6	1/03/18		64
7	03/03/18		65
8	4/03/18		54
9	10/03/18		55
10	11/03/18		48
11	15/03/18		49
12	16/03/18		52
13	22/03/18		60
14	23/03/18		60
15	29/03/18		55
16	30/03/18		55
17	05/04/18		55
18	06/04/18		56
19	12/04/18		58
20	13/04/18		60
21	19/04/18		58
22	20/04/18		62
23	26/04/18		60
24	27/04/18		55

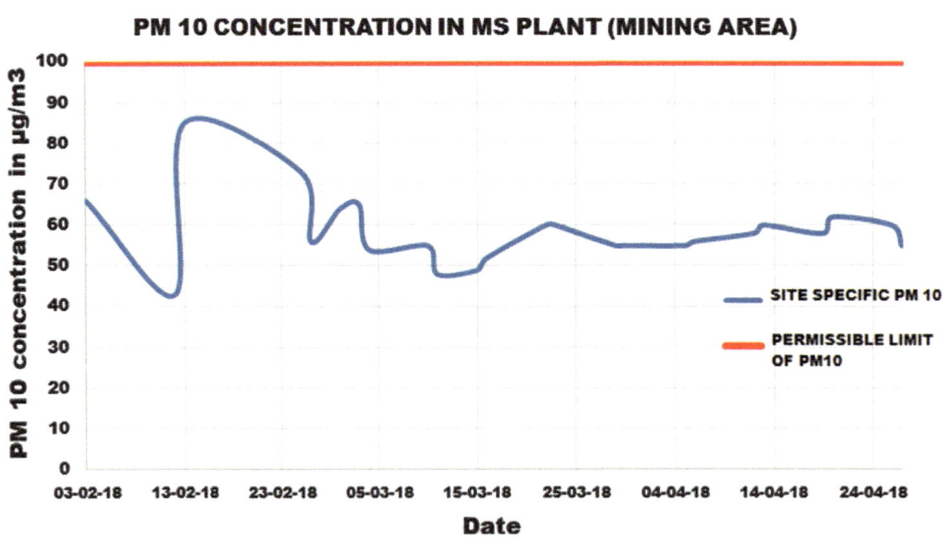

FIGURE 6.3 PM_{10} concentration in mining area.

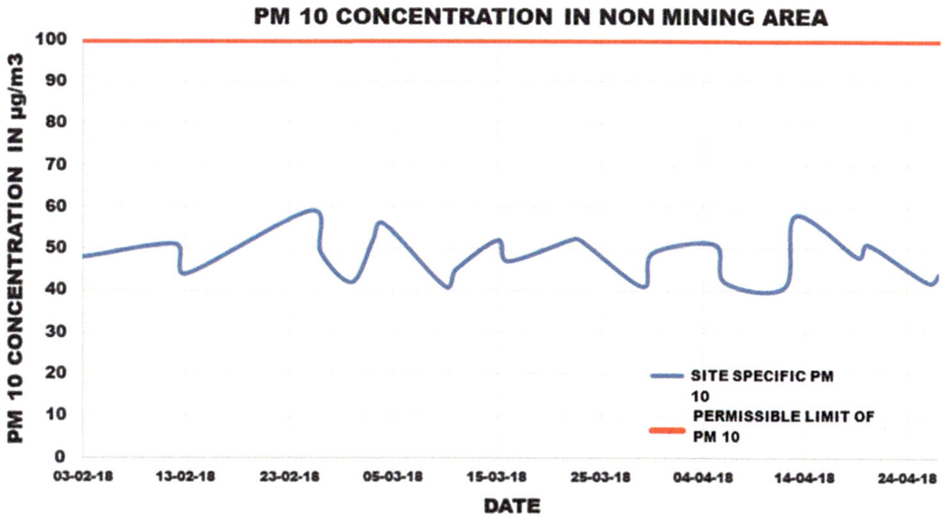

FIGURE 6.4 PM_{10} concentration in nonmining area.

chemical contamination, while the ratio $Cl^-/(CO_3^{2-}+HCO^{3-})$ showed that the mining area was prone to saltwater contamination. The entire study was done as a comparative study between a mining area and nonmining area, which revealed whether the mining activity was boosting the impact on that region. GIS tools were used for quantifying the impacts and establishing the area-wise impact of mining activity. GIS interpolation showed that the area where the inland mining activities occurred showed 100% saltwater contamination, but in the nonmining area, only 12% of area was affected by saltwater contamination. Further, the study was able to identify the source of saltwater contamination; it was seen that the mining area contained mine pits that were left open after mining, and these pits were accelerating the intrusion. It was also found that mining activities are not contributing any major impact to the air quality of the area, but truck traffic is causing minor dust emission during its operation. The study evaluated major techniques that can be adopted to minimize the observed impacts, as discussed next.

4.1 Mitigation measures to reduce impact

Mitigation measures/techniques for the impacts observed are classified into two categories: water environment and air environment.

4.1.1 Water environment

- Immediate refilling of the mine pits in site A as well water samples collected near these pits show higher saltwater contamination
- Mining and refilling should go hand to hand, which will help reduce the impact on GW.
- Supplying drinking water to Project Affected Peoples
- Monitoring GW samples from reclaimed land in postmonsoon and premonsoon seasons

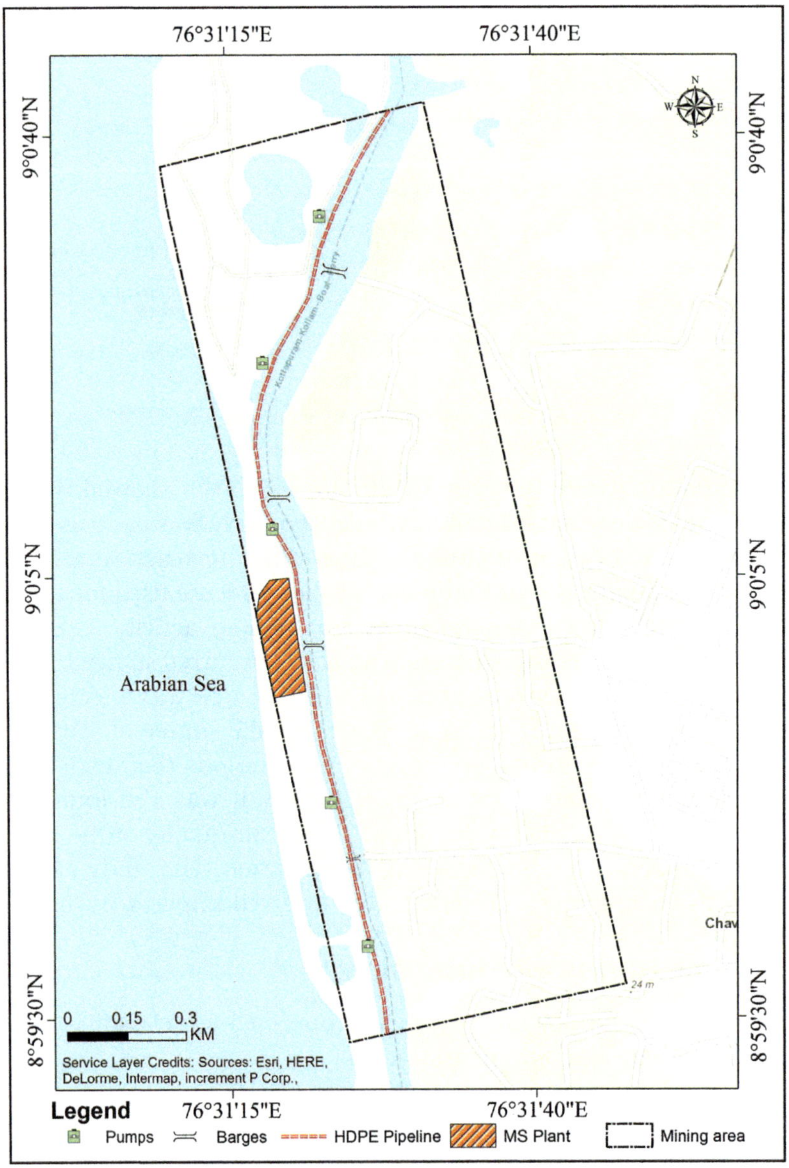

FIGURE 6.5 Alternate transportation methods using barge and HDPE pipeline system.

4.1.2 Air environment

- ❖ Usage of HDPE pipe lines with intermittent pumping stations to bring heavies to the MSP. Raw sand in the form of slurry can be pumped to the MSP through the pipeline system, which minimizes the truck travel distance (refer to Fig. 3.4).
- ❖ Barge/country boat deployment along the TS canal for transport of heavies. The operators can be chosen from areas that will provide a revenue to them (refer to Fig. 6.5).

Acknowledgments

Kashyap. V is thankful to CSIR, Govt. of India, for providing the SRF fellowship. Authors thank the Director, CSIR-National Institute for Interdisciplinary Science and Technology (CSIR-NIIST), Thiruvananthapuram, Kerala, India, for extending the laboratory facilities.

References

Concentration of Minerals, 2016. Comparative Analysis of Various Types of Mineral Raw Materials for Rare Earth Elements, Google Sites. Viewed 18 October 2018. https://sites.google.com/site/concentrationofminerals/home/the-practice-of-separation-of-minerals/comparative-analysis-of-various-types-of-mineral-raw-materials-for-rare-earth-elements.

DMG Kerala, n.d., Mineral Resources, C-Dit. Viewed 4 November 2018. http://dmg.kerala.gov.in/index.php?option=com_content&view=article&id=55&Itemid=61.

Environmental System Research Institute, 2016. How Kriging Works. Viewed 29th July 2019. http://desktop.arcgis.com/en/arcmap/10.3/tools/spatial-analyst-toolbox/how-kriging-works.htm.

IREL, 2017. About Us. Viewed 4 November 2018. http://www.irel.co.in/.

IS 3025-1, 1987. Methods of Sampling and Test (Physical and Chemical) for Water and Wastewater Part 1 – Sampling [CHD 32: Environmental Protection and Waste Management].

Klassen, J., Aleen, D.M., Kirtse, D., 2014. Chemical Indicators of Saltwater Intrusion for the Gulf Islands, British Columbia. Final Report. Department of Earth Sciences, Simon Rfaser University.

KMML, 2008a. A Brief History, Net Bios. Viewed 18 October 2018. http://kmml.com/php/showContent.php?linkid=10&partid=1.

KMML, 2008b. Applications, Net Bios. Viewed 23 July 2019. http://kmml.com/php/showContent.php?linkid=18&partid=2.

Kuswantoro, M., Al-Amri, N.S., Amro, M., Elfeki, M., 2012. Geostatistical Analysis Using GIS for Mapping Groundwater Quality: Case Study in the Recharge Area of Wadi Usfan, Western Saudi Arabia.

NAAQS, 2009. Guidelines for Sampling and Measurement of Notified Ambient Air Quality Parameters Issued by Central Pollution Control Board –India-Volume I.

National Mineral policy, 2019. Ministry of Mines, Government of India.

Olea, R.A., 1999. Geostatistics for Engineers and Earth Scientists. Kluwer Academic, Boston.

Revelle, R., 1941. Criteria for recognition of the sea water in ground-waters. Eos Trans. Am. Geophys. Union 22 (3), 593–597.

Sarada, D.P., Bhushanavathi, D.P., November 2015. $Cl^-/(CO_3{}^{2-} + HCO_3{}^-)$ ratio to evaluate salt water intrusion: a case study of Gnanapuram Area of Visakhapatnam, AP, India. Int. J. Sci. Res. 4 (11), 1311–1313 (#ijsrnet). https://www.ijsr.net/archive/v4i11/NOV151320.pdf.

Sen, A., Ahammed, Y.N., Banerjee, T., Chatterjee, A., Choudhuri, A.K., Das, T., et al., 2016. Spatial variability in ambient atmospheric fine and coarse mode aerosols over Indo-Gangetic plains, India and adjoining oceans during the onset of summer monsoons, 2014. Atmos. Pollut. Res. 7 (3), 521–532.

Todd, D.K., 1959. Ground Water Hydrology. John Wiley and Sons. Inc., United States, pp. 277–294.

Further reading

Shabani, M., 2012. Determining the Most Suitable Interpolation Method for Groundwater Chemical Characteristics Mapping.

Xiao, Y., Gu, X., Yin, S., Shao, J., Cui, Y., Zhang, Q., 2016. Geostatistical Interpolation Model Selection Based on ArcGIS and Spatio-Temporal Variability Analysis of Groundwater Level in Piedmont Plains, Northwest China.

7

Evaluation of coastal sediments: an appraisal of geochemistry using ED-XRF and GIS techniques

R.A. Renjith[1], R.G. Rejith[1,2], M. Sundararajan[1,2]

[1]MINERALS SECTION, MATERIALS SCIENCE AND TECHNOLOGY DIVISION, NATIONAL INSTITUTE FOR INTERDISCIPLINARY SCIENCE AND TECHNOLOGY (CSIR-NIIST), COUNCIL OF SCIENTIFIC & INDUSTRIAL RESEARCH, THIRUVANANTHAPURAM, KERALA, INDIA; [2]ACADEMY OF SCIENTIFIC AND INNOVATIVE RESEARCH (ACSIR), GHAZIABAD, UTTAR PRADESH, INDIA

Chapter outline

1. Introduction

The coastal sediments of India bestowed the predominant quantity of heavy mineral deposits of ilmenite, rutile, garnet, zircon, monazite, sillimanite, among others. Heavy minerals are considered strategic because of their contribution to the nation's economy (Raju et al., 2005; Rejith and Sundararajan, 2018). In India, major heavy mineral deposits are found at Chavara (Kerala), Manavalakurichi (Tamil Nadu), Sreekakulam (Andhra Pradesh), and Chathrapur (Odisha). The beach sands of Kerala, especially the Chavara deposit in Kollam district, is world-famous for heavy mineral concentration. The shaping of the coast is responsible for factors like physical, biological, geological, and anthropogenic processes (Noujas et al., 2019). The coastal sediments contain organic and inorganic materials such as detritus and nondetrital material in different proportions (Selvaraj et al., 2010). Mohanty et al. (2003) studied the geochemistry of monazite sands of Chathrapur beach placer deposit of Orissa. The geochemistry study identifies the

Remote Sensing of Ocean and Coastal Environments. https://doi.org/10.1016/B978-0-12-819604-5.00007-X

impact of anthropogenic activity on beach sediments and soil (Szefer, 2002). DíazRizo et al. (2015) studied the concentration Ni, Cu, Zn, and Pb levels in beach and dune sands from resorts in Havana, Cuba. In beach sands, the trace metals were present due to the silicate minerals, which are formed by weathering of the metamorphic rocks (Yalcin et al., 2015). Facchinelli et al. (2001) studied the effect of heavy metal contamination in soil, and their variability and origin.

Geochemistry of soils provides solid information regarding the weathering and provenance of parent rocks (Wei et al., 2004). Detailed geochemical investigation on REE, major, and minor element content of beach sands, and the nearby rock types helps to improve the provenance study of sediments (Armstrong-Altrin et al., 2012). Bernárdez et al. (2012) studied the variations in mineralogy and elemental geochemistry of marine sediments in the Northern Rias. Dubey et al. (2016) derived the weathering pattern and genesis of Precambrian rocks in parts of Kottayam in Kerala using the geochemistry of ferruginous soils collected over lateritic cover.

The beach sands of Varkala and the adjacent area contains a high concentration of heavy minerals such as ilmenite, rutile, zircon, monazite, sillimanite, and garnet. The black beach near the north cliff of Varkala contains a high concentration of ilmenite. Rejith and Sundararajan (2018) studied the heavy mineral present in the Kappil-Varkala coast, and this region contains 67.3% total heavy mineral (THM) out of this ilmenite (45.55%) as the major mineral. Silpa et al. (2016) investigated the influence of different morphology of high energy regime in the quartz grains of the Poovar-Varkala coast. Microtexture studies helped to identify the depositional environments like marine, Aeolian, fluvial, and glacial. The characteristics of hydrodynamics and coastal processes of the Kerala coast were studied by Noujas et al., (2019). Thankappan et al. (2018) investigated the effect of coastal morphology and shoreline changes along the Thiruvananthapuram coast. The effect of shoreline morphological changes to break-water construction in Veli-Varkala were studied by Noujas et al. (2014). Tertiary sedimentary coastal cliffs of Varkala consist of a composite fall-slippage model (Sajinkumar et al., 2017). Singh et al. (2016) identified the minerals present in the Warkalli formation in Kerala using the most advanced spectral techniques. According to Singh et al. (2016) the phyllosilicates exhibited as kaolinite and jarosite as natro-jarosite. Divya et al. (2018) studied the distribution and enrichment of radionuclide concentration in the Varkala region. The radioactivity of Varkala sand is high compared to the world average; it may be because of a high concentration of uranium in the sand. Gandhi et al. (2015) evaluate the distribution and abundance of foraminifera, which is a bioindicator for coastal pollution from Neendakara to Varkala beach. The large number of foraminifera population is found in Neendakara and Varkala beach sediments.

The objective of the present study is to examine geochemistry and structural characterization of the beach, cliff, and soil samples collected along the coast of the Varkala region using energy dispersive X-ray fluorescence spectroscopy (ED-XRF), X-ray powder diffraction (XRD), and Raman spectroscopy. Further, the distribution of major and trace elements were also analyzed using the ordinary kriging (OK) interpolation technique available in the GIS software environment.

2. Study area

The Varkala coast is situated in the southernmost district Thiruvananthapuram with a total 8 km coastline along the Arabian Sea on the west 8° 47′ 20.3028" to 8° 43′ 15.438″ north latitudes and 76° 40′ 23.2176″ to 76° 43′ 19.7148″ east longitudes (Fig. 7.1). The high energy waves, steep beach face, and medium-sized sand are characterized by the Thiruvananthapuram coast. The major morphological features of the coast are coastal plains, barrier beaches, pocket beaches, and cliff (Thankappan et al., 2018). The Varkala coast is situated between the famous heavy mineral deposit Chavara, and Manavalakurichi, which is particularly known for the ilmenite mineral. The total heavy mineral content of the Kappil-Varkala coast is high, about 51.06% (0.07 million tons). The heavy mineral-rich beach is influenced by factors such as hinterland geology, coastal geomorphology, subtropical to a tropical climate (Rejith and Sundararajan, 2018). The cliffs present in the Varkala coast belong to the Warkali formation, which consists of mio-pliocene placed over the Precambrian crystallines of the Kerala Khondalite Belt. The Varkala cliff consists mainly of arkosic sand, kaolinitic sandy clay, carbonaceous clay, and peat with plant remains (Singh et al., 2016).

3. Materials and method

Twenty-four samples were collected in a Varkala region to analyze the major and minor elements present in this area. It consists of 12 soil samples (within 1 km buffer area from the coastline) and 6 beach and cliff samples. The samples were collected in a polythene bag and dried in the sunlight. About 10 g of samples were finely powdered and subjected to characterization techniques like ED-XRF, XRD, and Raman spectroscopy analysis. X-ray fluorescence (XRF) spectrometry is the analytical technique to find the elements present in the sample from magnesium to uranium. The 5 g of sample is taken in pellets and running in a PANalytical XRF Epsilon 3 instrument. Ominon software is used to interpret the data. The XRD patterns of the beach, cliff, and soil samples were studied using the PANalyticalX'Pert Pro diffractometer. The resultant peaks were interpreted using X'PertHighScore Plus software (PANalytical) to identify the mineral phases. Raman spectroscopy is an important technique to find the minerals present in the sample. Confocal Raman microscope (WI-Tec, Inc., Germany, alpha 300R) is used to study the Raman shifts of the samples. The obtained data from the samples were processed by using the WI-Tec Project plus (v 2.1) software package. Pearson's correlation coefficient matrices were carried out to find the interrelationships and behavior of the elements present in the sample. The computer package IBM SPSS statistics 20 was used for statistical analysis.

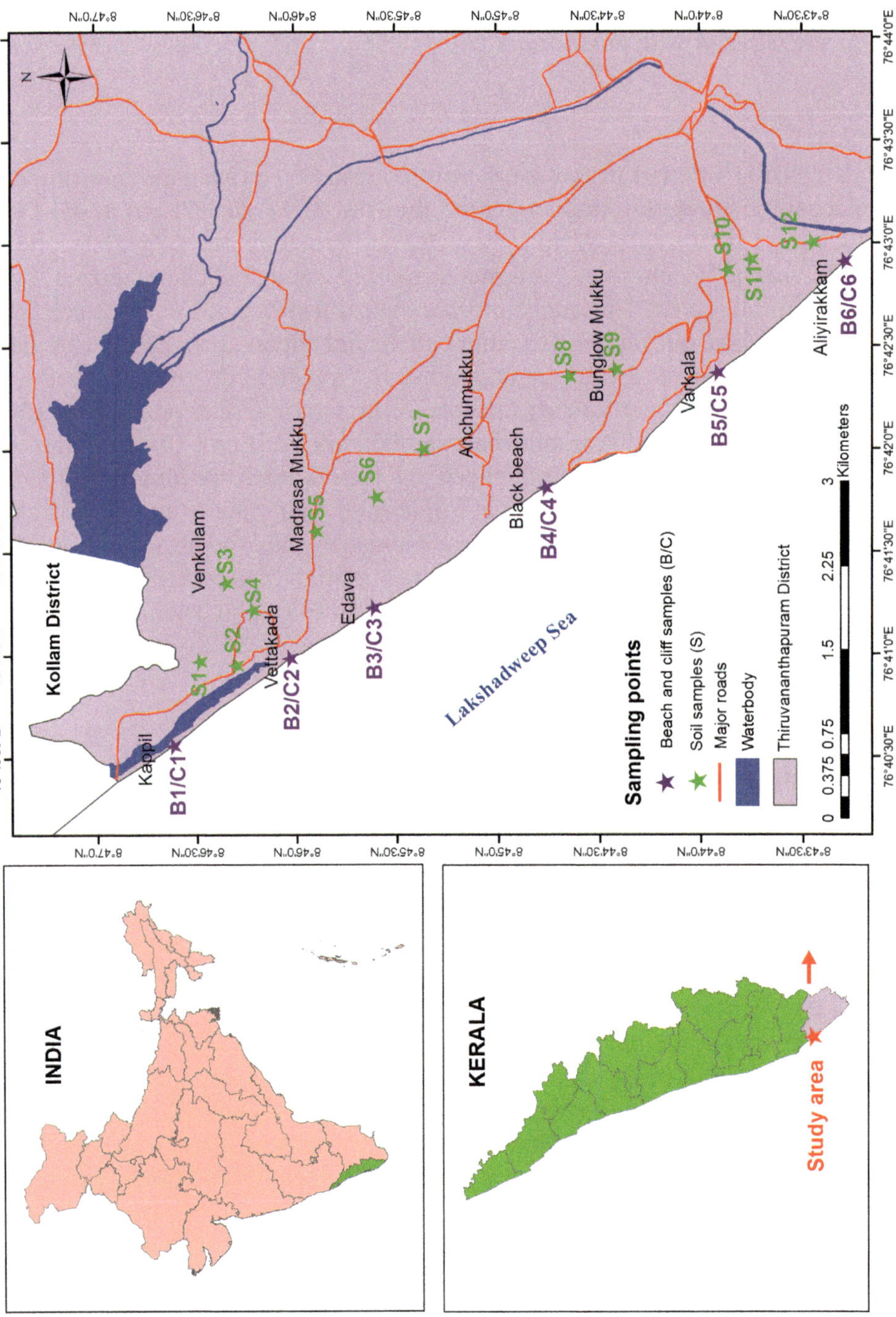

FIGURE 7.1 Study area.

The XRF results obtained for the 18 samples (6 beach samples and 12 soil samples) were used for generating surface maps showing the variation in geochemical components along the coast of Varkala. The OK always shows the least error for the prediction values generated using the geochemistry of soil samples (Tziachris et al., 2017; Panday et al., 2018). Zou et al. (2010) successfully proved that the OK shows the best accuracy for study areas with small sample data. Since only 18 samples were used for the present study, the OK interpolation technique with a spherical model available in the ArcGIS 10.2 software environment is adopted for the present study.

4. Results and discussion

XRF analysis of the beach sand, cliff, and soil samples are shown in Table 7.1, Table 7.2 and Table 7.3. Along the beach, TiO_2, and SiO_2 present as a major, which implies the presence of ilmenite and silica sand. The maximum TiO_2 content was seen in Kappil, Vettakada, Edava, and Black beach (B1–B4) in the range of 44.5%–53.8%. The Vettakada (B2) beach shows high TiO_2 content of 53.813%. The concentration of Fe_2O_3 content follows the same trend as TiO_2. The maximum Fe_2O_3 content was seen in Kappil, Vettakada, Edava, and Black beach (B1–B4) in the range of 28.6%–33.91%. The Vettakada (B2) beach shows high Fe_2O_3 content of 33.918%. The high concentration of Fe_2O_3 also indicates the presence of ilmenite mineral. The highest and lowest concentrations of SiO_2 were found in Aliyirakkam beach (73.831%) and Vettakada beach (4.613%), respectively. The highest concentration of SiO_2 reveals low presence of heavy minerals and high content of quartz. The highest concentration of SiO_2 was found in Varkala and Aliyirakkam beach in the range of 65%–73.8%. The ZrO_2 concentration was observed in the range of 9.85%–1.27%, in which the Black beach (B4) shows the highest concentration (9.857%) and lowest concentration at Aliyirakkam beach (1.271%). These values suggest the presence of zircon mineral in the sample. The highest and lowest concentration of Al_2O_3 was seen at Varkala beach (3.558%) and Edava beach (0.999%), respectively, which indicates sillimanite concentration. The radioactive elements of uranium and thorium are also present in the sample. The highest concentration of ThO_2 is seen in Black beach (0.788%). The highest concentration of ThO_2 found at Edava, Black beach, and Varkala beach (B3–B5) in the range of 0.26%–0.79%. The high concentration of uranium is present at Black beach (452.4 ppm). The highest concentration of lead and cadmium are present at Black beach (379.502 ppm) and Kappil beach (703.256 ppm). The distribution of major and trace elements were shown in Fig. 7.2A–G. The distribution maps clearly depict the presence of low heavy mineral content in the inland areas of Varkala within 1 km from the coastline.

XRF result of the cliff shows the highest concentrations SiO_2, Fe_2O_3, and Al_2O_3. The analysis of the cliff sample revealed that the amount of iron is high in the entire sample.

Table 7.1 XRF results of beach sand samples.

Compound	B1	B2	B3	B4	B5	B6
TiO_2 (%)	45.543	53.813	44.881	44.525	14.655	11.955
Fe_2O_3 (%)	28.601	33.918	29.169	29.151	9.167	7.732
SiO_2 (%)	16.662	4.613	11.908	10.026	65.081	73.831
ZrO_2 (%)	4.011	3.094	9.513	9.857	3.683	1.271
Al_2O_3 (%)	2.580	2.043	0.999	1.036	3.558	2.336
P_2O_5 (%)	0.583	0.494	0.615	1.161	1.250	1.067
V_2O_5 (%)	0.449	0.517	0.627	0.788	0.289	0.124
MgO (%)	0.279	0.343	0.303	0.264	0.141	0.047
CaO (%)	0.279	0.204	0.268	0.305	0.668	0.527
MnO (%)	0.177	0.214	0.163	0.098	0.044	0.054
Nb_2O_5 (%)	0.171	0.204	0.159	0.161	0.049	0.047
Cr_2O_3 (%)	0.139	0.139	BDL	BDL	BDL	0.046
HfO_2 (%)	0.130	0.113	0.266	0.289	0.099	0.037
ThO_2 (%)	0.071	0.016	0.370	0.788	0.261	0.051
Cl (%)	BDL	BDL	BDL	BDL	0.517	0.678
Nd_2O_3 (%)	BDL	BDL	0.578	1.312	0.382	0.056
Eu_2O_3 (%)	0.119	0.110	BDL	BDL	BDL	0.000
ZnO (%)	0.045	0.055	0.045	0.040	0.015	0.013
U (ppm)	BDL	36.500	231.500	452.400	156.300	BDL
Br (ppm)	BDL	BDL	BDL	BDL	BDL	11.700
As (ppm)	BDL	BDL	BDL	BDL	BDL	10.907
Se (ppm)	BDL	BDL	31.478	BDL	BDL	BDL
Rb (ppm)	6.216	6.307	BDL	BDL	BDL	BDL
Sr (ppm)	42.700	48.270	31.055	0.000	26.245	29.030
Y (ppm)	94.724	62.441	242.126	348.268	146.614	50.236
Cd (ppm)	703.256	198.967	144.170	297.532	281.250	345.238
Yb (ppm)	BDL	BDL	177.408	BDL	BDL	4.555
Pt (ppm)	40.279	BDL	BDL	BDL	BDL	BDL
Pb (ppm)	243.409	283.234	287.597	379.502	130.802	68.975
Bi (ppm)	186.042	255.203	195.013	176.085	26.103	9.598
Re (ppm)	0.200	5.700	BDL	BDL	BDL	BDL
K (ppm)	BDL	BDL	BDL	BDL	386.103	458.409
Ga (ppm)	BDL	BDL	BDL	BDL	BDL	75.269
Ni (ppm)	89.509	76.228	BDL	119.921	58.310	65.226

The maximum Fe_2O_3 concentration was seen at Black beach cliff (C4) of about 59.442%. The Fe_2O_3 concentration was found in the range 15.8%–59.44% and it is evident that the cliff contains mainly laterite. The other major element present in the cliff sample is SiO_2. The highest and lowest concentration of SiO_2 is present at Vettakada (67.977%) and Black beach cliff (19.095%). The SiO_2 concentration lies in the range of 19%–67.9%. The highest Al_2O_3 concentration is observed at Aliyirakkam beach (27.164%) and minimum at Vettakada (2.354%). The percentage of TiO_2 was low in all the samples (C1–C6), and it

Table 7.2 XRF results of cliff samples.

Compound	C1	C2	C3	C4	C5	C6
SiO_2 (%)	24.294	67.977	29.674	19.095	59.493	43.953
Fe_2O_3 (%)	45.807	24.032	40.355	59.442	15.792	26.058
Al_2O_3 (%)	27.132	2.354	26.549	17.583	21.923	27.164
MgO (%)	BDL	1.320	BDL	0.144	BDL	BDL
Cl (%)	0.102	1.074	0.274	0.111	0.044	0.102
CaO (%)	0.148	1.003	0.188	0.191	0.183	0.232
P_2O_5 (%)	0.722	0.987	0.818	0.879	1.124	0.941
TiO_2 (%)	1.170	0.906	1.509	1.164	1.032	1.140
CdO (%)	0.159	0.118	BDL	0.197	BDL	BDL
ZrO_2 (%)	0.035	0.055	0.075	0.463	0.058	0.097
K_2O (%)	0.028	0.048	0.170	0.075	0.203	0.173
Cr_2O_3 (%)	0.098	0.029	0.109	0.144	0.039	0.026
V_2O_5 (%)	0.100	0.012	0.092	0.140	0.044	0.029
Mn (ppm)	175.496	26.255	133.984	BDL	69.625	136.152
Ni (ppm)	172.574	84.086	181.847	204.794	114.028	90.530
Zn (ppm)	70.373	46.192	69.007	80.093	54.627	55.270
As (ppm)	11.740	BDL	BDL	BDL	36.052	BDL
Se (ppm)	BDL	BDL	BDL	22.906	BDL	BDL
Rb (ppm)	58.958	9.689	211.974	137.477	14.991	22.943
Sr (ppm)	48.354	156.034	51.899	31.392	18.143	26.076
Y (ppm)	BDL	BDL	BDL	BDL	BDL	17.165
Nb (ppm)	BDL	31.108	BDL	BDL	27.683	37.679
Eu (ppm)	BDL	492.821	BDL	BDL	BDL	BDL
Yb (ppm)	BDL	BDL	BDL	129.583	4.783	36.666
Pb (ppm)	529.335	85.128	583.550	1086.149	65.912	117.620
Bi (ppm)	18.479	BDL	68.622	BDL	BDL	BDL
Re (ppm)	9.300	5.300	10.400	16.500	BDL	BDL
Ga (ppm)	152.688	BDL	138.172	57.796	104.570	160.215
Br (ppm)	BDL	41.100	BDL	BDL	BDL	BDL
Cu (ppm)	97.460	BDL	83.640	135.006	78.687	BDL
In (ppm)	BDL	146.518	427.466	BDL	BDL	333.656
Os (ppm)	BDL	BDL	BDL	BDL	3.074	BDL
Ta (ppm)	BDL	BDL	25.878	BDL	BDL	BDL
Mo (ppm)	14.000	BDL	BDL	BDL	BDL	BDL
Ba (ppm)	524.082	BDL	BDL	1530.886	BDL	BDL

was found in the range of 1.5%–0.9%. The maximum TiO_2 concentration is present at the Edava cliff (C3). The rare earth distribution of beach and soil was shown in Fig. 7.3A–C using the GIS technique. On comparing with inland areas, the rare earth element concentration is high in beach sediments.

The XRF analysis of soil samples revealed that the concentration of the SiO_2 is very high in all the samples, and it was seen in the range of 82.57%–46.26%. The highest and

Table 7.3 XRF results of soil samples.

Compound	S1	S2	S3	S4	S5	S6	S7	S8	S9	S10	S11	S12
SiO_2 (%)	74.914	82.576	63.921	61.769	46.696	59.746	64.896	66.880	67.407	46.267	57.347	77.198
Al_2O_3 (%)	11.670	10.622	15.553	21.695	25.339	22.059	15.968	15.204	18.648	28.261	28.191	13.619
Fe_2O_3 (%)	9.698	2.871	13.243	10.351	22.438	11.229	9.852	12.424	8.384	20.952	10.671	5.018
TiO_2 (%)	1.637	1.619	3.918	1.608	2.925	4.284	1.656	2.086	2.039	1.940	1.548	1.572
P_2O_5 (%)	0.858	0.851	1.165	1.361	1.100	1.200	1.042	1.150	1.133	1.070	1.201	1.257
CaO (%)	0.633	1.004	1.026	2.095	0.611	0.310	3.630	1.092	1.274	0.557	0.444	0.782
K_2O (%)	0.307	0.198	0.372	0.663	0.391	0.284	2.272	0.725	0.590	0.523	0.340	0.319
ZrO_2 (%)	0.115	0.136	0.346	0.095	0.186	0.409	0.099	0.183	0.165	0.135	0.082	0.105
MgO (%)	BDL	BDL	0.130	0.132	BDL	0.117	0.289	BDL	0.154	BDL	BDL	BDL
Cr_2O_3 (%)	0.052	0.035	0.076	0.080	0.101	0.125	0.093	0.073	0.045	0.078	0.029	0.033
V (ppm)	236.500	119.762	390.881	218.127	489.301	385.055	199.978	265.685	200.258	356.151	149.451	145.922
Mn (ppm)	174.102	147.460	317.612	259.681	216.930	194.935	242.643	178.051	146.763	153.423	166.202	96.732
Ni (ppm)	77.957	101.061	67.976	195.049	183.497	212.024	178.782	114.342	128.094	175.403	94.774	73.713
Zn (ppm)	57.519	37.355	90.778	84.672	93.107	87.323	109.656	107.969	152.394	102.747	59.126	61.777
As (ppm)	22.419	4.014	30.978	25.979	64.001	32.947	17.496	44.308	23.404	55.518	BDL	17.875
Rb (ppm)	17.642	13.985	20.932	32.907	32.084	20.841	140.951	40.494	31.444	34.369	23.857	14.991
Sr (ppm)	32.152	20.844	44.051	58.903	47.932	28.354	134.430	61.097	76.371	77.553	90.549	37.300
Y (ppm)	BDL	BDL	23.543	16.772	19.843	28.740	29.291	16.142	18.504	20.472	49.370	BDL
Nb (ppm)	41.734	38.168	110.661	40.196	72.772	110.311	40.336	54.387	49.423	49.144	38.518	40.755
Eu (ppm)	BDL	BDL	902.038	BDL	BDL	627.374	745.252	724.873	545.391	852.015	656.786	302.455
Hf (ppm)	BDL	40.109	90.308	BDL	BDL	101.840	BDL	BDL	BDL	BDL	BDL	BDL
Pb (ppm)	40.939	16.060	61.363	51.058	113.721	55.607	59.970	51.522	52.172	113.814	91.255	21.352
Bi (ppm)	BDL	BDL	3.767	BDL	BDL	BDL	BDL	BDL	BDL	BDL	BDL	BDL
Th (ppm)	40.513	65.735	120.221	39.986	BDL	108.358	60.726	74.963	54.662	BDL	35.328	42.798
Re (ppm)	BDL	1.300	BDL	BDL	2.300	0.500	BDL	BDL	BDL	BDL	4.600	1.800
Ga (ppm)	74.731	38.441	91.129	109.677	167.742	141.667	94.892	110.215	106.183	173.925	125.806	82.796
Br (ppm)	6.600	8.900	7.300	13.000	16.900	14.900	17.500	BDL	BDL	22.500	BDL	BDL
Cu (ppm)	47.931	20.371	41.221	59.434	79.885	52.724	49.049	55.121	51.686	83.799	48.410	34.271
In (ppm)	BDL	BDL	211.315	BDL	BDL	84.139	BDL	BDL	BDL	BDL	BDL	BDL
Ir (ppm)	BDL	BDL	BDL	BDL	4.782	BDL	BDL	2.450	6.299	0.700	1.166	0.933
Os (ppm)	BDL	BDL	BDL	BDL	8.018	BDL	BDL	BDL	BDL	BDL	BDL	BDL

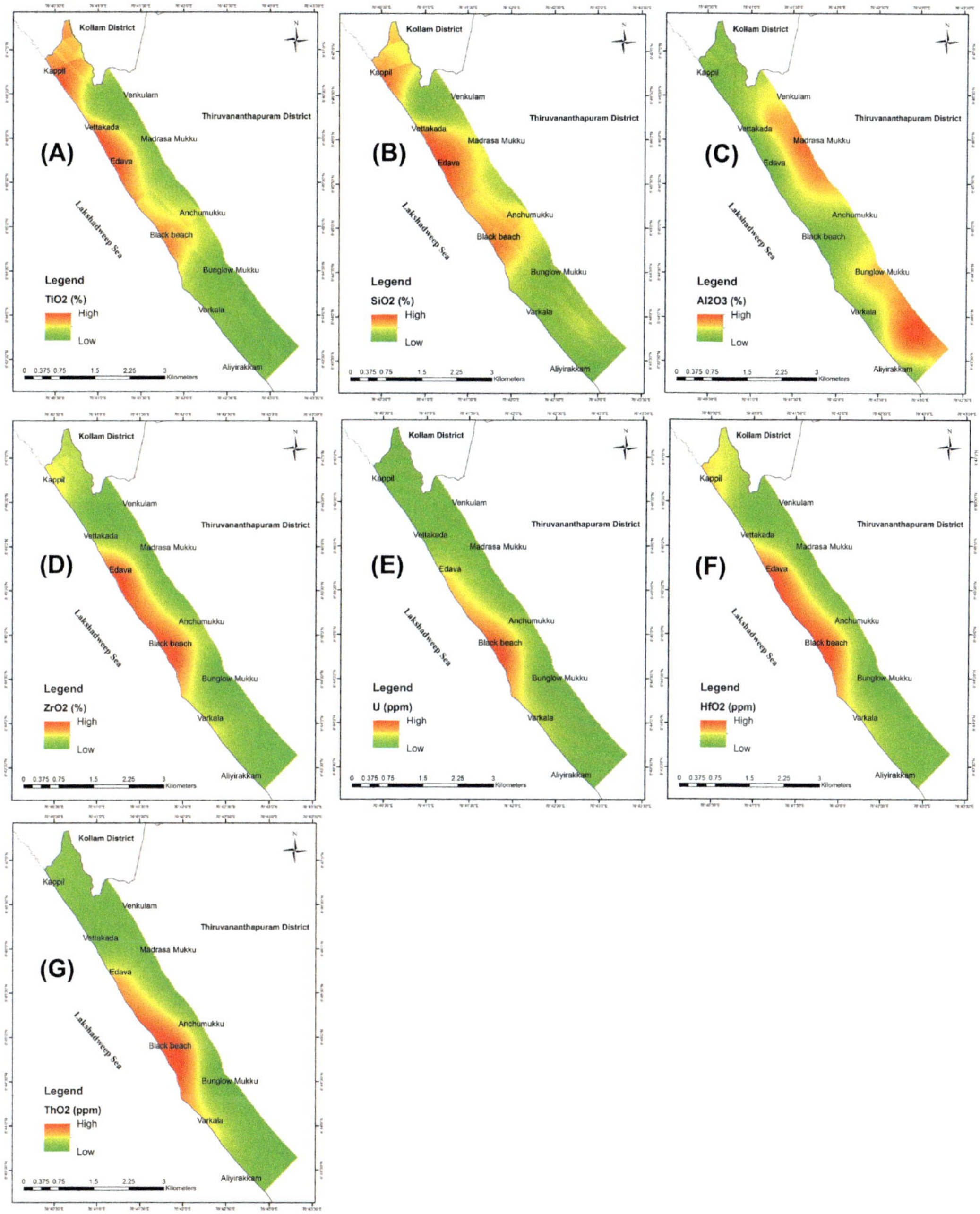

FIGURE 7.2 Map showing the distribution of major and trace elements (A) TiO_2, (B) SiO_2, (C) Al_2O_3, (D) ZrO_2, (E) U, (F) HfO_2, and (G) ThO_2.

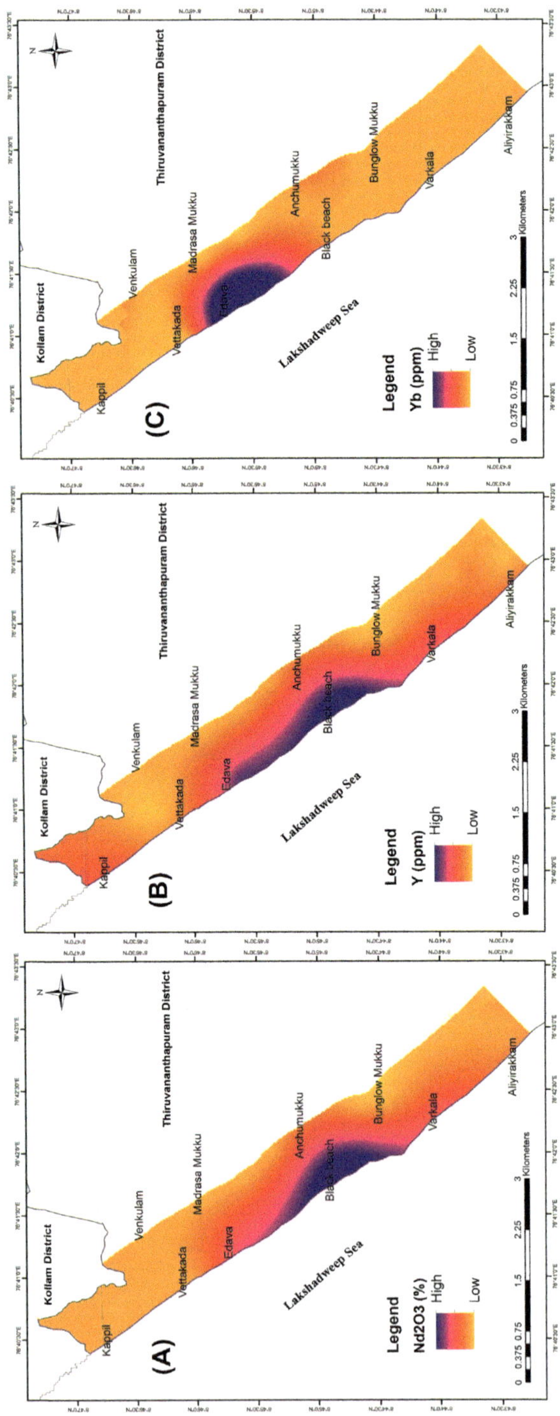

FIGURE 7.3 Map showing the distribution of rare earth elements (A) Nd_2O_3, (B) Y, and (C) Yb.

lowest concentration of SiO_2 was observed at Vettakada (S2) and Kilithattumukku (S10). The other major compound observed in the soil sample was Al_2O_3 and Fe_2O_3. The highest and lowest concentration of Al_2O_3 is present in Kilithattumukku (28.261%) and Vettakada (10.622%). The concentration of Fe_2O_3 is varied from in the range 22.43%–2.875%, with the highest concentration found in Jawahar higher secondary school (S5) and the lowest one in S2. The concentration of TiO_2 is low in all the samples in the range 4.28%–1.54%. The highest concentration of thorium is present in the Sreekrishna swami temple (120.221 ppm). The highest concentration of Pb is found at Kilithattumukku (113.814 ppm) and lowest at S2 (16.060 ppm). The statistical measures such as average, maximum, minimum, and standard deviation of beach sediment, cliff, and soil samples are shown in Tables 7.4– 7.6.

XRD of the beach (B2), cliff (C2), and soil (S2) sample were shown in Fig. 7.4A–C. The beach sand shows ilmenite as the major mineral and SiO_2 is the minor phase (Nair et al., 2009). The XRD study of cliff reveals that SiO_2 is the major phase, and Al_2O_3 and FeO is the minor phase. The soil sample contains mainly SiO_2 peaks. The Raman spectroscopy is an important analytic technique to find heavy minerals under a polarizing microscope. The Raman spectroscopy of beach (B2), cliff (C2), and soil (S2) are shown in Fig. 7.5A–C. The main Raman peak corresponds at 155 cm^{-1} due to the TiO_2 vibration. This peak is due to the oxygen moves strictly perpendicular around the titanium, which results in fewer changes for Ti–O and O–O bond lengths. As per the published Raman spectra of the ilmenite structure and the assignments made of their peaks, we attribute the Raman peak of $FeTiO_3$ ilmenite, ~615 cm^{-1}, to Ag symmetry as the symmetric stretching vibration of $Ti^{4+}O_6$ octahedra. That peak also has Ag symmetry and likely is produced by the translational motion of the Fe cation in the crystal lattice. The weaker Raman peaks are also observed at ~303 and ~192 ilmenite grains (Wang et al., 2004). The peaks observed for cliff and soil are almost the same. The peak observed at 106 and 167 cm^{-1} is due to the SiO_2 vibration, and the peak observed at 243 cm^{-1} corresponds to Fe–O vibration (Arrieta et al., 2011; Saksena, 1940). The Raman spectroscopy gives structural evidence of the minerals present in the sample.

The Pearson's correlation coefficient matrix for beach sand is shown in Table 7.7. The result indicates there is strong positive correlation between Pb and TiO_2, Fe_2O_3, and HfO_2 (R = 0.868, R = 0.884, and R = 0.865, respectively; $P < .05$). The significant correlation is observed between uranium, ZrO_2, and HfO_2 (R = 0.871, and R = 0.854, respectively; $P < .05$). SiO_2 is highly negatively correlated with both TiO_2 and Fe_2O_3 (R = −0.991, and R = −0.995; $P < .01$). ThO_2 is seen to be significantly correlated with uranium (R = 0.988; $P < .01$). The correlation matrix for soil sample is shown in Table 7.8. There is significant positive correlation between TiO_2 and ZrO_2, and HfO_2 (R = 0 0.973, and, R = 0.842, respectively; $P < .01$). Meanwhile there is good positive

Table 7.4 Statistical measures such as average, maximum, minimum, and standard deviation of beach sample.

Compound	Avg	Max	Min	SD
TiO_2 (%)	35.895	53.813	11.955	17.853
Fe_2O_3 (%)	22.956	33.918	7.732	11.410
SiO_2 (%)	30.354	73.831	4.613	30.659
ZrO_2 (%)	5.238	9.857	1.271	3.574
Al_2O_3 (%)	2.092	3.558	0.999	0.976
P_2O_5 (%)	0.862	1.250	0.494	0.334
V_2O_5 (%)	0.466	0.788	0.124	0.237
MgO (%)	0.230	0.343	0.047	0.112
CaO (%)	0.375	0.668	0.204	0.181
MnO (%)	0.125	0.214	0.044	0.070
Nb_2O_5 (%)	0.132	0.204	0.047	0.067
Cr_2O_3 (%)	0.054	0.139	0.000	0.068
HfO_2 (%)	0.156	0.289	0.037	0.100
ThO_2 (%)	0.260	0.788	0.016	0.293
Cl (%)	0.199	0.678	0.000	0.313
Nd_2O_3 (%)	0.388	1.312	0.000	0.510
Eu_2O_3 (%)	0.038	0.119	0.000	0.059
ZnO (%)	0.035	0.055	0.013	0.017
U (ppm)	175.340	452.400	0.000	180.485
Br (ppm)	1.950	11.700	0.000	4.777
As (ppm)	1.818	10.907	0.000	4.453
Se (ppm)	5.246	31.478	0.000	12.851
Rb (ppm)	2.087	6.307	0.000	3.234
Sr (ppm)	29.550	48.270	0.000	16.793
Y (ppm)	157.402	348.268	50.236	116.752
Cd (ppm)	328.402	703.256	144.170	197.317
Yb (ppm)	30.327	177.408	0.000	72.078
Pt (ppm)	6.713	40.279	0.000	16.444
Pb (ppm)	232.253	379.502	68.975	113.509
Bi (ppm)	141.341	255.203	9.598	99.686
Re (ppm)	0.983	5.700	0.000	2.312
K (ppm)	140.752	458.409	0.000	219.248
Ga (ppm)	12.545	75.269	0.000	30.728
Ni (ppm)	68.199	119.921	0.000	39.879

Table 7.5 Statistical measures such as average, maximum, minimum, and standard deviation of cliff sample.

Compound	Avg	Max	Min	SD
SiO_2 (%)	40.748	67.977	19.095	19.826
Fe_2O_3 (%)	35.248	59.442	15.792	16.197
Al_2O_3 (%)	20.451	27.164	2.354	9.641
MgO (%)	0.244	1.320	0.000	0.530
Cl (%)	0.285	1.074	0.044	0.394
CaO (%)	0.324	1.003	0.148	0.334
P_2O_5 (%)	0.912	1.124	0.722	0.140
TiO_2 (%)	1.154	1.509	0.906	0.201
CdO (%)	0.079	0.197	0.000	0.090
ZrO_2 (%)	0.131	0.463	0.035	0.164
K_2O (%)	0.116	0.203	0.028	0.074
Cr_2O_3 (%)	0.074	0.144	0.026	0.049
V_2O_5 (%)	0.070	0.140	0.012	0.049
Mn (ppm)	90.252	175.496	0.000	69.213
Ni (ppm)	141.310	204.794	84.086	51.475
Zn (ppm)	62.594	80.093	46.192	12.602
As (ppm)	7.965	36.052	0.000	14.539
Se (ppm)	3.818	22.906	0.000	9.351
Rb (ppm)	76.005	211.974	9.689	81.845
Sr (ppm)	55.316	156.034	18.143	51.016
Y (ppm)	2.861	17.165	0.000	7.008
Nb (ppm)	16.078	37.679	0.000	17.904
Eu (ppm)	82.137	492.821	0.000	201.193
Yb (ppm)	28.505	129.583	0.000	51.544
Pb (ppm)	411.282	1086.149	65.912	402.729
Bi (ppm)	14.517	68.622	0.000	27.517
Re (ppm)	6.917	16.500	0.000	6.449
Ga (ppm)	102.240	160.215	0.000	62.659
Br (ppm)	6.850	41.100	0.000	16.779
Cu (ppm)	65.799	135.006	0.000	54.654
In (ppm)	151.273	427.466	0.000	188.796
Os (ppm)	0.512	3.074	0.000	1.255
Ta (ppm)	4.313	25.878	0.000	10.565
Mo (ppm)	2.333	14.000	0.000	5.715
Ba (ppm)	342.495	1530.886	0.000	618.783

Table 7.6 Statistical measures such as average, maximum, minimum, and standard deviation of soil sample.

Compound	Avg	Max	Min	SD
SiO_2 (%)	64.135	82.576	46.267	11.037
Al_2O_3 (%)	18.902	28.261	10.622	6.145
Fe_2O_3 (%)	11.428	22.438	2.871	5.617
TiO_2 (%)	2.236	4.284	1.548	0.955
P_2O_5 (%)	1.116	1.361	0.851	0.149
CaO (%)	1.122	3.630	0.310	0.921
K_2O (%)	0.582	2.272	0.198	0.557
ZrO_2 (%)	0.171	0.409	0.082	0.103
MgO (%)	0.069	0.289	0.000	0.095
Cr_2O_3 (%)	0.068	0.125	0.029	0.030
V (ppm)	263.089	489.301	119.762	116.342
Mn (ppm)	191.211	317.612	96.732	59.788
Ni (ppm)	133.556	212.024	67.976	52.364
Zn (ppm)	87.035	152.394	37.355	30.572
As (ppm)	28.245	64.001	0.000	19.022
Rb (ppm)	35.375	140.951	13.985	34.306
Sr (ppm)	59.128	134.430	20.844	32.006
Y (ppm)	18.556	49.370	0.000	14.272
Nb (ppm)	57.200	110.661	38.168	26.712
Eu (ppm)	446.349	902.038	0.000	362.267
Hf (ppm)	19.355	101.840	0.000	37.707
Pb (ppm)	60.736	113.814	16.060	31.296
Bi (ppm)	0.314	3.767	0.000	1.088
Th (ppm)	53.608	120.221	0.000	36.488
Re (ppm)	0.875	4.600	0.000	1.428
Ga (ppm)	109.767	173.925	38.441	38.584
Br (ppm)	8.967	22.500	0.000	7.979
Cu (ppm)	51.992	83.799	20.371	17.382
In (ppm)	24.621	211.315	0.000	63.575
Ir (ppm)	1.361	6.299	0.000	2.112
Os (ppm)	0.668	8.018	0.000	2.315

correlation exhibited between Pb, Fe_2O_3, and Al_2O_3 with correlation coefficient values 0.955 and 0.876 ($P < .01$), respectively. Furthermore significant correlation is observed between HfO_2 and TiO_2, ZrO_2, and ThO_2 (R = 0.842, R = 0.898, and R = 0.777, respectively; $P < .01$). In the case of SiO_2 and Fe_2O_3 strong negative correlation is observed. The result shows that some elements show positive correlations, which may be due to similar origin.

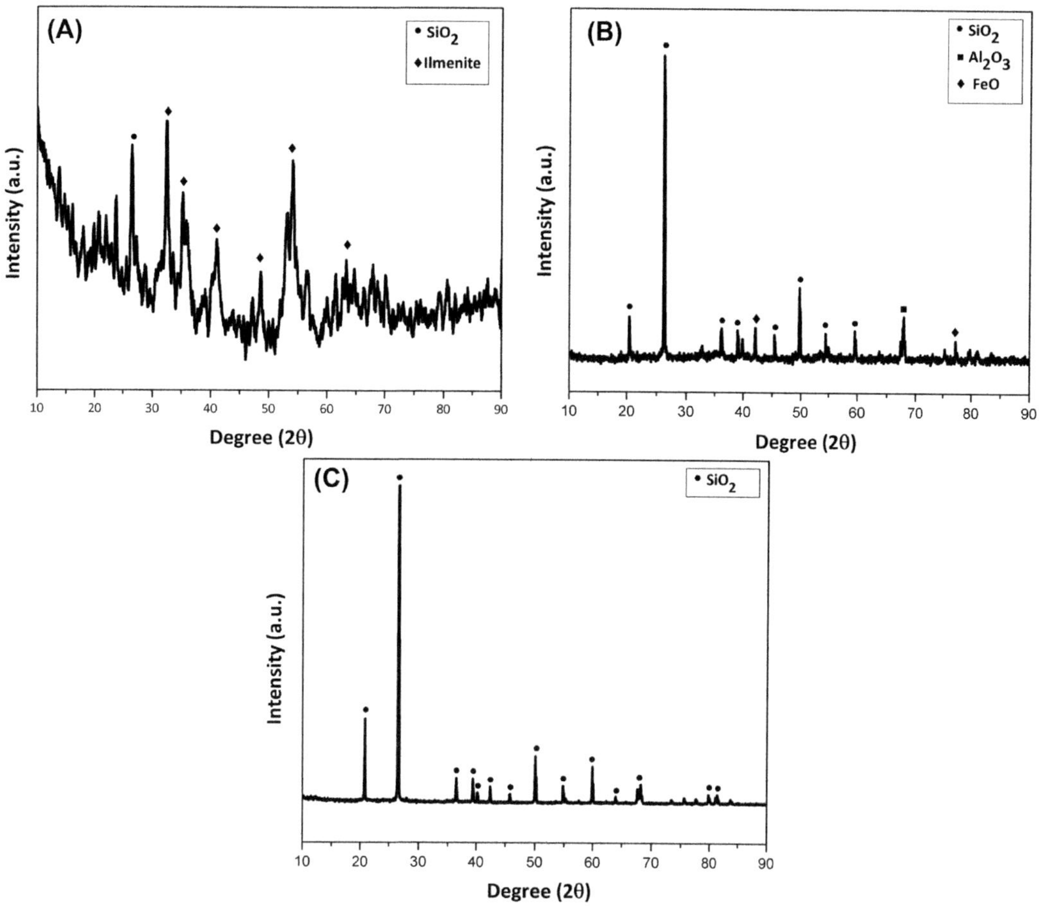

FIGURE 7.4 XRD of (A) beach, (B) cliff, (C) soil.

5. Conclusion

The depletion of TiO_2, ZrO_2, P_2O_5, U, and ThO_2 from beach samples to soil samples suggests the low presence of heavy minerals such as ilmenite, zircon, and monazite in inland areas of Varkala. The concentration of heavy metals such as Mn, Cr, Pb, Zn, Ni, and Cd in beach sediments are high compared to soil samples, which suggests the pollution of marine environments. The rare earth element concentration is also high in beach sediments compared to soil sediments. The rare earth elements are present in the beach sands such as Nd_2O_3, Eu_2O_3, Y, and Yb, but soil samples contain only Eu_2O_3 and Y. Integration of results obtained from advanced geochemical characterization with GIS techniques delivers solid information regarding the distribution of major, minor, and rare earth elements, and also, the heavy metal pollutants exist in marine environments.

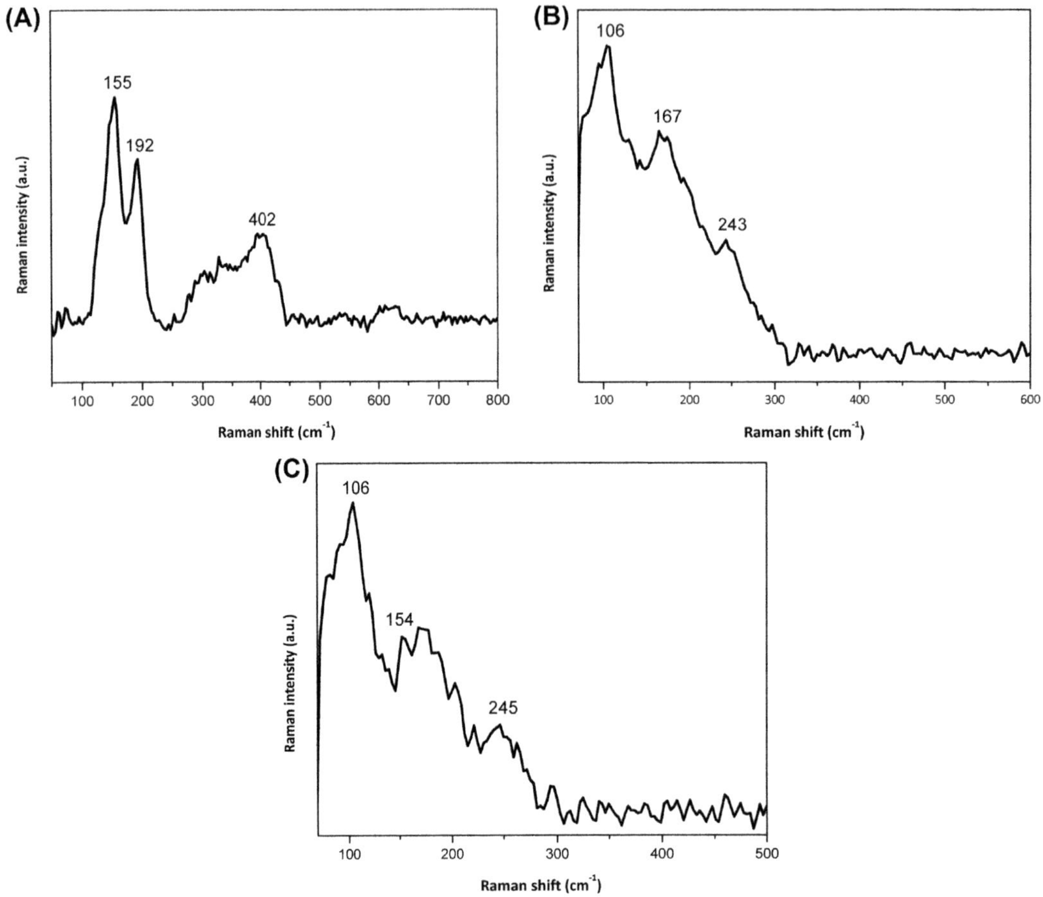

FIGURE 7.5 Raman spectrum of (A) beach, (B) cliff, and (C) soil.

Table 7.7 Correlation coefficient matrix for selected elements in beach sand.

	TiO_2	Fe_2O_3	SiO_2	ZrO_2	Al_2O_3	P_2O_5	ThO_2	HfO_2	U	Pb	Cd
TiO_2	1										
Fe_2O_3	0.999[b]	1									
SiO_2	−0.991[b]	−0.995[b]	1								
ZrO_2	0.489	0.523	−0.597	1							
Al_2O_3	−0.598	−0.630	0.653	−0.762	1						
P_2O_5	−0.741	−0.723	0.668	0.007	0.280	1					
ThO_2	0.166	0.202	−0.288	0.857[a]	−0.565	0.472	1				
HfO_2	0.581	0.613	−0.682	0.993[b]	−0.789	−0.065	0.838[a]	1			
U	0.209	0.245	−0.330	0.871[a]	−0.578	0.429	0.988[b]	0.854[a]	1		
Pb	0.868[a]	0.884[a]	−0.921[b]	0.800	−0.736	−0.332	0.632	0.865[a]	0.663	1	
Cd	−0.008	−0.028	0.051	−0.321	0.373	−0.100	−0.274	−0.290	−0.400	−0.151	1

[a]Correlation is significant at the 0.05 level (two-tailed).
[b]Correlation is significant at the 0.01 level (two-tailed).

Table 7.8 Correlation coefficient matrix for selected elements in soil sample.

	TiO_2	Fe_2O_3	SiO_2	ZrO_2	Al_2O_3	P_2O_5	ThO_2	HfO_2	Pb
TiO_2	1								
Fe_2O_3	0.348	1							
SiO_2	−0.327	−0.913[b]	1						
ZrO_2	0.973[b]	0.189	−0.166	1					
Al_2O_3	0.149	0.691[a]	−0.905[b]	0.014	1				
P_2O_5	0.195	0.113	−0.342	0.142	0.443	1			
ThO_2	0.573	−0.414	0.403	0.693[a]	−0.477	0.108	1		
HfO_2	0.842[b]	−0.090	0.068	0.898[b]	−0.132	0.031	0.777[b]	1	
Pb	0.230	0.905[b]	−0.961[b]	0.063	0.876[b]	0.136	−0.484	−0.135	1

[a]Correlation is significant at the 0.05 level (two-tailed).
[b]Correlation is significant at the 0.01 level (two-tailed).

Acknowledgments

The authors thank the Director, CSIR-National Institute for Interdisciplinary Science and Technology, Trivandrum, Kerala for extending the laboratory facilities.

References

Armstrong-Altrin, J.S., Lee, Y.I., Kasper-Zubillaga, J.J., Carranza-Edwards, A., Garcia, D., Eby, G.N., et al., 2012. Geochemistry of beach sands along the Western Gulf of Mexico, Mexico: implication for provenance. Chemie der Erde Geochemistry 72 (4), 345–362.

Arrieta, N., Goienaga, N., Martínez-Arkarazo, I., Murelaga, X., Baceta, J.I., Sarmiento, A., Madariaga, J.M., 2011. Beachrock formation in temperate coastlines: examples in sand-gravel beaches adjacent to the Nerbioi-Ibaizabal estuary (Bilbao, Bay of Biscay, North of Spain). Spectrochim. Acta Mol. Biomol. Spectrosc. 80 (1), 55–65.

Bernárdez, P., Prego, R., Giralt, S., Esteve, J., Caetano, M., Parra, S., Francés, G., 2012. Geochemical and mineralogical characterization of surficial sediments from the Northern Rias: implications for sediment provenance and impact of the source rocks. Mar. Geol. 291, 63–72.

Divya, P.V., Kaliprasad, C.S., Narayana, Y., Prakash, V., 2018. Distribution of natural radionuclides and assessment of excess lifetime cancer risk along coastal areas of Varkala in Kerala. J. Radioanal. Nucl. Chem. 1–7.

Dubey, P.N., Bhaskar, B.P., Chandran, P., Singh, B., Mishra, B.K., 2016. Geochemistry of some ferruginous soils of Kerala, India. J. Appl. & Nat. Sci. 8 (1), 196–207.

Facchinelli, A., Sacchi, E., Mallen, L., 2001. Multivariate statistical and GIS-based approach to identify heavy metal sources in soils. Environ. Pollut. 114 (3), 313–324.

Gandhi, M.S., Jeshma, P., Jisha, K., Rao, N.R., 2015. Neendakara and Varkala areas, Southern Kerala. J. Geosci. Res. 45–51.

Mohanty, A.K., Das, S.K., Vijayan, V., Sengupta, D., Saha, S.K., 2003. Geochemical studies of monazite sands of Chhatrapur beach placer deposit of Orissa, India by PIXE and EDXRF method. Nucl. Instrum. Methods Phys. Res. Sect. B Beam Interact. Mater. Atoms 211 (1), 145–154.

Nair, A.G., Babu, D.S., Damodaran, K.T., Shankar, R., Prabhu, C.N., 2009. Weathering of ilmenite from Chavara deposit and its comparison with Manavalakurichi placer ilmenite, Southwestern India. J. Asian Earth Sci. 34 (2), 115–122.

Noujas, V., Thomas, K.V., Nair, L.S., Hameed, T.S., Badarees, K.O., Ajeesh, N.R., 2014. Management of Shoreline Morphological Changes Consequent to Breakwater Construction.

Noujas, V., Kurian, N.P., Thomas, K.V., 2019. Comparison of coastal hydrodynamics in different energy regime coasts along Southwest coast of India. J. Earth Syst. Sci. 128 (4), 93.

Panday, D., Maharjan, B., Chalise, D., Shrestha, R.K., Twanabasu, B., 2018. Digital soil mapping in the Bara district of Nepal using kriging tool in ArcGIS. PloS One 13 (10), e0206350.

Raju, R.D., Ravi, G.S., Shivkumar, K., Reddy, L.S.R., Rohatgi, S., 2005. WDXRFS method for quantification of heavy minerals in sand samples. J. Geol. Soc. India 66 (4), 401.

Rejith, R.G., Sundararajan, M., 2018. Combined magnetic, electrostatic, and gravity separation techniques for recovering strategic heavy minerals from beach sands. Mar. Georesour. Geotechnol. 36 (8), 959–965.

Rizo, O.D., González, F.B., López, J.O.A., 2015. Assessment of Ni, Cu, Zn and Pb levels in beach and dune sands from Havana resorts, Cuba. Mar. Pollut. Bull. 100 (1), 571–576.

Sajinkumar, K.S., Kannan, J.P., Indu, G.K., Muraleedharan, C., Rani, V.R., 2017. A composite fall-slippage model for cliff recession in the sedimentary coastal cliffs. Geosci. Front. 8 (4), 903–914.

Selvaraj, K., Parthiban, G., Chen, C.T.A., Lou, J.Y., 2010. Anthropogenic effects on sediment quality offshore Southwestern Taiwan: assessing the sediment core geochemical record. Continent. Shelf Res. 30 (10–11), 1200–1210.

Saksena, B.D., 1940. Analysis of the Raman and infra-red spectra of α-quartz. Proc. Math. Sci. 12 (1), 93–139.

Silpa, B.L., Srinivas, R., Likhil, A.C., Aneesh, T.D., Prasad, K., Sajan, K., 2016. Microtextures on Quartz Grains in the Beach Sediments of a High Energy Regime, Kerala, Southwest Coast of India.

Singh, M., Rajesh, V.J., Sajinkumar, K.S., Sajeev, K., Kumar, S.N., 2016. Spectral and chemical characterization of jarosite in a palaeolacustrine depositional environment in Warkalli formation in Kerala, South India and its implications. Spectrochim. Acta Mol. Biomol. Spectrosc. 168, 86–97.

Szefer, P., 2002. Metals, Metalloids and Radionuclides in the Baltic Sea Ecosystem, vol. 5. Elsevier.

Thankappan, N., Varangalil, N., Varghese, T.K., Philipose, K.N., 2018. Coastal morphology and beach stability along Thiruvananthapuram, South-west coast of India. Nat. Hazards 90 (3), 1177–1199.

Tziachris, P., Metaxa, E., Papadopoulos, F., Papadopoulou, M., 2017. Spatial modelling and prediction assessment of soil iron using kriging interpolation with pH as auxiliary information. ISPRS Int. J. Geo-Inf. 6 (9), 283.

Wang, A., Kuebler, K.E., Jolliff, B.L., Haskin, L.A., 2004. Raman spectroscopy of Fe-Ti-Cr-oxides, case study: martian meteorite EETA79001. Am. Mineral. 89 (5–6), 665–680.

Wei, G., Liu, Y., Li, X.H., Shao, L., Fang, D., 2004. Major and trace element variations of the sediments at ODP Site 1144, South China Sea, during the last 230 ka and their paleoclimate implications. Palaeogeogr. Palaeoclimatol. Palaeoecol. 212 (3–4), 331–342.

Yalcin, M.G., Setti, M., Karakaya, F., Sacchi, E., Ilbeyli, N., 2015. Geochemical and mineralogical characteristics of beach sediments along the coast between Alanya and Silifke (Southern Turkey). Clay Miner. 50 (2), 233–248.

Zou, G., Xue, K., Huang, D., Su, C., Sun, J., August 2010. The comparison and study of small sample data spatial interpolation accuracy. In: 2010 Sixth International Conference on Natural Computation, vol. 4. IEEE, pp. 1897–1900.

8

Assessing the impact of aquafarming on landscape dynamics of coastal West Bengal, India using remotely sensed data and spatial metrics

Asit Kumar Roy[1], Mrinmoyee Naskar[1,2], Trisha Chakraborty[1], Sohini Neogy[1], Debajit Datta[1]

[1]*DEPARTMENT OF GEOGRAPHY, JADAVPUR UNIVERSITY, KOLKATA, WEST BENGAL, INDIA;* [2]*DEPARTMENT OF GEOGRAPHY, BARUIPUR COLLEGE, BARUIPUR, WEST BENGAL, INDIA*

Chapter outline

Remote Sensing of Ocean and Coastal Environments. https://doi.org/10.1016/B978-0-12-819604-5.00008-1

1. Introduction

Natural resource exploitation has become one of the most conspicuous phenomena in the epoch of Anthropocene, the geological timeframe in which mankind exerts the prime controlling forces on the environment, owing to unbridled growth in demographic structure worldwide, especially in the global south (Holling and Meffe, 1996; Kolbert et al., 2017). In general, a systematic positive relationship has been observed between resource utilization and the increasing population pressure as well as their means of living and livelihood generation (Sambrook et al., 1999). Due to the interplay of several factors that promote livability, such as provision of water transportation, broad coastal plains, availability of both terrestrial as well as marine resources, people mostly prefer to settle near the coast. However, in most instances, drastic alteration of landscapes driven by anthropogenic stressors are creating a menace, which has in turn proved to be detrimental for maintaining ecological equilibrium in and around the sensitive coastal tracts (Alonso-Pérez et al., 2003; Partecke et al., 2006).

In India, approximately 26% of the total demographic strength lives within 100 km of the coastline and nearly 20 million of the populace are directly dependent on coastal environment–related livelihood opportunities (D'Souza and Ishwar, 2012). The 7517 km long coastline of the Indian subcontinent is enriched with diverse ecologically sensitive coastal wetlands (i.e., mangroves, salt marshes, mudflats, interdunal swales, and coral reefs) and fertile coastal plains, which play a crucial role in pursuance of subsistence-based livelihood by the economically marginal coastal communities (D'Souza and Ishwar, 2012; Rajashekara, 2018). Local populace along with traditional fishermen have notable dependence on this ecologically rich coastal landscape through paddy-cum-fish cultivation, fresh and brackish water aquafarming, and deep-sea fishing. However, the negligible amount of production and unprofitability from intensive farming and fresh water fishing somehow diverted the interest of local farmers toward extensive monospecific brackish water aquafarming. Nationwide, approximately 0.18 million ha of land has already been acquired for this primarily monospecific shrimp cultivation (Ayyappan, 2006).

West Bengal and Odisha are such states of India, where 54,666 ha of land were allocated for tiger shrimp (*Penaeus monodon*) cultivation between 2012 and 2013, and the cumulative productivity was 3.34 metric ton/ha/year (Ayyappan, 2006). These statistics undoubtedly exhibited the predilection of the local populace toward shrimp aquafarming and portrayed the existence of capitalist magnates, influencing the regional socioeconomies very rapidly. However, the ecological sensitivity of the regional coastal environment was clearly neglected or not integrated with any aspects of the production system in this aquaindustrial boom (Chakraborty, 2010).

Along the Digha-Junput coastal stretch of the southwestern tract of littoral West Bengal, brackish water aquafarming has become the most visible spatial feature at present, which had only been implemented after massive as well as recurring alteration of natural landscapes (Chakraborty, 2010; Raja et al., 2012). Considering the local as well as international market demand of crustacean resources (*P. monodon, Scylla serrata, Mercenaria mercenaria*) usually found along this coastal tract, hectares of coastal lowlands were acquired without any proper planning or environmental assessment of the impact of monospecific shrimp cultivation (Ayyappan, 2006; Raja et al., 2012). Since shrimp cultivation is a capital-intensive industry, it is monetarily lucrative only for large investors, found fewer in numbers in this region, whereas the majority of the coastal populace, dependent on subsistence-based agroecosystem practices, stay far away from taking economic advantage of this ongoing aquacultural boom. Subsequently, this monopolistic practice has caused alarming degradation of the coastal wetland habitat along with the consequent obliteration of existing sustainable agroecosystems of this region.

Variegated research works have already been done on this region highlighting the land use and land cover (LULC) transformation, aquacultural growth, and coastal biodiversity assessment, but no such work has been done yet in integrating the spatiotemporal transformation of regional landscapes with the dominating posture of aquafarming (Chakraborty, 2010; Raja et al., 2012; Dutta et al., 2016). Thus, keeping in mind the noteworthy research gap in this context, this study primarily aims to estimate the spatiotemporal growth pattern of aquaculture through geospatial mapping, thereby establishing its association with other LULCs of this region. Moreover, the changing patterns of spatial concentration of aquafarms from coast to inland over the temporal span of 1988–2018 has also been analyzed. Accordingly, a probable future trend of extensive aquafarming has been exhibited, depending on the principles of spatial landscape ecology and community appraisals of the local coastal populace.

2. Methodology

The methodology of this study entails demarcation of area of study, acquisition of data, preprocessing and classification of satellite images, buffer-based zonation from coast to inland, change detection analysis, and application of landscape metrics/generation of landscape statistics (Fig. 8.1).

2.1 Area of study

The area of study is a 43 km long coastal stretch along the Medinipur littoral tracts of West Bengal comprising four major sectors, Digha, Chandipur, Dadanpatra, and Junput (Fig. 8.2). Wedged between 87°35′E to 87°50′E and 21°40′N to 21°50′N approximately, this mesotidal tract is characterized by Chenier plains, barriers islands, sparse mangrove stands, and a series of coastal sand ridges with intervening clayey tidal flats (Chakrabarti and Nag, 2015; Roy and Datta, 2018). Geologically, this region is a product of Holocene transgression and comprised of recent alluvial deposits with a number of intersecting rivers (Niyogi, 1975). The area experiences a subtropical humid climate with a mean maximum temperature of 27°C in December rising to about 41°C in May with average annual rainfall of 1192–1956 mm (Roy and Datta, 2018). The coast experiences

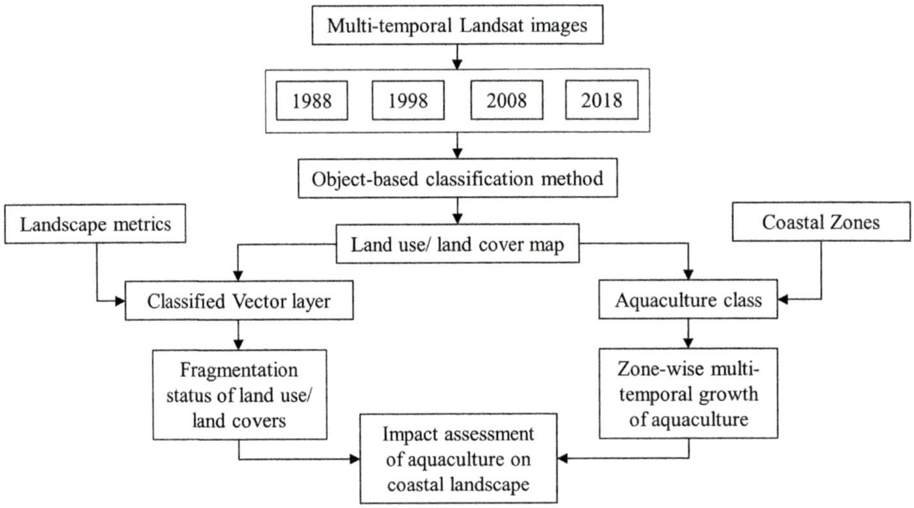

FIGURE 8.1 Methodological framework of the study.

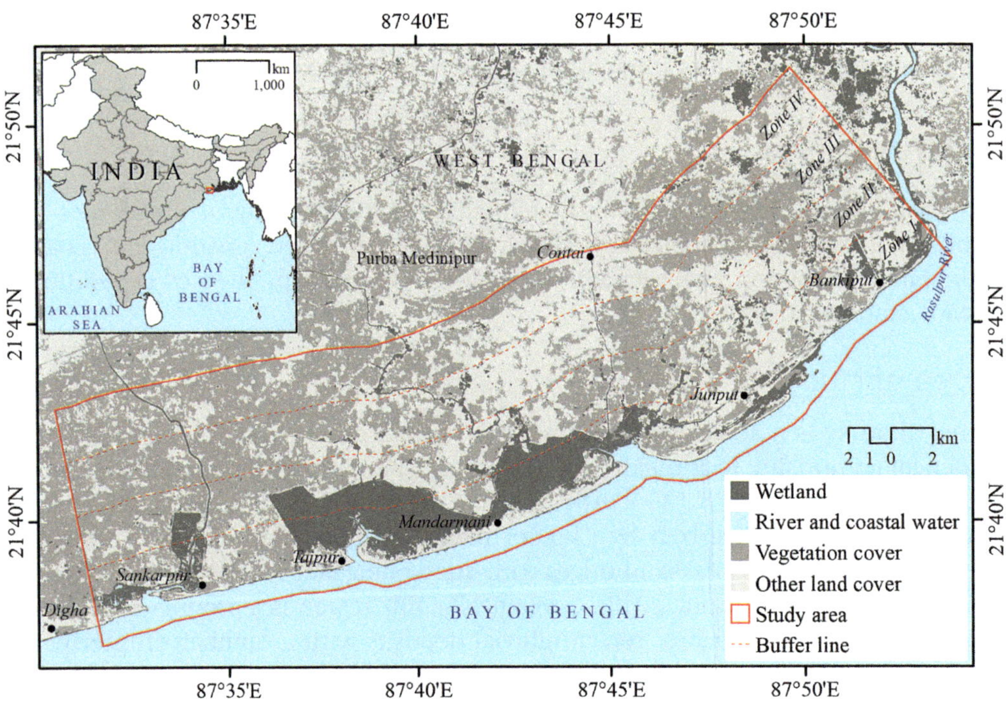

FIGURE 8.2 Location map of the selected coastal stretch.

semidiurnal tidal fluctuations and gets saturated with saline water (Barman et al., 2016). Chiefly associated with aquafarming practices, this halophytic extent is a storehouse of several mangrove species including *Acanthus ilicifolius, Ceriops decandra, Avicennia marina, Excoecaria agallocha, Porteresia coarctata, Kandelia candel,* and *Pandanus tectorius.* Nonmangrove species like *Casuarina equisetifolia* and *Eucalyptus* plantations can also be found in abundance as evidences of anthropogenic afforestation (Chaudhari et al., 2009). In this region, *Cynodon dactylon, Indigofera dendroides, Ipomoea pes-caprae, Phyla nodiflora, Evovulus nummularius,* and *Glycosmis pentaphylla* are the main sand dune stabilizer plants. Important fish species like *Tenualosa ilisha, Chelon parsia, Sillago sihama,* and *Harpodon nehereus* are widely found here (Kar et al., 2017).

2.2 Data acquisition and preprocessing

Four multitemporal cloud-free Landsat images from 1988 to 2018 encompassing this coastal stretch were downloaded from the open source archive of the United States Geological Service Glovis (http://glovis.usgs.gov) website for the present purpose (Hansen and Loveland, 2012; Roy and Datta, 2018). These were Landsat Thematic Mapper (TM) data (Path 139, Row 45; dated February 1, 1988, February 28, 1998, and February 24, 2008) and Landsat Operational Land Imager/Thermal Infrared Sensor data (Path 139, Row 45; dated February 3, 2018). All the images were orthorectified through UTM projection and WGS 84 datum. Both geometric and atmospheric corrections were carried out on all images for better accuracy in image classification (Hansen and Loveland, 2012).

2.3 LULC classification

Considering the proficiency of classifying a group of adjacent pixels under a specific LULC as a smallest classification unit in contrast with the differentiation of each single pixels based on their digital numbers, the object-based classification method was adopted for generating multitemporal LULC classification datasets. This classification comprises segmentation of raster datasets using the object-based image analysis software eCognition Developer 9.01 (Benz et al., 2004; Yu et al., 2016). It offers a highly suitable multiresolution segmentation technique that follows one of the most sophisticated segmentation algorithms from the fundamental aspect of the fractal net evolution approach and was accordingly chosen for the present classification (Blaschke et al., 2000; Kindu et al., 2013). It follows the bottom-up approach, where in the initial stage, each pixel was separated as individual objects following that these objects were merged afterward depending on the homogeneity criteria (e.g., color, shape, smoothness, and compactness of homogenous group of pixels) and produced bigger segmented forms. The derived segmented output was further exported into vector data format (.shp) to make it compatible with the topological vector space of ArcGIS 10.3. Thereafter, numerous vector segments were merged sequentially in accordance with diverse existing LULC classes in the ArcGIS platform using the collected ground control points and *a priori* knowledge of the investigators. Four chronological LULC maps had been prepared for 1988, 1998, 2008, and 2018 at an interval of 10 years each by identifying several LULC classes—aquaculture, bare earth and sandy beach, built-up, *Casuarina* plantation, cropland, mangrove vegetation, other vegetation, and river and coastal water.

2.4 Buffer based zonation from coast to inland

Littoral environment has only been witnessed in areas that have experienced maximum coastal water penetration without any hindrance (Chakraborty, 2010). Generally, the coastal influence has gradually been decreasing from backshore toward inland areas, which are prominently reflected in the floral and faunal communities with respect to changing edaphic and aquatic characteristics (Fenu et al., 2013). This prevailing environmental gradient was further simplified by segmenting it into four consecutive elongated zones parallel to the shoreline. The 43 km long coastline, extending 10 km inland, including the littoral tract, has been subdivided into four consecutive zones: Zone I, Zone II, Zone III, and Zone IV, each having 2.5 km width. Using the ArcGIS buffer tool, segmentation had been done through buffer generation (Ferguson, 2008). The prime objective of this spatial segmentation was to analyze the pattern of zonal dynamics of aquafarms over time, developed under the influence of a brackish water environment.

2.5 Change detection analysis

The post–classification change detection analyses were conducted in terms of areal dynamics and changing LULC patterns between 1988 and 2018 (Coppin et al., 2004; Roy and Datta, 2018). To know about the temporal statistics of interclass areal transformation, the geometric intersection tool was applied in ArcGIS platform to produce the transformation matrices for the three sequential time periods of 1988–1998, 1998–2008, and 2008–2018 (Castro et al., 2011; Datta and Deb, 2012). The prime LULC classes were identified from these to determine the major changes of regional LULC pattern.

2.6 Application of landscape metrics

A large set of landscape metrics were derived from the chronologically ordered LULC raster datasets using the spatial pattern analysis program, FRAGSTATS 4.2 (McGarigal and Marks, 1995; Wu et al., 2002; Matsushita et al., 2006; Tolessa et al., 2016). To assess the stress of aquaculture on the other LULC classes, eight metrics were selected in this study from class level analysis among numerous diversified landscape metrics: *viz.*-number of patches (NP), patch density (PD), largest patch index (LPI), landscape shape index (LSI), area-weighted mean shape index (SHAPEAM), area weighted mean fractal dimension index (FRACAM), clumpiness (CLUMPY), and interspersion juxtaposition index (IJI) (Supplementary Table 8.1) (Li et al., 2004; Su et al., 2011). These metrics were derived for every single LULC class with respect to the four assessment years.

These metrics were chosen based on appropriate review of literature in the related field of research to depict the temporal status of aggregation, dominance, and fragmentation of the studied landscape. Among these, NP, PD, SHAPEAM and FRACAM showed the fragmentation status, where higher value indicates more disintegration of landscape in class level. Conversely, the LSI and CLUMPY showed the aggregation scenario of individual LULC classes in which higher value is associated with compactness of landscape. Furthermore, the LPI represented the dominance of a specific LULC class, which has the largest patch share in the entire landscape area. In addition, the IJI depicted the level of evenness in the distribution of adjacent individual classes (Crews-Meyer, 2004; Li et al., 2004; Matsushita et al., 2006).

3. Results

3.1 LULC transformation scenario

During the span of 30 years, the entire study region experienced enormous amount of land use change along with complete alteration of land covers including cropland, aquaculture, bare earth and sandy beach, and other vegetation (Fig. 8.3).

In 1988, around 2643.52 and 20,453.05 ha land were occupied by aquafarming and cropland, respectively (Table 8.1). The major portion of land had been utilized for agricultural purposes. However, a notable portion of land amounting to 14,056.59 ha was under the coverage of other vegetation classes. In this period, only 126.48 ha of land was under the built-up concentration. However, settlements were not limited to this class only, rather they were spread throughout the region and merged with the other vegetated areas.

During 1998, the areal coverage of aquaculture was 3512 ha, which had increased considerably by 32.85% in comparison to 1988. Furthermore, the area of cropland and other vegetation were 21,078.86 and 13,357.23 ha, respectively, between which cropland slightly increased by 3.06% and other vegetation experienced a considerable decrease of 4.98% from those of 1988. In this phase, built-up coverage had increased by 78.38% (225.60 ha). In between 1988 and 1998, around 3828.32 ha of land under other vegetation was transformed into cropland, indicating a noticeable inclination of local populace toward agriculture (Table 8.2A). However, the 32.85% of increase in aquaculture was chiefly due to the transformation of bare land (550.65 ha), cropland (182.52 ha), and other vegetated areas (206.37 ha) into aquafarms as an emerging prospective livelihood option in this region.

The LULC statistics showed that the total aquaculture area was 3784.36 ha in 2008 with a 7.76% growth compared to 1998. In reality, a major portion of land (431.25 ha) was transformed into aquafarms through replacement of bare land (68.24 ha), cropland (234.43 ha), and other vegetated land (128.58 ha). However, the share of cropland to aquacultural transformation was considerably higher among all other classes in this period (Table 8.2B).

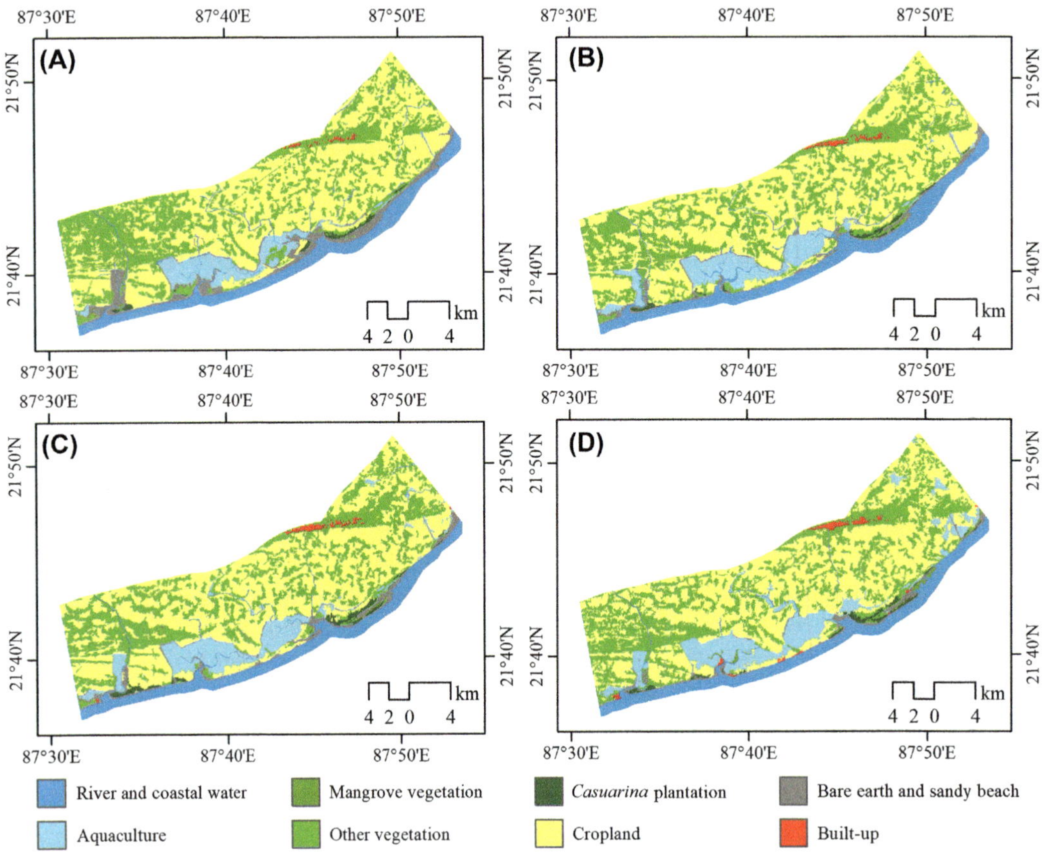

FIGURE 8.3 Spatial distribution of different LULC.

Table 8.1 LULC statistics of the study area from 1988 to 2018.

LULC class	1988 Area (ha)	1998 Area (ha)	(1988–1998) % Change	2008 Area (ha)	(1998–2008) % Change	2018 Area (ha)	(2008–2018) % Change
Aquaculture	2643.52	3512.00	32.85	3784.36	7.76	4724.69	24.85
Bare earth and sandy beach	2722.55	1293.36	−52.49	1496.12	15.68	884.73	−40.86
Built-up	126.48	225.60	78.38	285.66	26.62	371.73	30.13
Casuarina plantation	315.46	385.75	22.28	599.55	55.43	527.39	−12.04
Cropland	20453.05	21078.86	3.06	21264.24	0.88	20447.17	−3.84
Mangrove vegetation	1.40	1.27	−9.36	5.67	346.12	205.17	3518.60
Other vegetation	14056.59	13357.23	−4.98	12873.75	−3.62	13140.56	2.07
River and coastal water	4913.43	5378.40	9.46	4923.11	−8.47	4931.04	0.16
Total	45232.47	45232.47	–	45232.47	–	45232.47	–

Table 8.2A LULC transformation matrix from 1988 to 1998 (ha).

	1998							
1988	**Aquaculture**	**Bare earth and sandy beach**	**Built-up**	***Casuarina* plantation**	**Cropland**	**Mangrove vegetation**	**Other vegetation**	**River and coastal water**
Aquaculture	2515.42	42.55	0.00	0.63	40.68	0.00	9.47	34.77
Bare earth and sandy beach	550.65	922.88	0.55	103.41	167.90	1.01	348.11	628.05
Built-up	0.00	0.00	111.31	0.00	0.00	0.00	15.16	0.00
Casuarina plantation	0.00	4.77	0.00	219.15	13.24	0.00	78.21	0.09
Cropland	182.52	97.22	0.27	18.46	16924.47	0.00	3152.54	77.58
Mangrove vegetation	1.40	0.00	0.00	0.00	0.00	0.00	0.00	0.00
Other vegetation	206.37	124.39	113.47	44.01	3828.32	0.00	9695.29	44.73
River and coastal water	55.64	101.55	0.00	0.09	104.25	0.26	58.45	4593.17

Table 8.2B LULC transformation matrix from 1998 to 2008 (ha).

	2008							
1998	**Aquaculture**	**Bare earth and sandy beach**	**Built-up**	***Casuarina* plantation**	**Cropland**	**Mangrove vegetation**	**Other vegetation**	**River and coastal water**
Aquaculture	3252.03	60.32	0.00	1.44	139.31	0.27	28.45	30.18
Bare earth and sandy beach	68.24	826.65	5.94	60.03	153.48	5.31	63.42	110.31
Built-up	0.02	0.00	193.64	0.00	1.73	0.00	30.21	0.00
Casuarina plantation	1.17	24.10	0.00	313.95	17.46	0.00	28.53	0.54
Cropland	234.43	78.10	3.50	52.20	17467.79	0.00	3183.10	59.73
Mangrove vegetation	0.00	1.27	0.00	0.00	0.00	0.00	0.00	0.00
Other vegetation	128.58	101.13	82.58	166.17	3343.39	0.09	9504.01	31.28
River and coastal water	99.92	404.54	0.00	5.76	141.08	0.00	36.03	4691.07

A high amount of change was observed in the LULC statistics of 2018. Since aquaculture had experienced a boom after 2000, the total aquaculture area increased by 24.85% in 2008–2018 and occupied 4724.69 ha of land here. Consequently, the area of cropland decreased 3.84% and remained as 20,447.17 ha. Between 2008 and 2018, around 191.28 ha, 833.79 ha, and 122.38 ha of lands were transformed into aquafarms from bare land, cropland, and other vegetated areas, respectively, which were found to be the largest conversions into aquaculture compared to the previous phases (Table 8.2C). Furthermore, the concentration of built-ups was also increased by 30.13% compared to 2008 as a consequence of increasing population throughout this region for the prospective aquaculture and tourism-based economy.

Through generation of error matrices, the classification accuracy was measured for four consecutive LULC maps, where the derived Kappa statistics were 0.6287, 0.6892, 0.7459, 0.7810, and classification accuracies were 68%, 73%, 78%, 81% for the years 1988, 1998, 2008, and 2018, respectively (Supplementary Table 8.2A–D).

3.2 Spatiotemporal growth pattern of aquaculture and impact on environment

The buffer-wise zonation and its implication in assessment of the aquacultural expansion show a systematic pattern of spatiotemporal growth, which is intricately related to the land use dynamics of the entire study region from a broader perspective. In just 3 decades, sharp growth was detected in the amount of aquacultural practices (Fig. 8.4). In the initial stage of 1988, the major concentration of aquacultural farms was

Table 8.2C LULC transformation matrix from 2008 to 2018 (ha).

	2018							
2008	**Aquaculture**	**Bare earth and sandy beach**	**Built-up**	***Casuarina* plantation**	**Cropland**	**Mangrove vegetation**	**Other vegetation**	**River and coastal water**
Aquaculture	3486.61	24.71	2.77	2.88	149.80	35.01	58.35	24.23
Bare earth and sandy beach	191.28	587.82	76.23	81.30	89.70	92.17	131.10	246.52
Built-up	0.00	0.18	213.53	0.00	2.15	0.00	69.80	0.00
Casuarina plantation	4.36	48.21	0.00	362.08	25.29	0.58	139.33	19.71
Cropland	833.79	61.03	10.64	40.94	17177.99	6.26	3091.75	41.84
Mangrove vegetation	0.00	0.36	0.00	0.03	0.00	5.28	0.00	0.00
Other vegetation	122.38	57.64	68.56	37.55	2939.19	0.00	9620.74	27.70
River and coastal water	86.27	104.78	0.00	2.61	63.04	65.88	29.49	4571.05

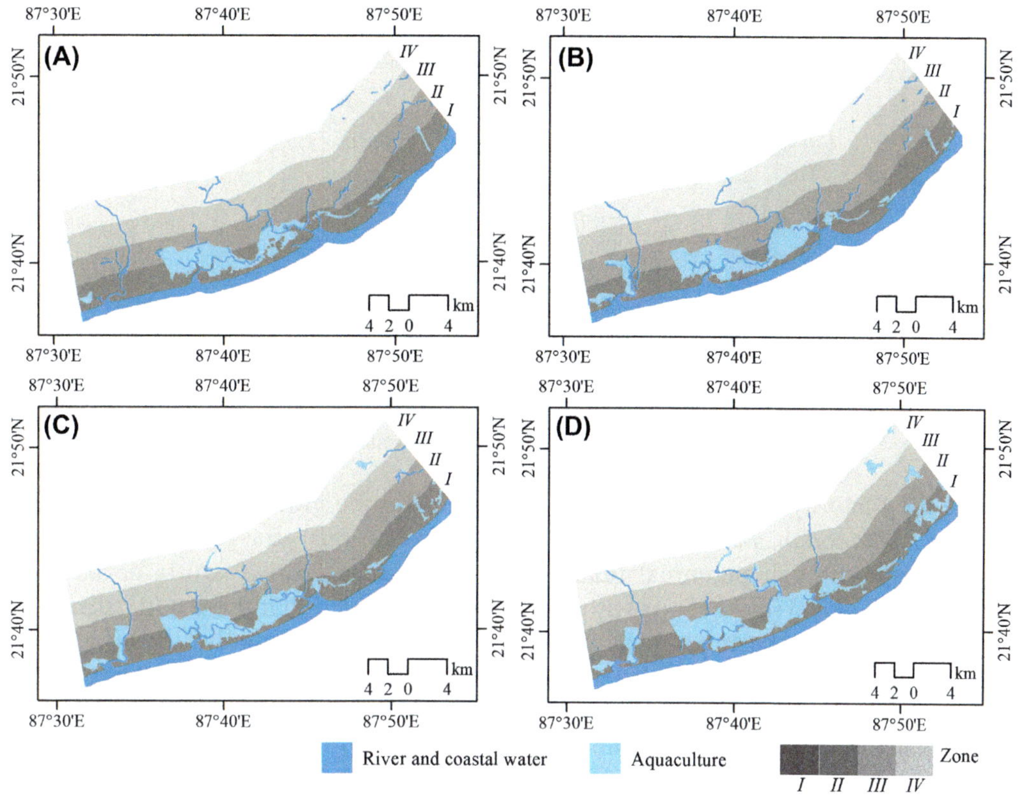

FIGURE 8.4 Zone wise growth of brackish water aquafarms.

identified around the coastal lowlands of Digha, Tajpur, and Mandarmani, which mainly fall in Zone I and Zone II. During this phase, the total area of aquaculture under these two zones were 1748.54 and 894.99 ha, respectively. In 1998, although the areal extent of aquaculture increased very sharply, spatially it was still limited into Zone I and Zone II primarily for utilizing the coastal brackish water sources. In the course of expansion of aquafarms, small creeks were embanked extensively, perennial water flows were obstructed instead of being diverted, and site-specific modifications in the channel morphology of creeks were done. In 2008, scattered patches of aquafarms started to grow on either side of the tidal creeks forwarding further inland. Following this trend, aquaculture gradually flourished away from the shoreline along Zone III and Zone IV. In 2008, zone-wise areal coverage of aquacultural farms was 2507.53, 1157.86, 39.78, and 79.20 ha, respectively, from Zone I to Zone IV (Fig. 8.5). Similarly, large patches of aquafarms situated in Zone I and Zone II were further intensified between 2008 and 2018, whereas the scattered aquacultural patches of Zone III and Zone IV expanded gradually and at the same time merged into each other toward converting into major inland brackish aquafarms, completely replacing the perennial creek networks. Subsequently, the zone-wise areal coverage of aquacultural farms was 3054.78, 1347.63, 171.16, and 151.13 ha, respectively, from Zone I to Zone IV in 2018 (Supplementary Table 8.3).

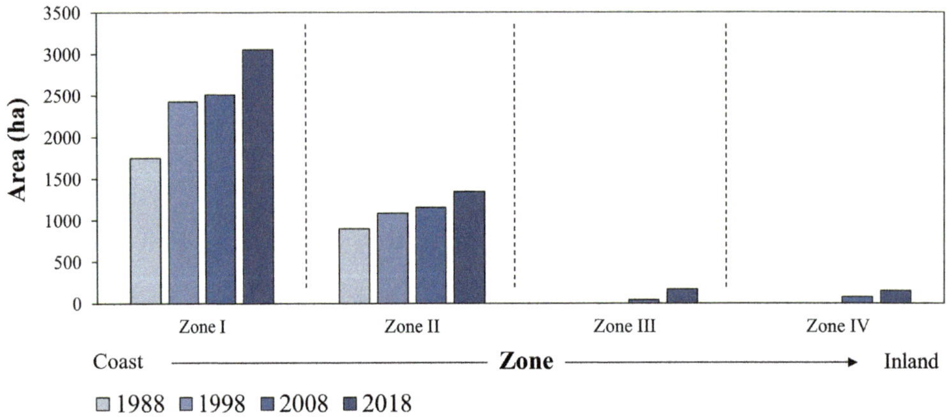

FIGURE 8.5 Spatiotemporal growth pattern of aquafarming from 1988 to 2018.

Since the majority of these aquafarms were found to be growing euryhaline species only (i.e., *P. monodon, Peneus indicus, Penaeus semisulcatus*), the recurrent supply of brackish water through perennial tidal creeks became imperative (Dutta et al., 2016; Mitra, 2019). Personal interactions with local farmers revealed that these aquafarms may be responsible for increasing salinity in the top soil layer and groundwater storage as a result of diffusion of dissolved salts (chlorides and sulfates of Na, Mg, and Ca) through leakage or seepage of brackish water (Xu et al., 2016). Gradually the increasing soil alkalinity has proven to be unfavorable for sustained crop yields, which further intensified the motivation of local marginal farmers and fishermen to adopt aquaculture as their new prospective livelihood option as an alternative to their hitherto practiced seasonal subsistence cropping. Accordingly, hectares of croplands and coastal wetlands were encroached and thereby converted into extensive aquafarms.

3.3 Change scenario of class-level landscape metrics

The outputs of class level metrics of four consecutive timeframes were compared for depicting the interclass adjustments and dynamicity of regional LULC patterns (Table 8.3). This was based on the major LULC classes—cropland, aquaculture, other vegetation, bare earth, and sandy beach, which act as the major drivers of landscape modification and intensification of aquacultural practices within the study region.

Both NP and PD have a close relationship in which increase in the number of individual patches indicates higher PD or more concentration of patches within per sq. km area. Here, NP and PD of aquacultural class increased gradually by 29 and 0.07/km^2, respectively, in the 1988–2018 period. However, a sharp growth was witnessed from 2008 as a result of partial switch over of local farmers from paddy monocropping to brackish water shrimp farming. Correspondingly, the bare earth and sandy beach, cropland, and other vegetated land experienced a sharp decrease in their NP and PD values owing to the conversion of those lands into aquacultural plots. As an indicator of land parcel shape, the SHAPEAM value of aquaculture was found to be highest (4.27) in 1988 and gradually experienced a decrease (unit value 0.33) as a result of getting

Table 8.3 Comparison of landscape metrics at class level from 1988 to 2018.

		Landscape metrics							
LULC class	Year	Number of patches (NP)	Patch density (PD)	Largest patch index (LPI)	Landscape shape index (LSI)	Area-weighted mean shape index (SHAPEAM)	Area weighted mean fractal dimension index (FRACAM)	Clumpiness (CLUMPY)	Interspersion juxtaposition index (IJI)
Aquaculture	1988	14	0.03	2.99	8.32	4.27	1.18	0.95	70.31
	1998	19	0.04	3.28	7.69	3.27	1.15	0.96	70.27
	2008	39	0.09	3.30	9.30	3.20	1.14	0.96	70.45
	2018	43	0.10	6.83	10.64	3.94	1.16	0.95	68.05
Bare earth and sandy beach	1988	31	0.07	1.76	15.56	5.14	1.21	0.91	76.33
	1998	30	0.07	0.92	20.08	5.86	1.24	0.83	77.91
	2008	23	0.05	1.24	18.79	6.01	1.24	0.86	82.05
	2018	22	0.05	0.39	17.52	5.53	1.24	0.83	91.34
Built-up	1988	17	0.04	0.09	5.91	2.23	1.13	0.86	0.00
	1998	11	0.02	0.17	5.79	2.55	1.14	0.90	2.72
	2008	10	0.02	0.36	5.70	2.78	1.15	0.91	17.39
	2018	24	0.05	0.32	7.11	2.25	1.12	0.90	59.69
Casuarina plantation	1988	9	0.02	0.38	5.60	2.89	1.14	0.92	56.79
	1998	12	0.03	0.36	7.53	2.96	1.15	0.90	55.48
	2008	13	0.03	0.59	8.41	3.62	1.18	0.91	65.01
	2018	16	0.04	0.41	8.68	2.89	1.15	0.90	63.47
Cropland	1988	184	0.41	26.10	27.60	12.72	1.27	0.90	27.63
	1998	92	0.20	25.32	27.67	12.92	1.27	0.90	28.54
	2008	106	0.23	31.23	26.61	13.95	1.27	0.90	32.03
	2018	159	0.35	28.55	32.78	15.50	1.28	0.88	27.73
Mangrove vegetation	1988	1	0.00	0.00	1.00	1.00	1.02	1.00	0.00
	1998	1	0.00	0.00	1.38	1.38	1.09	0.83	33.69
	2008	1	0.00	0.01	1.38	1.38	1.06	0.95	32.14
	2018	17	0.04	0.12	7.40	2.15	1.12	0.86	65.70
Other vegetation	1988	264	0.58	7.02	33.11	6.36	1.21	0.88	30.07
	1998	189	0.42	4.31	33.72	6.01	1.21	0.88	27.88
	2008	165	0.36	4.64	32.70	6.14	1.21	0.88	28.57
	2018	176	0.39	6.59	38.93	8.99	1.25	0.86	24.86
River and coastal water	1988	31	0.07	10.07	13.16	7.92	1.24	0.94	68.16
	1998	22	0.05	11.24	11.57	7.12	1.22	0.95	68.43
	2008	19	0.04	10.40	10.54	6.69	1.22	0.95	70.22
	2018	19	0.04	10.51	9.16	6.08	1.21	0.96	80.13

uniformity in the patch shape during the later years. A diametrically opposite pattern was observed for the SHAPEAM values of cropland (15.50) and other vegetation (8.99) classes, which experienced a rise of 2.77 and 2.63, respectively, in 2018. Due to presence of erstwhile isolated and smaller irregular patches of aquaculture class, its SHAPEAM value was found to be higher in 1988. However, the same was notably low due to their uniformity in shapes for cropland and other vegetation patches at that period. Similarly, the FRACAM value shows that other than aquaculture, the shape complexities were higher for bare earth and sandy beach, cropland and other vegetation class, and had increased by 0.03, 0.01, and 0.04, respectively, in 1988–2018. Although the FRACAM value of aquaculture decreased by 0.02 in 1988–2018, it actually depicted the gradual elimination of shape complexity over time.

Regarding landscape metrics of aggregation, the LSI statistics showed increasing trends in the aquaculture (2.32), cropland (5.18), and other vegetation (5.82) classes for the entire study period, which referred to a sequential pattern of disaggregation of these LULC patches as a result of continuous separation of larger patches into smaller amorphous patches through intrusion of other LULCs. Along with that, the average CLUMPY value was found to be highest for aquaculture (0.96) compared to the bare earth and sandy beach (0.86), crop land (0.89), and other vegetation (0.88) classes for the same period. Since the CLUMPY value of aquacultural class was close to 1, it depicted a larger assemblage of aquafarming plots that were not commonly detectable in the other LULC classes of the region.

As per the landscape association metrics, the cropland generally showed the largest LPI value (27.80%) during the entire study period (1988–2018). However, the LPI of aquaculture showed a sudden rise of 3.84% in 2018. Eventually, aquafarms engulfed the majority of other land use parcels and created some unified larger patches. Except cropland and aquaculture, other LULC classes had negligible values of LPI. Based on the patch adjacencies of unique classes, steady IJI values of aquaculture were found from 1988 until 2008 but it suddenly decreased by 2.4% in 2018 (Fig. 8.6). This eventually depicted that the class diversity in the adjacent patches of aquacultural farms had decreased and become alike after 2008 as a result of unimodal trend of regional LULC pattern. This similar pattern was also observed at the adjacent patches of other vegetation classes, since the neighboring patches of other vegetation classes were gradually consumed for the purpose of either aquaculture or subsistence agricultural practices.

4. Discussion

4.1 Aquaculture as a driver of LULC change

Assessing the geospatially derived LULC transformation matrices and class metrics, it was found that the natural coastal landscape of the study region had faced incessant stress from land acquisition for the purpose of brackish water aquaculture between 1988 and 2018. This ultimately led to the complete abolition of widespread terrestrial vegetation cover, barren wastelands, and farmland agroecosystems. Since the regional subsistence agriculture was no longer proven to be beneficial for economic prosperity

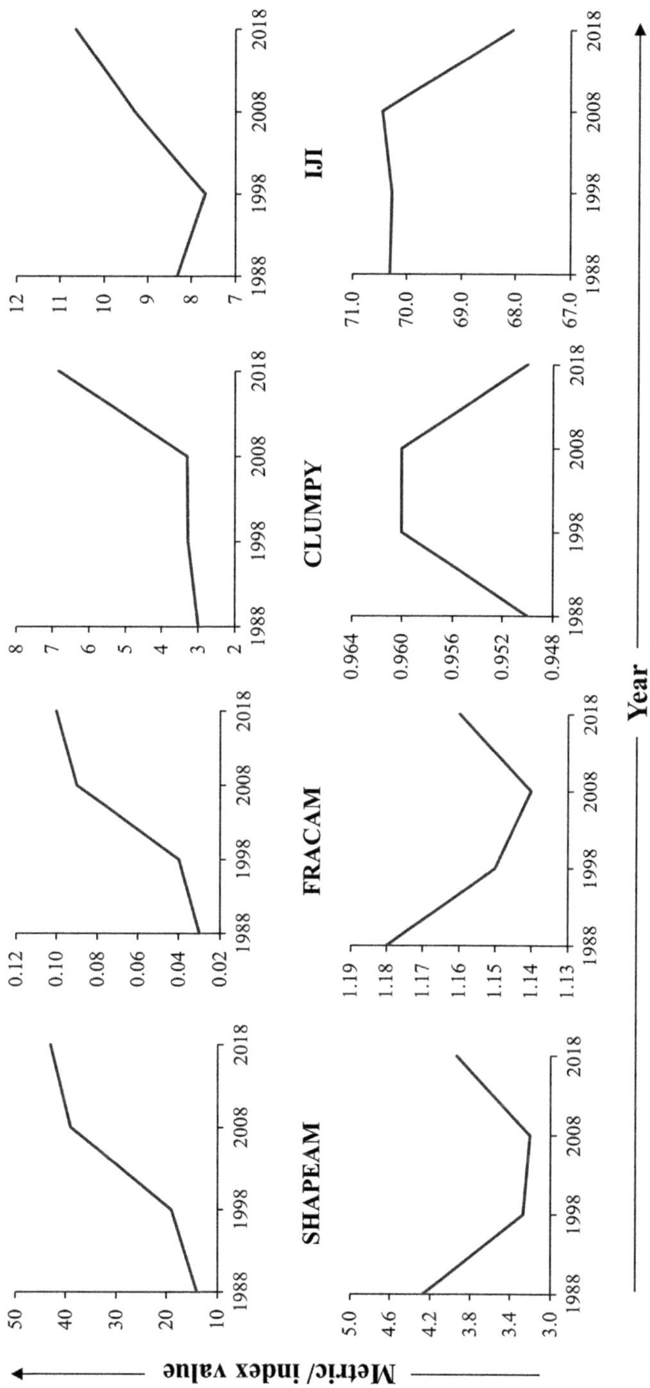

FIGURE 8.6 Changes in the statistics of landscape metrics for aquaculture.

due to its rudimentary approach, the local populace gravitated toward extensive aquafarming run under local magnates.

It has been found that with a massive areal expansion of 32.85% in the 1988–1998 period, these fisheries had started to concentrate just within 2.5 km of high tide level and gradually spread out to its maximum extent. Eventually, through the conversion of adjacent LULCs, specifically, the croplands, bare lands, and open vegetated areas, the massive patches of aquafarms were created. These were clearly reflected in the greater LPI value (6.83%) of 2018. Alternatively, it can also be said that other neighboring LULC patches of the aquafarms were progressively fragmented as well as pushed away through the manifestation of newly grown aquacultural patches.

Noticeably, around 24.85% growth of aquafarms were observed after 2008, triggered by the mass alteration of LULCs giving birth to a widespread vista of artificial wetlands within the region. In recent times, the dominance of aquaculture had clearly been reflected by means of the landscape metrics, where the higher values portrayed monodominance of a particular land use over the whole landscape in the form of continuous aquafarms.

4.2 Environmental impact of aquacultural spread

Coastal brackish water aquafarms are designed by modifying the lowlands through saline water intrusion as well as excavating and embanking them for artificial storage. Since these natural coastal depressions played a crucial role in groundwater recharging, their utilization for brackish water fisheries had resulted in the infiltration of dissolved salts in groundwater aquifers, causing an increase in pH of surface as well as groundwater. This artificially augmented salinity of the groundwater table may have eventually created an unfavorable condition for the coastal populace to use this water for drinking and irrigation purposes. In addition, the enormous amount of brackish water storage in the aquafarms also triggered the deteriorating state of physicochemical properties of the top soil layers of associated LULCs. Consequently, assemblage of salt-tolerant varieties of floral (i.e., *P. coarctata, Sesuvium portulacastrum, A. ilicifolius,* and *P. tectorius*) and faunal (i.e., *S. serrata, Uca rosea, Oxudercinae,* and *Paguroidea*) species have started to colonize around these aquafarms, even far away from the coast in some instances.

Formerly, croplands dominated this coastal region. At present, it is mostly utilized for paddy (*Boro*) cultivation during dry periods and partially used for paddy-cum-fish cultivation during wet periods. Although agriculture was the prime source of livelihood in this region since time immemorial, it still follows a traditional approach of production, neglecting the objectives of profit maximization and commercialization. Moreover, excessive groundwater irrigation has resulted in higher alkalinity and salinized surface soil, as a negative consequence of brackish water aquafarms, for acting together to degrade the natural environmental condition and decrease the agricultural yield rapidly (Dutta et al., 2016). Thus, the deteriorating hydroedaphic condition of this entire region minimizes the prospect of having sustainable agroecosystem and transforms it into a less synergetic ecosystem with minimum biodiversity.

5. Conclusions

Unidirectional conversion of LULCs from vegetated land, agricultural land, and bare surface, to aquafarms, as a direct product of commercialization, has posed a grave threat to the biophysical properties of the regional environment and has proven to be detrimental for maintaining sustainability in imminent years. During field verification and former explorations, it was observed that aquafarms had already succumbed to the problem of salinization in both surface and groundwater. Thus, it is affecting the plant succession and species richness simultaneously to reduce the above- and belowground biomasses, which are actually extensive repositories of carbon. Having disregarded the need for adoption of protective measures for efficient and optimal use of natural resources, the degree of landscape modification has already maximized between 1988 and 2018 as well as a complete takeover of the remaining fragments of agricultural patches, greeneries, and moist stretches of coastal lowlands is being apprehended. If this trend continues, then the entire coastal stretch will no longer be able to generate A necessary livelihood of the local populace and its deprecated ecological health will negatively impact the socioeconomy of West Bengal. Highlighting the concentrated stress on the coastal landscape in the form of expanding monospecific brackish water aquafarming, this study becomes a pioneering one in depicting the spatiotemporal growth of aquafarms in contrast with other prevailing LULCs and for analyzing the class-specific dynamics of the LULC patches using the principles of landscape ecology. In this context, this study achieved uniqueness in showing the sequential trend of land use dynamics under the influence of monopolistic economy generated from aquafarming and portraying a small glimpse of future socioeconomic turmoil and ecological disruptions as a consequence of promoting counterproductive livelihood options across the study region. However, the result of this study is site-specific and the depicted temporal dynamics of geospatial matrices and landscape metrics may not be generally applicable for any other coastal environment without necessary modifications.

Based on the study results and field-based experiences of the researchers, a few protective measures needed to be implemented here to combat the deteriorating environment, designed as follows:

1. Through integration of both agriculture and aquaculture, a unified food production system needs to be initiated in this region, which should hold a diversified agroecology.
2. Largescale monospecific aquafarming should not be encouraged throughout the region. Rather, integrated small mangrove-cum-shrimp farms can be introduced.
3. Fresh-water aquafarming should be introduced in this region for its contribution toward recharging fresh water into the aquifers.
4. Tidal creeks should not be blocked during the construction of fisheries; rather they should be kept perennial as a continuous source of brackish water.

5. Local government should encourage the coastal populace to adopt the mixed farming and multicropping method through canal irrigation and rain water harvesting toward sustainable resource management.

In summation, both the stakeholders and governing bodies need to be more aware regarding the environmental carrying capacity by keeping the anthropogenic modification up to its optimal limit and introducing integrated plans to follow the path of sustainable land management in this fragile coastal environment.

Supplementary data related to this chapter can be found online at https://doi.org/10.1016/B978-0-12-819604-5.00008-1.

The following are the Supplementary data related to this chapter:

Supplementary data

Supplementary Table 1 Detailed description of the selected class level landscape metrics

Class metrics	Unit	Range	Description
Number of patches (NP)	None	≥1	NP = 1 means when the LULC class contains only a single patch
Patch density (PD)	Number per 100 hectares	>0	PD is the number of patches in a specific LULC class per hectare area
Largest patch index (LPI)	Percentage	0 to 100%	LPI approaches 0 when the areal coverage of largest patch of a specific LULC class is extremely small. LPI is 100%, when the entire landscape consists of a single LULC patch
Landscape shape index (LSI)	None	≥1	LSI is 1, when the LULC class have a compact square shaped patch. Here, higher LSI value indicates towards amorphous shaped and fragmented patches of LULC class
Area-weighted mean shape index (SHAPEAM)	None	≥1	A perfect square shape of LULC patch refers to the lowest SHAPEAM value 1, whereas the higher SHAPEAM value is the indicator of irregularity in the patch shape for a particular LULC class
Area weighted mean fractal dimension index (FRACAM)	None	1 to 2	FRACAM value >1 refers to the departure from Euclidean geometry for a two dimensional class patch
Clumpiness (CLUMPY)	None	1 to -1	Positive CLUMPY value refers to the cluster formation of class patches and the negative value indicates towards the disaggregation of class patches
Interspersion juxtaposition index (IJI)	Percentage	0 to 100	To measure the adjacencies in respect to a specific LULC class, if the patches of all other LULC class are distributed equally then IJI approaches 100

Supplementary Table 2A Accuracy report of the classified image of 1988

Class name	Reference total	Classified total	Number correct	Producer's accuracy	User's accuracy	Kappa (*k*)
Aquaculture	26	20	17	65.38	85.00	0.80
Bare earth and sandy beach	17	11	8	47.06	72.73	0.67
Built-up	12	14	8	66.67	57.14	0.51
Casuarina plantation	10	11	8	80.00	72.73	0.70
Cropland	9	12	7	77.78	58.33	0.54
Mangrove vegetation	11	14	10	90.91	71.43	0.68
Other vegetation	6	9	4	66.67	44.44	0.41
River and coastal water	9	9	6	66.67	66.67	0.63
Column total				Overall classification accuracy = 68.00%		Overall Kappa = 0.6287

Supplementary Table 2B Accuracy report of the classified image of 1998

Class name	Reference total	Classified total	Number correct	Producer's accuracy	User's accuracy	Kappa (*k*)
Aquaculture	20	20	16	80.00	80.00	0.75
Bare earth and sandy beach	13	11	7	53.85	63.64	0.58
Built-up	13	14	10	76.92	71.43	0.67
Casuarina plantation	8	11	8	100.00	72.73	0.70
Cropland	9	12	7	77.78	58.33	0.54
Mangrove vegetation	12	14	10	83.33	71.43	0.68
Other vegetation	13	9	8	61.54	88.89	0.87
River and coastal water	12	9	7	58.33	77.78	0.75
Column total				Overall classification accuracy = 73.00%		Overall Kappa = 0.6892

Supplementary Table 2C Accuracy report of the classified image of 2008

Class name	Reference total	Classified total	Number correct	Producer's accuracy	User's accuracy	Kappa (*k*)
Aquaculture	21	20	18	85.71	90.00	0.87
Bare earth and sandy beach	10	11	7	70.00	63.64	0.60

Built-up	13	14	11	84.62	78.57	0.75
Casuarina plantation	9	11	8	88.89	72.73	0.70
Cropland	11	12	10	90.91	83.33	0.81
Mangrove vegetation	15	14	11	73.33	78.57	0.75
Other vegetation	11	9	7	63.64	77.78	0.75
River and coastal water	10	9	6	60.00	66.67	0.63
Column total				Overall classification accuracy = 78.00%		Overall Kappa = 0.7459

Supplementary Table 2D Accuracy report of the classified image of 2018

Class name	Reference total	Classified total	Number correct	Producer's accuracy	User's accuracy	Kappa (*k*)
Aquaculture	19	20	18	94.74	90.00	0.88
Bare earth and sandy beach	12	11	8	66.67	72.73	0.69
Built-up	15	14	13	86.67	92.86	0.92
Casuarina plantation	11	11	8	72.73	72.73	0.69
Cropland	10	12	8	80.00	66.67	0.63
Mangrove vegetation	13	14	12	92.31	85.71	0.84
Other vegetation	10	9	6	60.00	66.67	0.63
River and coastal water	10	9	8	80.00	88.89	0.88
Column total				Overall classification accuracy = 81.00%		Overall Kappa = 0.7810

Supplementary Table 3 Spatio-temporal statistics of aquacultural growth from 1988 to 2018 (ha)

	Year			
Zone ID	**1988**	**1998**	**2008**	**2018**
I	1748.54	2424.76	2507.53	3054.78
II	894.99	1087.25	1157.86	1347.63
III	0.00	0.00	39.78	171.16
IV	0.00	0.00	79.20	151.13

Acknowledgments

We are highly obliged to all of those villagers who provided valuable information during field investigations. For conducting this research, the first author received financial assistance (Fellowship Reference No. 3261/NET-JUNE 2015) from the University Grants Commission, Ministry of Human Resource Development, India.

References

Alonso-Pérez, F., Ruiz-Luna, A., Turner, J., Berlanga-Robles, C.A., Mitchelson-Jacob, G., 2003. Land cover changes and impact of shrimp aquaculture on the landscape in the ceuta coastal lagoon system, Sinaloa, Mexico. Ocean Coast Manag. 46 (6–7), 583–600.

Ayyappan, S., 2006. National Aquaculture Sector Overview: India. National Aquaculture Sector Overview Fact Sheets. Food and Agriculture Organization of the United Nations, Rome.

Barman, N.K., Chatterjee, S., Paul, A.K., 2016. Coastal Morphodynamics: Integrated Spatial Modeling on the Deltaic Balasore Coast, India. Springer, Switzerland.

Benz, U.C., Hofmann, P., Willhauck, G., Lingenfelder, I., Heynen, M., 2004. Multi-resolution, object-oriented fuzzy analysis of remote sensing data for GIS-ready information. ISPRS J. Photogramm. Remote Sens. 58 (3–4), 239–258.

Blaschke, T., Lang, S., Lorup, E., Strobl, J., Zeil, P., 2000. Object-oriented image processing in an integrated GIS/remote sensing environment and perspectives for environmental applications. Environ. Inf. Plan. Polit. & Public 2, 555–570.

Castro, M., Iglesias, L., Sánchez, J.A., Ambrosio, L., 2011. Sight distance analysis of highways using GIS tools. Transport. Res. C Emerg. Technol. 19 (6), 997–1005.

Chakrabarti, P., Nag, S., 2015. Rivers of West Bengal Changing Scenario. Department of Science and Technology, Government of West Bengal, Kolkata.

Chakraborty, S.K., 2010. Coastal environment of Midnapore, West Bengal: potential threats and management. J. Coast. Environ. 1 (1), 27–40.

Chaudhari, S., Devi Prasad, K.V., Shanker, K., 2009. Impact of Casuarina Plantations on Olive Ridley Turtle Nesting along the Northern Tamil Nadu Coast, India. Ashoka Trust for Research in Ecology and the Environment, Bangalore.

Coppin, P., Jonckheere, I., Nackaerts, K., Muys, B., Lambin, E., 2004. Digital change detection methods in ecosystem monitoring: a review. Int. J. Remote Sens. 25 (9), 1565–1596.

Crews-Meyer, K.A., 2004. Agricultural landscape change and stability in northeast Thailand: historical patch-level analysis. Agric. Ecosyst. Environ. 101 (2–3), 155–169.

D'Souza, N., Ishwar, N.M., 2012. Coastal Sustainability: Learnings from MFF (India) Projects. IUCN India Country Office, New Delhi.

Datta, D., Deb, S., 2012. Analysis of coastal land use/land cover changes in the Indian Sunderbans using remotely sensed data. Geo Spat. Inf. Sci. 15 (4), 241–250.

Dutta, D., Das, C.S., Kundu, A., 2016. A geo-spatial study on spatio-temporal growth of brackish water aquaculture along the coastal areas of West Bengal (India). Model. Earth Syst. & Environ. 2 (2), 61.

Fenu, G., Carboni, M., Acosta, A.T., Bacchetta, G., 2013. Environmental factors influencing coastal vegetation pattern: new insights from the Mediterranean Basin. Folia Geobot. 48 (4), 493–508.

Ferguson, D., 2008. GIS for wilderness search and rescue. In: 2008 ESRI Federal User Conference. In: http://proceedings.esri.com/library/userconf/feduc08/papers/gis_for_wilderness_search_and_rescue.pdf. Accessed 04.05.2019.

Hansen, M.C., Loveland, T.R., 2012. A review of large area monitoring of land cover change using Landsat data. Remote Sens. Environ. 122, 66–74.

Holling, C.S., Meffe, G.K., 1996. Command and control and the pathology of natural resource management. Conserv. Biol. 10 (2), 328–337.

Kar, A., Raut, S.K., Bhattacharya, M., Patra, S., Das, B.K., Patra, B.C., 2017. Marine fishes of West Bengal coast, India: diversity and conservation preclusion. Reg. Stud. Mar. Sci. 16, 56–66.

Kindu, M., Schneider, T., Teketay, D., Knoke, T., 2013. Land use/land cover change analysis using object-based classification approach in Munessa-Shashemene landscape of the Ethiopian highlands. Remote Sens. 5 (5), 2411–2435.

Kolbert, E., Wilson, E.O., Lovejoy, T.E., 2017. Living in the Anthropocene: Earth in the Age of Humans. Smithsonian Institution, Washington.

Li, Z., Li, X., Wang, Y., Ma, A., Wang, J., 2004. Land-use change analysis in Yulin prefecture, northwestern China using remote sensing and GIS. Int. J. Remote Sens. 25 (24), 5691–5703.

Matsushita, B., Xu, M., Fukushima, T., 2006. Characterizing the changes in landscape structure in the Lake Kasumigaura Basin, Japan using a high-quality GIS dataset. Landsc. Urban Plann. 78 (3), 241–250.

McGarigal, K., Marks, B.J., 1995. FRAGSTATS: Spatial Pattern Analysis Program for Quantifying Landscape Structure. Forest Science Department, USA.

Mitra, A., 2019. Estuarine Pollution in the Lower Gangetic Delta. Springer, Cham.

Niyogi, D., 1975. Quaternary geology of the coastal plain of West Bengal and Orissa. Indian J. Earth Sci. 2 (1), 51–61.

Partecke, J., Schwabl, I., Gwinner, E., 2006. Stress and the city: urbanization and its effects on the stress physiology in European blackbirds. Ecology 87 (8), 1945–1952.

Raja, R.A., Panigrahi, A., Kumar, S., 2012. Epidemiological investigation of brackish water culture systems in West Bengal, India. J. Appl. Aquacult. 24 (1), 49–59.

Rajashekara, S., 2018. Need for monitoring and conservation of marine floral and faunal biodiversity in the coastal region of Karnataka, India. Indian J. Geo Mar. Sci. 47 (1), 7–14.

Roy, A.K., Datta, D., 2018. Analyzing the effects of afforestation on estuarine environment of river Subarnarekha, India using geospatial technologies and participatory appraisals. Environ. Monit. Assess. 190 (11), 1–16.

Sambrook, R.A., Pigozzi, B.W., Thomas, R.N., 1999. Population pressure, deforestation, and land degradation: a case study from the dominican republic. Prof. Geogr. 51 (1), 25–40.

Su, S., Jiang, Z., Zhang, Q., Zhang, Y., 2011. Transformation of agricultural landscapes under rapid urbanization: a threat to sustainability in Hang-Jia-Hu region, China. Appl. Geogr. 31 (2), 439–449.

Tolessa, T., Senbeta, F., Kidane, M., 2016. Landscape composition and configuration in the central highlands of Ethiopia. Ecol. & Evol. 6 (20), 7409–7421.

Wu, J., Shen, W., Sun, W., Tueller, P.T., 2002. Empirical patterns of the effects of changing scale on landscape metrics. Landsc. Ecol. 17 (8), 761–782.

Xu, S., Xu, Y., Fu, Y., Wang, Q., 2016. Soil salinization and mitigation measures in land reclamation regions. In: Larramendy, M., Soloneski, S. (Eds.), Soil Contamination.

Yu, W., Zhou, W., Qian, Y., Yan, J., 2016. A new approach for land cover classification and change analysis: integrating backdating and an object-based method. Remote Sens. Environ. 177, 37–47.

9

Applications of geostationary satellite data in the study of ocean and coastal short-term processes: two cases in the East China Sea

Wenbin Yin[1,*], Daji Huang[1]

[1]STATE KEY LABORATORY OF SATELLITE OCEAN ENVIRONMENT DYNAMICS, SECOND INSTITUTE OF OCEANOGRAPHY, MINISTRY OF NATURAL RESOURCES, HANGZHOU, CHINA
[*]CORRESPONDING AUTHOR.

Chapter outline

1. Introduction

Oceanography is an observation-based discipline. For more than a century, oceanographers had to go to sea in ships to carefully sample the detailed vertical profile of water properties beneath the sea surface. Dr. Gifford Ewing, known as the father of space

Remote Sensing of Ocean and Coastal Environments. https://doi.org/10.1016/B978-0-12-819604-5.00009-3

oceanography, noted that the traditional field investigation forces oceanographers to consider "the class of problems that derive from the vertical distribution of properties at stations widely separated in space and time" (Wilson et al., 2008).

Since the first ocean-viewing satellite *Seasat* was launched into space on June 28, 1978, the satellite remote sensing shows great advantages in collecting the synoptic information of mid- and long-term variation processes at or near the surface of the global ocean, and therefore dramatically promoted the progress of oceanography. With rapid advances in computer hardware and software, oceanographers can easily acquire and analyze oceanic variables through satellite, such as sea surface temperature (SST), sea surface salinity (SSS), surface vector winds, variations in sea surface height associated with large-scale ocean circulation systems, and regional variations in ocean biology. The huge amount of high spatio-temporal satellite images permit a unique visualization of near-surface processes and therefore play an important role in the studies of climate change (Yang et al., 2013), material and energy transport in the global ocean (Chelton et al., 2007), and changes in marine ecosystems (Klemas, 2010).

Although the traditional polar-orbit ocean satellite is well suited for the study of medium- and long-term processes in the ocean, it encountered a great challenge when oceanographers turned their eyes to the short-term variations that prevail in the ocean due to the limit in temporal frequency of overpasses and the cover of clouds. The period of the short-term process in the ocean is usually from hours to days, thus the polar-orbit satellites that usually transit once or twice a day cannot capture the full spatial structure and complete evolution process. Because the orbit lies in Earth's equatorial plane, geostationary satellites can continuously observe the same surface area, therefore the interference of clouds can be eliminated as much as possible (Fig. 9.1). Compared to traditional polar-orbit satellites, the new generation geostationary satellites with high temporal-spatial resolution are uniquely capable of monitoring short-term and regional oceanic phenomena under the complex dynamic environment.

In this chapter, we will give two study cases about the short-term process in the cross-shelf transport of the East China Sea (ECS) to illustrate the great potential of geostationary satellites in observing the short-term process in the ocean and coastal dynamic environment. The first case (Yin and Huang, 2016) uses the world's first geostationary satellite ocean color sensor, the geostationary ocean color imager (GOCI), to explore the complete evolutionary process of submesoscale frontal waves in the Zhejiang coastal front zone (Fig. 9.1B). The second case (Yin and Huang, 2019) uses the new geostationary meteorological satellite Himawari-8 to explore the whole short-term variation in the surface upwelling off northeastern Taiwan (Fig. 9.1D).

2. Geostationary satellite data

2.1 GOCI data

GOCI is the world's first ocean color observation satellite placed in a geostationary orbit (Table 9.1). It was launched in June 2010 for near real-time monitoring of marine

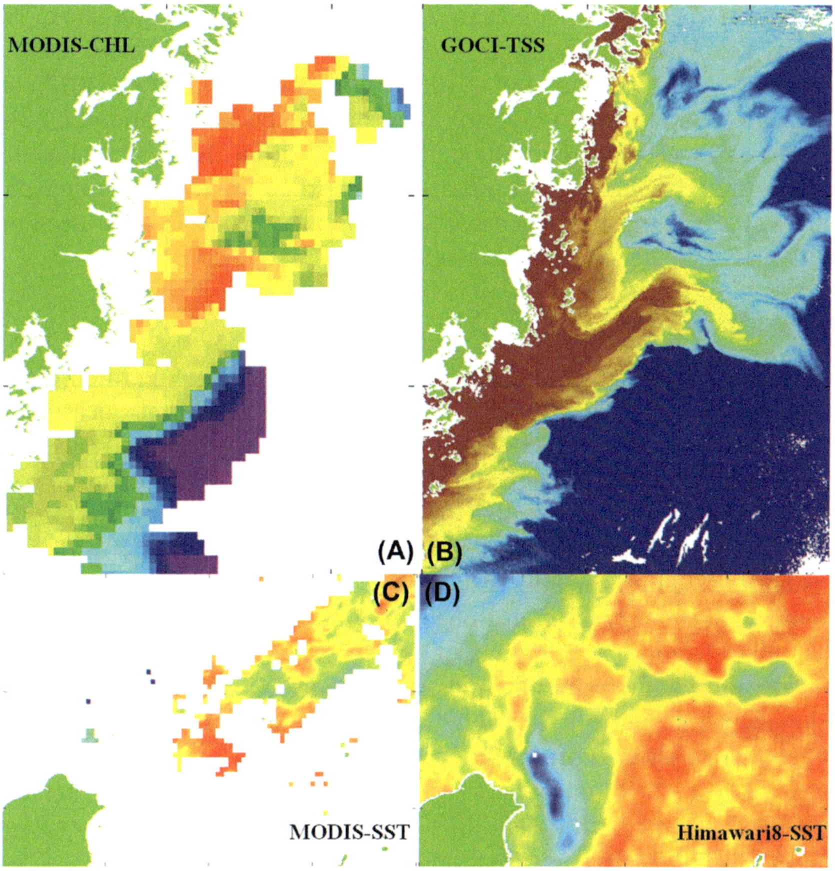

FIGURE 9.1 The distribution of MODIS-CHL (A) and GOCI-TSS (B) on July 15, 2016, and the daily SST distribution of MODIS (C) and Himawari-8 (D) on July 19, 2016.

environments in northeast Asia with a high spatial resolution (500 m) and a very high temporal resolution (1 h, eight times per day from 08:15 to 15:45 local time, UTC+8). It has six visible bands and two near-infrared bands. The observing area is about 2500 km× 2500 km centered at 130°E and 36°N, covering the coasts of Eastern China, the Korean peninsula, Japan, along with the corresponding shelves and open oceans (Choi et al., 2012; He et al., 2013). The derived ocean data products are mainly the water leaving radiance (Lw), chlorophyll concentration (CHL), concentration of dissolved organic material or yellow substance (CDOM), and concentration of suspended particles (TSS) in the near-surface zone of the sea. The GOCI-derived ocean color data were atmospherically corrected by the modified Management Unit of the North Sea Mathematical Models (MUMM) algorithm, which can be applied in highly turbid coastal waters (Lee et al., 2012; Wang et al., 2013). Ruddick had made a comparison of corresponding reflectance and TSS products between GOCI and MODIS_AQUA in the Bohai Sea, a turbid water region like the Zhejiang coastal zone. The result shows that the TSS spatial distributions of GOCI corresponding to previous studies of this region, and the GOCI data can already be used for study in the turbid waters (Ruddick et al., 2012). Therefore GOCI provides an effective tool to monitor the short-term ocean phenomenon in coastal waters such as surface tidal current (Hu et al., 2016), diurnal changes of biogeochemical properties (Lou and Hu, 2014), and resuspension processes of suspended particulate matter (Choi et al., 2012). In the first case, the GOCI Level-2 TSS data was used to analyze the instability waves of the Zhejiang coastal front.

2.2 Himawari-8 data

The Himawari-8 is a new geostationary meteorological satellite operated by the Japan Meteorological Agency since July 7, 2015 (Table 9.1). This satellite is located at 140.7°E and observes Earth from 80°E to 160°W between 60°N and 60°S. The Advanced Himawari Imager is a major instrument onboard the Himawari-8. It has 16 observation bands from visible to infrared wavelengths, and their spatial resolution is 0.5 or 1 km for visible and near-infrared bands and 2 km for infrared bands (Bessho et al., 2016). The scan frequency is around 10 min for Full Disk, and 2.5 min for Japan area and target areas. The high temporal-spatial resolution images of Himawari-8 provide new levels of capacity for the identification and tracking of rapidly changing ocean surface phenomena and for the derivation of quantitative products. In the second case, the daily SST, which was operationally retrieved from 10.4, 11.2 and 8.6 μm data of Himawari-8 was used to explore the complete short-term variation process of surface upwelling northeast of Taiwan. Kurihara et al. (2016) have made a comparison between this SST data from June to September 2015 and the drifting and tropical moored buoy data. The result showed a root-mean-square difference of ~0.59 K and a bias of ~−0.16 K.

Table 9.1 The detailed information of two geostationary satellites that were used in this chapter.

Platform		GOCI	Himawari-8
Bands		6 VIS, 2 NIR	3 VIS, 3 NIR, 10 IR
Coverage area		2500 km × 2500 km (center: 130°E, 36°N)	Longitude: 80°E–160°W Latitude: 60°S–60°N
Product		Water leaving radiance, normalized water leaving radiance, chlorophyll-a, colored dissolved organic matter, total suspended sediments	Aerosol property, cloud property, sea surface temperature, short wave radiation, chlorophyll-a, wild fire
Resolution	Temporal	Hourly (00:15–07:45 GMT)	10-minute, hourly, daily, monthly
	Spatial	0.5 km	1 , 2, 5 km

3. Applications: two cases in the East China Sea short-term cross-shelf process

The ECS is a marginal sea east of China and is part of the Pacific Ocean (Fig. 9.2A). It has a wide and shallow continental shelf and is a region of high marine-life productivity. Through the cross-shelf transport, the huge amount of nutrients from the land source and the outer sea provide strong support for maintaining high productivity in the ECS shelf. Therefore, the study of the cross-shelf transport process in the ECS has been the focus of scientists. Generally, the cross-shelf transport process in the ECS mainly includes the offshore transport of coastal waters off the coast of eastern China (*the green arrow* in Fig. 9.2A) and the inshore invasion of the Kuroshio waters near the shelf-break off northeastern Taiwan (*the magenta arrow* in Fig. 9.2A). The previous studies have focused more on the long-term variations in the cross-shelf process, such as the seasonal variation of Changjiang diluted water distributions (Chang and Isobe, 2003) and the seasonal variability of Kuroshio intrusion off north Taiwan (Tang et al., 2000). For the short-term variations in the cross-shelf process, there is little research due to the lack of effective observation data. With the rapid advance of science and technology, the new generation of geostationary satellites with high temporal-spatial resolution provides an effective research tool.

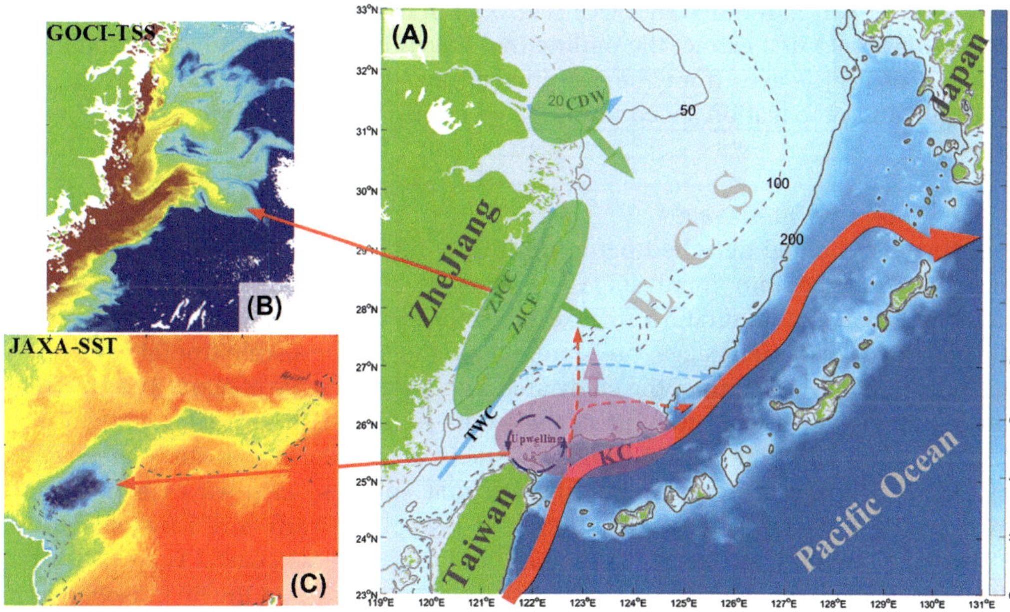

FIGURE 9.2 The circulation (*arrows*) and topography (background color) distribution in ECS shelf (A), the TSS image of frontal waves in Zhejiang coastal front zone (B), and the SST image of surface upwelling off the northeastern Taiwan (C). The *green arrows* in panel (A) represent the offshore transport of coastal waters, and the *magenta arrows* represent the inshore invasion of Kuroshio waters.

3.1 A complete evolutionary process of submesoscale frontal waves off the southeast coast of China

The Zhejiang coastal front (ZJCF) off the southeast coast of China is an interface between two currents (Fig. 9.2A), the nearshore Zhejiang coastal current (ZJCC) and the offshore Taiwan warm current (TWC). In situ measurements have demonstrated that the TWC has high salinity and low turbidity and flows northward along the 50–100 m isobaths all year round (Su and Pan, 1987). The ZJCC, characterized by low salinity and high turbidity, flows southward along the shoreline (Su and Yuan, 2005; Sun, 2006). The conventional wisdom holds that there is no direct passage of coastal waters from the coast to the outer shelf due to the barrier of front and the limitation of bottom topography. Recent studies have shown that the large scale cross-shelf penetrating fronts (Fig. 9.2B) that developed from small-scale frontal waves are universal in the Zhejiang coastal front area (Yuan et al., 2005; He et al., 2010). It can bring coastal waters across the front and transport to the outer shelf massively, and therefore play an important role in cross-front exchange of heat, effluent, and land-derived materials in the ECS (Yuan et al., 2010). However, the observation of features and understanding of the evolutionary process remain incomplete due to the lack of effective investigation techniques.

In this study, images of TSS from the GOCI were used to analyze ZJCF waves. The TSS images taken over 3 years from 2011 to 2013 showed that frontal waves are frequent in the ZJCF zone. In the period of September 2–11, 2013, a complete evolutionary process of frontal waves was monitored by GOCI (Fig. 9.3). In order to quantitatively analyze the whole evolution, we first extract the outline (*red line* in Fig. 9.3) and ridge line (*blue line* in Fig. 9.3) of front waves from the TSS images, and then calculate the wave height, wavelength, and propagation speed (for detailed information, see Section 2.2 of Yin and Huang, 2016).

3.1.1 Evolution of frontal waves

Using the rolled up and detached pattern of the frontal waves, the whole evolutionary process was divided into three stages: September 2–3 as the development stage, September 4–5 as the maturation stage, and September 6–11 as the decay stage.

During the development stage, on September 2, the submesoscale frontal waves first appeared in the northern part of the study area with a spatial scale of about 10 km. While in the southern part of the study area, the frontal outline still formed a straight line (Fig. 9.3A). Therefore, the coastal front was stable across the entire study area. After 1 day, the TSS image of September 3 showed that submesoscale waves had already formed south of 28.5°N, with a wave height of less than 10 km and a wavelength of 10–20 km (Fig. 9.3B). That means the equilibrium of the whole coastal front had been broken. Meanwhile, the whole front moved seaward, approaching the 50 m isobaths.

During the mature stage, the frontal waves rolled up but were still attached to the mother front. Meanwhile, the wave heights increased to 20–40 km with a mean wavelength of 33 km. The specific features were as follows. On September 4, 11 roll-up waves

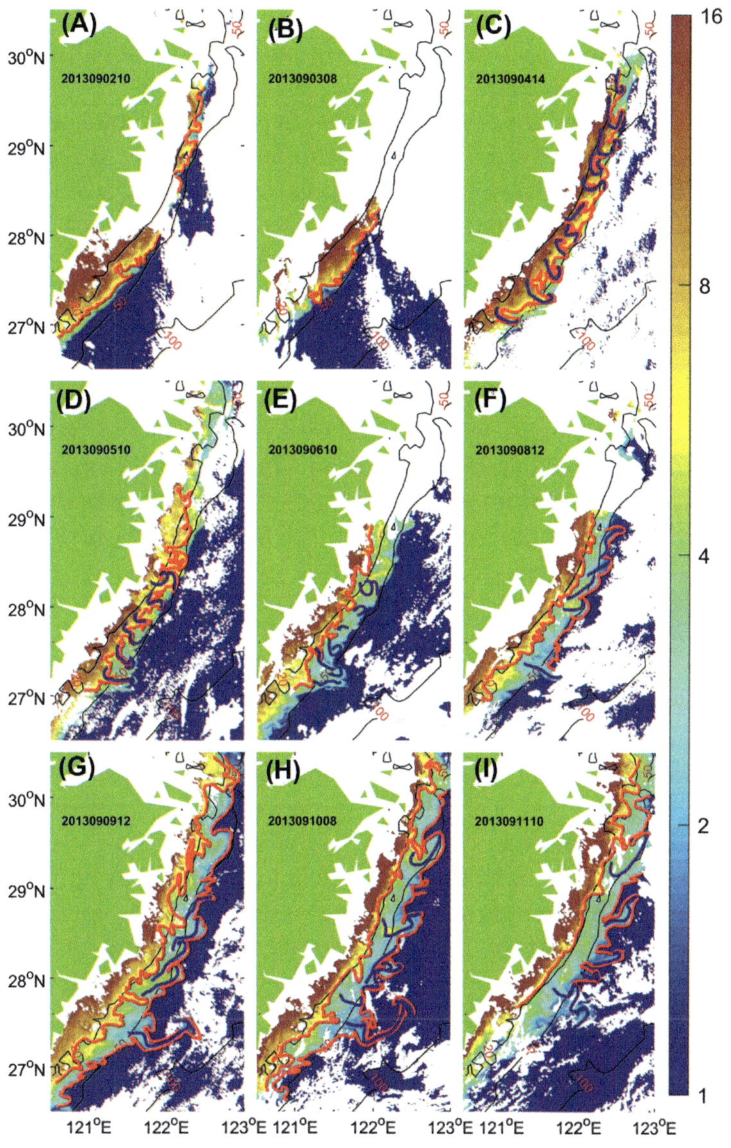

FIGURE 9.3 TSS images of complete evolutionary cycle of frontal waves between September 2 and 11, 2013.

were present in the study area. As can be seen, the wave height of the frontal waves increased from the north to south, the height of all frontal waves was 10–50 km, and the wavelength was 10–40 km with an average of 33 km (Fig. 9.3C). All 11 frontal waves were located around the 20–50 m isobaths. The spatial pattern of all the waves showed a cyclonic roll-up structure. On September 5, only seven frontal waves were visible. The wave height reduced to 20–40 km as the waves continued to curl or partially separate, with a wavelength of 15–35 km (Fig. 9.3D). Almost all these roll-up waves were still attached to the mother front and close to or beyond the 50 m isobaths with southward propagating speeds of 0.10–0.25 m/s.

During the decay stage, the roll-up waves became detached, extended further offshore, and moved northeastward, while the former mother front zone gradually stabilized into a straight line. The specific features of this stage were as follows. On September 6, all front waves started to partially or completely detach from the mother front (Fig. 9.3E). The wavelength still remained at 10–40 km, but the wave height was reduced to 10–20 km as isolated patches became partly detached. The spatial structure of these isolated patches was very complex and mainly characterized by a cyclonic vortex filament pattern. In the following days (Fig. 9.3F–I), the unstable frontal waves that separated from the coastal front had evolved into isolated patches and expanded to the northeast. All the isolated patches were located between the 50 and 100 m isobaths with a horizontal scale of 30–80 km, and some showed a mushroom-like pattern. Meanwhile, the mother front transformed from a wavy line into a straight line as the front gradually stabilized to an equilibrium state. At this point, the frontal waves had undergone a full life cycle.

3.1.2 Tidal oscillation of frontal waves

In addition to the daily variation, we also identified an hourly migration of the coastal front waves. The eight TSS images taken within a day show that the ZJCF makes a tidal-like oscillatory movement every day. The images from September 9 were selected to illustrate this movement as the good spatial coverage and small deformation of the frontal outline on that day (Fig. 9.4A). An onshore motion of the front with a speed of 0.3–0.9 m/s can be seen between 8:00 and 11:00, shifting to an offshore motion at 0.25–0.60 m/s between 12:00 and 14:00. The oscillatory range was approximately 6 km. Because the oscillatory movement of the front is similar to the local tidal motion, a simple analysis of the effect of the tidal current on front location was carried out. The results show that the front moved in phase with the tidal current (Fig. 9.4B). The mean ratio between the distance of hourly tidal movement and hourly frontal movement was 0.55. This suggests that the tidal-like oscillation of the front was caused primarily by the advection of the tidal current, although the movement was complex and was influenced by a range of factors.

3.1.3 The discussion of mechanism

The horizontal scale (L<20 km) and e-folding time (T<2 days) of the unstable frontal waves in the development stage suggested that this might be a submesoscale process in the upper mixed layer (Boccaletti et al., 2007). To confirm this conjecture, we calculated the Rossby number ($R_O = U/fL$), the horizontal scale ($L = NH/f$), and the growth rate $\left(\tau = \sqrt{\frac{54}{5}}\frac{\sqrt{1+Ri}}{f}\right)$ of the baroclinic modes in the coastal front zone based on

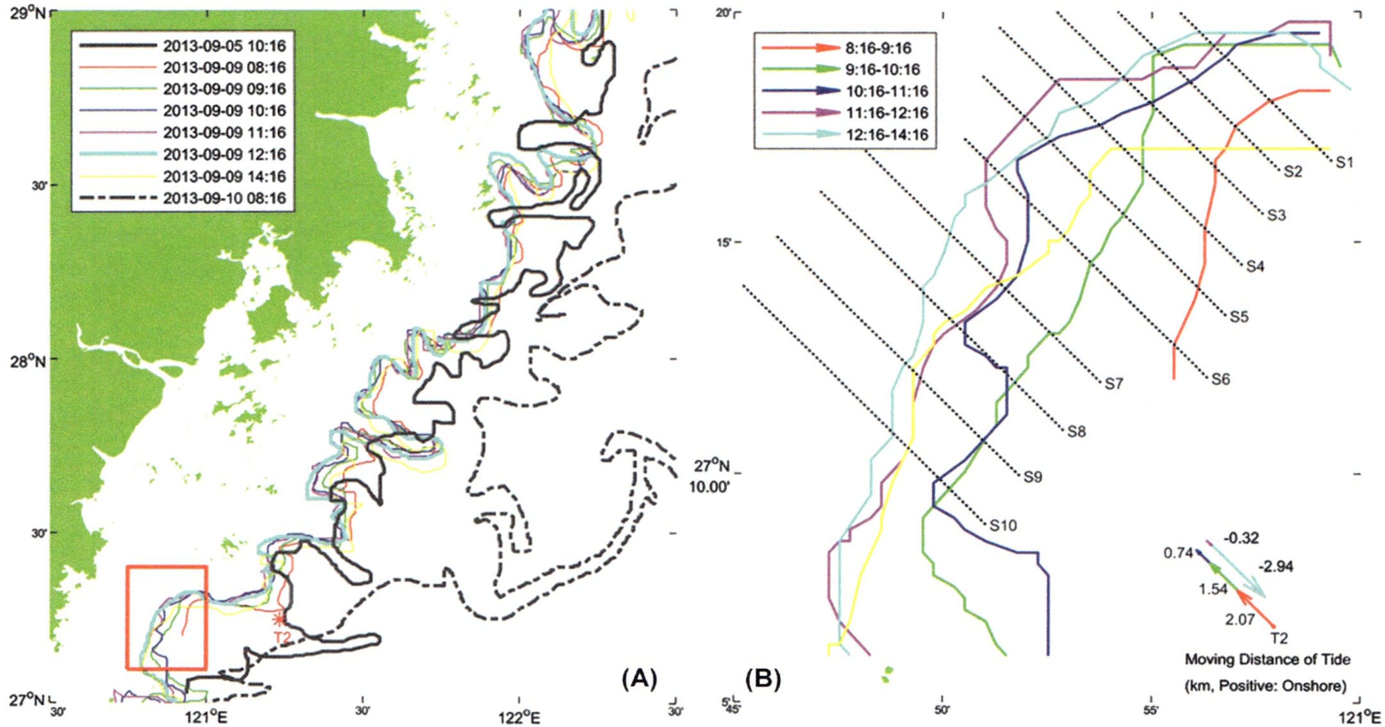

FIGURE 9.4 (A) Frontal outlines on September 5, 9, and 10, 2013 and the (B) section of frontal outlines used to compare with tidal current data. The thin gray lines in panel (B) S1–S10 are the transect lines used to estimate the hourly movement of the front. The arrows in the lower right corner of panel (B) show the movement in kilometer of tidal currents within the same time period.

submesoscale dynamics. The results showed that the theoretical estimates were close to the observations in the development stage. This indicated the frontal waves in the development stage represented a baroclinic submesoscale process.

Several mechanisms are known to be active in the generation of submesoscale processes in the surface layer of the ocean (Thomas et al., 2008): (1) frontogenesis, (2) unforced instabilities, and (3) forced motion. After checking the wind and temperature data over the same period, we found that the collective effect of forced motion and unforced instability caused by the transit of a tropical storm (the Kong-Rey) was the most likely cause of submesoscale instability. Initially, the wind in the front zone was weak, and the front was close to geostrophic balance. When the storm hit the area, the wind increased from 3 to 10 m/s and transited from upwelling-favorable to downwelling-favorable (down-front) wind rapidly (Fig. 9.5A). This change led a reduction in the potential vorticity and thus provided favorable conditions for submesoscale phenomena (Thomas and Lee, 2005). At the same time, the strong storm and the accompanying cold air cooled the sea surface and resulted in a sharp reduction of SST (Fig. 9.5B). This surface cooling induced the loss of buoyancy in the surface layer, and strengthened the lateral buoyancy gradient that can trigger submesoscale instability (Thomas et al., 2008). Additionally, the unforced instability caused by the restratification in the upper mixed layer may also play an important role in the submesoscale process since the Ri numbers was about $Ro^{-(1/2)}$ and close to O(1) (Boccaletti et al., 2007; Thomas et al., 2008).

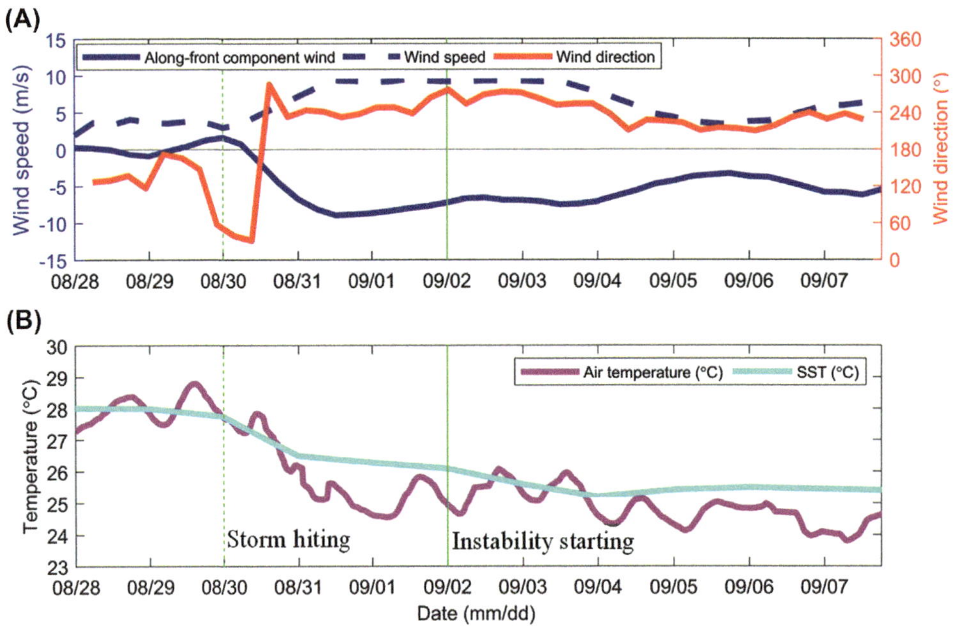

FIGURE 9.5 The magnitude and direction of area-averaged 6-h CCMP wind (A), the moored-buoy monitored 15-min air temperature and area-averaged daily AVHRR SST (B) before and after the onset of frontal unstable processes. The direction of the along-front component in panel (A) is toward northeast.

3.2 The short-term variation of surface upwelling off northeastern Taiwan

The upwelling off northeastern Taiwan was caused by the collision of the northward Kuroshio current against a nearly zonal-running continental shelf break of the ECS. It is an important pathway for shelf-ocean exchanges of dissolved and particulate material between the ECS and the Kuroshio current, and therefore has attracted much attention (Chen, 1996; Su, 1998; Wong et al., 2000). Multiple observational and numerical results have shown that this upwelling persists year-round (Liu et al., 1992) and exhibits significant seasonal variations, particularly in the upper layer (Cheng et al., 2009). The surface upwelling occurs more frequently in summer than in winter under the influence of the Kuroshio current and the wind (Tang et al., 2000; Chang et al., 2009). In addition to the seasonal variation, several studies have reported that the upwelling also has short-term biweekly variations (Lin et al., 1992; Gopalakrishnan et al., 2013). However, due to limitations in the spatiotemporal resolution and coverage of the observed data, the lifecycle and associated dynamics are poorly understood.

In this study, daily SST data from Himawari-8 were used to explore the detailed features of the short-term variations in surface upwelling. In order to demarcate the surface upwelling and quantify its features, we proposed a gradient-based edge detection algorithm. After that, the temperature difference between upwelled and surrounding water was used as an index *UPI* to measure the intensity of upwelling, and the area of upwelling $S_C$was also calculated (for detailed information, see Yin and Huang, 2019).

3.2.1 The whole short-term variation of surface upwelling

The high spatial-temporal resolution JAXA-SST images taken from 2015 to 2017 show that surface upwelling is frequent off northeastern Taiwan. In the period of July 18–August 3, 2016, a complete evolutionary process of surface upwelling was monitored by Himawari-8 (Fig. 9.6). Based on the variations of *UPI* and S_C (Fig. 9.7), the whole process can be divided into two stages: July 18–25 as the intensification stage, and July 26–August 3 as the decay stage.

During the intensification stage, between July 18 and July 25, the *UPI* intensified quickly from 1.1°C to 2.5°C, and the core area S_C expanded by over six times over 8 days, from 950 to 5900 km^2 (Fig. 9.7A). The center of the upwelling was located quasi-stationary near the 200 m isobath off northeastern Taiwan throughout that stage (Fig. 9.7B). The AVISO surface flow field and the JAXA-SST spatial pattern (e.g., July 20 and 21 in Fig. 9.6) indicate that there is a cyclonic circulation in the upwelling area during this stage. After July 22, the core area was stretched northeastward by the advection of the Kuroshio current and formed an elongated cold-water area (Fig. 9.6).

After the intensity peaked on July 25, the upwelling switched to the decay stage until August 3. The *UPI* decreased quickly from 2.5 to 0.4°C, and the core area S_C decreased by nearly 1/5 over 9 days, from 5900–970 km^2 (Fig. 9.7A). Due to the advection of the Kuroshio current, the center of the upwelling started to move northeastward with an

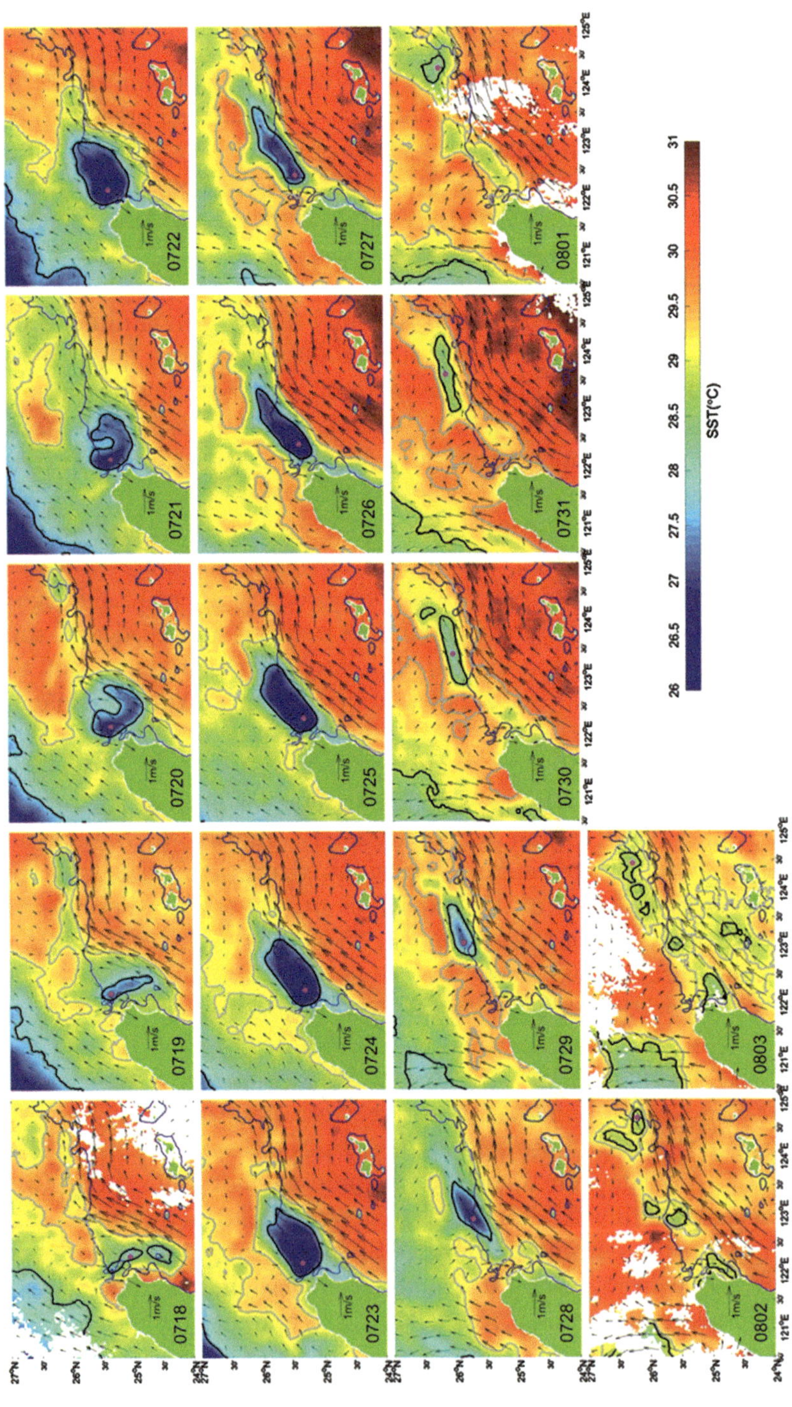

FIGURE 9.6 SST images of a complete short-term surface upwelling process between July 18 and August 3, 2016.

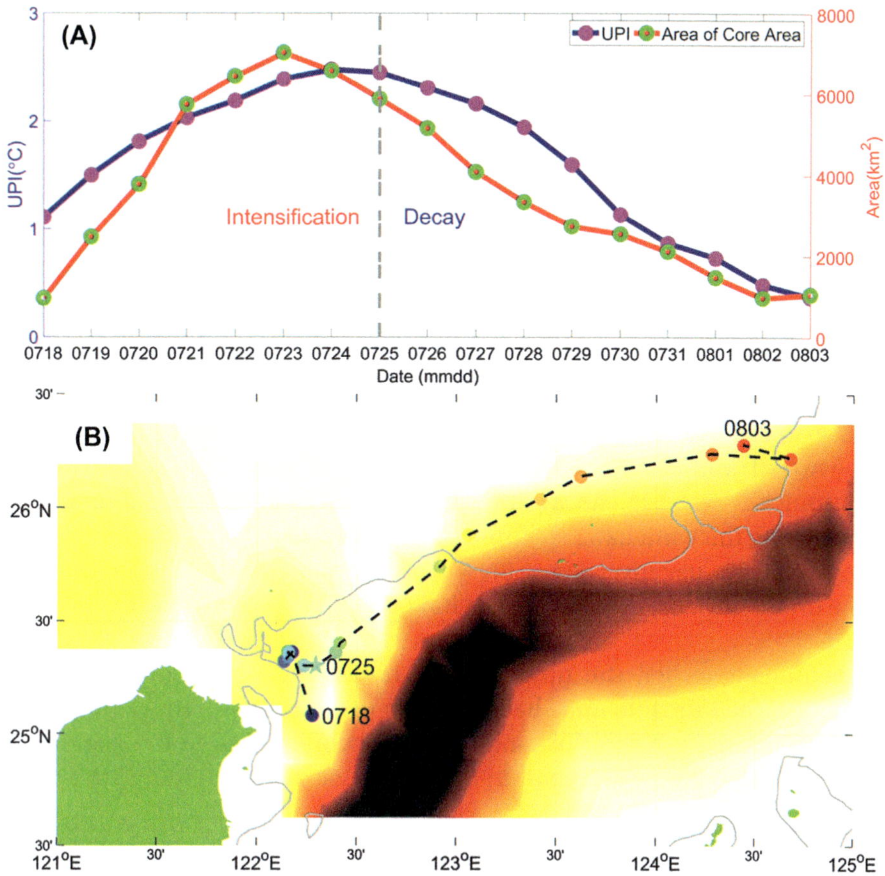

FIGURE 9.7 (A) The UPI and core area and (B) the daily position of the upwelling center over the entire process. The background colors in panel (B) indicate the mean kinetic energy of the altimeter-derived geostrophic current over the lifecycle of the process. The *solid gray line* in panel (B) indicates the 200 m isobath.

average speed of 0.30 m/s, and crossed over the 200 m isobath from the deep area to the shelf area on July 28 (Fig. 9.7B). The shape of the core area continued to be stretched northeastward, and the surface upwelling changed from an elliptical shape to a long-strip shape and finally split into several small low-temperature patches (Fig. 9.6).

3.2.2 A preliminary discussion of mechanism

Several mechanisms (e.g., tides, the Kuroshio current, the wind-stress curl, typhoons, and mesoscale eddies) have been invoked to explain the variations in the surface upwelling in previous studies (Chang et al. 2008, 2009; Jan et al., 2011; Shen et al. 2011, 2017; Tsai et al., 2013). Considering the matching of time scales (~2 weeks), the direct impacts of mesoscale eddies (~100 days) and tides (~0.5–1 day) on short-term processes are not considered. Since no typhoon passed through Taiwan during the same period, the effect of typhoon is also not considered.

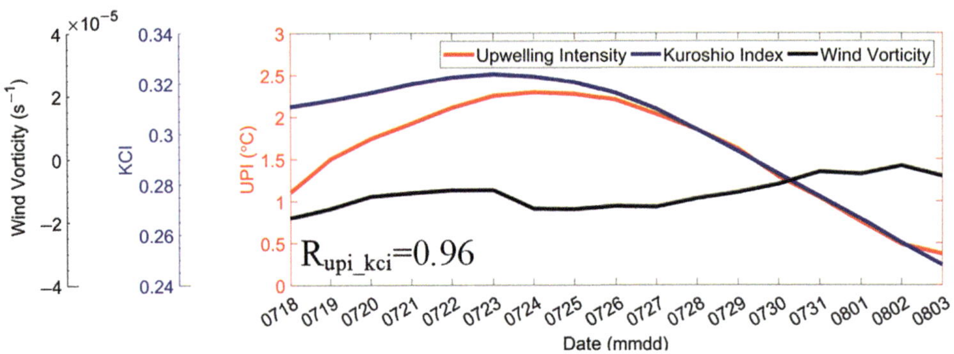

FIGURE 9.8 The time series of the *UPI*, *KCI*, and core area average wind curl between July 18 and August 3, 2016.

A positive/negative wind curl indicates that divergent/convergent flow drives upward/downward vertical motion and intensifies/weakens the surface upwelling. Therefore, a quick reversal in the wind curl can result in a short-term variation in the UPI. After checking the wind data, we found that the averaged wind curl in the core area of upwelling is always negative (Fig. 9.8). A negative value of the wind curl indicates a convergence on the surface; therefore, the wind curl cannot be responsible for this short-term process.

As we mentioned earlier, the upwelling off northeastern Taiwan is caused by the collision between the Kuroshio and shelf break, so the intensity of upwelling *UPI* should be proportional to the Kuroshio flux, and inversely proportional to the angle θ between the Kuroshio and topography gradient. In this study, we define an index that reflects the tendency of the Kuroshio invasion: $KCI = \frac{KVT}{KAA}$, where the *KVT* is the Kuroshio volume transport, and *KAA* is the Kuroshio axis angle that is related to the θ. It's obvious that *KCI* should have a positive correlation with *UPI* if the short-term process was mainly affected by the Kuroshio. Our correlation analysis result shows that *UPI* has good synchronicity with *KCI* and that the correlation coefficient is approximately 0.96 (Fig. 9.8). This means that this short-term process was likely dominated by short-term fluctuations in the Kuroshio current.

4. Conclusion

In this chapter, two cases were used to show the applications of geostationary satellite data in observing the short-term process in the ECS cross-shelf process. Compared to traditional polar-orbit satellites, the results show that the geostationary satellite has a great advantage in the time resolution and spatial coverage of data, and allows us to capture the details of rapid and continuous variation in the ocean. We think that the geostationary ocean satellite will greatly promote the study of short-term processes in marine science just as the contribution of polar-orbit satellites to mid- and long-term process did in previous decades.

References

Bessho, K., Date, K., Hayashi, M., Ikeda, A., Imai, T., Inoue, H., Kumagai, Y., Miyakawa, T., Murata, H., Ohno, T., Okuyama, A., Oyama, R., Sasaki, Y., Shimazu, Y., Shimoji, K., Sumida, Y., Suzuki, M., Taniguchi, H., Tsuchiyama, H., Uesawa, D., Yokota, H., Yoshida, R., 2016. An introduction to Himawari-8/9—Japan's new-generation geostationary meteorological satellites. J. Meteorol. Soc. Japan. Ser. II 94 (2), 151–183.

Boccaletti, G., Ferrari, R., Fox-Kemper, B., 2007. Mixed layer instabilities and restratification. J. Phys. Oceanogr. 37 (9), 2228–2250.

Chang, P.H., Isobe, A., 2003. A numerical study on the Changjiang diluted water in the yellow and East China Seas. J. Geophys. Res. Oceans 108 (C9), 1–17.

Chang, Y.-L., Wu, C.-R., Oey, L.-Y., 2009. Bimodal behavior of the seasonal upwelling off the northeastern coast of Taiwan. J. Geophys. Res. 114 (C3), 13.

Chang, Y., Liao, H.-T., Lee, M.-A., Chan, J.-W., Shieh, W.-J., Lee, K.-T., Wang, G.-H., Lan, Y.-C., 2008. Multisatellite observation on upwelling after the passage of typhoon Hai-Tang in the Southern East China Sea. Geophys. Res. Lett. 35 (3), 5.

Chelton, D.B., Schlax, M.G., Samelson, R.M., de Szoeke, R.A., 2007. Global observations of large oceanic eddies. Geophys. Res. Lett. 34 (15).

Chen, C.T.A., 1996. The Kuroshio intermediate water is the major source of nutrients on the East China Sea continental shelf. Oceanol. Acta 19 (5), 523–527.

Cheng, Y.H., Ho, C.R., Zheng, Z.W., Lee, Y.H., Kuo, N.J., 2009. An algorithm for cold patch detection in the sea off Northeast Taiwan using multi-sensor data. Sensors 9 (7), 5521–5533.

Choi, J.-K., Park, Y.J., Ahn, J.H., Lim, H.-S., Eom, J., Ryu, J.-H., 2012. GOCI, the world's first geostationary ocean color observation satellite, for the monitoring of temporal variability in coastal water turbidity. J. Geophys. Res. Oceans 117 (C9).

Gopalakrishnan, G., Cornuelle, B.D., Gawarkiewicz, G., McClean, J.L., 2013. Structure and evolution of the cold dome off Northeastern Taiwan: a numerical study. Oceanography 26 (1), 66–79.

He, L., Li, Y., Zhou, H., Yuan, D., 2010. Variability of cross-shelf penetrating fronts in the East China Sea. Deep Sea Res. Part II Top. Stud. Oceanogr. 57 (19), 1820–1826.

He, X.Q., Bai, Y., Pan, D.L., Huang, N.L., Dong, X., Chen, J.S., Chen, C.T.A., Cui, Q.F., 2013. Using geostationary satellite ocean color data to map the diurnal dynamics of suspended particulate matter in coastal waters. Remote Sens. Environ. 133, 225–239.

Hu, Z., Wang, D.-P., Pan, D., He, X., Miyazawa, Y., Bai, Y., Wang, D., Gong, F., 2016. Mapping surface tidal currents and Changjiang plume in the East China Sea from geostationary ocean color imager. J. Geophys. Res. Oceans 121 (3), 1563–1572.

Jan, S., Chen, C.C., Tsai, Y.L., Yang, Y.J., Wang, J., Chern, C.S., Gawarkiewicz, G., Lien, R.C., Centurioni, L., Kuo, J.Y., 2011. Mean structure and variability of the cold dome Northeast of Taiwan. Oceanography 24 (4), 100–109.

Klemas, V., 2010. Remote sensing techniques for studying coastal ecosystems: an overview. J. Coast Res. 27 (1), 2–17.

Kurihara, Y., Murakami, H., Kachi, M., 2016. Sea surface temperature from the new Japanese geostationary meteorological Himawari-8 satellite. Geophys. Res. Lett. 43 (3), 1234–1240.

Lee, B., Young-Je, P., Jae Hyun, A., Joo-Hyung, R., Sang-Wan, K., 2012. Improvement of atmospheric correction for extremely turbid waters using modified MUMM atmospheric correction algorithm. In: International Symposium on Remote Sensing, Incheon, Korea.

Lin, C.-Y., Shyu, C.-Z., Shih, W.-H., 1992. The Kuroshio fronts and cold eddies off northeastern Taiwan observed by NOAA-AVHRR imageries. Terr. Atmos. Ocean Sci. 3 (3), 225–242.

Liu, K.-K., Gong, G.-C., Lin, S., Yang, C.-Y., Wei, C.-L., Pai, S.-C., Wu, C.-K., 1992. The year-round upwelling at the shelf break near the northern tip of Taiwan as evidenced by chemical hydrography. Terr. Atmos. Ocean Sci. 3 (3), 243–275.

Lou, X., Hu, C., 2014. Diurnal changes of a harmful algal bloom in the East China Sea: observations from GOCI. Remote Sens. Environ. 140, 562–572.

Ruddick, K., Vanhellemont, Q., Yan, J., Neukermans, G., Wei, G.M., Shang, S.L., 2012. Variability of suspended particulate matter in the Bohai Sea from the geostationary ocean color imager (GOCI). Ocean Sci. J. 47 (3), 331–345.

Shen, D.L., Li, X.F., Pietrafesa, L., Bao, S.W., IEEE, 2017. Geostationary Satellite Observations and Numerical Simulation of Typhoon-Induced Upwelling to the Northeast of Taiwan. In: 2017 IEEE International Geoscience and Remote Sensing Symposium. IEEE, New York, pp. 3552–3555.

Shen, M.L., Tseng, Y.H., Jan, S., 2011. The formation and dynamics of the cold-dome off Northeastern Taiwan. J. Mar. Syst. 86 (1–2), 10–27.

Su, J., 1998. Circulation dynamics of the China Seas north of 18°N. Sea 11, 483–505.

Su, J.L., Pan, Y.Q., 1987. On the shelf circulation north of Taiwan. Acta Oceanol. Sin. 6 (Suppl. 1), 1–20.

Su, J.L., Yuan, Y.L., 2005. Physical Oceaography in Chinese Coastal Oceans. China ocean press, Beijing, China.

Sun, X.P., 2006. Regional Oceanography of Chinese Coastal Oceans. China ocean press, Beijing, China.

Tang, T.Y., Tai, J.H., Yang, Y.J., 2000. The flow pattern north of Taiwan and the migration of the Kuroshio. Continent. Shelf Res. 20 (4–5), 349–371.

Thomas, L.N., Lee, C.M., 2005. Intensification of Ocean fronts by down-front winds. J. Phys. Oceanogr. 35 (6), 1086–1102.

Thomas, L.N., Tandon, A., Mahadevan, A., 2008. Submesoscale processes and dynamics. In: Hecht, M.W., Hasumi, H. (Eds.), Ocean Modeling in an Eddying Regime. American Geophysical Union, Washington, DC, pp. 17–38.

Tsai, Y.L., Chern, C.S., Jan, S., Wang, J., 2013. Numerical study of cold dome variability induced by typhoon morakot (2009) off Northeastern Taiwan. J. Mar. Res. 71 (1–2), 109–131.

Wang, M., Ahn, J.H., Jiang, L., Shi, W., Son, S., Park, Y.J., Ryu, J.H., 2013. Ocean color products from the Korean geostationary ocean color imager (GOCI). Opt. Express 21 (3), 3835–3849.

Wilson, W.S., Lindstrom, E.J., Willis, J.K., 2008. Satellite oceanography, history and introductory concepts. In: John Steele, S.T., Karl, T. (Eds.), Encyclopedia of Ocean Sciences. Academic Press, pp. 65–79.

Wong, G.T.F., Chao, S.-Y., Li, Y.-H., Shiah, F.-K., 2000. The Kuroshio edge exchange processes (KEEP) study — an introduction to hypotheses and highlights. Continent. Shelf Res. 20 (4–5), 335–347.

Yang, J., Gong, P., Fu, R., Zhang, M., Chen, J., Liang, S., Xu, B., Shi, J., Dickinson, R., 2013. The role of satellite remote sensing in climate change studies. Nat. Clim. Change 3 (10), 875–883.

Yin, W., Huang, D., 2016. Evolution of submesoscale coastal frontal waves in the East China Sea based on geostationary ocean color imager observational data. Geophys. Res. Lett. 43 (18), 9801–9809.

Yin, W., Huang, D., 2019. Short-term variations in the surface upwelling off Northeastern Taiwan observed via satellite data. J. Geophys. Res. Oceans 124 (2), 939–954.

Yuan, D., Li, Y., He, L., Zhou, H., Li, R., Wang, F., Lei, H., Hu, D., 2010. An observation of the three-dimensional structure of a cross-shelf penetrating front off the Changjiang mouth. Deep Sea Res. Part II Top. Stud. Oceanogr. 57 (19), 1827–1834.

Yuan, D.L., Qiao, F.L., Su, J., 2005. Cross-shelf penetrating fronts off the southeast coast of China observed by MODIS. Geophys. Res. Lett. 32 (19).

10

Evaluation of heavy metals in coastal aquifers and seawater: an appraisal of geochemistry using ICPMS and remote sensing

K. Shravanraj[1], R.G. Rejith[1,2], M. Sundararajan[1,2]

[1]MINERALS SECTION, MATERIALS SCIENCE AND TECHNOLOGY DIVISION, NATIONAL INSTITUTE FOR INTERDISCIPLINARY SCIENCE AND TECHNOLOGY (CSIR-NIIST), COUNCIL OF SCIENTIFIC & INDUSTRIAL RESEARCH, THIRUVANANTHAPURAM, KERALA, INDIA; [2]ACADEMY OF SCIENTIFIC AND INNOVATIVE RESEARCH (ACSIR), GHAZIABAD, UTTAR PRADESH, INDIA

Chapter outline

1. Introduction

In current trends and growing interest in environmental issues, monitoring of water pollution has high importance, in which seawater pollution assessment is a prime concern. Each year, seawater is polluted by the dumping of ballast water (10 billion tons),

Remote Sensing of Ocean and Coastal Environments. https://doi.org/10.1016/B978-0-12-819604-5.00010-X

sewage water (10,000 million gallons), oil (3.25 million metric tons), and solid wastes (millions of tons). Eighty percent of nonorganic marine pollution comes from land-based intrusions only. All rivers carry pollutants from the catchment areas into the sea. It may consist of sedimentation, agricultural runoff, sewage, solid waste, chemicals, metals, radioactive substances, oil, and biological wastes.

Heavy metal pollution sources include (1) acidification, (2) eutrophication, (3) plastic debris, and (4) toxins. The increase in the level of atmospheric CO_2 tends to make oceans more acidic than the tolerable level. The increased acidic nature dissolves the calcium carbonate structures in the sea. The increase in the amount of nutrients in the sea due to eutrophication will lead to the excessive growth of marine plant species and organisms and subsequent decrease in the oxygen level. Eighty percent of marine debris is plastic, which does not degrade quickly; it will last for a long time. This will lead to the intrusion of plastic wastes in the marine food cycle and it will gradually kill the organisms. Some toxins like polychlorinated biphenyls (PCB), dichlorodiphenyltrichloroethane, tributyltin, pesticides, furans, dioxins, phenols, and radioactive wastes do not undergo disintegration very easily. Such persistent toxins are creating severe problems in the marine system. Heavy metals like Hg, Pb, Ni, As, and Cd can vary the biochemical stability of marine water, and it will affect the living organisms. Through bioaccumulation, these toxins will enter into the tissues of living beings in the sea (Voica et al., 2012).

Heavy metal contamination is critical since it affects aquatic as well as terrestrial organisms and causes ecological variations and problems. These pollutants have a tendency to settle in the bottom sediments and are very large in seaports and industrialized coastal areas. Seawater will contain heavy metals in low concentration. Human activities, like mine wastes, oil and gas exploration, pesticides, paints, leather, textiles, fertilizers, pharmaceutical industries, domestic effluents, agricultural runoff, acid rain, and so forth, increase the amount of heavy metal deposition in the sea. Heavy metals are bioavailable and indestructible, and may cause toxic effects on living beings if they exceed a certain threshold level (Ridgway et al., 2002). Some heavy metals are essential for life, some are beneficial, and a large number are highly toxic. The availability of heavy metals in biomass helps to predict the effect and possible human health risks. Fifteen heavy metals such as Be, V, Cr, Mn, Fe, Co, Ni, Cu, Sn, As, Se, Mo, Cd, Sb, and Pb have been selected and prior to analysis detailed study on those selected metals was done.

Beryllium (Be) is found as feldspar mineral deposits. Beryllium in soil can easily enter into the food cycle through plants. Therefore, the weathering of soil and rocks will lead to beryllium mixing in water bodies. Burning of fossil fuels and atmospheric fallout on rain harvesting systems can also account for beryllium pollution (Walsh, 2009). Vanadium (V) is usually found in phosphate rocks and certain ores and is present as organic complexes in some crude oils (Baroch, 2006; Lide, 2008). Soil erosion and leaching as well as water erosion of land surfaces increase vanadium concentration in water bodies. Processing of ores and vanadium-enriched slags, and urban sewage are the major human-initiated sources of vanadium pollution (Byerrum, 1991; Byerrum et al., 1974). Most rocks and soils contain small amounts of chromium (Cr). Hexavalent chromium compounds are carcinogenic and the guideline value is 0.05 ppm (Barceloux and Barceloux, 1999). Chromium is present in the environment in trace amounts, but human activities lead the rise in concentration level and it causes serious health issues (Islam et al., 2018; Katz and Salem, 1993). Manganese (Mn) is a common element present in almost every soil type, and it is easily deposited at water bodies. Some manganese compounds are soluble and can suspend in air. Manganese will be present as suspended particles in the atmosphere because of soil erosion, industrial emissions, volcanic emissions, and burning of fossil fuels (IPCS, 1999). Forest fires, ocean spray, and volcanic activities are the major natural sources of manganese intrusion in water bodies (Howe et al., 2004; Corathers, 2009).

Iron (Fe) holds fourth place in Earth's crust and is essential for almost every living being, from microorganisms to humans and animals. Through many natural pathways, a large amount of iron is depositing in water bodies. Soil and rock erosion and underground streams are the major natural sources of iron pollution in water. Apart from that, increased human activities also lead to pollution (Ityel, 2011). Most of the cobalt (Co) found in Earth's crust is as chemically combined forms. Cobalt occurs naturally in the environment in soil, rocks, air, water, plants, and animals. Also, extra concentration may be increased through soil runoff into water bodies and deposition of suspended cobalt particles. Coal combustion and mining processes are also responsible for cobalt-induced pollution (Hawkins, 2001). Generally, nickel (Ni) is a nontoxic element but some nickel compounds show carcinogenic character (Greenwood and enEarnshaw, 2002). Power plants release large amounts of nickel particles into the air and they settle into the surface water and ground through raindrops. Wastewater streams also responsible for the intrusion of nickel into sea and water bodies (Gianguzza, 2002). Due to industrialization a large increase in copper (Cu) production has been seen over the last decades. Consequently copper concentration in water bodies also increased. As compared to other elements, variation in the concentration of copper is not as harmful, but may interfere with the intended uses of water. Urban sewage flow is one of the major reasons for higher copper concentration in water bodies (Emsley, 2011). Zinc (Sn) naturally occurs in water. Average zinc concentration in seawater is 0.6–5 ppb and in the case of rivers, it is 5–10 ppb. The major reason for zinc deposition in water bodies is aerial deposition. Approximately 11,000 to 60,000 metric tons of zinc particles fall from air to sea per year and 6000 tons from river flow (Jickells, 1995; Prasad, 2008).

Arsenic (As) is found in many minerals as a combination of sulfur and metals but also in pure crystalline form. Leaching and percolation of arsenic in soil and leaching of arsenic in ground water from sediments containing arsenic-containing minerals are the major natural sources of arsenic pollution. Arsenite (As (III)) is the most toxic form of arsenic and causes acute toxicity. Arsenic trioxide is widely used in the production of pesticide, herbicides, and insecticides. Apart from agrochemicals, wood preservatives, industrial sources, mineral processing, burning of fossil fuels, and such also cause the percolation of arsenic into water bodies (Ravenscroft et al., 2009). Selenium (Se) is found in elemental stage and it will occur naturally in the environment. E-wastes can be the primary source for the pollution caused by the selenium. Since selenium is one of the constituents in fertilizers, fertilized agroland soils will contain significant amounts of it. The amount of selenium in soil and water will increase due to wastewater runoff as well as settled down from air. Selenium exposure can occur through water or food (Xu, 2007).

Molybdenum (Mo) occupies various oxidation states in different minerals. Usually molybdenum compounds are less water-soluble materials, even though when it encounters air or water it will form a soluble molybdate ion. Coal combustion is the major source of molybdenum pollution. In the pigments industry and in the field of catalysis this element has significant applications. Cadmium (Cd) found in zinc ores and zinc production will deposit cadmium as a by-product. Since cadmium is water soluble, it will easily bioaccumulate in water bodies and living beings. Forest fire, volcanic emissions, and weathering of soil and rocks are the natural ways of cadmium percolation into water bodies. While considering anthropogenic reasons, the list will show metal smelting and refining, fuel combustion, paper production, metal processing, and wastewater treatment (Rieuwerts, 2017). Antimony (Sb) is a gray metalloid mainly found as sulfide mineral stibnite. Since it has no biological function in nature, it is considered a toxic substance. Rock weathering and soil run off are the only known natural sources for antimony and its intrusion into water bodies (Filella et al., 2002). Lead (Pb) occurs naturally in our environment, but most of it comes because of human activities. Natural sources like weathering of soil, forest fires, and volcanoes will deposit lead in nature. Still anthropogenic activities are higher than natural intrusions. Discharge of ammunition, leaded fuel in light aircraft, combustion of coal/wood, and metal production are the major human activities that increase the concentration of lead in atmosphere. Lead paints, lead water pipes, and leaded fuels also contribute to this. Household wastes, waste management, and sewage treatment plants also may contain a significant amount of lead particles. These will easily be transferred to marine systems and other water bodies. Since lead is not an essential element, it will cause harmful effects through intake (Verma and Dwivedi, 2013).

The advent of remote sensing techniques made a breakthrough in the research activities of oceans and coastal environments (Jones, 2014; Rejith et al., 2018). Advances in satellite sensor technology, production of satellite imageries with increased spatial and temporal resolutions, and improved software and hardware computations opened up new areas of applications for remote sensing and GIS in coastal environments (Davis et al., 2019). From the very early days, the Landsat multispectral remote sensing dataset has been successfully used for retrieving solid information regarding physical and chemical parameters of surface water bodies (Sydor, 1978; Ren et al., 2019). Usually the correlations between elemental concentrations and remote sensing reflectance are studied for deriving the relation that can be further used for developing predictive models for identifying potential heavy metal contamination in water bodies (Rostom et al., 2017). Remote sensing reflectance values are either derived from satellite data or measured in situ using spectral radiometers. Chen et al. (2012) derived the correlation of heavy metals with the remote sensing reflectance values derived from EO-1 Hyperion and the Landsat data. Swain and Sahoo (2017) mapped the heavy metals such using moderate-resolution imaging spectroradiometer and Landsat data.

The main objective of the present study is to determine the concentration of heavy metal pollutants in seawater and groundwater along the Kozhikode coast in Calicut district, Kerala, India. Heavy metals such as Be, V, Cr, Mn, Fe, Co, Ni, Cu, Zn, As, Se, Mo, Cd, Sb, and Pb were analyzed for the water samples. Further, the correlation coefficient (r) between the Landsat bands and heavy metal concentrations were also analyzed.

2. Study area

Calicut is one of the coastal districts in Kerala sharing boundaries with Kannur on the north, Wayanad on the east, Malappuram on the south, and Lakshadweep Sea on the west side. The study area is located at approximately11°19′9.02″ N to 11°13′52.87″ N latitudes and 75°44′43.52″ E to 75°46′40.17″ E longitudes (Fig. 10.1).The whole district comes under toposheets 58A/49M and the areal distribution is exactly 2344 sq km. According to census (2011), 3,089,543 is the total population. The average rainfall is 3063 mm per year. Climate conditions are classified into four seasons including summer, southwest tropical monsoon, northeast tropical monsoon, and winter. Physiographically there are three divisions in Calicut district, high land (>76 m amsl), mid-land (7.6–76 m amsl) and low land (‹ 7.6 m amsl). Alluvial, laterite, and forest loam are the three types of soil found in the district. Alluvial soil occupies riverbanks and coastal plain. Groundwater occurs in the weathered portions of crystalline and alluvial formations in the district (Salaj et al., 2018). The basement rocks were found to be crystalline so dug wells are suitable for this area. As per the records of Central Ground Water Board Department, government of India, published in the year 2013, the depth to groundwater level lies between 0.73 and

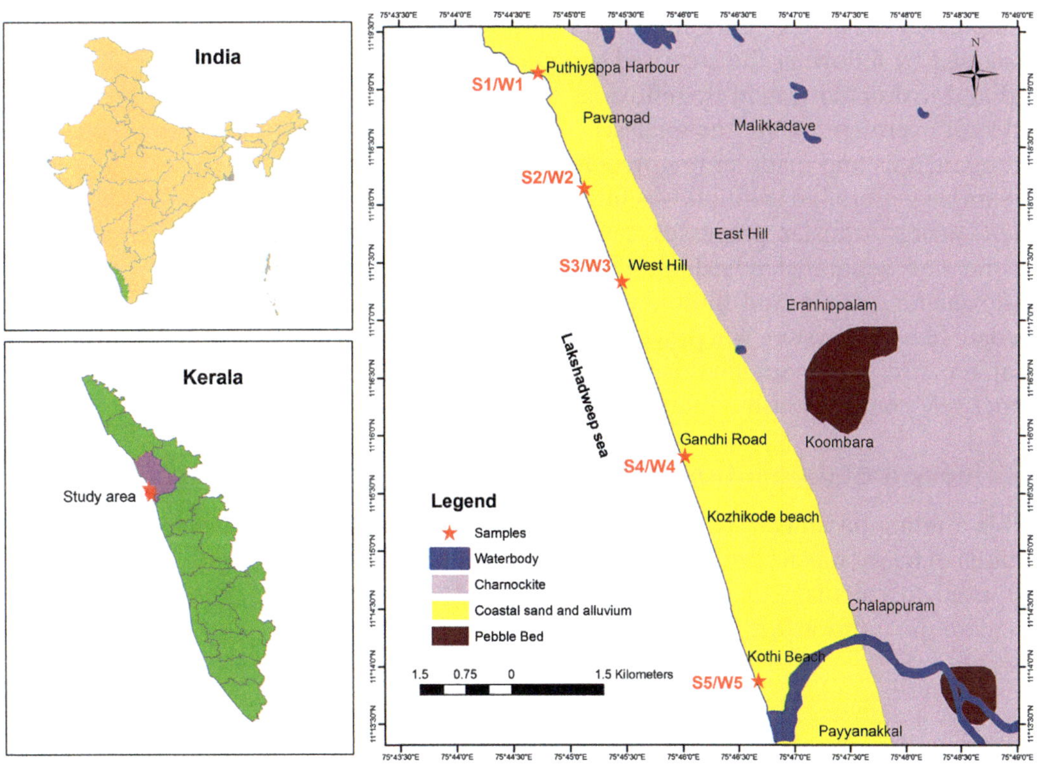

FIGURE 10.1 Study area.

16.11 m during premonsoon and 0.26–10.62 m in postmonsoon. The study area falls in the lowland stretch including Payyanakkal, Kothi beach, Calicut beach, Gandhi road, West hill, and Pavangad and Puthiyappa settlements. Narrow beach, estuaries, backwater tidal flats, and such comprise the study area limit. The coastal stretches have sandy, lateritic, weathered rock, and fractured rock types of aquifers.

3. Materials and methods

The proposed methodology adopted for the present study includes the following systematic procedures: (1) sampling and laboratory analysis, (2) heavy metal pollution index (HPI), and (3) principal component analysis (PCA) of Landsat imagery.

3.1 Sampling and laboratory analysis

High-density polyethylene bottles (100 mL capacity) were used for water sample collection. Bottles were rinsed three times with the sample and all samples were acidified with 1% nitric acid at the collection point itself. To avoid solid particles, all samples were filtered prior to filling and water was filled to the top of the sample bottle. A membrane filter of 0.1 μm was used to filter. Collected samples were stored at 4°C in the dark, which will improve the preservation time of the sample (it will last up to 4 weeks).

A portion of homogenous and thoroughly mixed sample is accurately weighed and processed by following the standard protocols. The sample was solubilized with nitric acid and hydrochloric acid by refluxing. The sample was made to the required volume and kept overnight before the analysis. The sample is treated with ultrapure nitric acid before analysis and made to the predetermined volume.

Sample material in solution was introduced by nebulization into a plasma generator, where energy transfer processes cause ionization. The ions are extracted from the plasma and separated based on their mass-to-charge ratio by a quadrupole mass spectrometer. An electron multiplier or Faraday detector detects ions transmitted and the data handling system will process the ion information. Instrumental drift caused by a sample matrix were corrected using internal standards. The Thermo Scientific iCAP RQ inductively coupled plasma mass spectrometer was used for analysis.

3.2 Heavy metal pollution index

HPI is a rating parameter, which gives the quality of water with respect to the concentrations of heavy metals present in the sample. Arithmetic average of the concentrations was used to calculate the overall HPI. Eq. (10.1) (Mohan et al., 1996) was used to calculate the index value.

$$\mathrm{HPI} = \frac{\sum_{i=1}^{n} W_i Q_i}{\sum_{i=0}^{n} W_i} \tag{10.1}$$

where

Si = maximum permissible limit for drinking water given by BIS (2012).

Wi = reciprocal value of Si

n = number of parameters considered

Qi = subindex of i'th parameter

$$Q_i = \sum_{i=0}^{n} \frac{Mi}{Si} \times 100 \tag{10.2}$$

Mi = monitored value of heavy metals

If the measured concentration of any metal exceeds its permissible value, then the quality of the water will be worse.

3.3 Principal component analysis of Landsat imagery

The Landsat 8 imagery of the study area acquired on 2019-04-05 was used for deriving the correlation between the heavy metal concentration and the reflectance values. Empirical models developed using reflectance calibrated Landsat imageries and water parameters have opened up new areas of satellite remote sensing applications in the characterization of surface water. González-Márquez et al. (2018) uses Landsat 8 for mapping the water quality parameters such as turbidity, phosphates, total suspended solids, and electrical conductivity of surface water. Hereher et al. (2010) uses Landsat (ETM+) for studying chlorophyll-a, ammonia nitrogen, temperature, and so on. The Landsat 8 imagery consists of 12 bands as shown in Table 10.1 in which the visible- near-infrared (NIR)- short-wave infrared (SWIR) bands such as bands 1–5 and band 7 were used for the present study. Reflectance calibration and atmospheric correction were carried out using fast line-of-sight atmospheric analysis of spectral hypercubes modules available in the software package of ENVI version 5.1 (Rejith and Sundararajan, 2019). PCA is one of the most widely used multivariate statistical techniques, which linearly

Table 10.1 Band designations of Landsat eight imagery.

Band no.	Bands	Wavelength(μm)	Resolution(m)
1	Coastal aerosol	0.43–0.45	30
2	Blue	0.45–0.51	30
3	Green	0.53–0.59	30
4	Red	0.64–0.67	30
5	Near infrared (NIR)	0.85–0.88	30
6	SWIR 1	1.57–1.65	30
7	SWIR 2	2.11–2.29	30
8	Panchromatic	0.50–0.68	15
9	Cirrus	1.36–1.38	30
10	Thermal infrared (TIRS) 1	10.6–11.19	100
11	Thermal infrared (TIRS) 2	11.50–12.51	100

transforms the correlated multivariate datasets to a set of uncorrelated image bands. The newly generated uncorrelated sets of image bands are called principle (PC) bands (Pour and Hashim, 2011). The highest data variance was observed in the first PC band and then the value decreases. The last PC band with low data variance and least information content can be removed from further steps, thereby reducing the spectral redundancy in multispectral and hyperspectral datasets (Chuvieco, 2016). Braga et al. (1993) uses the PCA of Landsat 5 bands for analyzing the water quality parameters and manganese and iron content. Batur and Maktav (2018) use PCA data fusion of Landsat 8 OLI, Sentinel 2A, and Göktürk-2 data for analyzing chlorophyll-a, pH, Secchi disk, dissolved oxygen, and so on of surface water.

4. Results and discussion

Collected seawater and groundwater were analyzed for heavy metal concentrations using inductively coupled plasma mass spectrometry (ICP-MS). Fe, Mn, Zn, Cu, Pb, Cr, Ni, V, Co, As, Mo, Se, Sb, Cd, and Be concentrations were obtained in ppb range.

4.1 ICPMS results

The ICPMS results of Fe, Mn, Zn, Cu, Pb, Cr, Ni, V, Co, As, Mo, Se, Sb, Cd, and Be are shown in Tables 10.2 and 10.3 and Fig. 10.2. Arithmetic mean was calculated for each heavy metal (Tables 10.4 and 10.5), and it follows an order Fe > Cu > Zn > Mo > V > Ni > Cr > Mn > Pb > As > Se > Co > Cd > Sb > Be in the case of seawater and Fe > Mn > Zn > Cu > Pb > Cr > Ni > V > Co > As > Mo > Se > Sb > Cd > Be in the case of

Table 10.2 ICPMS results for groundwater in ppb level.

Ground water	W1	W2	W3	W4	W5
Fe	82.572	63.031	173.362	62.443	143.163
Mn	32.013	35.463	10.426	27.475	11.222
Zn	27.588	17.028	12.899	14.606	16.939
Cu	3.814	2.399	3.142	1.985	1.224
Pb	2.858	1.155	0.841	0.838	0.693
Cr	2.228	2.040	2.070	1.892	1.866
Ni	2.151	2.174	1.615	2.116	1.555
V	1.080	0.615	2.267	0.965	2.343
Co	0.770	0.805	0.270	0.701	0.241
As	0.659	0.281	1.581	0.681	1.382
Mo	0.513	0.302	1.236	0.501	0.670
Se	0.300	0.258	0.300	0.251	0.410
Sb	0.217	0.134	0.143	0.125	0.218
Cd	0.121	0.041	0.023	0.036	0.036
Be	0.027	0.026	0.012	0.020	0.009

Table 10.3 ICPMS results for seawater in ppb level.

Sea water	S1	S2	S3	S4	S5
Fe	70.844	106.249	74.550	85.334	72.503
Cu	3.690	4.697	5.894	6.703	7.247
Zn	1.902	2.309	2.017	3.237	3.103
Mo	1.307	1.322	1.330	1.387	1.373
V	0.628	0.710	0.658	0.686	0.707
Ni	0.463	0.553	0.369	0.521	0.466
Cr	0.439	0.732	0.498	0.759	1.275
Mn	0.386	0.621	0.392	0.669	0.509
Pb	0.277	0.177	0.255	0.387	0.306
As	0.135	0.155	0.133	0.137	0.165
Se	0.060	0.101	0.059	0.099	0.229
Co	0.050	0.073	0.032	0.049	0.036
Cd	0.036	0.023	0.020	0.043	0.039
Sb	0.019	0.021	0.018	0.027	0.027
Be	0.012	0.014	0.012	0.011	0.005

groundwater. In both cases iron occupies the first position and beryllium occupies the last, while all other metals are showing variations. Now let us discuss each metal separately—their sources, chances of pollution, and consequences.

In water, chemicals will react with beryllium and this results in an insoluble form. Thus dissolution of beryllium contents will be less. That is why in almost every location beryllium level is very low. In seawater as well as groundwater, beryllium is the least concentrated element. In case of groundwater, locations follows the order W1 > W2 >W4 >W3 >W5, while in seawater S2 >S1≥S3 >S4 >S5 is the concentration trend. Beryllium pollution is less studied all over the world. Since these selected locations have some small-scale industries as well as ports, there is a chance of oil and chemical spills in the sea. This may cause the increase in beryllium concentrations. However, in our case study, the beryllium concentration is not complicated. Locations 1 and 2 are industrial areas, with some small-scale industries as well as S1 having a mini harbor along with the sea.

In our case study, concentration of vanadium in ground water follows the order W5>W3>W1> W4>W2 and it is S5> S2> S4> S3> S1 in seawater. Highest concentration comes in location five. While considering the geography of that location, we found that there is a riverfront opening to the sea. It carries sewage water as well as domestic/industrial wastes. Comparing with all other locations, this area was the most polluted one, and as we discussed above, urban sewage is the main human-made source of vanadium into the sea. The effect of this pollution is evident in the groundwater result too.

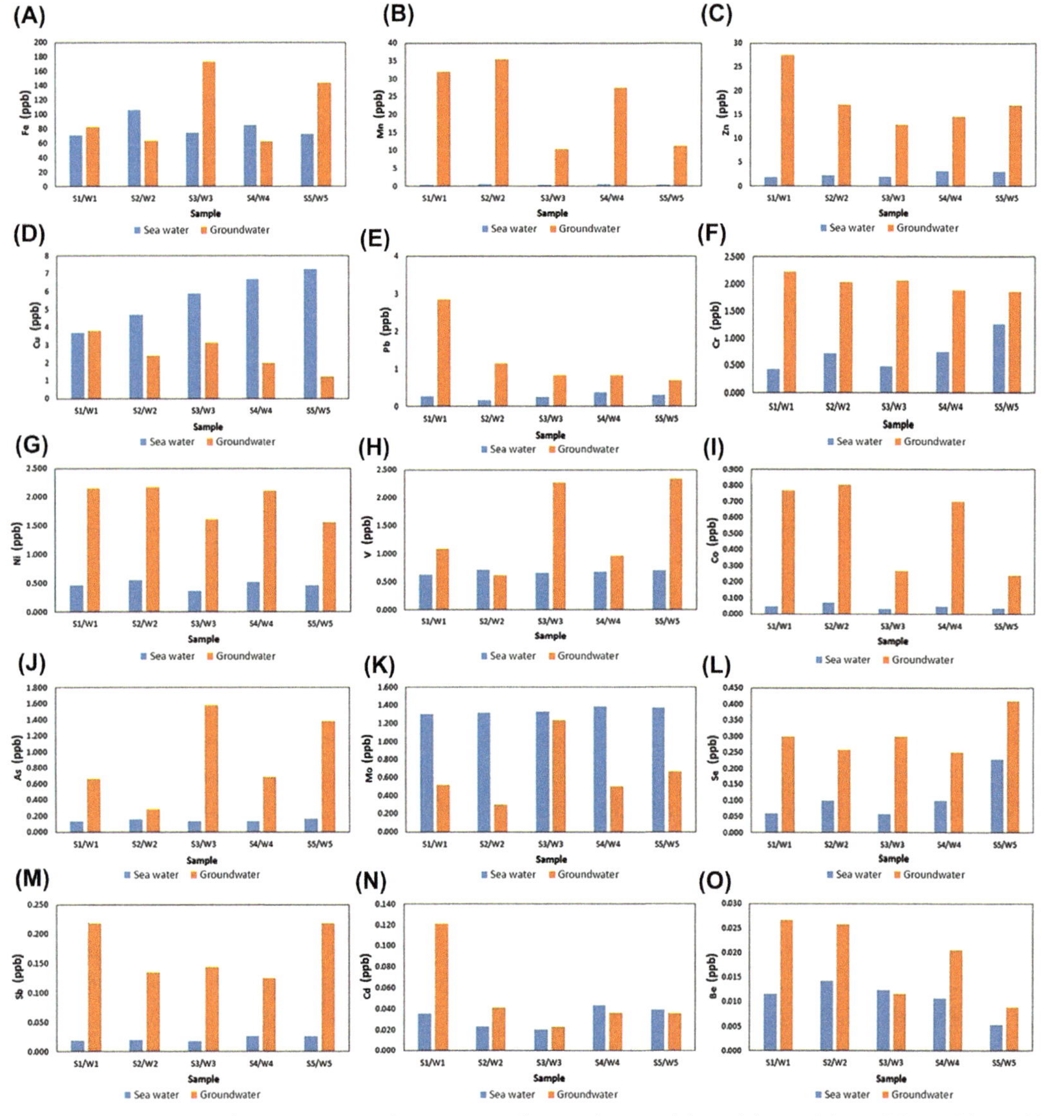

FIGURE 10.2 Heavy metal concentration of seawater and groundwater. (A) Fe; (B) Mn; (C) Zn; (D) Cu; (E) Pb; (F) Cr; (G) Ni; (H) V; (I) Co; (J) As; (K) Mo; (L) Se; (M) Sb; (N) Cd; (O) Be.

S5> S4> S2> S3> S1 and W1> W3> W2> W4> W5 are the results we got for chromium from ICPMS data. In the case of chromium also, the final location S5 has a high concentration in seawater. This may due to the same reason, of the presence of the riverfront opening into the sea. Rather than a riverfront, it has become a sewage pathway from urban areas. The desirable limit of chromium in drinking water is 50 ppb. In the

Table 10.4 Statistical measures such as minimum, maximum, average, and standard deviation of the seawater results in ppb.

	Seawater			
Elements	**Max**	**Min**	**Average**	**SD**
Fe	106.249	70.844	81.896	14.741
Cu	7.247	3.690	5.646	1.455
Zn	3.237	1.902	2.513	0.619
Mo	1.387	1.307	1.344	0.034
V	0.710	0.628	0.678	0.035
Ni	0.553	0.369	0.474	0.070
Cr	1.275	0.439	0.741	0.330
Mn	0.669	0.386	0.516	0.129
Pb	0.387	0.177	0.280	0.076
As	0.165	0.133	0.145	0.014
se	0.229	0.059	0.109	0.070
Co	0.073	0.032	0.048	0.016
Cd	0.043	0.020	0.032	0.010
Sb	0.027	0.018	0.022	0.004
Be	0.014	0.005	0.011	0.003

Table 10.5 Statistical measures such as minimum, maximum, average, and standard deviation of the groundwater results in ppb.

	Groundwater			
Elements	**Max**	**Min**	**Average**	**SD**
Fe	173.362	62.443	104.914	50.511
Mn	35.463	10.426	23.320	11.757
Zn	27.588	12.899	17.812	5.731
Cu	3.814	1.224	2.513	1.005
Pb	2.858	0.693	1.277	0.900
Cr	2.228	1.866	2.019	0.147
Ni	2.174	1.555	1.922	0.309
V	2.343	0.615	1.454	0.796
Co	0.805	0.241	0.557	0.278
As	1.581	0.281	0.917	0.544
Mo	1.236	0.302	0.645	0.356
se	0.410	0.251	0.304	0.064
Sb	0.218	0.125	0.168	0.046
Cd	0.121	0.023	0.051	0.040
Be	0.027	0.009	0.019	0.008

samples, no one exceeds the limit and is actually far below values. Regarding chromium, groundwater in those areas are safe to drink. As seen in Tables 10.1 and 10.2, manganese concentrations follow the pattern S4 > S2 > S5 > S3 > S1 and W2 > W1 > W4 > W5 > W3. Manganese is the second highest concentrated element in collected groundwater samples. Since it is a common element in nature, through soil erosion it will mix with groundwater streams. The maximum limit according to BIS standards is 100 ppb. The highest value we got for groundwater in our study area is 35.46 ppb. It is safe to drink concerning manganese. The average value we got for manganese is 0.516 ppb in seawater. Robin et al. (2012) reported manganese concentrations from Cochin, Allepy, and Kayamkulam coastal zones are 6.61, 7.51, and 6.17 ppb, respectively. Compared to those values, the values in Calicut are very low.

Since iron is the most used metal around us, there is no doubt regarding the first position that iron occupied in our studies. Iron is a highly concentrated metal in both seawater and ground water in our study area. W3 > W5 > W1 > W2 > W4 and S2 > S4 > S3 > S5 > S1 are the orders of test results. The highest iron concentration we got from the W3 location was 173.36 ppb. Water is safe for drinking with iron concentration up to 300 ppb level. In seawater studies, S2 shows the highest value. This location is highly populated among the selected locations, and has large number of housing colonies as well as apartments. Because their best and only option for waste disposal is the nearby sea, it may cause variation on metal concentrations in the marine environment.

S2 > S1 > S4 > S3 > S5 and W2 > W1 > W4 > W3 > W5 are the location strategy according to cobalt concentration resulting from the ICPMS test. Overall, in the case of groundwater, cobalt occupies the 9th position and the 12th position in seawater results. Since in both cases concentrations are below 1 ppb, we can conclude that there is no trouble with cobalt metal around the study area. However, the study area has many small-scale industries including paint and glass. The proper waste disposal is essential for avoiding future problems.

The average concentration of nickel is 1.922 and 0.474 ppb in groundwater and seawater, respectively. In both the cases, location 2 showed maximum concentration. Since up to 20 ppb of nickel in groundwater is safe to drink, there are no worries related to the obtained results. Results are much lower than the limiting value. From Udayakumar et al. (2011), it is clear that in the southwest coastal zone nickel can have values in between 0.56 and 13.83 ppb.

In seawater analysis, copper occupies the second most concentrated metal position. The maximum obtained value is 7.274 ppb at location 5. It shows a regular order that in going from location 5 to 1, the values are gradually decreasing. Since urban sewage flow is the major polluting agent, the downstream present at location 5 may significantly influence this pattern. The average copper concentration for seawater obtained is around 5.646, compared with the value reported by Robin et al. (2012) for Cochin, Allepy, and Kayamkulam, which are lower than this. However, the total average (0.18–6.2 ppb) they reported will occupy our value ranges. The reported values for copper in the southwest coast by Udayakumar et al. (2011) is 3.88–13.06 ppb. Groundwater has good quality according to copper concentration; 3.814 ppb is the largest concentration we got. It can be up to 50 ppb in drinking water (Bureau of Indian Standards-BIS).

From the results, 2.513 ppb is the average zinc concentration in seawater; 3.237 ppb is the highest value we got for zinc, which falls under the low zinc category when compared to the other coastal zones. Thus it does not come under the zinc-polluted zone. The average value reported for the southwest coastal zone is 0.89–17.18 ppb (Robin et al., 2012). According to Udayakumar et al. (2011), 3.93–35.58 ppb is the range for zinc in the southwest coast. So in our study area pollution caused by zinc is negligibly low. In case of groundwater, 17.812 ppb is the highest obtained value. Since it is an essential element, the amount of zinc in the groundwater is not above its dangerous limit.

Since arsenic can cause dangerous toxic effects (above some certain level), we should give a prime level concern regarding this metal. The maximum permissible limit in drinking water is 10 ppb. Above that it will cause serious health issues. In our study, the average arsenic value obtained is 0.917 ppb. It is very low compared to the limit. In seawater arsenic also shows very low concentration. In the present study, no location attained above 2 ppb arsenic concentration.

While considering selenium levels, locations shows a trend of S5 $>$ S4 $>$ S1 $>$ S3 $>$ S2 in seawater and W5 $>$W1 $\geq$ W3$>$ W2$>$ W4 in groundwater. The desirable limit for drinking purposes is 10 ppb per liter, and in the present study the values show beyond the limit. The S5 seawater sample is the highest among the five locations. This also accounted for the presence of downstream present in that area. It will carry most of the agricultural runoff and it will definitely contain significant amount of selenium. The values 0.645 ppb (groundwater) and 1.344 ppb (seawater) are the average molybdenum concentrations obtained from the results. The limit for drinking water is 70 ppb and is safe here. The low concentration can be explained concerning the low solubility in water exhibiting by the molybdenum compounds. It won't easily dissolve in water.

In both types of water sample studies, cadmium occupies the last positions in the table. The average values are less than 1 ppb. While considering the maximum limit of 3 ppb, the obtained values are safe. Since most of the wastewater streams ended in the marine area, we should be very careful regarding these kinds of elements having a small range of permissible limits. Wastewater runoffs are the major source of cadmium type metals into the sea. The average concentrations obtained for cadmium in seawater is 0.032 ppb. Compared to the reported values for Cochin and Alappuzha (Robin et al., 2012), it was 0.17 and 0.12 ppb, respectively, and the cadmium concentration in the selected study is very low. When Udayakumar et al. (2011) studied seasonal changes of dissolved metals in surface coastal waters of southwest India, they covered almost all the

Kerala coastal zones. Thus obtained values for cadmium fall between 0.10 and 4.60 ppb. The obtained values are very low for antimony compared to others. Since this metal has no significant biological activity in the living species, even the smaller values are not essential in the drinking water. In our results, there is a contradiction that the seawater results are lower than the antimony values in the groundwater. This may due to soil runoff or weathering. Antimony groups among the heavy metals are not well discussed in terms of pollution and safety concerns.

W1> W2> W3> W4> W5 and S4> S5> S1> S3> S2 are the orders obtained for lead from the results. Lead is not at all acceptable in drinking water, even though some amount of lead will be there due to improper waste management. In seawater, no value exceeds 0.3 ppb but in the case of groundwater it is larger compared to seawater. The highest concentration is 2.858 ppb (groundwater), and it belongs to location 1. But the maximum permissible limit is 10 ppb in drinking water. This area includes the largest port among the whole study area, and it is always occupied with motor boats. The fuel from the boats contain significant amounts of lead. While considering the Cochin coastal zones, the reported value is 1.99 ppb (Robin et al., 2012), which is very large compared to the present seawater results. It may be because the Calicut coast is not well established as an industrial giant compared to Cochin. As per Udayakumar et al. (2011), the southwest coastal zone has an average lead concentration between 0.05 and 5.33 ppb.

4.2 Heavy metal pollution index

Obtained values, both seawater as well as groundwater, have been compared as each location can separately compare with the standard values. Statistical analyses were done with the help of SPSS software (Table 10.6). HPI calculation was helpful to analyze the quality of the groundwater available in those locations. The calculated HPI value is 5.459 and comparing with the standard HPI characterization values given in Table 10.7, it falls in the low HPI category. That means the heavy metal intrusion and consequent pollution is very low in the selected location zones. It can be concluded that there is not much influence of the sea on groundwater streams at those locations.

4.3 Correlation coefficient

Pearson's correlation coefficient matrices for obtained metal concentrations are given in Table 10.8 (seawater) and Table 10.9 (groundwater). In seawater, correlation between Ni and Fe, and Mn and Fe, may possibly have the same source of contamination with coefficients of 0.739 and 0.722, respectively. Cr–Cu (0.735), Sb–Cu (0.757), As–Cr (0.873) and Sb–Cr (0.812) show significant correlation in seawater analysis. All are significant at the level 0.05. In the case of groundwater, Fe–V (0.957) and Mo–Fe (0.903) show significant correlation coefficients. This may be the result of same pollution source. Rather than this pair, Cd–Mn (0.521), Pb–Mn (0.547), and Zn–Mn (0.505) show the best

Table 10.6 Heavy metal Pollution Index (HPI) calculations for groundwater based on mean heavy metal concentration.

			Groundwater		
Metals	**Si μg/L**	**Wi = 1/Si**	**Mi μg/L**	**Qi = Mi/Si*100**	**Wi*Qi**
Cr	50	0.020	2.019	4.038	0.081
Mn	100	0.010	23.32	23.320	0.233
Fe	300	0.003	104.914	34.971	0.117
Ni	20	0.050	1.922	9.610	0.481
Cu	50	0.020	2.513	5.026	0.101
Zn	5000	0.000	17.812	0.356	0.000
As	10	0.100	0.917	9.170	0.917
Se	10	0.100	0.304	3.040	0.304
Mo	70	0.014	0.645	0.921	0.013
Pb	10	0.1	1.277	12.77	1.277
Cd	3	0.333	0.051	1.700	0.567
Co	10	0.1	0.557	5.57	0.557
Sb	—	—	—	—	—
Be	—	—	—	—	—
V	—	—	—	—	—
		Σwi = 0.851			Σwi*Qi = 4.646
HPI for groundwater = 5.459					

Table 10.7 Categorization of heavy metal pollution index.

HPI	Class
<15	Low
15–30	Medium
>30	High

correlations in groundwater. In both cases, the least correlations between metals are also spotted. This indicates different sources of pollution in each case. Compared to seawater, groundwater correlations are less. This may be because groundwater may have various pollution sources in each location.

4.4 Monitoring heavy metal concentration using Landsat imagery

Reflectance calibrated FCC images of Landsat data are shown in Fig. 10.3. The Landsat 8 data is used to retrieve and map heavy metal concentrations in seawater in the study areas. Each element falls under different bands and wavelength regions. Tables 10.10 and 10.11 show the correlation between heavy metal concentration and Landsat data. Among

Table 10.8 Correlation coefficients for parameters of seawater samples.

	Fe	Cu	Zn	Mo	V	Ni	Cr	Mn	Pb	As	Se	Co	Cd	Sb	Be
Fe	1														
Cu	−0.172	1													
Zn	0.065	0.826	1												
Mo	−0.073	0.907*	0.965**	1											
V	0.602	0.580	0.661	0.546	1										
Ni	0.739	−0.104	0.415	0.200	0.546	1									
Cr	−0.003	0.735	0.777	0.683	0.770	0.267	1								
Mn	0.722	0.365	0.707	0.597	0.744	0.828	0.396	1							
Pb	−0.481	0.567	0.677	0.768	−0.097	−0.031	0.235	0.224	1						
As	0.271	0.371	0.455	0.283	0.798	0.407	0.873	0.337	−0.238	1					
Se	−0.109	0.671	0.694	0.598	0.680	0.190	0.987**	0.260	0.200	0.880*	1				
Co	0.867	−0.533	−0.119	−0.319	0.305	0.839	−0.157	0.553	−0.486	0.191	−0.212	1			
Cd	−0.340	0.349	0.695	0.641	0.031	0.340	0.428	0.317	0.831	0.115	0.426	−0.151	1		
Sb	0.033	0.757	0.986**	0.920*	0.636	0.467	0.812	0.675	0.661	0.519	0.747	−0.079	0.766	1	
Be	0.550	−0.711	−0.621	−0.618	−0.312	0.177	−0.836	0.058	−0.471	−0.576	−0.883*	0.606	−0.554	−0.668	1

*Correlation is significant at the 0.05 level (two tailed).

**Correlation is significant at the 0.01 level (two tailed).

Table 10.9 Correlation coefficients for parameters of groundwater samples.

	Fe	Mn	Zn	Cu	Pb	Cr	Ni	V	Co	As	Mo	Se	Sb	Cd	Be
Fe	1														
Mn	−0.937*	1													
Zn	−0.362	0.505	1												
Cu	−0.040	0.322	0.544	1											
Pb	−0.351	0.547	0.951*	0.769	1										
Cr	−0.085	0.396	0.688	0.958*	0.853	1									
Ni	−0.944*	0.980**	0.458	0.359	0.534	0.380	1								
V	0.957*	−0.981**	−0.340	−0.240	−0.401	−0.274	−0.983**	1							
Co	−0.942*	0.992**	0.488	0.357	0.552	0.398	0.997**	−0.985**	1						
As	0.961**	−0.982**	−0.396	−0.146	−0.402	−0.233	−0.951*	0.981**	−0.965**	1					
Mo	0.903*	−0.827	−0.428	0.198	−0.294	0.047	−0.762	0.804	−0.787	0.892*	1				
Se	0.626	−0.683	0.064	−0.434	−0.164	−0.308	−0.768	0.775	−0.738	0.646	0.268	1			
Sb	0.253	−0.206	0.683	0.055	0.495	0.224	−0.286	0.376	−0.249	0.257	−0.033	0.770	1		
Cd	−0.376	0.521	0.985**	0.642	0.981**	0.741	0.506	−0.370	0.524	−0.393	−0.366	−0.050	0.599	1	
Be	−0.840	0.970**	0.606	0.539	0.694	0.602	0.958*	−0.927*	0.970**	−0.910*	−0.693	−0.698	−0.143	0.642	1

*Correlation is significant at the 0.05 level (two tailed).

**Correlation is significant at the 0.01 level (two tailed).

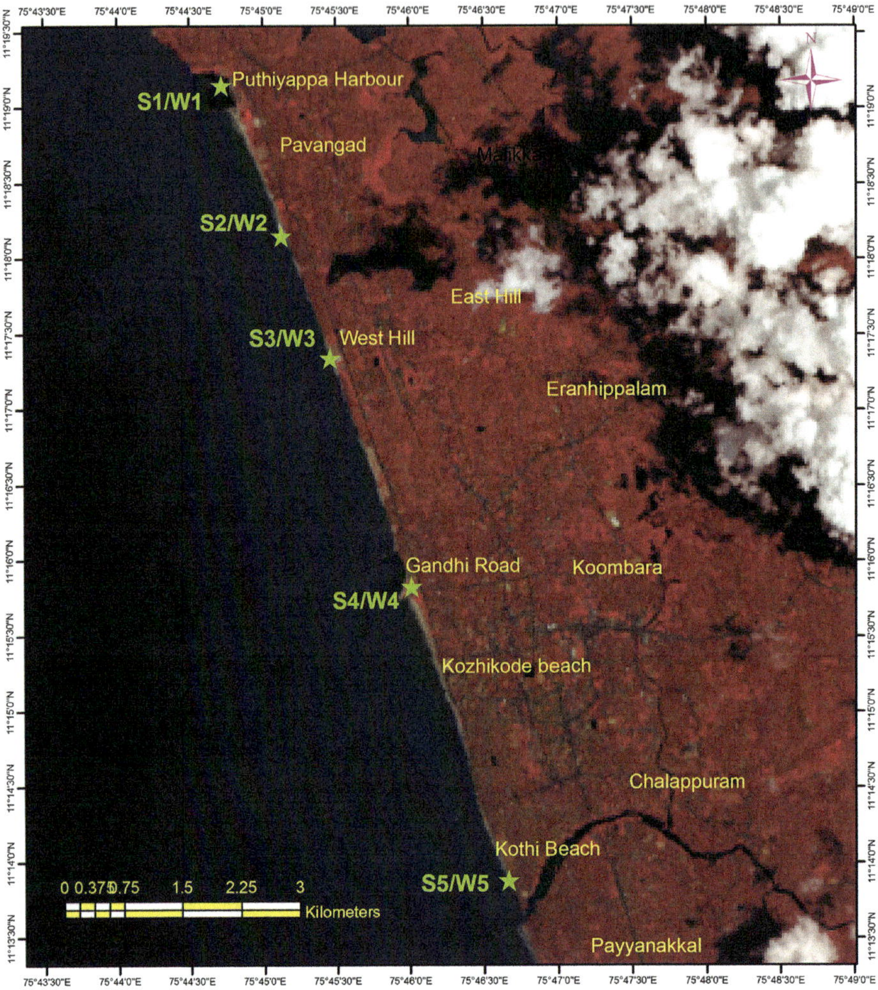

FIGURE 10.3 Reflectance calibrated FCC image of Landsat data.

all the heavy metals Co/band 5 (Fig. 10.4) and Cu/PCB4 (Fig. 10.5) show the best correlations. The elements Fe, V, Ni, Mn, and Co have the best correlation with the B5 band (NIR region) while Cu, Zn, Mo, Pb, Cd, and Sb show the best correlation with the B1 (coastal aerosol) band, and Cr, Se, Be in B7 (SWIR 2) and As with the B4 (red) band. The metal and its best correlation bands are given in Tables 10.7 and 10.8. In the case of principle component bands, Fe, Ni, and Co fall in PCB1 and Cu, Cr, As, Se, Cd, Sb, and Be show the best performance with PCB4. PCB2 has Zn, V, and Mn while PCB3 has only two metals in best correlation with Mo and Pb.

Table 10.10 Correlation between heavy metal concentration and Landsat data.

Heavy metals	Landsat bands	R2
Fe	B5	0.5821
Cu	B1	0.795
Zn	B1	0.323
Mo	B1	0.5675
V	B5	0.0934
Ni	B5	0.8432
Cr	B7	0.1644
Mn	B5	0.3499
Pb	B1	0.3602
As	B4	0.1007
Se	B7	0.1748
Co	B5	0.9353
Cd	B1	0.0286
Sb	B1	0.2041
Be	B7	0.4634

Table 10.11 Correlation between heavy metal concentration and PC bands of Landsat data.

Heavy metals	PC bands	R2
Fe	PCB1	0.2414
Cu	PCB4	0.8894
Zn	PCB2	0.5979
Mo	PCB3	0.7834
V	PCB2	0.4168
Ni	PCB1	0.4482
Cr	PCB4	0.5466
Mn	PCB2	0.6765
Pb	PCB3	0.4777
As	PCB4	0.1153
Se	PCB4	0.5207
Co	PCB1	0.7195
Cd	PCB4	0.2538
Sb	PCB4	0.5473
Be	PCB4	0.7609

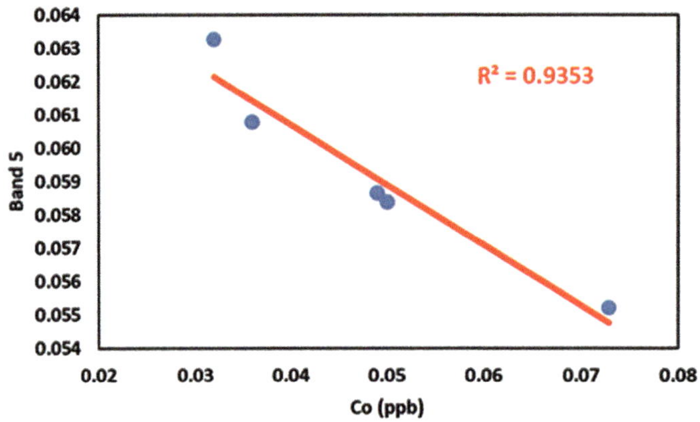

FIGURE 10.4 Correlation between band 5 and Co concentration.

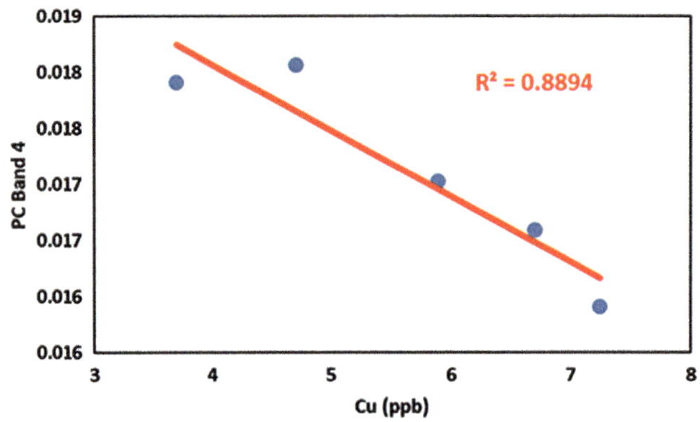

FIGURE 10.5 Correlation between PC band 4 and Cu concentration.

5. Conclusion

From the results, it is clear that the study area is not as polluted as other industrialized and well-developed areas along the coasts of Kerala. Statistical calculations also account for this conclusion. It shows that the strong correlation between two heavy metals may be due to the same kind of pollution-causing factors. The correlations with the spectral bands of Landsat remote sensing data help to identify the potential sites of heavy metal contamination, and this relation can be used for further monitoring of heavy metal pollutants in the future.

Acknowledgments

The authors thank the Director, CSIR-National Institute for Interdisciplinary Science and Technology, Trivandrum, Kerala for extending the laboratory facilities.

References

Barceloux, D.G., Barceloux, D., 1999. Chromium. Clin. Toxicol. 37 (2), 173–194.

Baroch, E.F., 2006. Vanadium and vanadium alloys. In: Kirk-Othmer Encyclopedia of Chemical Technology. John Wiley & Sons, Inc. May 12, 2009.

Batur, E., Maktav, D., 2018. Assessment of surface water quality by using satellite images fusion based on PCA method in the Lake Gala, Turkey. IEEE Trans. Geosci. Remote Sens. 57 (5), 2983–2989.

Byerrum, R.U., Eckardt, R.E., Hopkins, L.L., et al., 1974. Vanadium. National Academy of Sciences, Washington, DC, pp. 19–45.

Braga, C.Z.F., Setzer, A.W., de Lacerda, L.D., 1993. Water quality assessment with simultaneous Landsat-5 TM data at Guanabara Bay, Rio de Janeiro, Brazil. Remote Sens. Environ. 45 (1), 95–106.

Byerrum, R.U., 1991. Vanadium. In: Merian, E. (Ed.), Metals and Their Compounds in the Environment. VCH, Weinheim, Germany, pp. 1289–1297.

Chen, C., Liu, F., Tang, S., July 2012. Estimation of heavy metal concentration in the pearl river estuarine waters from remote sensing data. In: 2012 IEEE International Geoscience and Remote Sensing Symposium. IEEE, pp. 2575–2578.

Chuvieco, E., 2016. Fundamentals of Satellite Remote Sensing: An Environmental Approach. CRC press.

Corathers, L.A., 2009. Mineral Commodity Summaries 2009: Manganese. United States Geological Survey. Retrieved 30 April 2009.

Davis, E., Wang, C., Dow, K., 2019. Comparing Sentinel-2 MSI and Landsat 8 OLI in soil salinity detection: a case study of agricultural lands in coastal North Carolina. Int. J. Remote Sens. 40 (16), 6134–6153.

Emsley, J., 2011. Nature's Building Blocks: An AZ Guide to the Elements. Oxford University Press.

Filella, M., Belzile, N., Chen, Y.W., 2002. Antimony in the environment: a review focused on natural waters: I. Occurrence. Earth Sci. Rev. 57 (1–2), 125–176.

Gianguzza, A., Pelizzetti, E., Sammartano, S. (Eds.), 2002. Chemistry of Marine Water and Sediments. Springer, Berlin.

González-Márquez, L.C., Torres-Bejarano, F.M., Rodríguez-Cuevas, C., Torregroza-Espinosa, A.C., Sandoval-Romero, J.A., 2018. Estimation of water quality parameters using Landsat 8 images: application to Playa Colorada Bay, Sinaloa, Mexico. Appl. Geomat. 10 (2), 147–158.

Greenwood, N.N., enEarnshaw, A., 2002. Chemistry of the Elements, second ed. Butterworth-Heinemann, Oxford.

Hawkins, M., 2001. Why we need cobalt. B. Appl. Earth Sci. 110 (2), 66–70.

Hereher, M., Salem, M., Darwish, D., 2010. Mapping water quality of Burullus Lagoon using remote sensing and geographic information system. J. Am. Sci. 7 (1), 138–143.

Howe, P., Malcolm, H., Dobson, S., 2004. Manganese and its Compounds: Environmental Aspects. World Health Organization.

IPCS, W., 1999. Environmental Health Criteria 210-Principles for the Assessment of Risks to Human Health from Exposure to Chemicals. Geneva WHO Available. http://www.inchem.org/documents/ehc/ehc/ehc210.htm (Accessed 17 January 2016).

Islam, M., Karim, M., Zheng, X., Li, X., 2018. Heavy metal and metalloid pollution of soil, water and foods in Bangladesh: a critical review. Int. J. Environ. Res. Publ. Health 15 (12), 2825.

Ityel, D., 2011. Ground water: dealing with iron contamination. Filtrat. Separ. 48 (1), 26–28.

Jickells, T., 1995. Atmospheric inputs of metals and nutrients to the oceans: their magnitude and effects. Mar. Chem. 48 (3–4), 199–214.

Jones, R., 2014. Quantifying extreme weather event impacts on the Northern Gulf Coast using landsat imagery. J. Coast Res. 31 (5), 1229–1240.

Katz, S.A., Salem, H., 1993. The toxicology of chromium with respect to its chemical speciation: a review. J. Appl. Toxicol. 13 (3), 217–224.

Lide, D.R., 2008. CRC Handbook of Chemistry and Physics, eighty eighth ed. FL: CRC Press, Boca Raton. 4-40, 4-90, 4-92, 4-98.

Mohan, S.V., Nithila, P., Reddy, S.J., 1996. Estimation of heavy metals in drinking water and development of heavy metal pollution index. J. Environ. Sci. & Health Part A 31 (2), 283–289.

Pour, A.B., Hashim, M., 2011. Spectral transformation of ASTER data and the discrimination of hydrothermal alteration minerals in a semi-arid region, SE Iran. Int. J. Phys. Sci. 6 (8), 2037–2059.

Prasad, A.S., 2008. Zinc in human health: effect of zinc on immune cells. Mol. Med. 14 (5), 353–357.

Ravenscroft, P., Brammer, H., Richards, K., 2009. Arsenic Pollution: A Global Synthesis, vol. 28. John Wiley & Sons.

Rejith, R.G., Sundararajan, M., 2019. Mapping of mineral resources and lithological units: a review of remote sensing techniques. Int. J. Image & Data Fusion 10 (2), 79–106.

Rejith, R.G., Sundararajan, M., Kaliraj, S., 2018. A geo-processing modeling of deltaic suspended sediment variability. J. Geovisualization & Spat. Anal. 2 (2), 13.

Ren, C., Wang, Z., Zhang, Y., Zhang, B., Chen, L., Xi, Y., et al., 2019. Rapid expansion of coastal aquaculture ponds in China from Landsat observations during 1984–2016. Int. J. Appl. Earth Obs. Geoinf. 82, 101902.

Ridgway, K.R., Dunn, J.R., Wilkin, J.L., 2002. Ocean interpolation by four-dimensional weighted least squares—application to the waters around Australasia. J. Atmos. Ocean. Technol. 19 (9), 1357–1375.

Rieuwerts, J., 2017. The Elements of Environmental Pollution. Routledge.

Robin, R.S., Muduli, P.R., Vishnu Vardhan, K., Ganguly, D., Abhilash, K.R., Balasubramanian, T., 2012. Heavy metal contamination and risk assessment in the marine environment of Arabian Sea, along the Southwest Coast of India. Am. J. Chem. 2 (4), 191–208.

Rostom, N.G., Shalaby, A.A., Issa, Y.M., Afifi, A.A., 2017. Evaluation of Mariut lake water quality using hyperspectral remote sensing and laboratory works. Egypt. J. Remote Sens. & Space Sci. 20, S39–S48.

Salaj, S.S., Ramesh, D., Babu, D.S., Kaliraj, S., 2018. Impacts of Urbanization on Groundwater Vulnerability along the Kozhikode Coastal Stretch, Southwestern India Using GIS Based Modified DRASTIC-U Model.

Swain, R., Sahoo, B., 2017. Mapping of heavy metal pollution in river water at daily time-scale using spatio-temporal fusion of MODIS-aqua and Landsat satellite imageries. J. Environ. Manag. 192, 1–14.

Sydor, M., 1978. Analysis OF suspended solids IN lakes using landsat multispectral data. Can. J. Spectrosc. 23 (3), 91–97.

Udayakumar, P., Chandran, A., Jose, J.J., Rajesh, B.R., Babu, K.N., Ouseph, P.P., 2011. Seasonal dynamics of dissolved metals in surface coastal waters of Southwest India. B. Environ. Contam. Tox. 87 (6), 662–668.

Verma, R., Dwivedi, P., 2013. Heavy metal water pollution-A case study. Recent Res. Sci. & Technol. 5 (5).

Voica, C., Kovacs, M.H., Dehelean, A., Ristoiu, D., Iordache, A., 2012. ICP-MS determinations of heavy metals in surface waters from Transylvania. Rom. J. Phys. 57 (6–7), 1184–1193.

Walsh, K.A., 2009. In: Vidal, E.E., et al. (Eds.), Beryllium Chemistry and Processing. ASM International, Materials Park, OH.

Xu, Z., 2007. In: Devillanova, F.A. (Ed.), Handbook of Chalcogen Chemistry: New Perspectives in Sulfur, Selenium and Tellurium. Royal Society of Chemistry, ISBN 978-0-85404-366-8, p. 460.

11

Remote sensing for exploring heavy mineral deposits: a case study of Chavara and Manavalakurichi deposits, southwest coast of India

R.G. Rejith[1,2], M. Sundararajan[1,2], S. Venkatesan[3], M.A. Mohammed-Aslam[4]

[1]*MINERALS SECTION, MATERIALS SCIENCE AND TECHNOLOGY DIVISION, NATIONAL INSTITUTE FOR INTERDISCIPLINARY SCIENCE AND TECHNOLOGY (CSIR-NIIST), COUNCIL OF SCIENTIFIC & INDUSTRIAL RESEARCH, THIRUVANANTHAPURAM, KERALA, INDIA;* [2]*ACADEMY OF SCIENTIFIC AND INNOVATIVE RESEARCH (AcSIR), GHAZIABAD, UTTAR PRADESH, INDIA;* [3]*DEPARTMENT OF GEOLOGY, NATIONAL COLLEGE, TRICHY, TAMIL NADU, INDIA;* [4]*DEPARTMENT OF GEOLOGY, SCHOOL OF EARTH SCIENCES, CENTRAL UNIVERSITY OF KARNATAKA, GULBARGA, KARNATAKA, INDIA*

Chapter outline

1. Introduction

The major remote sensing applications in geology and mineral exploration include mapping of faults, fractures, and lithological units associated with ore deposit and prospecting potential zones of mineral occurrences using spectral signatures of minerals (Rejith and Sundararajan, 2019). The advent of multispectral satellite datasets like the Landsat series made a breakthrough in the field of geological remote sensing and mineral exploration. Since 1972, remote sensing data products of the Landsat series have

Remote Sensing of Ocean and Coastal Environments. https://doi.org/10.1016/B978-0-12-819604-5.00011-1

been used for deriving solid information regarding occurrences of mineral deposits on Earth's surface (Albert, 1976; Nath et al., 2019). Landsat data were successfully used for mapping alteration minerals (Banerjee et al., 2019; Eldosouky et al., 2017; Liu et al., 2017), Li-bearing pegmatites (Cardoso-Fernandes et al., 2019), iron ore deposits (Ghrefat et al., 2018), gold exploration (Yousefi et al., 2018), Zn–Pb sulfide mineralization (Beiranvand Pour et al., 2018), and so on. The advanced multispectral dataset like Advanced Spaceborne Thermal Emission and Reflection Radiometer also successfully delineate the occurrences of gold sulfide mineralizations (Rani et al., 2019), alteration minerals (Fereydooni and Mojeddifar, 2017), silicate and carbonate rocks (Kurata and Yamaguchi, 2019), mafic and ultramafic rock units (Rejith and Sundararajan, 2018b), limestone deposits (Basavarajappa et al., 2019), and sillimanite mineralization (Amer and El-Desoky, 2017). Increased advancements in satellite sensor technology have paved the way for production of remote sensing datasets with high spatial, spectral, and temporal resolutions that made a significant breakthrough in the field of mineral exploration. Hyperion, sentinel, and Worldview-2 help delineate alteration minerals (Liu et al., 2017; Abubakar et al., 2019), gold prospecting (Xu et al., 2019), and lithium (Li)-bearing pegmatites (Cardoso-Fernandes et al., 2018).

The nature of minerals and their occurrence play an essential role in selecting satellite data and also choosing the best method for image processing (Zeinelabedin and El Nadi, 2014). Today image processing software packages provide more versions and advanced capabilities for image processing, enhancement, and classification, which results in accurate subpixel-level mapping of minerals. Linear transformation techniques like minimum noise fraction (MNF), independent component analysis, and principle component analysis (PCA) reduce spectral redundancy and noise-dominated bands by converting these satellite bands to a set of uncorrelated image bands (Liu et al., 2014; Ge et al., 2018). The extreme spectrally pure pixels are extracted from each band and subjected to a selection of unique true end members or target members. The image spectra corresponding to these target end members are spectrally matched with spectral signatures of minerals either taken from spectral libraries or measured using spectroradiometers. Finally classification algorithms like mixture tuned match filtering, spectral angle mapper (SAM), support vector machine, and linear spectral unmixing were used for image classification, resulting in the generation of accurate mineral maps (dos Reis Salles et al., 2017; Rajendran and Nasir, 2019). Apart from these techniques, band rationing, color composites, and spectral indices were also successfully used for mapping potential zones of mineral occurrences based on their variation in spectral behavior (Sabin, 1997; Mahdevar et al., 2015).

The beach sands of the Indian coast are endowed with high concentration of heavy minerals (HMs), particularly ilmenite followed by other minerals like rutile, sillimanite, garnet, zircon, and monazite. All these minerals are strategically important due to their high economic value arising from their significant applications in various industries (Rejith and Sundararajan, 2018a; Amritha et al., 2019). The rutile and upgraded ilmenite with high TiO_2 content are used for the production of titanium products, which results in commercial employment of large quantities of ilmenite mineral (Sundararajan et al., 2009). Monazite is a rare-earth phosphate mineral that is mainly used for extracting rare earths (Lazo et al., 2018). In India, the major coastal placer deposits of HMs are Ratnagiri deposits (Maharashtra), Chavara (Kerala), Manavalakurichi (Tamil Nadu), Bhimunipatnam (Andhra Pradesh), and Chatrapur (Orissa) (Ali et al., 2001). Many researchers have completed detailed investigation on beach sands and their HM content of the Indian coast (Rajganapathi et al., 2013; Viveganandan et al., 2013; Loveson et al., 2014; Sajimol et al., 2017) particularly on Chavara and Manavalakurichi placer deposits (Babu et al., 1994; Ramakrishnan et al., 1997; Nair et al., 2009; Gayathri et al., 2017). In the context of remote sensing, limited studies were carried out on beach environments. Chandrasekar et al. (2011) uses the SAM algorithm for mapping the beach minerals of the Ovari coast in Tamil Nadu.

In the present study, an attempt has been made to investigate the strategic minerals in beach sediments of Chavara and Manavalakurichi placer deposits of India using standardized hyperspectral analysis techniques applied to Landsat data.

2. Study area

The coasts of Chavara in Kerala and Manavalakurichi in Tamil Nadu are selected for the present study (Fig. 11.1). The Chavara deposits having the length of 44.14 km tends from Azheekal to Paravur along the southwest coast of Kerala with latitude 9°8′3.282″ N to 8°48′41.0688″ N and longitude 76°27′39.7944″ E to 76°38′54.474″ E. The Manavalakurichi deposit having the length of 51.27 km tends from Neerody to Pallam along the southwest coast of Tamil Nadu with latitude 8°17′27.078″ N to 8°4′38.6076″ N and longitude 77°5′54.978″ E to 77° 33′53.2584″ E. The factors prevailing the formation of these beach placers includes provenance rocks, climate, west-flowing rivers, and coastal processes (Ali et al., 2001). The khondalites and charnockites form the provenance and flow of about nine west-flowing rivers forming the black sand deposits of Chavara deposits (Suresh Babu and Thrivikramaji, 1993). A humid tropical climate with an average annual temperature of 27.6°C and a mean annual rainfall of 1936 mm prevails in Chavara. Manavalakurichi is occupied mainly by quaternary fluvio-marine deposits, which receive an average rainfall of 846–1456 mm, and have an average temperature of about 22.8–33.6°C.

The Chavara in Kerala and Manavalakurichi in Tamilnadu are two main coastal placer deposits of India. The tsunami waves that hit in 2004 made a significant depletion of about 2% in HM concentration of Chavara (Prakash et al., 2016). The Chavara deposits that lie between Kayamkulam and Neendakara possess an average grade of 45% THM (Total Heavy Mineral) up to a depth of 7.5 m that consists of ilmenite (2.7 Mt-35% grade), sillimanite (2 Mt-7%), rutile (1 Mt-2.5%), zircon (0.9 Mt-2.5%), and monazite (0.17 Mt-0.5%) (Ali et al., 2001). The beach sediments collected along a multiberm beach at Chavara consist of mainly ilmenite (ranges 35%–64%) followed by zircon, sillimanite, rutile, monazite, and leucoxene (Unnikrishnan et al., 2000). The ilmenite mineral of

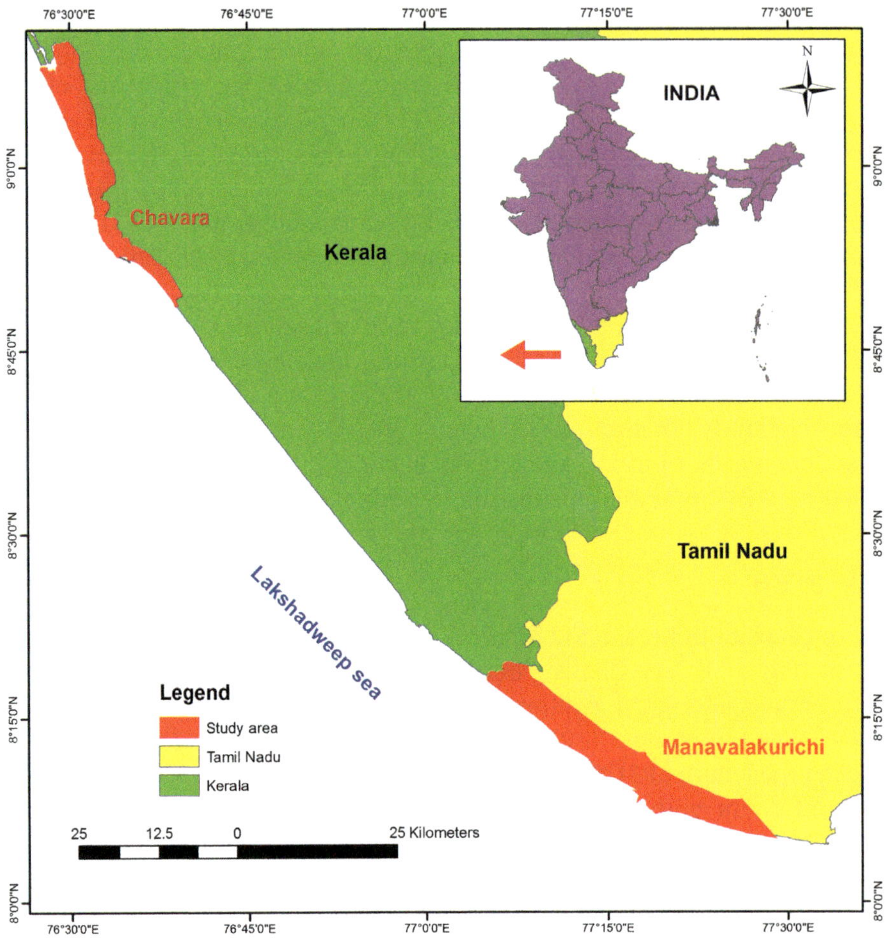

FIGURE 11.1 Study area map.

Chavara undergoes advanced alteration to pseudorutile or leucoxene compared to Manavalakurichi ilmenite (Nair et al., 2009). The Manavalakurichi deposits in Kanyakumari district of Tamilnadu shows 39% THM up to a depth of 7.5 m in which ilmenite accounts for 24% followed by garnet (5.5%), sillimanite (3.5%), zircon (2%), rutile (1.8%), monazite (1%), and leucoxene (0.9%) (Ali et al., 2001).

3. Materials and methods

The commonly used hyperspectral data processing steps were used for deriving the potential HM deposits from Landsat imagery. The processing steps adopted for the present study are shown in Fig. 11.2. It includes a series of six continuous processes such

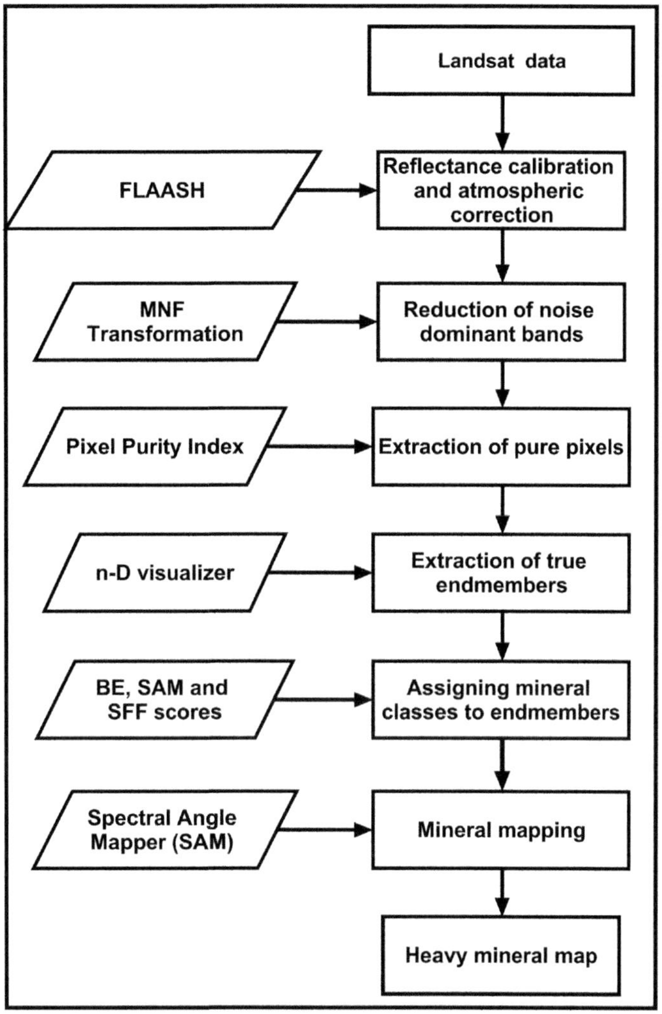

FIGURE 11.2 Flowsheet showing the steps of hyperspectral analysis.

as (1) reflectance calibration and atmospheric correction using FLAASH module, (2) removal of noise-dominant bands using MNF transform, (3) extraction of pure pixels using pixel purity index (PPI), (4) extraction of true endmembers using n-dimensional (n-D) visualizer, (5) identification of mineral classes using the scores of spectral matching techniques, and (6) image classification using SAM classifier (Chandrasekar et al., 2011). After FLAASH correction, the Landsat bands were linearly transformed using MNF transformation, where the noisy bands with less information were removed from further processing. The MNF bands having eigenvalues near zero can be removed using a threshold value. The PPI extracts the pure pixels and generates a PPI image with pure pixels shown as bright. The pure pixels are plotted in an n-D visualizer, where the spectrally unique pixels representing target end members form data clouds representing true end members. The commonly used spectral matching techniques sun as SAM, binary encoding (BE), and spectral feature fitting (SFF) are used for identifying the mineral classes (Pal et al., 2011). Each technique gives a score of 0–1, where 1 represents the best match of image spectra to the given reference spectra. A spectral library of HMs developed using the beach sands of Kerala is used as the reference spectra for the present study (Rejith et al., 2020). The image spectra showing the maximum score out of 3 will be assigned to that particular mineral class. Finally, these mineral classes are used for classifying the whole image using a suitable classification algorithm.

The SAM classifier is used for generating the mineral maps showing the potential occurrences of HM deposits. The SAM classifier has been successfully used for mapping ore minerals of Zn, Cu, apatite, marbles, and so on (Khatun et al., 2019); hydrothermal alteration minerals (Noori et al., 2019); and lithological mapping (Salaj et al., 2019). SAM is a widely used per-pixel image classification method that uses the spectral angle for measuring the similarity between training pixels (reference spectra) and the target spectra (image spectra) (De Carvalho and Meneses, 2000). Here the image spectra and reference spectra are taken as unit vectors in an n-D plane where the spectral angle between them is measured. The image spectra showing the smaller angle will be assigned to that particular class of minerals. Likewise, all the pixels of the image can be assigned to a particular mineral class, and finally, a classified image showing the distribution of mineral is obtained.

4. Results and discussion

The cloud-free Landsat 8 scenes of Chavara and Manavalakurichi acquired on 2019-06-01 and 2019-11-01 were used for the present study. The FLAASH atmospherically corrected, and reflectance calibrated seven bands (bands 1–5 VNIR and bands 6–7 SWIR) were subjected to MNF transformation. The eigenvalue plots of MNF bands are shown in Fig. 11.3. The eigenvalues obtained for the last band (band 7) of Chavara and Manavalakurichi Landsat scenes are 1.2472 and 1.4629. The least informative and noise dominant last MNF band (MNF band 7) was removed from further processing. The PPI images were generated using 100 iterations with a threshold value of 3. Usually, high threshold values are used for hyperspectral data. In case of multispectral datasets like Landsat, high threshold values may lead to the generation of nonextreme pixels, which affect adversely the accuracy of classification (Pal et al., 2011). Two end members of ilmenite and light (quartz) minerals derived from the Landsat data show the best match with the same minerals of reference spectra. The score obtained for SAM, SFF, and BE

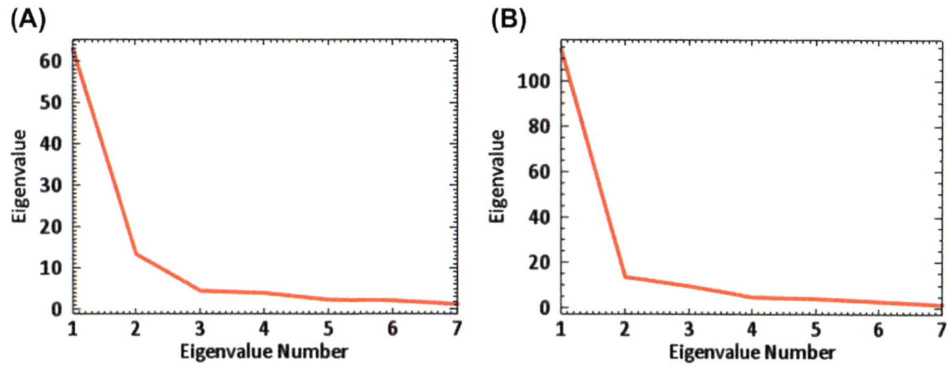

FIGURE 11.3 Eigenvalue plots of MNF bands. (A) Chavara; (B) Manavalakurichi.

are shown in Table 11.1. In case of Chavara Landsat scene, ilmenite and light minerals show the highest scores of 2.533 and 2.258, whereas in the case of Manavalakurichi, they show the highest scores of 2.430 and 2.486, out of 3. Thus the hyperspectral analytical approach facilitates the identification of the high concentration minerals in beach sands, out of which the ilmenite mineral is a strategic one. These identified mineral classes are used as the training spectra for the SAM classification. The SAM classified mineral maps showing the distribution of ilmenite mineral along the coast of Chavara and Manavalakurichi are shown in Figs. 11.4 and 11.5. Since the ilmenite mineral predominantly exists along the two study areas, the hyperspectral techniques were able to derive only the end member of ilmenite from the Landsat imagery. The low spectral resolution of Landsat imagery also plays an important role in getting these results. The transportation and concentration of ilmenite mineral and other HMs along the beach are controlled by factors such as grain size, the high specific gravity of minerals, strong winnowing action of waves, and so on (Rajganapathi et al., 2013). So pixels in the classified maps showing the presence of ilmenite mineral may also indicate the occurrence of other low concentration HMs. On applying the same techniques to hyperspectral imageries, more end members of low concentration minerals may be derived successfully.

Table 11.1 Results obtained for spectral matching techniques.

Study area	End member	SAM	SFF	BE	Score
Chavara	Ilmenite	0.770	0.906	0.857	2.533
	Light minerals (quartz)	0.544	1	0.714	2.258
Manavalakurichi	Ilmenite	0.728	0.845	0.857	2.430
	Light minerals (quartz)	0.635	0.994	0.857	2.486

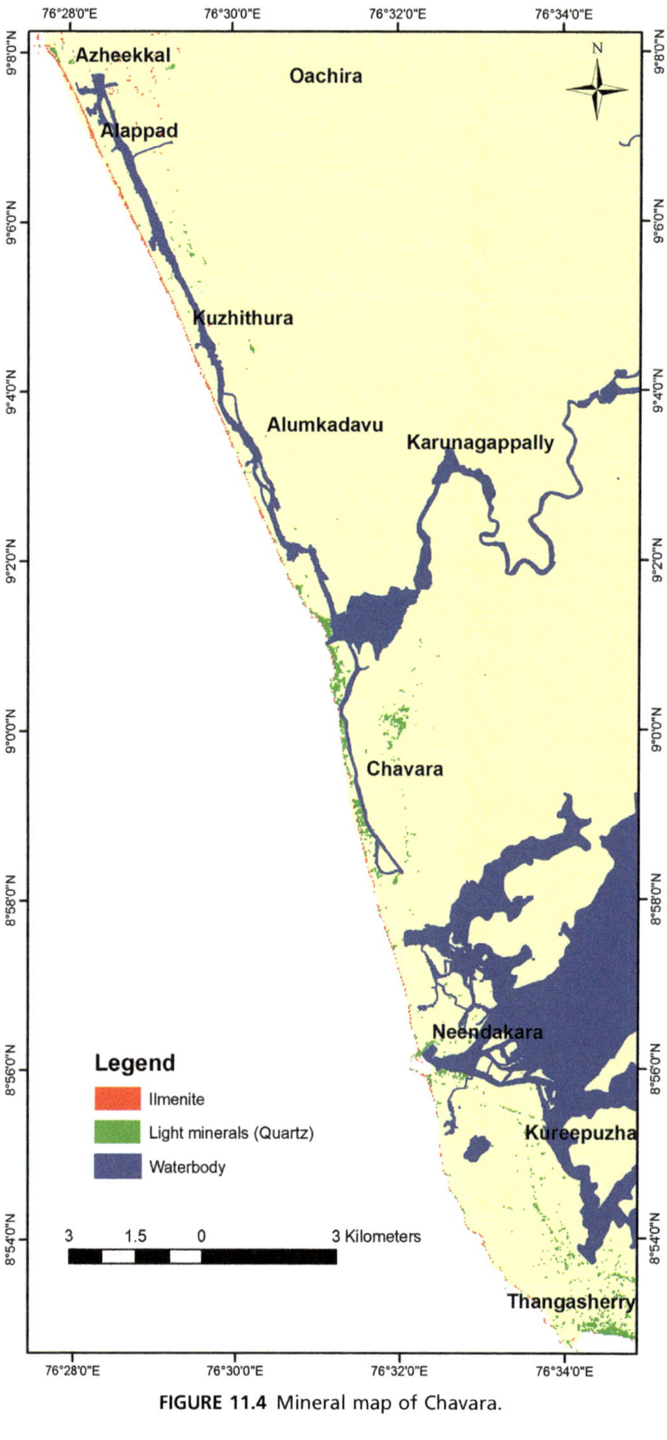

FIGURE 11.4 Mineral map of Chavara.

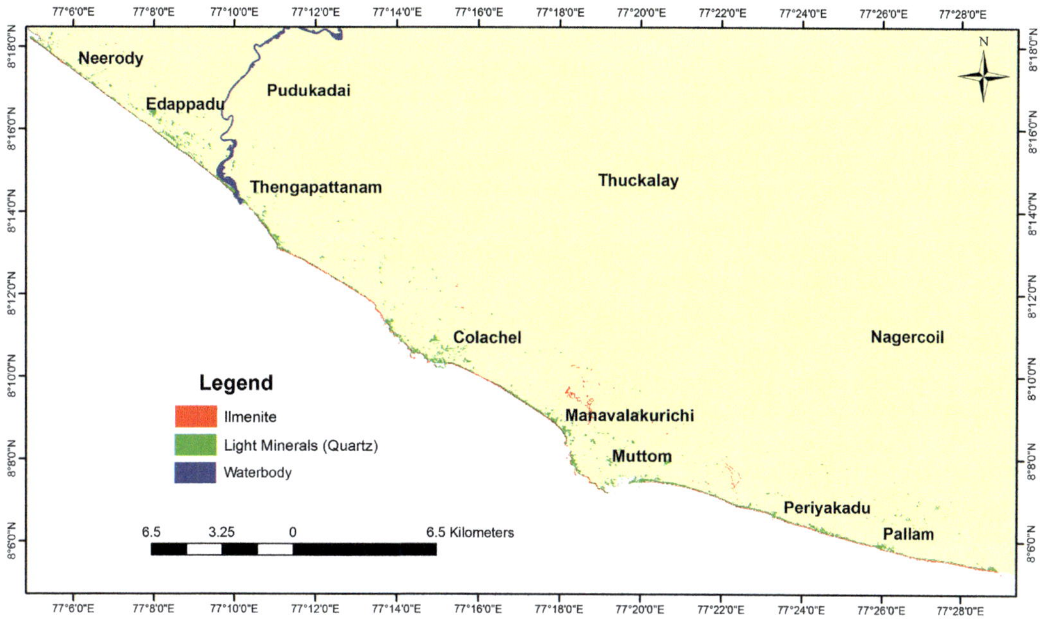

FIGURE 11.5 Mineral map of Manavalakurichi.

5. Conclusion

Effective mapping of strategic minerals in beach sands can be achieved using a combination of hyperspectral analysis techniques like MNF transformation, PPI, n-D visualizer, spectral matching techniques, and SAM classifier. All these techniques applied to the multispectral Landsat imagery help to derive the potential occurrences of ilmenite mineral along the coast of Chavara in Kerala and Manavalakurichi in Tamil Nadu. The ilmenite exists predominantly in Chavara and Manavalakurichi followed by other strategic minerals. The mineral spectra of ilmenite and light minerals (mainly quartz) show a strong correlation with the reference spectra, and thereby it significantly increases the accuracy of the SAM classified mineral maps. The present study using Landsat imagery has opened up new areas of applications in remote sensing and initiates the potential of hyperspectral datasets for mapping beach minerals.

Acknowledgments

The first author thanks the DST-INSPIRE Division, Department of Science and Technology (DST), Government of India, for providing the INSPIRE fellowship. The authors also thank the Director, CSIR-National Institute for Interdisciplinary Science and Technology (CSIR-NIIST), Trivandrum, Kerala, India for extending the laboratory facilities.

References

Abubakar, A.J.A., Hashim, M., Pour, A.B., 2019. Identification of hydrothermal alteration minerals associated with geothermal system using ASTER and hyperion satellite data: a case study from Yankari Park, NE Nigeria. Geocarto Int. 34 (6), 597–625.

Ali, M.A., Krishnan, S., Banerjee, D.C., 2001. Beach and inland heavy mineral sand investigations and deposits in India—an overview. Explor. Res. Atom Miner. 13, 1–21.

Albert, N.R.D., January 1976. Application of computer-enhanced Landsat imagery to mineral-resource assessment in south-central Alaska. Geophysics 41 (No. 2), 337–337. 8801 S YALE ST, TULSA, OK 74137: Soc Exploration Geophysicists.

Amer, R., El-Desoky, H., 2017. A remote sensing method for mapping sillimanite mineralization. J. Afr. Earth Sci. 134, 373–382.

Amritha, A., Sundararajan, M., Rejith, R.G., Mohammed-Aslam, M.A., 2019. La-Ce doped TiO_2 nanocrystals: a review on synthesis, characterization and photocatalytic activity. SN Appl. Sci. 1 (11), 1441.

Babu, D.S., Thomas, K.A., Das, P.M., Damodaran, A.D., 1994. Alteration of ilmenite in the Manavalakurichi deposit, India. Clays Clay Miner. 42 (5), 567–571.

Banerjee, K., Jain, M.K., Jeyaseelan, A.T., Panda, S., 2019. Landsat 8 OLI data for identification of hydrothermal alteration zone in Singhbhum Shear Zone using successive band depth difference technique-a new image processing approach. Curr. Sci. 116 (10), 00113891.

Basavarajappa, H.T., Jeevan, L., Rajendran, S., Manjunatha, M.C., 2019. Aster mapping of limestone deposits and associated lithounits of parts of Chikkanayakanahalli, southern part of Chitradurga Schist Belt, Dharwar Craton, India. J. Indian Soc. Remote Sens. 47 (4), 693–703.

Beiranvand Pour, A., Park, T.Y., Park, Y., Hong, J., Zoheir, B., Pradhan, B., Ayoobi, I., Hashim, M., 2018. Application of multi-sensor satellite data for exploration of Zn–Pb sulfide mineralization in the Franklinian basin, north Greenland. Remote Sens. 10 (8), 1186.

Cardoso-Fernandes, J., Lima, A., Teodoro, A.C., October 2018. Potential of Sentinel-2 data in the detection of lithium (Li)-bearing pegmatites: a study case. In: Earth Resources and Environmental Remote Sensing/GIS Applications IX, vol. 10790. International Society for Optics and Photonics, p. 107900T.

Chandrasekar, N., Mujabar, P.S., Rajamanickam, G.V., 2011. Investigation of heavy-mineral deposits using multispectral satellite data. Int. J. Remote Sens. 32 (23), 8641–8655.

Cardoso-Fernandes, J., Teodoro, A.C., Lima, A., 2019. Remote sensing data in lithium (Li) exploration: a new approach for the detection of Li-bearing pegmatites. Int. J. Appl. Earth Obs. Geoinf. 76, 10–25.

De Carvalho, O.A., Meneses, P.R., February 2000. Spectral correlation mapper (SCM): an improvement on the spectral angle mapper (SAM). In: Summaries of the 9th JPL Airborne Earth Science Workshop, JPL Publication 00–18, vol. 9. JPL Publication Pasadena, CA.

dos Reis Salles, R., de Souza Filho, C.R., Cudahy, T., Vicente, L.E., Monteiro, L.V.S., 2017. Hyperspectral remote sensing applied to uranium exploration: a case study at the Mary Kathleen metamorphic-hydrothermal U-REE deposit, NW, Queensland, Australia. J. Geochem. Explor. 179, 36–50.

Eldosouky, A.M., Abdelkareem, M., Elkhateeb, S.O., 2017. Integration of remote sensing and aeromagnetic data for mapping structural features and hydrothermal alteration zones in Wadi Allaqi area, south eastern Desert of Egypt. J. Afr. Earth Sci. 130, 28–37.

Fereydooni, H., Mojeddifar, S., 2017. A directed matched filtering algorithm (DMF) for discriminating hydrothermal alteration zones using the ASTER remote sensing data. Int. J. Appl. Earth Obs. Geoinf. 61, 1–13.

Gayathri, G.S., Rejith, R.G., Jeelani, S.H., Sundararajan, M., Aslam, M.M., Chidambaram, S., 2017. Heavy mineral resources in Tamil Nadu, India: an overview. Geochem. & Mineral. Coast. Sediments Tamil Nadu 110.

Ge, W., Cheng, Q., Jing, L., Armenakis, C., Ding, H., 2018. Lithological discrimination using ASTER and Sentinel-2A in the Shibanjing ophiolite complex of Beishan orogenic in Inner Mongolia, China. Adv. Space Res. 62 (7), 1702–1716.

Ghrefat, H.A., Al Zahrani, A.A., Galmed, M.A., 2018. Mapping oolitic iron ore deposits in the ashumaysi formation, western Saudi Arabia, using different remote sensing techniques applied to Landsat 7 enhanced thematic mapper plus data (ETM+). J. Indian Soc. Remote Sens. 46 (9), 1399–1411.

Khatun, M., Sharma, R.U., Chattoraj, S.L., 2019. Mineralogical mapping using field and image based spectra in parts of Delhi-Aravalli Fold Belt, Rajasthan, India. Int. J. Econ. & Environ. Geol. 8–14.

Kurata, K., Yamaguchi, Y., 2019. Integration and visualization of mineralogical and topographical information derived from ASTER and DEM data. Remote Sens. 11 (2), 162.

Lazo, D.E., Dyer, L.G., Alorro, R.D., Browner, R., 2018. Treatment of monazite by organic acids II: rare earth dissolution and recovery. Hydrometallurgy 179, 94–99.

Liu, L., Zhou, J., Han, L., Xu, X., 2017. Mineral mapping and ore prospecting using Landsat TM and hyperion data, Wushitala, Xinjiang, northwestern China. Ore Geol. Rev. 81, 280–295.

Liu, L., Zhou, J., Jiang, D., Zhuang, D., Mansaray, L.R., 2014. Lithological discrimination of the mafic-ultramafic complex, Huitongshan, Beishan, China: using ASTER data. J. Earth Sci. 25 (3), 529–536.

Loveson, V.J., Richa, K., Sundararajan, M., Gujar, A.R., 2014. Remote-sensing perspective and GPR subsurface perception on the growth of a recently emerged spit at Talashil coast, west coast of India. GISci. & Remote Sens. 51 (6), 644–661.

Mahdevar, M.R., Ketabi, P., Saadatkhah, N., Rahnamarad, J., Mohammadi, S.S., 2015. Application of ASTER SWIR data on detection of alteration zone in the Sheikhabad area, eastern Iran. Arab. J. Geosci. 8 (8), 5909–5919.

Nath, B., Niu, Z., Mitra, A.K., 2019. Observation of short-term variations in the clay minerals ratio after the 2015 Chile great earthquake (8.3 *M*w) using Landsat 8 OLI data. J. Earth Syst. Sci. 128 (5), 117.

Nair, A.G., Babu, D.S., Damodaran, K.T., Shankar, R., Prabhu, C.N., 2009. Weathering of ilmenite from Chavara deposit and its comparison with Manavalakurichi placer ilmenite, southwestern India. J. Asian Earth Sci. 34 (2), 115–122.

Noori, L., Pour, A.B., Askari, G., Taghipour, N., Pradhan, B., Lee, C.W., Honarmand, M., 2019. Comparison of different algorithms to map hydrothermal alteration zones using ASTER remote sensing data for polymetallic vein-type ore exploration: Toroud–Chahshirin Magmatic Belt (TCMB), north Iran. Remote Sens. 11 (5), 495.

Pal, S.K., Majumdar, T.J., Bhattacharya, A.K., Bhattacharyya, R., 2011. Utilization of Landsat ETM+ data for mineral-occurrences mapping over Dalma and Dhanjori, Jharkhand, India: an advanced spectral analysis approach. Int. J. Remote Sens. 32 (14), 4023–4040.

Prakash, T.N., Varghese, T.I., Prasad, R., Nair, L.S., Kurian, N.P., 2016. Erosion and heavy mineral depletion of a placer mining beach along the south-west coast of India: part II—sedimentological and mineralogical changes. Nat. Hazards 83 (2), 797–822.

Ramakrishnan, C., Mani, R., Babu, D.S., 1997. Ilmenite from the Chavara deposit, India: a critical evaluation. Mineral. Mag. 61 (405), 233–242.

Rani, K., Guha, A., Mondal, S., Pal, S.K., Kumar, K.V., 2019. ASTER multispectral bands, ground magnetic data, ground spectroscopy and space-based EIGEN6C4 gravity data model for identifying potential zones for gold sulphide mineralization in Bhukia, Rajasthan, India. J. Appl. Geophys. 160, 28–46.

Rajganapathi, V.C., Jitheshkumar, N., Sundararajan, M., Bhat, K.H., Velusamy, S., 2013. Grain size analysis and characterization of sedimentary environment along Thiruchendur coast, Tamilnadu, India. Arab. J. Geosci. 6 (12), 4717–4728.

Rajendran, S., Nasir, S., 2019. Mapping of hydrothermal alteration in the upper mantle-lower crust transition zone of the Tayin Massif, Sultanate of Oman using remote sensing technique. J. Afr. Earth Sci. 150, 722–743.

Rejith, R.G., Sundararajan, M., 2018a. Combined magnetic, electrostatic, and gravity separation techniques for recovering strategic heavy minerals from beach sands. Mar. Georesour. Geotechnol. 36 (8), 959–965.

Rejith, R.G., Sundararajan, M., 2018b. Mapping of maficultramafic rock units in late Archean Bhavani Complex, southern India using ASTER thermal bands. In: Shear Zones and Crustal Blocks of Southern India, Vol. 34.

Rejith, R.G., Sundararajan, M., 2019. Mapping of mineral resources and lithological units: a review of remote sensing techniques. Int. J. Image & Data Fusion 1–28.

Rejith, R.G., Sundararajan, M., Gnanappazham, L., Loveson, V.J., 2020. Satellite-based spectral mapping (ASTER and Landsat Data) of mineralogical signatures of beach sediments: a precursor insight. Geocarto Int. 1–24.

Salaj, S.S., Srivastava, S.K., Dugal, R., Upadhyay, R., Babu, D.S., Kaliraj, S., 2019. Application of ASTER remote sensing for lithological mapping in the Udaipur district of Rajasthan, India. In: Applications and Challenges of Geospatial Technology. Springer, Cham, pp. 99–119.

Suresh Babu, D.S., Thrivikramaji, K.P., 1993. Palaeogeographic interpretation of Kerala beach placers, SW coast of India. Indian J. Mar. Sci. 22, 203–208.

Sabin, F.F., 1997. Remote Sensing: Principles and Interpretation, Thrid ed. WH Freeman and Company, New York.

Sajimol, S., Rejith, R.G., Lakshumanan, C., Sundararajan, M., 2017. Sedimentology and geochemistry of heavy mineral deposits along the coast of Kanyakumari district, Tamil nadu, India. Geochem. & Mineral. Coast. Sediments Tamil Nadu 145.

Sundararajan, M., Bhat, K.H., Velusamy, S., Babu, N., Janaki, M.E.K., Sasibhooshanan, S., Das, P.M., 2009. Characterization of ilmenite from Kerala coastline, India: implications in the production of synthetic rutile. J. Miner. Mater. Char. Eng. 8 (06), 427.

Unnikrishnan, E., Dinesh, A.C., Nambiar, A.R., Gangadharan, A.V., 2000. Granulometric study of placer sands across a multi-berm beach at Chavara, southwest coast of India. Indian J. Mar. Sci. 29, 295–304.

Viveganandan, S., Lakshumanan, C., Sundararajan, M., Eswaramoorthi, S., Natesan, U., 2013. Depositional environment of sediments along the Cuddalore coast of Tamilnadu, India. Indian J. Geo Mar. Sci. 42 (3), 375–382.

Xu, Y., Meng, P., Chen, J., 2019. Study on clues for gold prospecting in the Maizijing-Shulonggou area, Ningxia Hui autonomous region, China, using ALI, ASTER and WorldView-2 imagery. J. Vis. Commun. Image Represent. 60, 192–205.

Yousefi, T., Aliyari, F., Abedini, A., Calagari, A.A., 2018. Integrating geologic and Landsat-8 and ASTER remote sensing data for gold exploration: a case study from Zarshuran Carlin-type gold deposit, NW Iran. Arab. J. Geosci. 11 (17), 482.

Zeinelabedin, K.A.E., El Nadi, A.H.H., 2014. The use of Landsat 8 OLI image for the delination of gossanic ridges in the Red Sea Hills of NE Sudan. Am. J. Earth Sci. 1 (3), 62–67.

12

Simulation studies about the role of Lakshadweep-Maldives ridge in determining tsunami characteristics along the Kerala coast of India

Praveen Sathee Sankar[1], Shafique Matin[2]

[1]DEPARTMENT OF PHYSICS, ST. THOMAS COLLEGE, KOZHENCHERRY, KERALA, INDIA;
[2]ENVIRONMENT PROTECTION AGENCY (EPA), WEXFORD, IRELAND

Chapter outline

1. Introduction

A tsunami is a sequence of long tidal waves produced by various geological events like eruption of volcanoes, underwater earthquakes, landslides, tectonic movements, or even in cases of an asteroid hit. Such events result in a rapid flooding of coastal areas, causing a fatal impact to coastal life (Bernard and Titov, 2015). Tsunamis can broadly be divided into local and distant types. In local tsunamis, people living in coastal areas feel the tremors quickly and have very limited time to react and before the flood water arrives. In distant tsunamis, people living in coastal areas may not get the tremor sensations and have enough time before tsunami flooding reaches their areas. The tsunami warning systems started in Japan in the late 1940s to alert coastal communities only in an event of

Remote Sensing of Ocean and Coastal Environments. https://doi.org/10.1016/B978-0-12-819604-5.00012-3

a local tsunami, followed by the United States using advanced technologies that include both local and distant warning systems for tsunamis. The distant tsunami that occurred on Boxing Day (i.e., December 26, 2004) attacked many parts of the Kerala coast. However, there were serious differences and deviations in the observed tsunami characteristics along the Kerala coast. The varying tsunami physiognomies along the coast of Kerala state can be attributed to several physical oceanographic processes and their nonlinear interactions. The various physical oceanographic processes as reported by Kowalik et al. (2005a,b), Murty et al. (2005a,b,c, 2006a,b,c), and Nirupama et al. (2005a,b, 2006) focused on the bathymetric variations, tsunami energy, reflections from land boundaries, resonance amplifications, tidal interactions, role of internal waves, several different oscillations, energy extraction, and dispersions and gaps in the continental shelves. Much less has been discussed about the role of ridges in determining tsunami characteristics, and nothing considering Lakshadweep-Maldives ridge (LMR).

The South-East Arabian Sea (SEAS) is a prominent zone beside the Kerala coast and encompassing the LMR (Matin et al., 2019). In this region the tsunami wave undergoes several severe wave transformation processes coupled with nonlinear interactions. The tsunami of December 2014 that originated from the Sumatra-Andaman earthquake of 9.0 magnitude occurred at 00:58:53 UTC. Tsunami waves arrived relatively earlier to the eastern coastline of India compared to the western coast. The Kerala coast, which is in the shadow zone (west coast) of the propagation direction of the epicenter, was not expected to be approached by the tsunami waves. The impact of this tsunami on the coast of Kerala and its late arrivals of the wave was due to the echoed waves from the LMR and east coast of the African continent. The role of LMR for causing delayed tsunami was mysterious and its characterization was badly required.

In this chapter a tsunami inundation modeling was simulated to understand the role of LMR in modifying tsunami characteristics along the Kerala coast. Two scenarios were considered: with the presence of LMR and without LMR (hypothetical), to understand the related uncertainties.

2. Methodology

2.1 Study area

The study area spans along the Kerala coast of Southwestern India. The Kerala coastline is 595 km long and accommodates 1.1 million people, directly and indirectly dependent on the cost and related business and covers about 15% of the total geographical area of the state. Geographically the Kerala coast is in the shadow zone of the tsunami propagation and assumed to be the first of this scale. The western coast of Kerala is relatively flat compared to the eastern districts. There is a dense network of interconnected brackish water channels and river estuaries. A network of rivers called Kayals in the Malayali language discharge into the coastal sea. These Kayals are separated from the sea by huge sandbars and form coastal lagoons. Monsoon brings most of the rainfall in the coastal zones and are relatively prominent in the southern districts with an average rainfall of 2923 mm annually. Gale force winds, storms, flash floods, cyclones, and sudden sea-level rise can be seen during monsoon.

2.2 Method

Different tsunami models are being used to estimate and understand tsunami-related phenomena like wave height, inundation height, extent of flooding, and time to arrival of waves in cases of distant tsunamis. In this investigation the TUNAMI N2 model, developed by the Japan-based university, Tohoku (Goto et al., 1997) was used. The model gives more prominence to fine grid data with good resolution as it is very important in shaping tsunami wave characteristics along the coast. This model was set up and calibrated earlier (Kurian et al., 2006; Praveen et al., 2011) for the southwest coast of India. The nested grid version of the TUNAMI N2 model is used here. Numerical simulations for estimating tsunami inundation require two types of data inputs: (1) point of event information, which includes its geology and seismology; and (2) bathymetry and topographical information of the coast under consideration. The model uses finite grids for the propagation of tsunami. The entire domain is divided into different grids, A to D (Fig. 12.1). Grid D is the coastal area for which the model stimulation was computed.

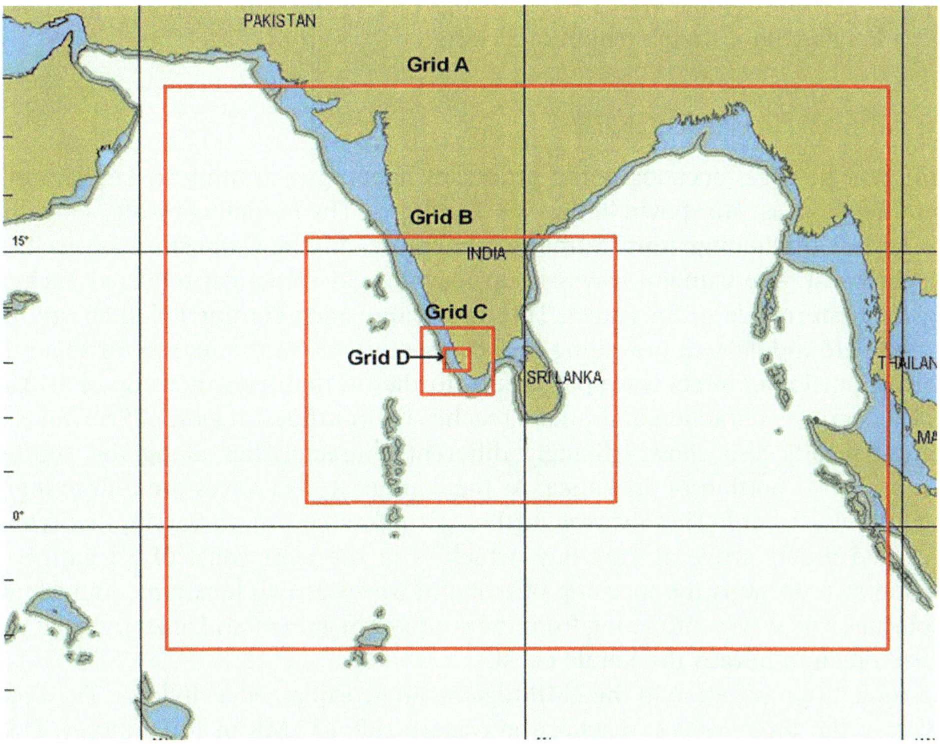

FIGURE 12.1 Data grids used as model inputs.

Starting from grid D, each outer grid will be expanded three times the equivalent inner grid. Nearly all of the Indian Ocean is taken as the outer grid A, and the three subgrids B, C, and D each lie within the outer one. The topographic data for grid D was obtained from field surveys. The sources such as GEBCO, NHO Chart, C-MAP, and SRTM were used for the preparation of topographic/bathymetric grids for the coarse grids A, B, and C.

After setting up the model, the simulation was carried out for 8 h starting from the source of tsunami generation to the final inundation at the coast for two different sources of tsunami generation i.e., Sumatra in 2004 and Makran in 1945). The output of the model in time-series format in an interval of 15 s was extracted and was merged to an animation frame for in-depth analysis and for identifying the characteristics in the SEAS. Particularly for understanding the role of LMR, numerical modeling was carried out, incorporating two different criteria, i.e., with LMR and without LMR. With LMR is the prevailing condition in the Arabian Sea in which the LMR resides in a perpendicular position with reference to Sumatra source and aligned in a north–south direction. Without LMR is a hypothetical condition in which the land mass of LMR was totally removed and was fed with bathymetric values as per the bathymetric conditions prevailing beside (Fig. 12.2). This study was done for two coasts in Kerala, one in Southern Kerala and other in Northern Kerala in the southwest coast of India. The sector areas are Neendakara-Thotapally in South Kerala with a stretch length of 44 kms and Mogral Thalappadi of Northern Kerala with a stretch length of 25 kms.

3. Results and discussion

The different physical oceanographic processes happening around Sri Lanka and the adjacent Kerala coast are shown in Figs. 12.3 and 12.4. The modeling results show that a tsunami wave originating from Sumatra undergoes different transformations before hitting the coast. The tsunami waves upon reaching Sri Lanka get reflected back after hitting the eastern side of Sri Lanka. The role of the steep continental shelf and deep bathymetry around 3000 m prevailing in those sectors forms the reason for this reflection. Also a portion of direct wave propagates further in to the northern tip of Sri Lanka where it undergoes refraction and finally reaches the northeast region of Sri Lanka. The simulation results also show strikingly different characteristics along the southeast portion and west portion of Sri Lanka. At the southeast, the waves are diffracting and propagating northward. These waves further get refracted due to varying bathymetric features and finally arrive at Palk Bay, which is in the west coast of Sri Lanka. The simulation also unravels the focusing of tsunami waves at two locations, Kanyakumari and Colachel. The waves diffracting from the southern region of Sri Lanka proceed in the northern direction toward the Kerala coast.

The tsunami propagation in the LMR region can be explained as follows. The first set of waves, or the direct waves, reaches the eastern side of LMR in 190 minutes. Due to bathymetric constraints no shoaling occurs and the reflection of waves are only partial.

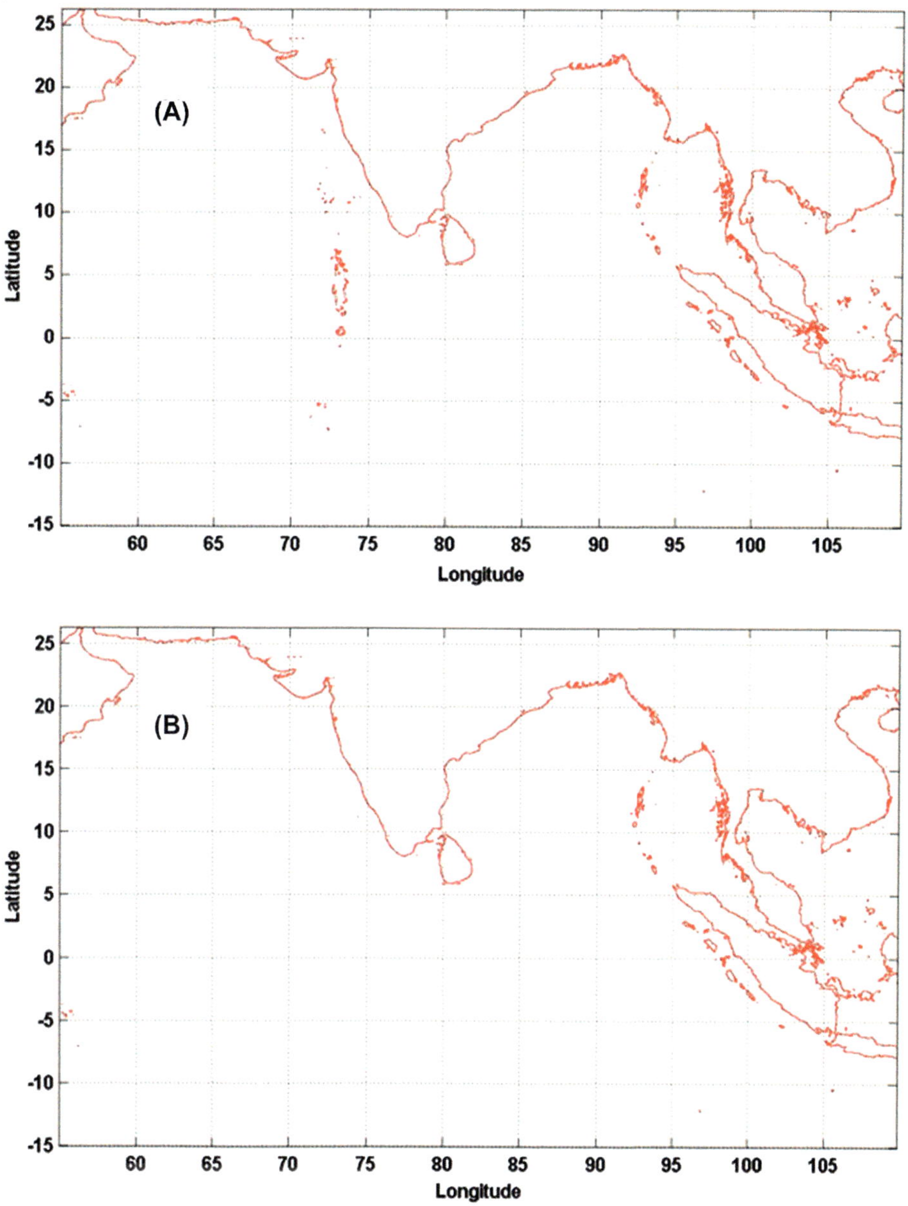

FIGURE 12.2 Grids used for LMR investigation: (A) with LMR, (B) without LMR.

The partially reflected waves move in two different trajectories, either to the south-western Sri Lanka coast or to the Kerala coast. Such tsunami waves can be regarded as the second set of waves arriving at the Kerala coast. These waves undergo wave interference with the reflected waves from LMR leading to higher run-up values at the coast.

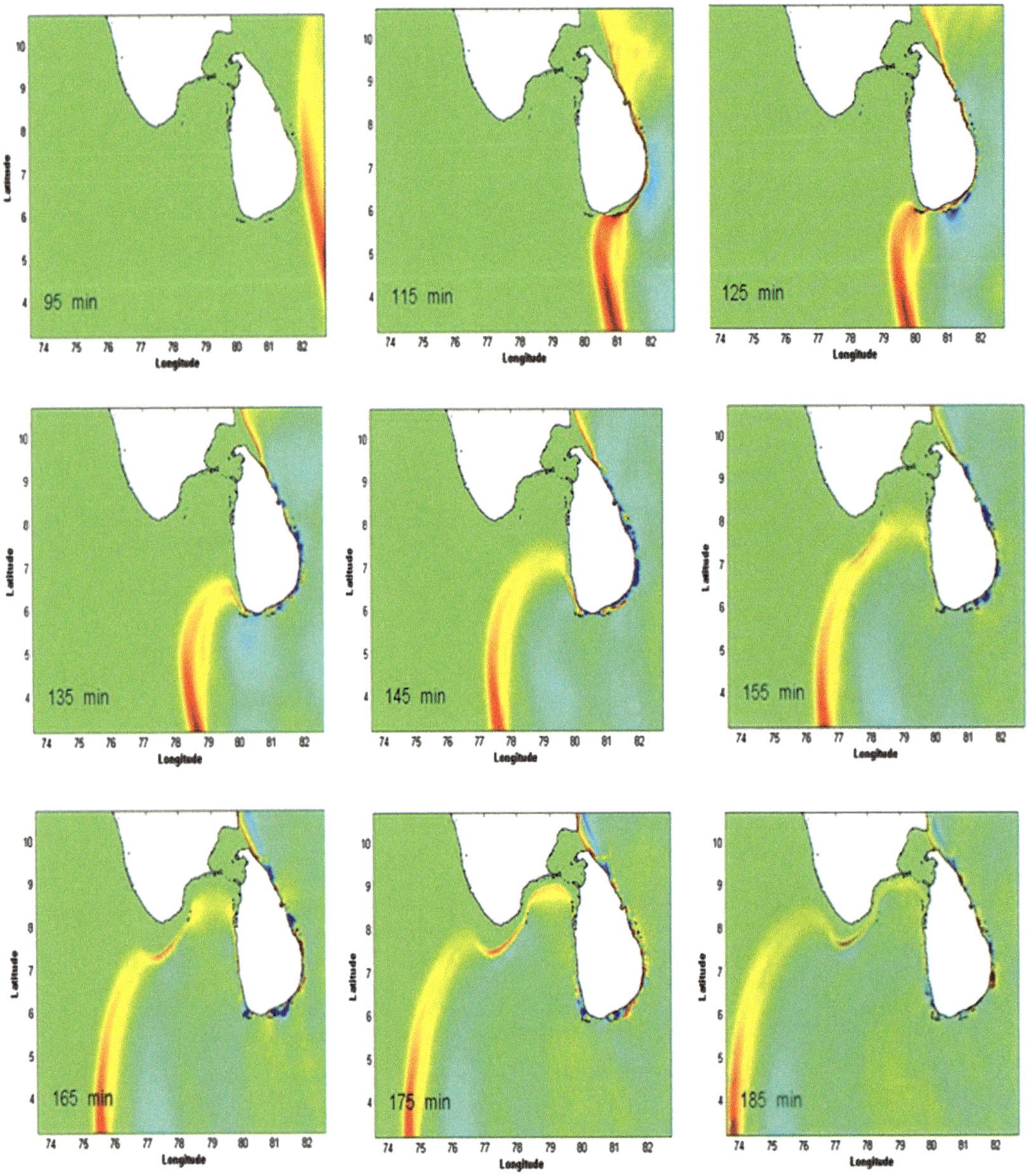

FIGURE 12.3 Modeling results showing reflection, refraction, and diffraction along the SEAS.

At the same time a portion of waves that are not reflected traverses the LMR and reaches the western portion of LMR. This will lead to tsunami wave amplification due to increased shoaling effect.

The waves that are moving westward undergo Total Internal Reflections (TIR) (Murty et al., 2008), mainly due to the propagation within two bathymetric values i.e., from

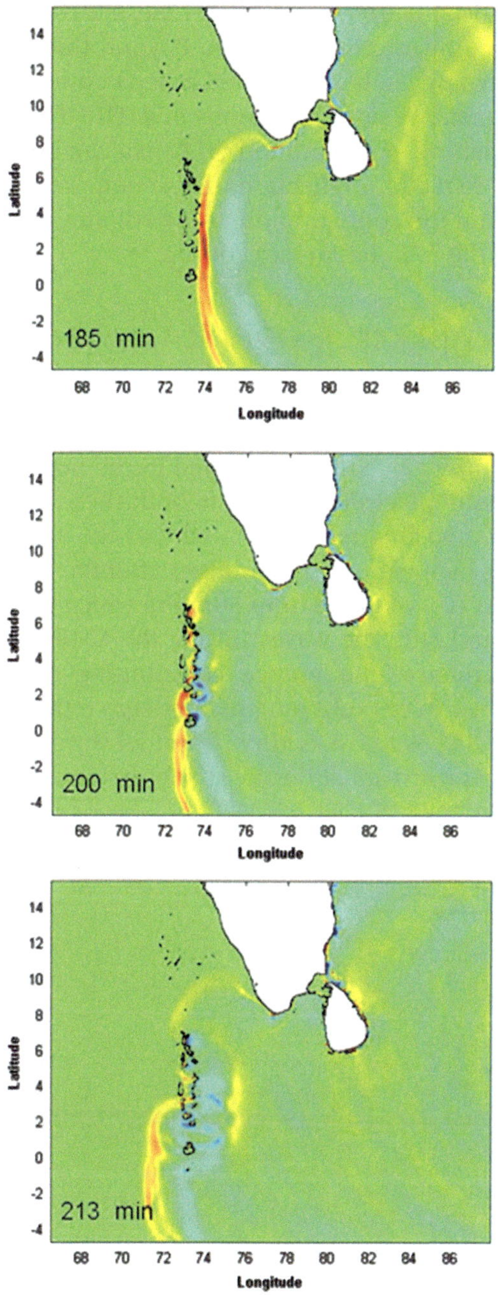

FIGURE 12.4 Propagation maps of reflection occurred east of LMR.

shallow bathymetry to deep bathymetry. Finally, these waves reach the southwest coast of Sri Lanka and the southwest coast of India. Beyond this any late arrival of tsunami waves along the Kerala coast can be mentioned as the boundary reflections from the east coast of Africa and Somalia besides reflections and TIR. The various physical oceanographic processes happening in and around Sri Lanka can be finally summarized as the effect of direct waves, refraction, reflection, diffraction, and TIR. The complexity prevailing in bathymetry and the configuration of the adjoining coast are also contributing factors responsible for the inundation and run-up.

4. The influence of LMR for 2004 Sumatra tsunami

To understand the effect of LMR, numerical modeling was carried out for two different scenarios, with and without LMR. For measuring the amplitudes of tsunami, representative locations at a depth of 1–2 m was selected in the northern and southern sectors. The run-up along different places pertaining to both these sectors were also simulated. In southern Kerala the modeling results showed that the arrival of the first wave is the same for both scenarios happening at 283 minutes after the generation of the initial wave at Sumatra (Fig. 12.5). Waves initiated from Sumatra which were propagating toward the west also encounter the LMR. The waves hitting the LMR will get redirected and will further undergo constructive interference contributing to higher amplitude values (Table 12.1). The reflected waves subsequently undergo refraction and will move toward the southern coast, arriving 60 minutes after the initial wave. This can be regarded as the second set of waves, which arrived at the coast after being deflected from Sri Lanka. The

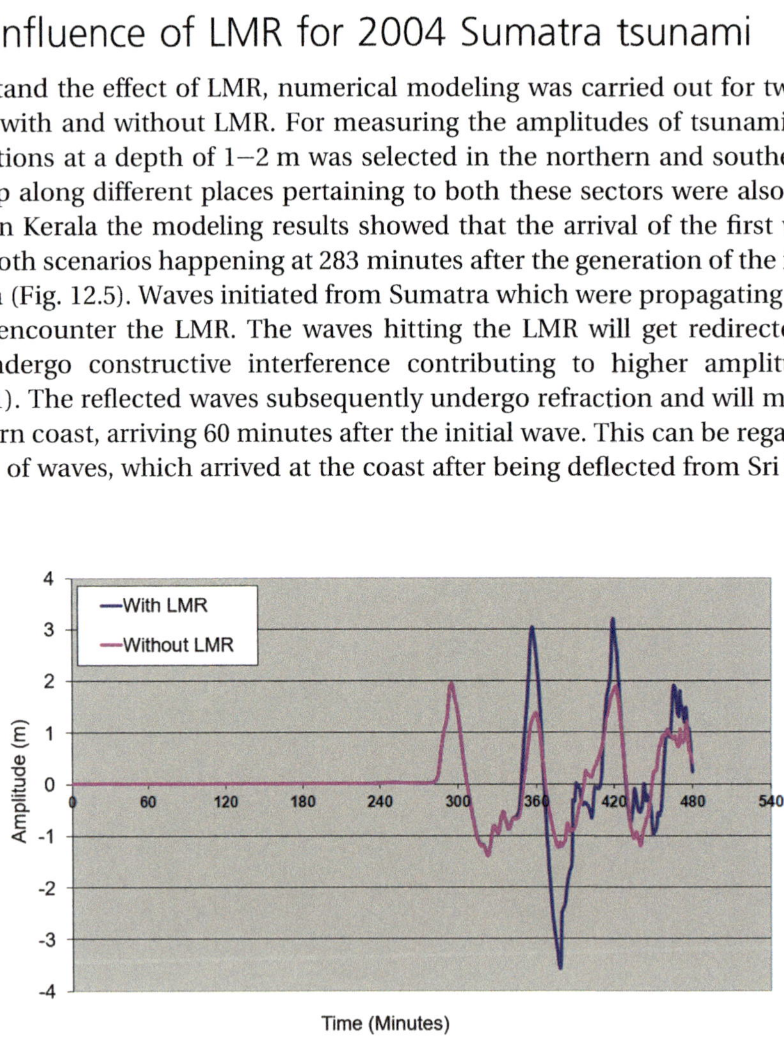

FIGURE 12.5 Tsunami simulations showing the effect of LMR for the southern Kerala coast.

Table 12.1 Showing results to compare the run-up for southern Kerala coast with and without LMR.

		Run-up (m)	
Sl. No.	**Location**	**With LMR**	**Without LMR**
1.	Neendakara north	2.5–3.0	1.5–2.0
2.	Chavara	2.5–3.0	2.0–2.5
3.	Pandarathuruthu	3.0–3.5	1.5–2.0
4.	Pannickarkadavu	3.0–3.5	1.5–2.0
5.	Alappad	4.0–4.5	2.0–2.5
6.	Azhekkal	4.5–5.0	2.0–2.5
7.	Valiyazhekkal	4.5–5.0	1.5–2.0
8.	Arattuppuzha	4.0–4.5	1.5–2.0
9.	Trikkunnapuzha	4.0–4.5	2.0–2.5
10.	Thotappally	2.5–3.0	2.0–2.5

waves also propagate between the gaps of LMR in the westward direction in two bathymetric conditions, leading to total internal reflection. However, at some locations on the southern coast this wave reached only after 2 h with reference to the initial wave.

For the scenario of without LMR, modeling results show that there is no reflection and the waves proceed toward the east coast of Africa and Somalia. It can be concluded that for a propagation scenario without LMR in the Arabian Sea there are only three waves: direct diffracted waves, refracted waves, and reflected waves from the eastern coast of Somalia and Africa. The time series plot of tsunami amplitudes for south and north Kerala are shown in Figs. 12.5 and 12.6.

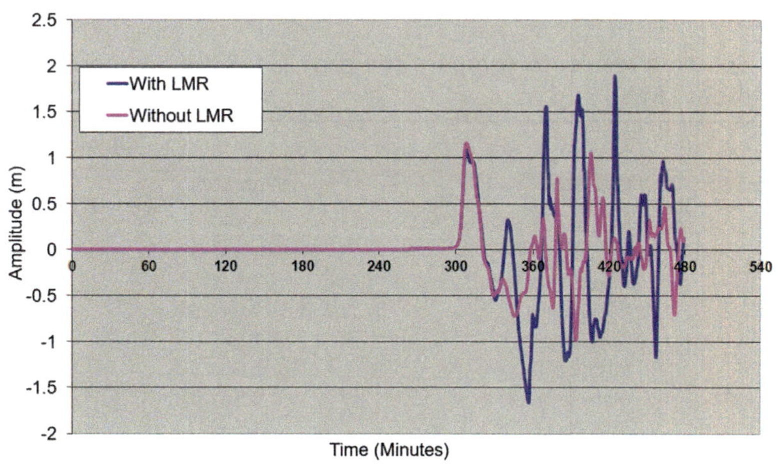

FIGURE 12.6 Simulation studies showing the effect of LMR for the northern Kerala coast.

The waves arriving at northern Kerala were also diffracted from Sri Lanka, arriving 308 minutes after the initial displacement of the tsunami wave. The model results showed a pertinent finding about the western side of LMR. The tsunami waves, if propagated through the western side would lead to a time delay in propagation. This can be attributed to the blockade imposed by LMR during the event of propagation when it is coming from the west. Moreover, the amplitude values of tsunami are much less. The bottom dissipation process and decrement in the energy of the wave due to dispersion are the factors responsible for the same.

The pattern of reflection is visible in the northern Kerala region too. The reflected waves arrive only after 370 minutes with reference to the initial wave arrival time at the coast. Waves that arrived after this time were also reflected waves arising from boundary reflections. Table 12.2 shows the different run-up values along the northern Kerala for the two scenarios, with LMR and without LMR. Here the run-up values are much less for the without-LMR scenario compared to that of the with-LMR scenario, clearly highlighting the effect of LMR and reflection. The variation in direct and reflected tsunami amplitudes comes in the array of 1.0–1.5 m. The boundary reflections from the east coast of Somalia and Africa were also simulated in this study. So it can be summarized that an increase in tsunami run-up heights will be high for the with-LMR scenario along both southern and northern Kerala, whereas the values will be high particularly for southern Kerala due to the proximity with the source and the magnitude of tsunamigenic earthquake coupled with directivity.

5. The effect of LMR for 1945 Makran-like tsunami

The impact of LMR in shaping tsunami run-up characteristics along the coast of Kerala for tsunamis coming from Makran source was also investigated. The with-LMR and

Table 12.2 Showing results to compare the run-up for the northern Kerala coast with and without LMR.

		Run-up (m)	
Sl. No.	**Location**	**With LMR**	**Without LMR**
1	Mogral	1.5–2.0	0.5–1.0
2	Kumbla north	1.0–1.5	0.5–1.0
3	Ankadi	1.5–2.0	0.5–1.0
4	Kukkaru	1.0–1.5	0.5–1.0
5	Uppala north	1.5–2.0	0.5–1.0
6	Hosangadi	1.0–1.5	0.5–1.0
7	Manjeswaram	0.5–1.0	0.5–1.0
8	Talappadi south	1.5–2.0	0.5–1.0

without-LMR scenarios of model run were also carried out for the Makran source of tsunami generation. The comparison of tsunami amplitudes for southern and northern Kerala in the nearshore locations are shown in Figs. 12.7 and 12.8 and are depicted in Tables 12.3 and 12.4.

The results for the Makaran tsunami numerical simulation shows major differences compared with that of Sumatra. Here the without-LMR scenario shows amplified run-up

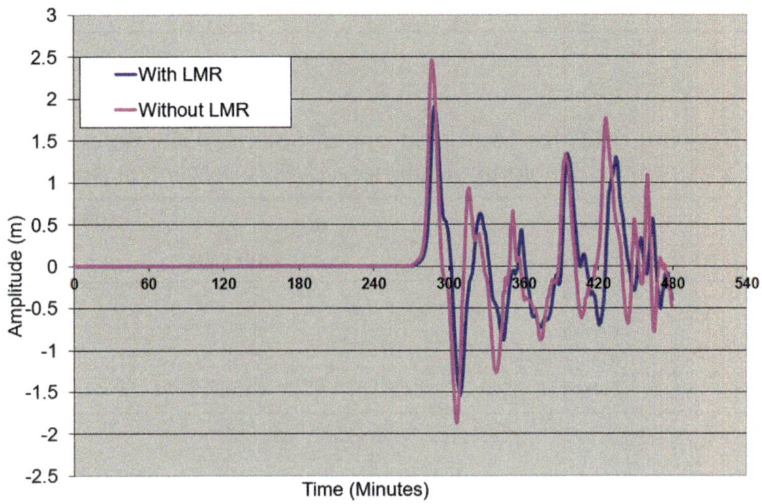

FIGURE 12.7 Simulated results of tsunami amplitudes without and with LMR for the south coast of Kerala with reference to Makran source.

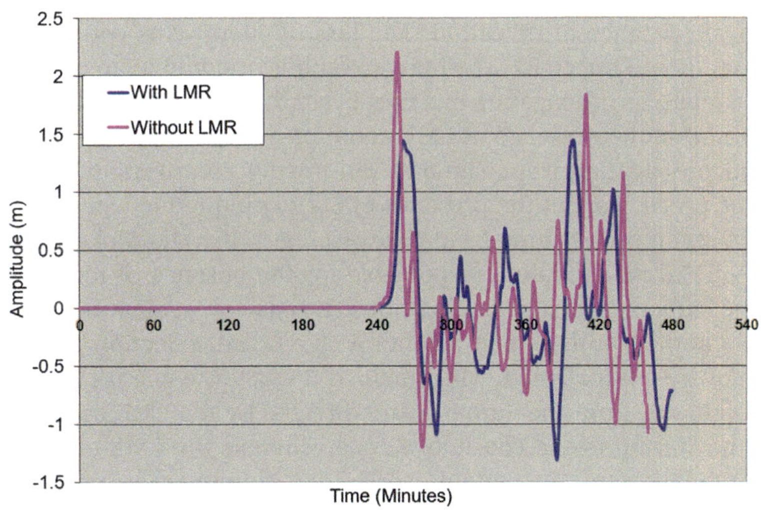

FIGURE 12.8 Simulated results of tsunami amplitudes with and without LMR for the north coast of Kerala with reference to Makran source.

Table 12.3 Showing cases for four locations of southern Kerala coast without and with LMR and its run-up in a condition where LMR is present or absent.

		Run-up (m)	
Sl. No.	**Location**	**With LMR**	**Without LMR**
1.	Neendakra	1.5–2.0	2.0–2.5
2.	Kovilthottam	1.0–1.5	1.5–2.0
3.	Cheriyazhekkal	0.5–1.0	1.0–1.5
4.	Valiyazhekkal	0.5–1.0	1.0–1.5

Table 12.4 Showing cases for four locations of northern Kerala coast without and with LMR and its run-up in a condition where LMR is present or absent.

		Run-up (m)	
Sl. No.	**Location**	**With LMR**	**Without LMR**
1.	Kumbla	1.5–2.0	2.0–2.5
2.	Uppala	1.5–2.0	2.0–2.5
3.	Manjeswaram	1.5–2.0	2.0–2.5
4.	Talappadi	1.5–2.0	2.0–2.5

values along both sectors of the Kerala coast. It shows the absence of a physical oceanographic phenomenon. The absence of reflection caused by LMR is evident here. It can be attributed mainly to the propagation trajectory of a tsunami from the west. The orientation of the Makran tsunamigenic earthquake fault and directivity of the propagation led to the absence of reflection. The loss of energy flux contributed toward the lesser run-up values at the coast. The Makran fault is oriented in an east–west direction, whereas the tsunami is propagating in a north–south direction. From these results it can be summarized that the presence of LMR can increase run-up along the coast of Kerala for a tsunami propagating from Sumatra due to the contribution of several physical oceanographic processes. But the presence of LMR could reduce the run-up to a certain extent for a tsunami propagating from the west coast, like the Makran source.

The fault of Makran is mainly responsible for the absence of reflection, which also resulted in the tsunami's change of direction. The east–west alignment of the fault gave the tsunami a north–south course without a significant reflection from the LMR. The LMR is also north–south aligned, which helped it further to change its course. Tsunami waves that originated from the Sumatra and directed by the LMR bring potentially high inundation and run-up to the coasts of Kerala, whereas the LMR obstructs the run-up and trailed inundation up to a certain degree for a tsunami that started in Makran.

6. Conclusions

The 2004 tsunami devastated many parts of Kerala with tremendous human loss. From the observed pattern of tsunami run-up and inundation at the coast it was understood that the Southeast Arabian Sea and the area encompassing it contributed significantly to the increased inundation and run-up along the entire coast of Kerala. Besides, the field observations on the arrival times of waves and increased amplitudes, it was evident that the LMR had a major role in shaping tsunami wave characteristics along the coast of Kerala. An investigation was carried out to understand the propagation characteristics of tsunami in SEAS and for studying the role of LMR in hiking tsunami run-ups along the coast. Numerical modeling was carried out for two scenarios, with LMR and without LMR, and model runs were initiated for two different sources of tsunami generation—one east of India and the other west of India with reference to historical tsunamigenic sources. The modeling results concluded that LMR could contribute toward increased run-ups along the coast of Kerala for a tsunami originating from Sumatra, whereas the simulation studies for Makran indicated the absence of reflection leading to the reduction in the run-up values along the coast.

Acknowledgments

The authors are thankful to Scientists of Former Marine Sciences Division, NCESS, Government of India, Thiruvananthapuram for support, guidance, and fruitful discussions. The authors are greatly indebted to National Center for Coastal Research (Formerly ICMAM PD), Government of India, Chennai for the support extended. The work emanates from the work done by National Center for Earth Science Studies (formerly CESS) under the national Project of Tsunami Inundation Modeling coordinated by ICMAM PD, Chennai.

References

Bernard, E., Titov, V., 2015. Evolution of tsunami warning systems and products. Philos. Trans. Ser. A Math. Phys. & Eng. Sci. 373 (2053), 20140371. https://doi.org/10.1098/rsta.2014.0371.

Goto, C., Ogawa, Y., Shuto, N., Imamura, F., 1997. Numerical Method of Tsunami Simulation with the Leap-Frog Scheme (IUGG/IOC Time Project). IOC Manual, UNESCO. No. 35.

Kowalik, Z., Knight, W., Logan, T., Whitmore, P., 2005a. Numerical modelling of the global tsunami: Indonesian tsunami of 26 December 2004. Sci. Tsunami Hazards 23, 40–56.

Kowalik, Z., Knight, W., Logan, T., Whitmore, P., 2005b. The tsunami of 26 December 2004: numerical modelling and energy considerations. In: Papadopoulos, G.A., Satake, K. (Eds.), Proc. Int. Tsunami Sym., pp. 140–150. Chania, Greece.

Kurian, N.P., Abilash, P.P., Rajith, K., Murali Krishnan, B.T., Kalaiarasan, P., 2006. Inundation characteristics and geomorphological impacts of December 2004 tsunami on Kerala coast. Curr. Sci. 90, 240–249.

Matin, S., Dash, S.K., Usha, T., 2019. A semi-analytical approach to understand remote sensing-based backscattering characteristics for Kerala coast using in situ observation. In: Kumar, P., Rani, M.,

Chandra Pandey, P., Sajjad, H., Chaudhary, B. (Eds.), Applications and Challenges of Geospatial Technology. Springer, Cham.

Murty, T.S., Rao, A.D., Nirupama, N., 2005a. Inconsistencies in travel times and amplitudes of the 26 December 2004 tsunami. J. Mar. Med. 7, 4–11.

Murty, T.S., Nirupama, N., Nistor, I., Rao, A.D., 2005b. Conceptual differences between the Pacific, Atlantic and Arctic tsunami warning systems for Canada. Sci. Tsunami Hazards 23, 39–51.

Murty, T.S., Nirupama, N., Rao, A.D., 2005c. Why the earthquakes of 26 December 2004 and the 27 March 2005 differed so drastically in their tsunami-genic potential. Newslett. Voice of the Pacific 21, 2–4.

Murty, T.S., Nirupama, N., Nistor, I., Hamdi, S., 2006a. Far field characteristics of the tsunami of 26 December 2004. ISET J. Earthq. Technol. 42, 213–217.

Murty, T.S., Nirupama, N., Nistor, I., Hamdi, S., 2006b. Why the Atlantic generally cannot generate trans-oceanic tsunamis. ISET J. Earthq. Technol. 42, 227–236.

Murty, T.S., Rao, A.D., Nirupama, N., Nistor, I., 2006c. Numerical modelling concepts for the tsunami warning systems. Curr. Sci. 90, 1073–1081.

Murty, T.S., Kurian, N.P., Baba, M., 2008. Roles of reflection, energy trapping and secondary undulations in the tsunami on Kerala coast. Int. J. Ecol. Dev. 10, 100–114.

Nirupama, N., Murty, T.S., Rao, A.D., Nistor, I., 2005a. The role of gravoid and elastoid modes in oscillations around Andaman and Nicobar Islands. In: Sundar, V. (Ed.), Indian Ocean Tsunami. Indian Institute of Technology, Madras, India, pp. 41–52.

Nirupama, N., Murty, T.S., Rao, A.D., Nistor, I., 2005b. Numerical tsunami models for the Indian Ocean countries and states. Indian Ocean Surv. 2, 1–14.

Nirupama, N., Murty, T.S., Nistor, I., Rao, A.D., 2006. The energetics of the tsunami of 26 December 2004 in the Indian Ocean: a brief review. Mar. Geodes. 29, 39–48.

Praveen, S.S., Reshmi, A.K., Dhanya, P., Arjun, S., Kalarani, K.,N.P., Ramana Murthy, M.V., Shahul Hameed, T.S., Prakash, T.N., 2011. Numerical simulation of 26 December 2004 tsunami on the southwest coast and Lakshadweep Islands of India. Mar. Geod. Spec. Issue on Tsunamis 34, 59–67.

13

Measuring the vulnerability of coastal ecosystems in a densely populated west coast landscape, India: a remote sensing perspective

M.K. Rafeeque, M. Rameshan, M.K. Sreeraj

NATIONAL CENTRE FOR EARTH SCIENCE STUDIES, THIRUVANANTHAPURAM, KERALA, INDIA

Chapter outline

1. Introduction

The Indian coastal zone, with a length about 7510 km, is one of the most densely populated coastal regions of the world. Of India's population of 1270 million, 1710

Remote Sensing of Ocean and Coastal Environments. https://doi.org/10.1016/B978-0-12-819604-5.00013-5

million people live along the coastal districts. The coastal lands in India have become the most sought-after zones for major economic activities and urbanization, making the coastal ecosystems vulnerable to reclamation and land use modifications. Added to this is the impact of sea level rise due to global warming. Sewage discharge and coastal erosion are the other major concerns of the coastal zones. Daily more than 3000 million liters of domestic sewage are regularly added to the coastal waters. Coastal ecosystems all over the world are subjected to various intensities of anthropogenic pressures largely due to industrialization and overexploitation by increasing population. Like the ancient civilization, modern industrialization and settlements are also concentrating on fragile landscapes like the coastal zone. Coastal belts and associated landscapes attract civilization causing high density of population in coastal zones all over the world. India, the second-largest populated country of the world, has an extensive coastline of over 7500 km, which supports a large coastal community dependent on the ocean and the coast for their subsistence. As the density, as well as modernization of population, increases, the dependent coastal ecology is modified or redefined accordingly.

Coastal zones consisting of mangroves, mudflats, coastal wetlands, estuaries, and coral reefs represent multifunctional, highly productive, and critical habitats for a wide variety of plants, fish, shellfish, and other coastal living things. The coastal ecologies are affected mainly by the economic growth and urbanization of the landward area close to the coastal waters, estuaries, and backwaters. Coastal wetlands such as mangrove ecosystems along with biologically active mudflats are nature's "kidneys," which provide water quality improvement through filtering of agricultural and industrial waste and recharge of aquifers. Mangroves (Fig. 13.1) are taxonomically diverse assemblages of trees and shrubs that form the dominant plant communities in tidal, saline wetlands along sheltered tropical and subtropical coasts and are well adapted to salt water and fluctuations of tide level (Tomlinson, 1986; Clough, 1982). The mangrove ecosystem is nature's support for coastal stabilization, protecting communities from cyclones, tsunamis, and storms, maintaining critical habitats for many common, threatened, and endangered species; fish production; and for generating economic opportunities and providing the community with products for subsistence and survival (Qasim and Kathiresan, 2005; Spalding et al., 2010; Kathiresan, 2011; Baconguis and Mauricio, 1991). Mudflats are equally important morphological and ecosystem units that are natural protectors of the coastal zone. Mangroves flourish in mudflats, tidal flats, and estuaries, and the sustainability and protection of these ecosystems become truly critical.

The southwest coast of India has an extensive coastline of about 590 km with a large network of estuaries and backwaters along the Kerala coast. The major urban areas such as Kochi, Kozhikode, Kannur, Kollam, and Thiruvananthapuram are located in the coastal zone. The estuaries and backwaters and the tidal-influenced water bodies of Kerala have extensive patches of mangroves and tidal/mudflats (Mohanan, 1997, 2004; Basha, 1991, 1992; Kurian et al., 1995; CESS, 1995, 1998; Anitha and Thomas, 2006; Mohanan and Thomas, 2011). These are under threat due to pollution and reclamation for tourism, real estate, industries, and port. Protection of these ecosystems are not

FIGURE 13.1 Mangrove ecosystem at Edakochi.

effective, even in the presence of regulations and restrictions introduced through acts and notifications like the Environmental Protection Act (1986), Coastal Regulation Zone (CRZ) notification (1991, 2011), Wetlands (Management and Conservation) Rules (2010), Kerala Conservation of Paddy Land and Wetland Act (2008), among others. Among these acts, rules and notifications, the CRZ (1991, 2011) is the most effective. The issues of conservation of coastal ecosystems with the CRZ notification are discussed with examples from the coastal zone of Kerala.

One of the proven approaches to ensure livelihood security to the fisher and other local communities and to protect endangered coastal morphology and ecosystems is setting aside spatial buffers around coastal ecosystems, which would preserve coasts for posterity, mitigate forces of the ocean, allow a zone for any future marine transgressions, ensure livelihood sustainability, and facilitate sustainable resource development. Accordingly, the Ministry of Environment and Forests (MoEF) established the Coastal Regulation Zone (CRZ) notification in 1991 and 2011 under the Environment Protection Act of 1986. The CRZ notification recognizes mangrove areas and biologically active mudflats as ecologically very sensitive and categorizes them as CRZ-IA, which puts them under the highest order of protection. As an additional protection to mangrove ecosystems, a 50 m buffer zone is provided, surrounding mangrove ecosystems with an area of coverage more than 1000 m^2.

2. Study area

The west coastal plain of India is mainly divided as the Kachch peninsular coast, Konkan coast, and Kerala (Malabar) coast from north to south, respectively. Kerala coastal plain of SW peninsular India is one of the highly populated states of India with a population density of 2931 per/km^2. Out of six municipal corporations, five are located in the coastal plain of the state. Kerala is a narrow elongated state in NW–SE direction with a maximum length of around 550 km and a varying width of 15–120 km. The elevation of Kerala state ranges from Mean Sea Level (MSL) to 2695 m in Anamudi Hills. The relief features are drastically changing from its coastal plain at MSL in the western side to a highly elevated Western Ghats mountain range in the east within a short distance of width. Out of 14 districts of Kerala, 10 districts comprise the coastal plain of the state; of this, Kottayam doesn't have any coastline in its territory (Fig. 13.2). The drainage pattern of Kerala includes 3 east-flowing rivers and 41 west-flowing rivers, 25 backwaters (including inlets), 2 freshwater lakes, and 7 lagoons. Backwaters and lagoons are the cradles of the coastal ecosystems such as mangroves and filtration ponds.

The coastal plain of the state is comprised of a young coastal plain of the Holocene period. Major geomorphic processes active in Kerala are marine, estuarine, fluvial, and denudational processes (Chattopadhyay and Mahamaya, 2007). Even though the Kerala coastline is a high energy coast, its coastal plain is more stable than the remaining part of the west coast of India.

3. Methodology

All the available information on mangroves and mudflats are on large-scale maps that do not support microlevel management. Microlevel information in cadastral (survey plot) level on mangrove coverage gives the scope for local-level management through local bodies. Accordingly, species composition and geospatial distribution of mangrove ecosystems and its associates along the Kerala coast have been studied in a cadastral level through field mapping using high precision GPS incorporating information from high resolution World View 2 satellite images. This has been done as part of the preparation of local level Coastal Zone Management Plan (CZMP) for the state of Kerala during 2012–2014. Geospatial analysis of the mangrove ecosystem is done on a GIS platform. Appropriate management strategies to protect and restore the mangrove and mudflat ecosystems under the provisions of CRZ (2011) are discussed.

3.1 Data processing

Toposheet and satellite images were georeferenced using Arc GIS 10.3 version. Cadastral maps were also georeferenced by ground-truthing the existing Ground Control Points (GCPs) of survey plots. The satellite image World View 2 has a very high spatial resolution with 0.46 m panchromatic and 1.84 m multispectral data set. To get very high resolution pan-sharpened Multi Spectral Scanner (MSS) images, these data sets were resampled into 50 cm spatial resolution in Erdas Imagine. The spatial extensions of field-verified mangroves and biologically active mudflats and other coastal ecosystems were demarcated from these satellite images by incorporating the information of these coastal morphologies surveyed using GPS. Spatial distribution of mangrove ecosystems and biologically active mudflats are presented in cadastral level maps.

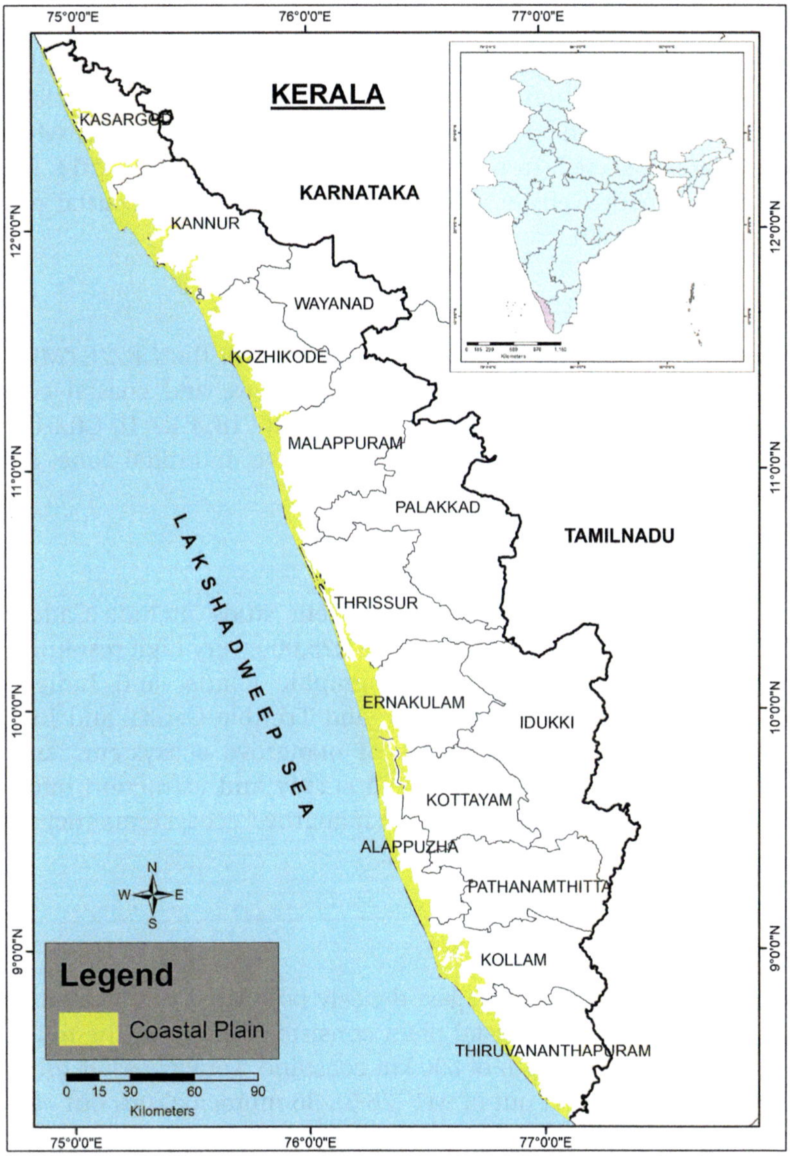

FIGURE 13.2 Location map of Kerala coastal zone.

3.2 Taxonomic investigation

Taxonomic studies were carried out for mangrove and its associated plant species. Morphological characters and floral characters were studied to identify the species. Fresh plant samples along with reproductive parts were collected for species-level identification. The plant species were grouped into respective classes based on standard classification (Tomlinson, 1986). Information on major threats to the existing mangroves was also collected during the field observations.

3.3 Coastal population study

The coastal plains and tidal-influenced water bodies were demarcated during the field survey. As per CRZ notification of Government of India (GOI), the salinity of 5 ppt is considered the tidal limit. The population and spatial extensions of listed local bodies of coastal plain were collected from the Census India report of 2011 and Local Self Government Department website. Accordingly, the density of coastal population was derived from the census 2011 reports.

3.4 Coastal regulation zone maps

The details on mangroves and mudflats are provided in the CRZ/CZMP maps, which have the high tide line, low tide line, CRZ categories, and coastal ecosystems and morphologies. The CRZ is categorized as CRZ IA, CRZ IB, CRZ II, CRZ III, and CRZ IV depending on whether it is ecologically very sensitive, intertidal zone, developed, undeveloped, or waterbody, respectively.

3.5 Data source and instruments

The principal data sources used for the present study include Cadastral maps in 1:4000/1:5000, Survey of India Toposheets in 1:25,000, very high resolution (submeter) satellite imageries (World View 2), hydrographic charts, and Indian tide tables. DISTOPRO laser meter, Brunton compass, and Trimble GeoXT and Juno GPSs were used for field mapping. The boundaries of mangrove ecosystems and biologically active mudflats were mapped through field survey and extraction information from satellite imageries. A detailed inventory of mangrove ecosystems including mudflats was prepared based on these observations.

4. Analysis and discussions

Kerala coastal plain is one of the longest densely populated rural–urban agglomerated coastal plains in India. Kerala coastal plain consists of 2.42% of the total geographical area of 38,863 km^2 with the length of 690 km coastline. Kerala coastal zone includes 246 Gramapanchayats (Table 13.1) out of 941 (26%), 36 municipalities out of 87 (42%), and out of 6 municipal corporations, 5 Municipal Corporations (MCs) (83%) (Table 13.2). Being one of the highly populated states of India, its low lying coastal areas (2.42% of its total geographical area) include 8.25% of its total population, with an average coastal density of 2931 per/km^2; 42% of urban municipalities, and 83% of municipal corporations are located in the coastal zone of the state (Table 13.3). The geomorphic features of Kerala Coastal Zone include denudational hills, headlands, and coastal cliffs beside the normal coastal landforms of coastal plains beaches, barrier reefs, bars, backwaters, and lagoons (Fig. 13.3).

Table 13.1 Local bodies of Kerala coastal zone districts.

District	Gramapanchayat	Municipality	Corporation
Kasaragod	Manjeshwar, Kumbla, Mangalpady, Meenja, Chemnad, Mogral-Puthur, Puthige, Chengala, Ajanur, Bedadukka, Muliyar, Pallikkara, Uduma, Kayyur-Cheemeni, Kinanur-Karinthalam, Pullur-Periya, Cheruvathur, Padanna, Trikkaripur, Valiyaparamba, West-Eleri	Kasaragod, Nileshwar, Kanhangad	Nil
Kannur	Karivellur-Peralam, Cheruthazham, Kunhimangalam, Madayi, Ramanthali, Chapparapadavu, Chengalayi, Kurumathur, Mayyil, Pariyaram, Cherukunnu, Ezhome, Kadannappally-Panapuzha, Kannapuram, Mattool, Pattuvam, Chokli, New Mahe, Thrippangottur, Chirakkal, Kolacheri, Narath, Pappinisseri, Azhikode, Kalliasseri, Valapattanam, Chembilode, Dharmadam, Eranholi, Kadambur, Muzhappilangad, Peralassery, Pinarayi, Vengad, Anjarakandy, Keezhallur, Kadirur	Payyannur, Thaliparambu, Thalassery, Anthoor, Panoor	Kannur
Kozhikode	Azhiyur, Chorode, Edacherry, Eramala, Onchiyam, Thuneri, Cheruvannur, Keezhariyur, Maniyur, Moodadi, Thikkodi, Thiruvallur, Thurayur, Velom, Atholi, Chemancherry, Chengottukavu, Naduvannur, Ulliyeri, Thalakulathur, Balusseri, Kottur, Chelannur, Kakkodi, Mavoor, Olavanna, Perumanna, Peruvayal, Kadalundi	Vadakara, Koilandi, Feroke, Payyoli, Ramanattukara	Kozhikode
Malappuram	Vazhakkad, Vazhayur, Chelembra, Thenchipalam, Vallikkunnu, Moonniyur, Cheriyamundam, Niramaruthur, Tanalur, Thalakkad, Vettom, Kaladi, Mangalam, Maranchery, Purathur, Triprangode, Veliyancode, Perumpadappu	Parappanangadi, Tirurangadi, Tanur, Tirur, Ponnani	Nil
Thrissur	Punnayoor, Punnayoorkulam, Engandiyoor, Kadappuram, Manalur, Mullassery, Orumanayoor, Pavaratty, Vadanappilly, Venkitangu, Edathiruthy, Kattoor, Thanniam, Anthikkad, Nattika, Thalikulam, Valappad, Kaipamangalam, Mala, Mathilakam, Padiyoor, Perinjanam, Puthenchira, Sree Narayanapuram, Vellangalur, Edavilangu, Eriyad, Poyya	Chavakkad, Guruvayoor, Kodungallur	Nil
Errnakulam	Chendamangalam, Chittattukara, Ezhikkara, Kottuvally, Kuzhappilly, Pallippuram, Puthenvelikara, Vadakkekkara, Edavanakkade, Elamkunnapuzha, Kadamakkudy, Mulavukade, Nayarambalam, Njarakkal, Varapuzha, Alangad, Karumalloor, Kunnukara, Cheranallur, Kadungallur, Chellanam, Kumbalam, Kumbalangy, Udayamperoor, Amballur	Paravur, Eloor, Thrikkakara, Kalamassery, Thrippunithara, Maradu	Kochi

Continued

Table 13.1 Local bodies of Kerala coastal zone districts.—cont'd

District	Gramapanchayat	Municipality	Corporation
Kottayam	Chempu, Maravanthuruthu, Udayanapuram, T.V. Puram, Thalayazham, Vechoor	Vaikom	Nil
Alappuzha	Devikulangara, Krishnapuram, Pathiyoor, Arookutty, Aroor, Perumpalam, Chennam-Pallippuram, Panavally, Ezhupunna, Kodamthuruth, Kuthiathode, Pattanakkad, Thuravoor, Thycattussery, Vayalar, Thanneermukkom, Cherthala South, Kadakkarapally, Mararikkulam North, Aryad, Mararikkulam South, Ambalappuzha North, Ambalappuzha South, Punnapra North, Punnapra South, Purakkad, Chingoli, Karthikapally, Thrikkunnappuzha, Arattupuzha, Kandalloor, Muthukulam	Kayamkulam, Cherthala, Alappuzha, Haripad	Nil
Kollam	East Kallada, Kunnathor, Mandrothuruthu, Pavithreswaram, Perayam, Sasthamcotta, West Kallada, Alappad, Chavara, Kulasekharapuram, Mynagappally, Panmana, Thevalkkara, Thodiyoor, Clappana, Adichanallur, Chathannur, Nedumpana, Panayam, Perinad, Neendakara, Thekkumbhagom, Thrikkadavoor, Thrikkaruva, Chirakkara, Mayyanad, Poothakkulam	Karunagapally, Paravoor	Kollam
Thiruvananthapuram	Anchuthengu, Edava, Vakkom, Vettoor, Azhoor, Cherunniyoor, Chirayinkeezhu, Kadakkavoor, Kadinamkulam, Kizhuvilam, Manamboor, Mangalapuram, Ottoor, Andoorkonam, Venganoor, Elakamon, Chenkal, Karode, Karumkulam, Kottukal, Kulathoor, Poovar, Thirupuram	Varkala, Attingal	Thiruvananthapuram

Table 13.2 Local bodies of Kerala coastal zone districts.

Sl. No:	District	Gramapanchayat	Municipality		Corporation	
1	Kasargod	21	3	3	0	0
2	Kannur	37	5	9	1	1
3	Kozhikode	29	5	7	1	1
4	Malappuram	18	5	12	0	0
5	Thrissur	28	3	7	0	1
6	Ernakulam	25	6	13	1	1
7	Kottayam	6	1	6	0	0
8	Alappuzha	32	4	6	0	0
9	Kollam	27	2	4	1	1
10	Thiruvananthapuram	23	2	4	1	1
TOTAL		**246**	**36**	**87**	**5**	**6**

Table 13.3 Population scenario of Kerala coastal zone.

Type of Local Body (LB)	Low population[a]	Moderate population[b]	High p[c]	Very high p[d]	NA	Total
Gramapanchayat	8	21	115	98	4	**246**
Municipality	0	0	5	23	8	**36**
Municipal Corporation	0	0	0	5	0	**5**
Total	**8**	**21**	120	126	**12**	287

[a]<500 = low population.
[b]501–1000 = moderate population.
[c]1001–2000 = high population.
[d]>2000 = very high population.

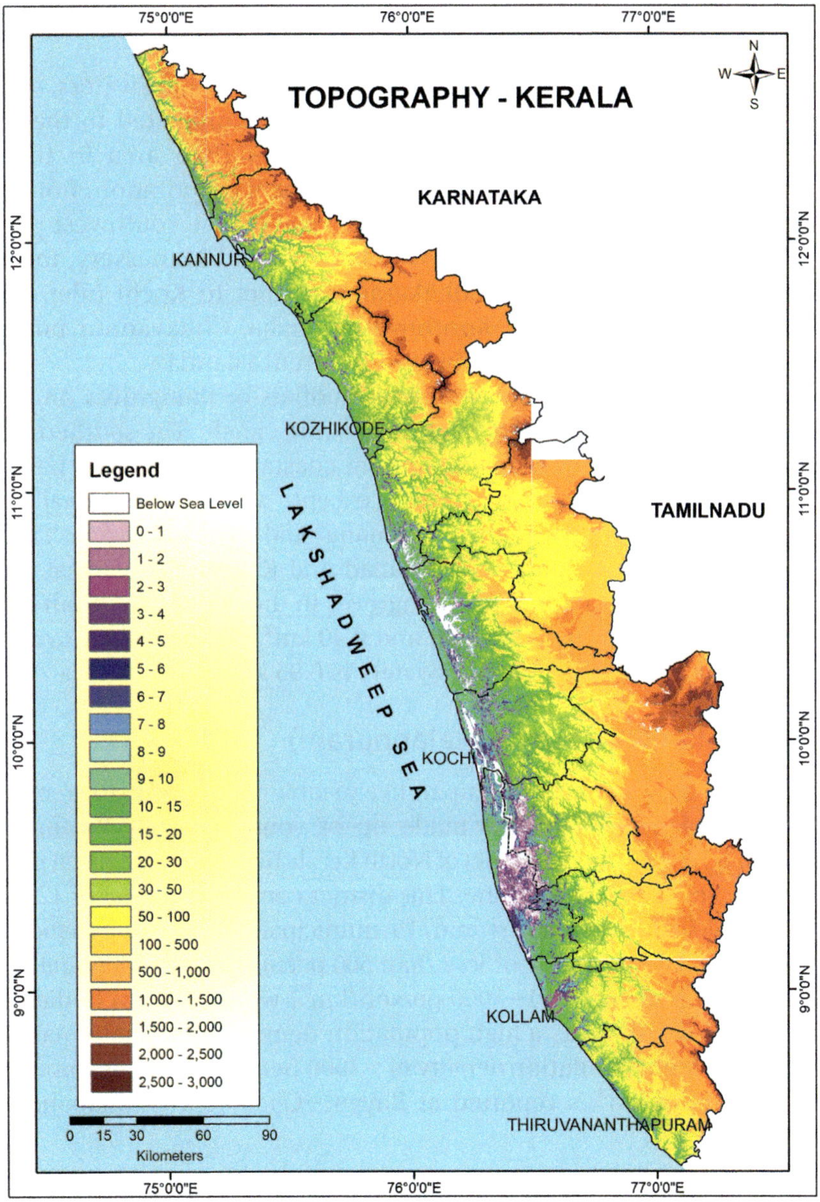

FIGURE 13.3 Physical setup of Kerala coastal zone.

The details of major mangrove ecosystems are given in Table 13.4. Being a high-energy coast, the seacoast of Kerala does not report any mangroves or mudflats. Valapattanam Kayal, Thaliparamba backwaters, Vembanad lake and associated waterbodies, and Ashtamudi Lake are the major coastal ecosystems of the area. The other important coastal ecosystems are Poyya, Aikady (Kumble), Madhuvahini, Chandragiri, Kuppam, Akalapuzha, Kadalundi, Chettuva, Pallippuram, Cherai–Nayarambalam, Ezhikkara–Mulavukadu, Vypin, Kayamkulam, Paravur Kayal, Edava Kayal, Anjuthengu–Kadinamkulam, Edayar (Karamana), and Poovar. There are number of small and large backwater islands and islets in the Kerala coastal region that provide a good habit for coastal ecosystems like Mulavukadu, Perumbalam, and Kumbalam in Vembanad Lake; Munroe Island in Ashtamudi Lake; and Kavvai island in Valiyaparamba Kayal.

Vembanad and associated waterbodies are a network of backwaters, estuaries, and interconnecting canals. These backwater systems get connected to the sea through Kochi tidal inlets. Kochi corporation is the largest urban area in the state, and Maradu municipality is a fast-growing suburb of Kochi corporation. Both consist of a number of backwater islands in Vembanad backwater. The southwest part of Kochi, consisting of Fort Kochi, Mattanchery, Mundamveli, and Manassery, forms part of a barrier island that extends from Andhakaranazhi inlet to Kochi inlet. Champakara canal, Nettur–Kundannur puzha, Konthuruthy puzha, Chilavannur puzha, Thevara puzha, and Ayini thodu divide Maradu into different islands.

The seacoast of Kerala does not have any mudflats or mangroves on its sea shore. Mangroves are confined to estuaries, backwaters, and canals. The southern sector of the southwest coast along the Thiruvananthapuram coastal zone does not have any significant mangrove and mudflat ecosystem except a few very small patches in Kadinamkulam and Veli kayals and Karamana and Neyyar rivers. The mangrove coverage along the southern part of Vembanad and Kayamkulam Kayal in Alappuzha district of Kerala is 1.43 km^2. Spread of mangroves in the two major urban centers Kochi and Maradu adjoining Vembanad is 0.48 and 0.40 km^2, respectively. Coverage of mangroves in Ashtamudi and adjoining backwaters is 0.95 km^2.

4.1 North Kerala (Kasaragod–Malappuram)

North Kerala or the Malabar coast is comparatively stable compared to the remaining part of Kerala. This coastal plain is mainly made up of young coastal alluvium and mixed denudational hills. The Kasargod district of North Kerala has a narrow coastal plain and is the least populated coastal belt of the state. This district comprises a total of 125 local bodies including 2 municipal corporations and 18 municipalities. 8 Gramapanchayats come under the low population density of less than 500 person/km^2 and 13 Gramapanchayats are of moderate population (501–1000 person/km^2) while 55 local bodies including 3 municipalities are classified as a high population density area. The remaining 49 local bodies have a very high population density of >2000 person/km^2. The lowest population density of 324 person/km^2 is reported at Kayyur–Cheemeni Gramapanchayat of the

Table 13.4 Major mangrove ecosystem potentials of Kerala coastal zone.

District	Backwaters and rivers	Major mangrove locations	Major mangrove species observed	Total area of mangroves (km^2)
Kasaragod	Manjeswar, Shiriya, Valiyaparamba backwaters	Poyya, Shiriya, Kumbala, Mogral, Madhuvahini, Chandragiri	*Excoecaria agallocha*, *Rhizophora mucronata*, *Avicennia officinalis*, *Acanthus ilicifolius*	1.22
Kannur	Valiyaparamba backwaters, Kavvayi backwaters, Kottappuram backwaters, Kuppam river, Valapattanam river, Anjarakkandi river, Mahi river	Elambachi, Kunhimangalam, Pattuvam, Arippambra-Kutyeri-Panakkad, Azhikode, Nadal, Melur, Dharmadam, Kuyyali, Mayyazhi,	*Excoecaria agallocha*, *Bruguiera gymnorrhiza*, *Barringtonia* sp., *Avicennia marina*, *R. mucronata*, *Sonneratia caseolaris*, *Acanthus ilicifolius*, *Lumnitzera racemosa*, *Kandelia candel*, *A. officinalis*	9.1
Kozhikode	Kuttiyadi river, Korappuzha, Kallayi river, Chaliyar river, Alakapuzha backwater, Mahi river	Kuttiyadi river, Akalapuzha, Korappuzha, Kalipoika, Kadalundi,	*Avicennia officinalis*, *R. mucronata*, *A. ilicifolius*, *Excoecaria agallocha*, *B. gymnorrhiza*, *K. candel*	1.12
Malappuram	Beeyam lake, Bharathappuzha, Tirur river, Kadalundi river	Keezhayil, Kootayi	*Avicennia officinalis*, *R. mucronata*, *A. ilicifolius*	0.65
Thrissur	Chettuva backwater, Enamavu lake, Karuvannur river, Periyar river	Chettuva, Methala, Athikadavu	*Rhizophora mucronata*, *A. ilicifolius*, *A. officinalis*, *A. marina*	1.32
Ernakulam	Vembanad lake, Veeranpuzha lake, Periyar river	Mulavukadu, Kadamakkudi, Vypin, Marad, Vembanad	*Avicennia officinalis*, *A. marina*, *R. mucronata*, *S. caseolaris*, *A. ilicifolius*, *E.agallocha*, *Br. gymnorrhiza*, *L. racemosa*, *K. candel*, *Barringtonia* sp.	5.56
Kottayam	Vembanad lake, Muvattupuzha river	Poothotta, Murinjapuzha, Vadakkemuri, Thalayazham	*Br. gymnorrhiza*, *L. racemosa*, *A. ilicifolius*, *Sonneratia alba*	0.01
Alappuzha	Kayamkulam kayal, Vembanad estuary-Vayalar kayal and Ezhupunna kayal	Puthuppally, Arattupuzha, Karthikappally, Kandalloor, Pallipuram, Thaikkattussery, Panavalli, Vaduthala, Arookutti, Aroor, southeast of Chellanam, Ezhupunna, Thuravur	*Avicennia officinalis*, *A. marina*, *R. mucronata*, *E.agallocha*, *B. gymnorrhiza*, *L. racemosa*, *Acanthus ilicifolius*, *S. alba*	1.29

Continued

Table 13.4 Major mangrove ecosystem potentials of Kerala coastal zone.—cont'd

District	Backwaters and rivers	Major mangrove locations	Major mangrove species observed	Total area of mangroves (km^2)
Kollam	Edava Nadayara kayal, Paravur kayal, Ashtamudi kayal, Vattakayal and Kayamkulam kayal (part) Ithikkara river, Kallada river, Pallikkal river	Ashtamudi kayal (Neendakara village and Thekkumbhagom), Paravur kayal (Adichanalloor area), south of Kayamkulam inlet of Vattakayal (Panmana area), the banks of Kayamkulam kayal	*Avicennia officinalis*, *R. mucronata*, *E. agallocha*, *Acanthus ilicifolius*	0.88
Thiruvananthapuram	Nadayar kayal, Anjuthengu kayal, Kozhithottam kayal, Mungottu kayal, Kadinamklam kayal and Veli-Akkulam lake Neyyatinkara river, Karamana Ar, Vamanpuram river, Killi Ar, Karichal canal and TS canal	Anjuthengu kayal and Vamanapuram river	*Avicennia officinalis*, *R. mucronata*	0.02

Kasargod district and the highest population density of North Kerala is recorded at Kannur municipal corporation.

Kannur district of North Kerala is described as the area of the highest coastal mangrove ecosystem of Kerala with a total mangrove spread of 9.1 km^2. And it is the only district with major mangrove potential in North Kerala. The important mangrove potentials of the sector are Kuppam–Valapattanam river basin, Payyannur–Perumba river basin, Dharmadam, Kumbala, and Akalapuzha backwaters. The other mangrove concentrations include Poyya, Shiriya, Mogral, Maduvahini, Elambachi, Azhikode, Kuyyali, Mayyazhi, Korapuzha, and Kalipoika.

4.2 Central Kerala (Thrissur–Alappuzha)

The central Kerala or Cochin coastal plain is the vast and most populated coastal sector of the state. It comprises the 106 local bodies including 1 municipal corporation and 14 municipalities. There is no local body classified as low population density. The lowest population density of 585 person/km^2 is reported at Vechoor Gramapanchayat of the Kottayam district and the highest population density of 5914 person/km^2 is at Kochi corporation of the Ernakulam district. Five Gramapanchayats are classified as moderately populated and 46 local bodies including 2 municipalities are grouped into highly populated areas while the remaining 55 local bodies including 1 municipal corporation and 12 municipalities are recorded with a population of very high density.

This coastal belt is known for its backwater ecosystems and long parallel barrier reefs. The great Vembanad backwater system, extending from the Periyar river basin in north Ernakulam to the Kuttanad region of the Alappuzha district is the cradle of the coastal ecosystem of the state, and it is designated as one of the Ramsar sites under the Ramsar Convention (Fig. 13.4). The total area of mangroves in Kochi corporation is 0.49 km^2, which is distributed along the backwaters, interconnecting canals, and filtration ponds. Edakochi has the largest coverage with 0.18 km^2 mangrove forest. The most abundant mangrove species in this patch of mangrove community is *Avicennia officinalis*. One of the rare species of mangrove, *Kandelia candel*, is also found in this patch. The backwaters of Willington Island, Vathuruthi area, Mattummal Backwater Island, Thanthonnithurth, and Mangalavanam area also has thick patches of mangrove forest. The mangrove ecosystems in Maradu cover an area of 0.41 km^2, which is distributed in the backwaters, its interconnecting canals, and in filtration ponds. The backwater island, Nettor, which is surrounded by Kundanoor thodu on the east, Nettor thodu on the south, and Vembanad backwater on the west and north side has a maximum mangrove coverage of 0.20 km^2, of which 51,595.83 m^2 is occupied in Kundannor thodu near Kaithamanakara. The major mangrove species identified in Maradu municipality are *Avicennia officinalis, Rhizophora mucranata, Acanthus ilicifolius, Bruguiera gymnorhiza, Lumnitzera racemosa, Bruguiera sexangula, Derris trifoliate*, and *Barringtonia racemosa*. Mangrove associates such as *Lannea coromandelica, Cerbera manghas, Premna serratifolia, Clerodendrum inerme, Thespesia populnea, Cayratia trifolia, Syzygium* sp., and *Acrostichum aureum* (fern) are also observed.

The mangroves are thickly distributed in the Alappuzha district, especially at the Puthuppally, Arattupuzha, Karthikappally, and Kandalloor regions of Kayamkulam Kayal; Pallipuram, Thyccattussery, Panavalli, Vaduthala, Arookutti areas of Vayalar Kayal; Aroor and southeast of Chellanam of Vembanad kayal; and in Ezhupunna, Thuravur regions of Ezhupunna Kayal (Fig. 13.4). Extensive mangrove patches are also seen in the active Pokkali fields of Kadakkarapally, Kodamthuruth, and Pattanakadu regions. Eight exclusive species of mangroves are observed in Alappuzha district. *Avicennia officinalis, Avicennia marina, Rhizophora mucronata, Excoecaria agallocha, B. gymnorrhiza, L. racemosa, Acanthus ilicifolius*, and *Sonneratia alba* are the mangrove species observed in Kayamkulam kayal, Vayalar kayal, and Vembanad kayal and its interconnecting canal systems in the district. *Excoecaria agallocha* and *Avicennia officinalis* are the most dominant and abundant mangrove species in the Alappuzha district. The major mangrove associates observed are *Acrostichum aureum, Pandanus odoratissimus, C. trifolia, Cerbera odollam, C. inerme, Calophyllum inophyllum, P. serratifolia*, and *Thespesia populnea*. The total area of mangroves is 1.43 km^2 with a protective buffer zone of 5.40 km^2. Fig. 13.5 gives an abstract of cadastral scale mapping with a few major mangrove and mudflat (intertidal zone) coverage areas of Vembanad Lake.

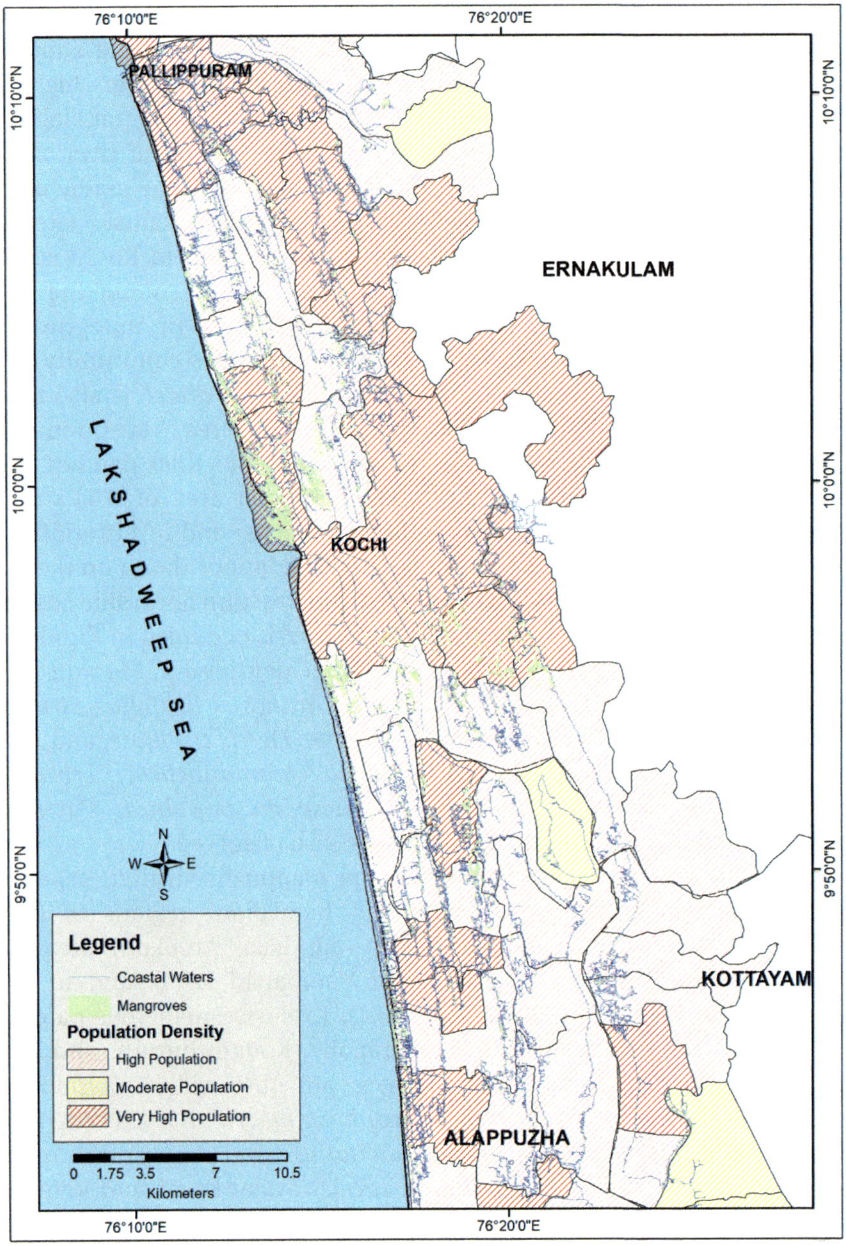

FIGURE 13.4 Significant mangrove ecosystems and population density along Vembanad backwater system.

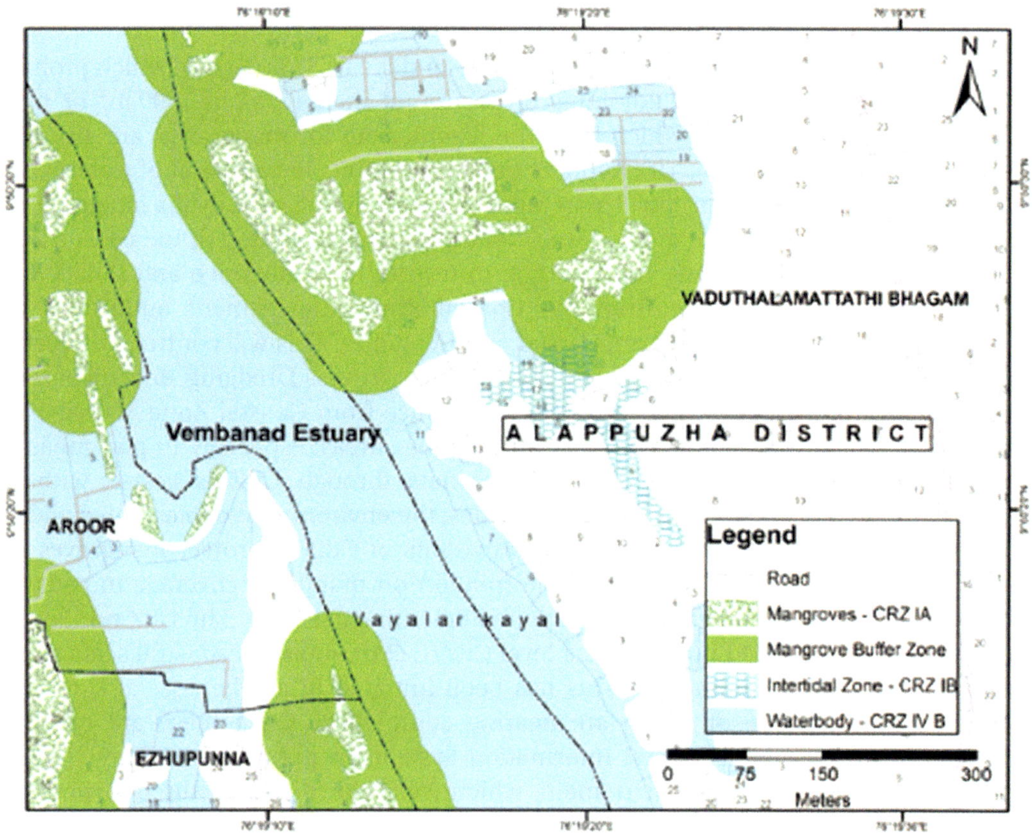

FIGURE 13.5 Mangroves and other coastal ecosystems mapping in cadastral scale—Vemband lake of Alappuzha district.

4.3 South Kerala (Kollam and Thiruvananthapuram)

The southernmost coastal belts of Kerala comprise the narrow young coastal plains and lateritic cliffs and a number of headlands. Out of 56 local bodies of this sector, 3 Gramapanchayats are listed in moderately populated classes and 19 Gramapanchayats are grouped as highly populated regions. The remaining 34 local bodies including 4 municipalities and 2 municipal corporations are in the very high population category. Mundrothuruth of Kollam district has the lowest population density of 706 person/km^2, and Kollam municipal corporation with a population density of 5885 person/km^2 is reported as the highest in South Kerala.

The spread of mangrove is mostly limited to lower stretches of the Ashtamudi estuary (Neendakara village and Thekkumbhagom) in Kollam district. Patches of mangroves are also observed on the banks of Paravur kayal (Adichanalloor area), Vattakayal (in Panmana) south of Kayamkulam inlet, and along the banks of Kayamkulam kayal. Small patches of mangroves are also observed at Manamboor and Kadakkavoor along the bed of Anchuthengu kayal and Vamanapuram river. The major mangrove species observed are *Avicennia officinalis* and *Rhizophora mucronata* in this sector.

4.4 CRZ notification and mangrove conservation

The CRZ notification has categorized mangroves under CRZ IA category, which prohibits all activities in this zone and adjoining 50 m around mangroves of 1000 m^2 or more. Biologically active mudflats, which form the ideal terrain for mangroves, are also categorized as CRZ IA (Fig. 13.6). The notification also prohibits reclamation of waterbodies, tidal flats, and mudflats, which will help regeneration of mangroves over a longer period. Discharge of untreated effluents and waste dumping are prohibited in the coastal zone. Construction of tidal barrages and bunding to restrict tidal influence and saline water flow now requires clearance from the Coastal Zone Management authorities and Ministry of Environment and Forests based on EIA studies. This will control the blocking of saline water for the gradual destruction of mangroves. Dredging and disposal of dredge spoil from backwaters also requires clearance from Coastal Zone Management authorities, which will help reclamation of mangrove areas, especially in port areas.

The management of mangroves and mudflats through CRZ is vested with the respective Coastal Zone Management authorities, the environment departments and the local bodies. The implementation of the provisions of CRZ to protect mangroves and mudflats are possible only if accurate information on mangrove coverage in cadastral scale (survey plotwise) is available for the implementing agencies. The CRZ notification gives specific direction to prepare local level CRZ/CZMP maps in cadastral scale for the use of implementing authorities. This has been initiated in all the coastal states and union territories, some of which are nearing completion. Once these are uploaded through the web-enabled Coastal Information System for public access, any violation will be scrutinized in the public domain, which will help to control the destruction of mangroves and mudflats.

5. Results

A current estimate of mangrove forests of the world is less than half of what it once was (Spalding et al., 1997; Spiers, 1999), and much of what remains is in a degraded condition. About 35% of mangroves were lost from 1980 to 2000 (MA, 2005), and the forests have been declining at a faster rate than inland tropical forests and coral reefs (Duke et al., 2007). The largest extent of mangroves is found in Asia (42%) followed by Africa (20%), North and Central America (15%), Oceania (12%), and South America (11%). Mangrove coverage has been showing a declining trend over the last few years along the southwest coast, especially along the Kerala coastal zone (Mohanan, 1999).

Mangroves have been reported to be present in different parts of the southwest coastal zone of India. The seacoast of Kerala is high energy and sandy (significant wave height generally > 0.50 m and average sediment size about 0.28 mm), which is not conducive for mangroves (Anitha and Thomas, 2006). In Kerala, the extent of mangroves is reported to be maximum in Kannur (Mohanan, 1999; Kurian et al., 1995). The total mangrove coverage has been reported as 50 km^2 by Mohanan (1999) while Kurian et al. (1995)

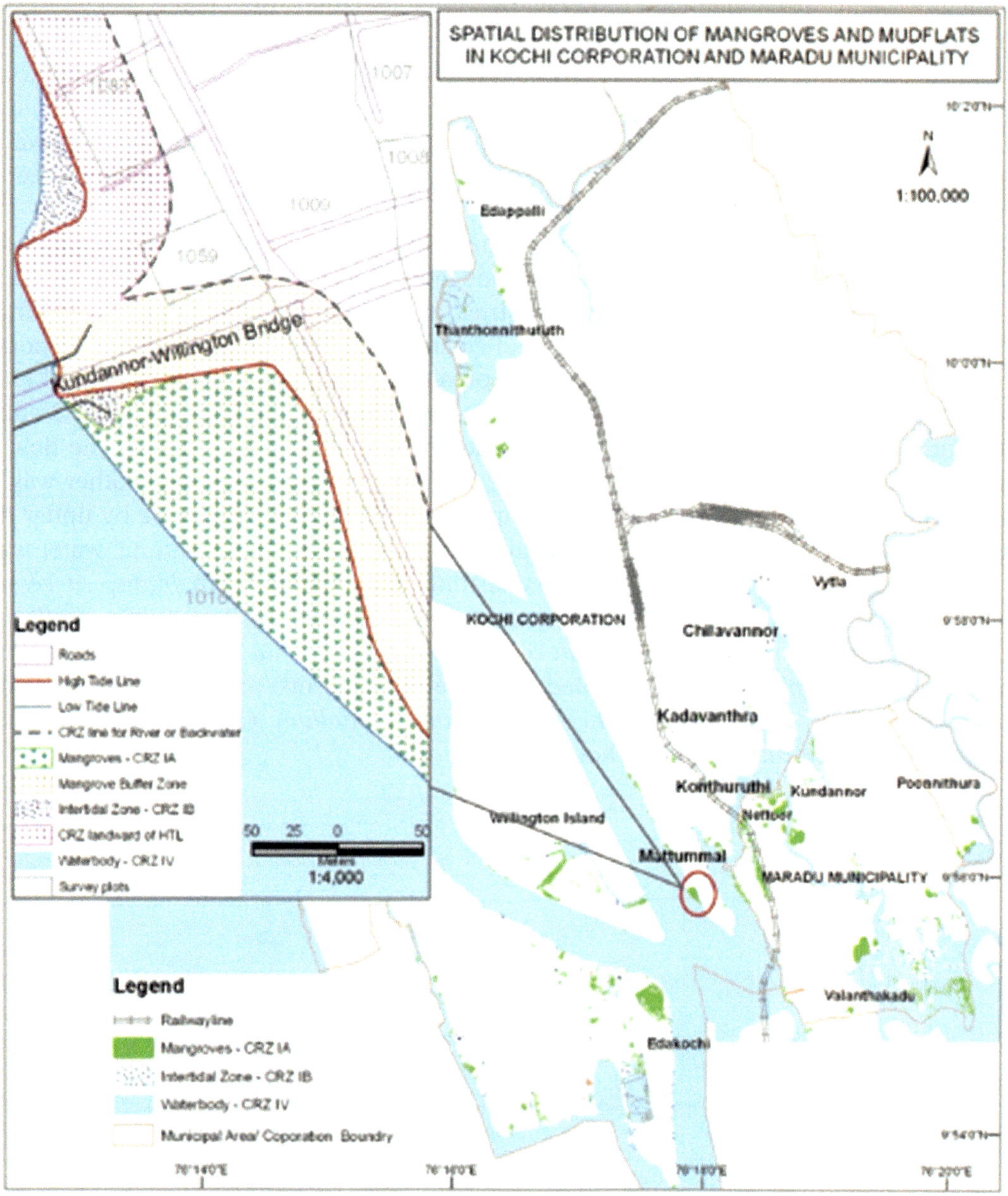

FIGURE 13.6 CRZ mapping and conservation of coastal ecosystems in Kochi corporation and Maradu municipality.

reported only 11 km^2. The CZMP of Kerala (CESS, 1995), which is prepared in a large scale, reported the area of mangrove coverage as 63 km^2. This huge mangrove potential of Kerala is reduced to 9.1 km^2 as per the current study. In the Ernakulam district, 260 ha of mangrove ecosystem was reported by Unni (1998). About 14 dominant species of mangroves have been identified in Kerala by Mohanan (1999), of which 7 species of true mangroves and 2 species of semimangroves have been reported in Kumbalam island in Ernakulam district (Arun and Shaji, 2013). One of the main reasons for the differences in the area is because of the different data sources and methods used and the scale of mapping (Anitha and Thomas, 2006). In the present study, the mapping of mangroves is done in cadastral level based on field mapping and submeter-level resolution imageries for the better result.

The sensitive mangrove ecosystems of the southwest coast of India are under constant flux due to both natural and anthropogenic forces. Large-scale reclamation of wetlands, discharge of effluents from industries, as well as dredged spoils (Fig. 13.7) from the Port Trust have increased significantly. Reclamation of mudflats and filtration ponds (Figs. 13.8 and 13.9) have been taking place on a large scale in the urban areas of Kerala coastal plain, especially in the Kochi–Maradu urban fringe of the central Kerala. Gradual destruction of mangroves is carried out without being noticed by blocking the flow of brackish water into the extensive mangrove vegetations in Puthuvype. Another way of removing mangroves unnoticed is filling up of mangrove and areas close by under the pretext of disposing dredged spoil from the shipping channels and national water way.

Ten exclusive mangrove species consisting of *Avicennia officinalis*, *Avicennia marina*, *Rhizophora mucronata*, *S. alba*, *Acanthus ilicifolius*, *Excoecaria agallocha*, *B. gymnorrhiza*, *L. racemosa*, *Kandelia candel*, and *Barringtonia* sp. are observed along southwest coast of India now studied. The major mangrove associates observed are *P. odoratissimus*, *C. trifolia*, *Acrostichum aureum*, *C. odollam*, *C. inerme*, *Calophyllum inophyllum*, *P. serratifolia*, and *Thespesia populnea*.

FIGURE 13.7 Disposing dredged spoils in mangrove ecosystem by Port Trust.

FIGURE 13.8 Reclamation of coastal ecosystems.

FIGURE 13.9 Reclamation of mudflats at Kundannor.

6. Conclusion

Mangrove forests around the globe are under immense pressure from clear-cutting, land-use change, hydrological alterations, chemical spill, and climate change, which are equally applicable to the southwest coast of India. In the future, the ongoing accelerated sea-level rise and overexploitation by population explosion in a nonsustainable manner could be the biggest threat to mangrove ecosystems. The geospatial distribution map in cadastral level generated will serve as a baseline to develop adaptive management strategies in anticipation of sea-level rise, set conservation priorities, monitor the reclamation of coastal wetlands, improve terrestrial carbon accounting, and quantify the ecological functions of mangrove forests. The current study encompasses the most comprehensive mapping of mangrove ecosystems and population distribution ever attempted in this part of the southwest coast of India. Detailed mapping of mangroves in the cadastral scale has brought out many mangrove areas that were hitherto not reported.

Lack of having an accurate and detailed database on the spread of mangrove ecosystem in an easily accessible format has been the major constraint in its conservation and management through CRZ. The detailed cadastral level mapping of coastal ecosystems as part of the CZMP for CRZ applications provides a detailed inventory of mangroves and mudflats in the coastal zone. Regardless of whether ownership is private or public, all the mangroves and biologically active mudflats are categorized as CRZ IA with an adjoining buffer zone. It gives maximum protection by prohibiting any activity within the mangrove area and in the adjoining 50 m buffer zone. At the same time, the private ownership of a major part of mangroves could result in conflicts of usage, which needs to be addressed for the sustenance of mangroves. The government or other agencies could think of acquiring privately owned mangrove areas and potential mangrove habitat such as mudflats and giving appropriate compensation. This could help in resolving conflict of interests and promote healthy growth of mangroves. The present study provides cadastral level information on mangroves and mudflats for a small part of the coastal zone of the southwest coast of India. The study is being further continued to cover the remaining areas. The web-enabled Coastal Information System under development is expected to provide easy access to decision-makers and the public all information on mangroves and mudflats, which will take us a long way toward its conservation.

Acknowledgments

Dr. P.C. Rao, Director, NCESS, is thanked for his support and valuable guidance. The active support and direction of Dr. K.V. Thomas, Dr. K.K Ramachandran, Dr. C.N Mohanan, Dr. K. Raju, and Dr. L. Sheela Nair are well remembered and acknowledged. The great effort done by Mr. M. Rameshkumar, Mr. S Mohanan, Mrs. Abhi Aravind, Mrs. T. Neelima, and Mrs. R.V. Viji were the foundation of this work and we will be thankful toward your effort. The authors are also thankful to all the administrative and project staff of the Coastal Processes Group of NCESS for their support.

References

Anitha, P.M., Thomas, K.V., 2006. Conservation of Mangroves through Coastal Regulation Zone for Coastal Hazard Management. Extended Abstracts, XVIII Kerala Science Congress, Thiruvananthapuram, pp. 249–251.

Arun, T.R., Shaji, C.N., 2013. Diversity and distribution of mangroves in Kumbalam island of Kerala, India. J. Environ. Sci. Toxicol. & Food Technol. 4 (4), 18–26.

Baconguis, S.R., Mauricio, R.A., 1991. Forage and livestock production in the mangrove forest. Canopy 16, 9–10.

Basha, S.C., 1991. Distribution of mangroves in Kerala. Indian For. 117 (6), 439–448.

Basha, S.C., 1992. Mangroves of Kerala: a fast disappearing asset. Indian For. 118 (3), 175–190.

Census India, Population Finder, 2011, 2019. http://www.censusindia.gov.in/pca/Searchdata.aspx.

CESS, 1995. Coastal Zone Management Plan of Kerala. Centre for Earth Science Studies, Thiruvananthapuram, p. 77.

CESS, 1998. Coastal Landuse and Resources. Centre for Earth Science Studies, Thiruvananthapuram, p. 176.

Chattopadhyay, S., Mahamaya, C., 2007. Field Workshop – IGCP. Geomorphology of Kerala: an Overview, vol. 514, pp. 64–71.

Clough, B.F., 1982. Mangrove Ecosystems in Australia - Structure, Function and Management. ANU Press.

Coastal Regulation Zone Notification, Ministry of Environment and Forests, Government of India, New Delhi, Gazette of India, Extraordinary, Part-II, Section 3, Sub-section (ii) of dated the 6th January, 2011, http://moef.gov.in/.

Coastal Regulation Zone Notifications, 19th February 1991. Government of India, New Delhi, The Gazette of India vide number S.O. 114(E). http://moef.gov.in/.

Duke, N.C., Meynecke, J.O., Dittmann, S., Ellison, A.M., Anger, K., Berger, U., Cannicci, S., Diele, K., Ewel, K.C., Field, C.D., Koedam, N., Lee, S.Y., Marchand, C., Nordhaus, I., Dahdouh-Guebas, F., 2007. A world without mangroves? Science 317 (5834), 41–42.

Kathiresan, K., 2011. Eco-biology of mangroves. In: Metras, J.N. (Ed.), Mangroves: Ecology, Biology and Taxonomy. Nova Science Publishers, New York, p. 386.

Kurian, N.P., Samsuddin, M., Ramachandran, K.K., Salim, 1995. Resource evaluation using remote sensing for aquaculture site selection. Proc. 6th Kerala Sci. Congr. 23–25.

MA (Millennium Ecosystem Assessment), 2005. Mille Ecosystems and Human Well-Being: Synthesis. Island Press, Washington, DC.

Mohanan, C.N., 1997. In: Thampi, K.B., Nayar, N.M., Nair, C.S. (Eds.), Mangroves. The National Resource of Kerala, pp. 149–158. Thiruvananthapuram.

Mohanan, C.N., 1999. Mangroves. In: Thampi, K.B., Nair, N.M., Nair, C.S. (Eds.), Natural Resources of Kerala, World Wide Fund for Nature – India, pp. 149–158.

Mohanan, C.N., 2004. Conservation of the fragile ecosystem diversity of Kerala. In: CESS, Earth System Science and Natural Resources Management. Centre for Earth Science Studies, Thiruvananthapuram, pp. 381–398.

Mohanan, C.N., Thomas, K.V., 2011. The status of mangroves in Kerala. In: Bhatt, J.R., Macintosh, D.J., Nayar, T.S., Pandey, C.N., Nilaratna, N.P. (Eds.), Towards Conservation and Management of Mangrove Ecosystems in India. IUCN, India, pp. 154–159.

Qasim, S.Z., Kathiresan, K., 2005. Biodiversity of Mangrove Ecosystems. Hindustan Publishing Corporation, New Delhi.

Spalding, M.D., Blasco, F., Field, C.D. (Eds.), 1997. World Mangrove Atlas. The International Society for Mangrove Ecosystems, Okinawa.

Spalding, M., Kainuma, M., Collins, L., 2010. World Atlas of Mangroves. Earth- scan Publications, London, p. 319.

Spiers, A.G., 1999. Review of international/continental wetland resources. In: Finlayson, C.M., Spiers, A. G. (Eds.), Global Review of Wetland Resources and Priorities for Wetland Inventory, Supervising Scientist Report 144, pp. 63–104. Canberra, Australia.

Tomlinson, P.B., 1986. The Botany of Mangroves. Cambridge University Press, Cambridge.

The Environment (Protection) Act, 1986. Ministry of Environment and Forests. Government of India, New Delhi. http://moef.gov.in/rules-and-regulations/environment-protection/.

The Kerala Conservation of Paddy Land and Wetland Act, 2008. Kerala Gazette, Extra ordinary. Government of Kerala, Thiruvananthapuram, 12th August 2008. http://sanitation.kerala.gov.in/wp-content/uploads/2017/07/the-kerala-conservation-of-paddy-land-and-wetland-act-2008.pdf.

Unni, P.N., 29-31 January 1998. Biodiversity of Mangrove Vegetation in Wetlands of Kerala. International Conference on Asian Wetlands, New Delhi and Bharatpur, India.

14

Modeling of coastal environmental vulnerability in South India: a multiple parametric approach using remote sensing and GIS

Kaliraj Seenipandi[1], K.K. Ramachandran[1], N. Chandrasekar[2]

[1]NATIONAL CENTRE FOR EARTH SCIENCE STUDIES (NCESS), THIRUVANANTHAPURAM, KERALA, INDIA; [2]CENTRE FOR GEOTECHNOLOGY, MANONMANIAM SUNDARANAR UNIVERSITY & FRANCIS XAVIER ENGINEERING COLLEGE, TIRUNELVELI, TAMIL NADU, INDIA

Chapter outline

Remote Sensing of Ocean and Coastal Environments. https://doi.org/10.1016/B978-0-12-819604-5.00014-7

1. Introduction

The coastal zones consist of a special set of environments where marine and terrestrial areas influence each other. Sixty percent of the world's major cities are located in coastal zones, and 40% of the global populations live within 100 km of the coastline (IPCC, 1991; Adger, 1996; Bossel, 1999; Cutter, 2003; Barnett et al., 2008). The coastal environment is the life-support system for the ecosystem within the transitional zone that is dynamically influenced by marine and terrestrial factors. The environmental systems have tightly intertwining relationships between humans and coastal resources, which are amplifying the terrestrial and marine environmental ecosystems within a transitional zone (IPCC, 1991). The coastal environments include sandy beaches, bays, backwater creeks, sheltered estuaries, exposed rocky headlands, dramatic cliffs, wild dunes, tidal wetlands, mangrove swamps, and inshore reef systems, and serve as critical habitats for a diverse range of flora and fauna for feeding, breeding, nursery, and resting areas (Peterson et al., 1998; Pethick and Crooks, 2000; Cutter, 2003; Gallopin, 2006). In recent decades, coastal environments across the world are under severe environmental stress due to a wide range of natural and anthropogenic factors, including land use/land cover change, placer mining, sea-level rise, global warming, and pollution. These factors are inducing massive impacts on coastal and marine environmental ecosystems (Pantin, 1997; Bossel, 1999; Lambert, 2001; Pratt et al., 2004; Esty et al., 2005; Kaliraj et al., 2016). Understanding the coastal environment is vital for preparing sustainable development and management plans subject to vulnerability, risk, and hazards. Environmental vulnerability is defined as the influencing factors that magnify or attenuate the effects of extreme natural or human-induced events that are deteriorating the environmental-ecosystem systems (Kaly et al., 1999; Alwang et al., 2001; Pratt et al., 2002; Parris and Kates, 2003; SOPAC, 2005). The coastal environment consists of a complex system of marine and terrestrial forces with the hierarchy of different ecosystems. During the past decades, many hotspots along the Indian coastline are adversely affected by human activities and coastal habitats are being threatened due to improper environmental management causing the failure of socioeconomic systems. The coastal environment is frequently exposed to a variety of risks and vulnerability to natural and anthropogenic hazards that trigger potential changes to the exposed systems (Briguglio et al., 1999; Pratt et al., 2002; Kasperson et al., 2005; Skondras et al., 2011; Kaliraj et al., 2017). Along the South Indian coastline, the environmental ecosystems are among the most vulnerable to physical damage resulting from human-induced impact and climate variability. Increasing adverse impacts will further exacerbate the environmental ecosystems by reducing adaptive capacity. Mapping of coastal vulnerability, especially environmental ecosystems on a country or regional scale, is an extremely complex task since the variability of parameters influencing a particular ecosystem cannot be exactly determined in space and time (Cutter, 1996; Campbell, 1997; Barnett et al., 2008).

The GIS-based vulnerability index model is necessary to focus on objectives like (1) mapping vulnerability to natural and the anthropogenic hazards, (2) improving adaptive capacity and building resilience of environmental ecosystems, and (3) reducing potential impact resulting from environmental changes. Integrated remote sensing and GIS provide such knowledge to enhance human ability for insight into multiparametric vulnerability analysis under specific events and guide environmental management options at any level toward a sustainable development path (Kaliraj and Chandrasekar, 2012). Nowadays, many researchers have proposed a variety of vulnerability indices such as the composite human vulnerability index and global vulnerability mapping based on spatial and nonspatial data analysis. The South Pacific Geoscience Commission (SOPAC) developed the environmental vulnerability index (EVI) model under the program of the United Nations Environment Program (UNEP). They have measured EVI for 243 nations and other geographies based on more than 50 parameters indicating various types of environmental issues (Kaly et al., 2002; SOPAC, 2005). According to the SOPAC report (2004) for global environmental vulnerability, the Indian coastal environment falls under extremely vulnerable with rank 216 and an estimated vulnerability index score of 385. Managing the coastal environment involves an assessment of vulnerability using different indices like CVI, EVI, and Habitat Vulnerability Index (HVI) by incorporating multiple geoenvironmental parameters; therefore, the remote sensing and GIS techniques have been used extensively to provide spatial analyses of the degree of vulnerability (Lambert, 2001; Kaly et al., 2004; Skondras et al., 2011; Kaliraj et al., 2015).

Integrated remote sensing and GIS techniques provide an effective platform for investigating the coastal hazard risk using vulnerability indices at national, local authority, and site levels. For example, the GIS-based EVI model uses multiple parameters concerned with resilience and susceptibility of the coastal environment to damage or degradation. The EVI model focuses on multiple factors that quantify potential vulnerability to environmental damage at a scale of country or region level (Turner et al., 2003; Pratt et al., 2004; Kaly et al., 2004), and it is considered as the factors interacting with environmental systems and human activities. The EVI model requires a framework with three aspects of vulnerability: (1) the mapping and assessment of the hazards and risks, (2) identifying sensitive factors (natural and anthropogenic factors) that cause environmental damage (or degradation), and (3) the innate ability of the environment to cope with the risks and resilience (National Research Council, 2000; Lambert, 2001; Kasperson and Kasperson, 2001).

The EVI model is used as a valuable tool for mapping environmental vulnerability and helpful for decision-making and environment management at national or international levels. The applications of the EVI model involve the various fields in earth and environmental sectors and provide basic information sources for ensuring vulnerability and resilience for people and organizations (United Nations, 1994; Briguglio et al., 1999;

Cutter, 2003; Esty et al., 2005). The EVI model examines levels of risk and conditions of the parameters (variables) affecting the environmental system and predicting how the environment is likely to cope with future events; for example, the changing of coastal vegetation and how it causes damage the environment in a particular area. Assessment of EVI in a particular area provides insights into the degree of vulnerability to environmental degradation and can be considered in allocating financial aid and projects. The advantage of a GIS-based EVI model is that it provides rapid and standardized methods for data collection, processing, and analysis for characterizing potential environmental vulnerability at site-specific scale due to a variety of natural and anthropogenic issues (IPCC, 1991; Kaly et al., 1999; Gallopin, 2006; Barnett et al., 2008). EVI modeling is a complex process of environmental functioning and resilience because of incorporating different indices and multiple indicators of various spatial, temporal, and hierarchical levels. The development of the EVI model could be calculated by the vulnerability index using multiple weighted parameters and providing a single-figure expression (vulnerability map) of their relative environmental vulnerabilities. Many researchers have developed different vulnerability indices for assessment of vulnerability and risk associated with sea-level rise, flooding, pollution, erosion, and other anthropogenic impacts (Villa and McLeod, 2002). The present study deals with the development of GIS-based EVI models using multiple geoenvironmental parameters that influence site-specific environmental ecosystems in the South Indian coastal zone.

2. The geographical profile of the study area

The southern coast of India dwells in a variety of environmental ecosystems along with many human settlements. Along the coastline, many hotspots found valuable environmental, aesthetic, and recreational resources, and they are noted under high vulnerability to physical damage and environmental degradation during the past decades (Kaliraj and Chandrasekar, 2012; Kaliraj et al., 2019). The GIS-based EVI model is formulated for assessing and mapping environmental vulnerability along the South Indian coastal stretch between Kanyakumari and Thengapattinam that covers the geographical extent of 77°9′49.23″E to 77°34′14.99″E longitude and 8°6′32.63″N to 8°14′15.30″N latitude. Fig. 14.1 shows the location of the South Indian coastal stretch. The coastal stretch comprises sandy beaches, dunes, estuaries, and other fluvio-marine landforms (Kaliraj et al., 2017). The coastal landforms are frequently faced with the dynamic action of wave and littoral currents. Three major drainage networks such as Pazhayar, Valliyar, and Thamirabarani and their tributaries are the primary sources of sediment discharge that are influencing the wetland environmental ecosystems. In the coastal stretch, the subtropical climate prevails with the rainfall range of 826–1456 mm, and the temperature range at 23.78–33.95°C. The coastal stretch is modifying the different landforms shapes and structures due to natural and anthropogenic forces causing physical damage to environmental ecosystems during past decades.

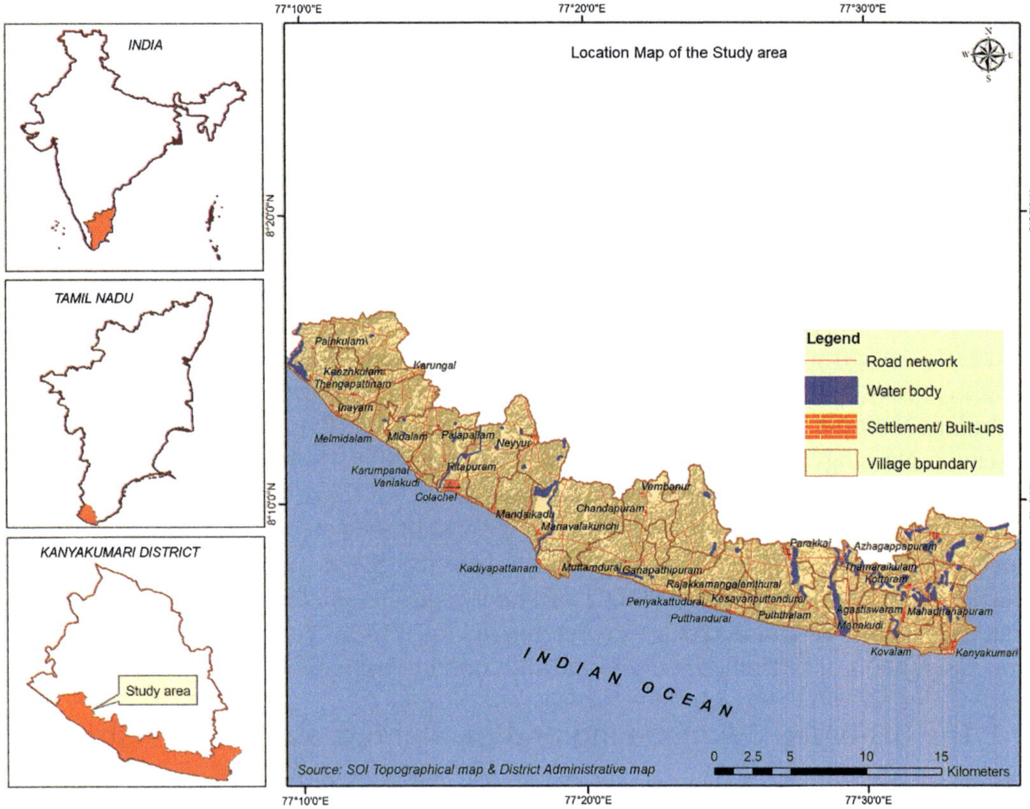

FIGURE 14.1 The geographical location of the South Indian coastal stretch.

3. Materials and methods

The SOPAC developed the EVI under the UNEP. The present study deals with the framework of GIS-based EVI model for the assessment of potential vulnerability to environmental damage in various landforms due to erosion, lack of sediments, storm surge, inundation, and seawater intrusion, and so on. In this study, the GIS-based EVI model is designed using site-specific parameters for mapping environmental vulnerability in the southern coast of India, and these nine geoenvironmental parameters mainly influence site-specific environmental issues. These multiple parameters are derived from different types of primary and secondary data sources using systematic GIS tools and functionalities. Satellite image products include Landsat ETM+ and OLI (30 m) images (30 m) and IKONOS (2.5 m), which have been used for extracting temporal shoreline change, land use/land cover, and geomorphic landforms. The volumetric change of sediment load is estimated using the digital elevation model (DEM) data like SRTM and ASTER DEMs. Using the Garmin GPS ETREX 30, the in situ survey was carried out for measuring shoreline, benchmark elevation, and demarcating high-tide line. Secondary data sources include Survey of India (SOI) topographical maps, Geological Survey of India (GSI) geology and geomorphology maps, National Hydrographic Office (NHO) bathymetry chart and tide gauge data, wave atlas, buoys data, climate data, and census data published by the authenticated departments and which government agencies have used for deriving various thematic layers. SOI topographical maps have been used to extract coastal relief and physiographic details like contour, drainage, and water bodies. GSI published maps were used for preparing geology and geomorphology layers. The spatial variability of sea level (mm/yr) in the South Indian offshore has been estimated using the long-term observed data source of USACE-Wave Information Study Atlas and Permanent Service for Mean Sea Level (PSMSL), UK. Similarly, the wave energy (kJ/km^2) and storm surge/wave run-up (m) were estimated based on incorporating different data sources like water height, gravity, wind speed, salinity, and sea surface temperature, observed through buoys of Indian National Centre for Ocean Information Services (INCOIS), National Institute of Ocean Technology (NIOT), NIO, and IMD and other MODIS-derived climatic data sources. Coastal geomorphic landforms have been extracted from the combined data sets of DEM and Landsat ETM+ and IKONOS images (Kaliraj et al., 2017). Land use and land cover (LULC) features were demarcated using Landsat ETM+ and OLI images using the maximum likelihood classifier (MLC) of supervised classification techniques. The thematic layers of EVI parameters are georeferenced to UTM-WGS 84 projection for mapping and assessment of coastal environmental vulnerability using GIS tools and functionalities.

3.1 The GIS-based EVI modeling and parameters assigned weights

The Central Geomatics Laboratory of the National Center for Earth Science Studies has formulated a new approach of GIS-based EVI model for mapping and assessment of environmental vulnerability using nine types of geoenvironmental and social parameters. The advantage of this model serves like as a global composite of the SOPAC EVI model, but it is composed of site-specific parameters influencing the environmental ecosystems. The GIS-based EVI model is applied on the South Indian coast due to physical, environmental, and anthropogenic conditions that affect the environmental stabilities and make it suitable for the assessment of environmental vulnerability. However, the accuracy of the EVI model depends on input of multiple geoenvironmental data sources published by various departments and agencies, and the algorithm of this index that may be further developed is conveyed. In this analysis, the model is mainly composed of nine types of weighted parameters to characterize the site-specific environmental vulnerability (Kaliraj et al., 2015). These EVI parameters are heterogeneous in qualitative measurements (linear, areal, and volume) and scale (scale of data sources with different ranges). The theory of the EVI scoring system identifies three aspects: (1) risk of hazards, (2) inherent resistance to damage, and (3) acquired vulnerability resulting from past damage. To be able to calculate an EVI value, the EVI parameters have assigned vulnerability weights a range of 1–5 based on the degree of magnitude to environmental vulnerability, damage, hazards, and adaptability. The lowest vulnerability score of 1 is attributed to parameters having the ability to resist and (or) recover environmental damage, while a parameter having more susceptibility or tendency to produce extreme vulnerability to the environment is assigned a higher vulnerability score of 5. Table 14.1

Table 14.1 The characteristics of EVI parameters and assigned vulnerability weights.

		EVI parameter characteristics and assigned vulnerability weights				
Sl.		Very low	Low	Moderate	High	Very high
no.	EVI parameter	1	2	3	4	5
1	Wave energy and shoreline impacts (kJ/km^2)	<2.5	2.5–5.5	5.5–8.5	8.5–12.5	>12.5
2	Groundwater table fluctuation (m)	>1.5	1.5–3.0	3.0–4.5	4.5–6.0	<6.0
3	Coastal sediment load change rate (m/km^2)	>10 (accretion)	5.0–10.0 (accretion)	5.0 and −5.0 (seasonal accretion/erosion)	−10.0 and −5.0 (erosion)	> −10.0 (erosion)
4	Storm surge or wave run-up (m)	<1.5	1.5–2.5	2.5–3.5	3.5–4.5	>4.5
5	Suspended sediments or river discharge rate (mg/L)	>122	92–122	62–92	32–62	<32
6	Coastal vegetation change rate (km^2)	<0.5	0.5–1.0	1.0–1.5	1.5–2.0	<2.0
7	Coastal mining and quarrying (km^2)	<0.25	0.25–0.50	0.50–0.75	0.75–1.0	>1.0
8	Seawater intrusion impacts (GALDIT score)	<4.45	4.45–5.45	5.45–6.45	6.45–7.45	>7.45
9	Coastal population growth rate (%)	<10	10.0–20.0	20.0–30.0	30.0–40.0	>40.0

shows the EVI parameter's characteristics and assigned vulnerability weights. In this EVI model, the functionalities are devised as per the user's decision to define a vulnerability score to the parameters based on potentiality to vulnerability and hazards. All the EVI parameters have transformed into a common scale based on their assigned weights that facilitate setting raster-pixel thresholds using ESRI-ArcGIS software. The GIS-based EVI model uses the empirical algorithm to accumulate the vulnerability scores assigned to each parameter and corresponding feature classes, and then the average vulnerability is calculated to generate the EVI value at pixel (10 * 10 m) scale. The EVI model's equation is expressed as, $EVI = \sum_{i=1}^{n}(WE * GW * SL * SW * SS * CV * MQ * SWI * PG)/9$where EVI is the EVI score or value; *WE* refers to wave energy released at the shoreline (kJ/km^2); *GW* refers to the depth of the groundwater table (m); *SL* is an estimated volume of coastal sediment load change rate (m/km^2); *SW* is the storm surge or wave run-up (m); *SS* is a measured value of suspended sediments or river discharge rate (mg/L); *CV* refers to coastal vegetation change rate (km^2); *MQ* is an estimated area of placer mining and beach quarrying (km^2); *SWI* refers seawater intrusion impacts (Groundwater occurrence (G), Aquifer hydraulicconductivity (A), Level of groundwater table above sea level (L), Distance from the shoreline (D), Impact of existing seawater intrusion (I), and Thickness of the aquifer (T) (GALDIT) score); and *PG* refers to coastal population density (persons/km^2) and population growth rate The output of the EVI model shows the map of environmental vulnerability with its potential vulnerability rate at site-specific scale.

4. Results and discussions

The GIS-based EVI model is composed of multiple parametric analyses for mapping and assessment of environmental vulnerability in the South Indian coastal stretch. The characteristics of EVI parameters and assigned vulnerability scores are described as follows.

4.1 Wave energy and shoreline impacts

The coastal dynamic processes depend on waves and currents that release energy to the shoreline and lead to changes in environmental ecosystems on a long-term scale. The coastal zones with prevailing high-wave energy are more susceptible to environmental ecosystems being disturbed (Kaliraj et al., 2014). Fig. 14.2 shows the spatial variability of wave energy releasing at various parts of the South Indian coastal stretch. The high-wave energy in Kanyakumari, Kovalam, Muttam, and Inayamputhenthurai coastal zones are in the range of 6.5–8.5 kJ/km^2, and these areas have been assigned vulnerability weights of 5. The interacting high-wave energy with the shoreline is triggering severe environmental

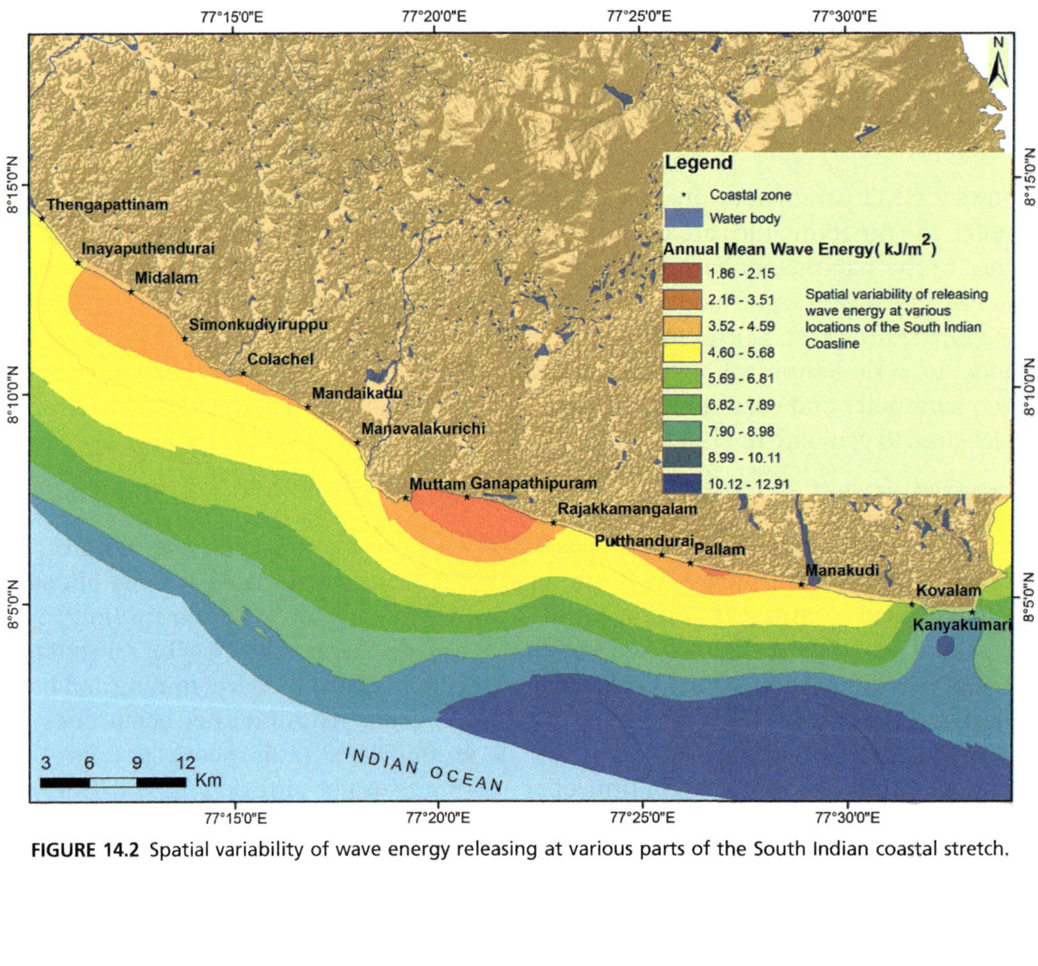

FIGURE 14.2 Spatial variability of wave energy releasing at various parts of the South Indian coastal stretch.

damage with the zones having deep bathymetry and steep shelves. Similarly, the coastal zones with high-wave energy are backwashing a larger amount of sediments in the Manavalakurichi–Mandaikadu and hence they are assigned a weighted value of 4. The coastal zones experience moderate vulnerability to environmental changes due to seasonally fluctuating wave-energy (2.5–4.5 kJ/km^2) in the coastal zones of Manavalakurichi and Mandaikadu, and they are assigned a weighted value of 3. Moreover, the coastal zones of Manakudi, Rajakkamangalam, Colachel, Simonkudiyiruppu, and Midalam are noted as lower wave-energy zones (2.5–3.5 kJ/km^2), whereas the formation of accreted landforms provides more suitability for environmental ecosystems, therefore these areas are assigned lower weight values ranging from 1 to 2. Along the South Indian coast, the wave energy operates the cyclic process of the marine environment and extends their life span significantly. The wave-energy releasing forces determine erosion or accretion processes that are triggering environmental vulnerability to erosion, flooding, and seawater intrusion (Kaliraj et al., 2013, 2014, 2016). It is noted that erosion landforms are highly susceptible to environmental vulnerability due to the removal of sediments and nutrients from beaches, dunes, cliffs, and wetlands that cause biodegradation of the coastal and marine environment.

4.2 Groundwater table fluctuation

Changing the level of the groundwater table depends on permeability and infiltration of runoff that affects environmental ecosystems by controlling the freshwater–saltwater interface at the transition zones (Kaliraj et al., 2019). Hydrodynamic processes of watering or dewatering from sea or land sources influence the rate of sediment wetness, soil moisture, and vegetation growth that cause environmental systems in an area subject to groundwater table level (Pantin, 1997; Pratt et al., 2004; Kaliraj et al., 2019). In the coastal stretch, the groundwater table is estimated using seasonally observed groundwater table data from open/dug wells for the years 2011 and 2018. Fig. 14.3 shows the spatial variability of the groundwater table level at various parts of the South Indian coastal stretch. The coastal zones with fluctuating higher groundwater table (>6.0 m) have appeared with dried sediments, causing backwashing or erosion and serving poor habitats for flora and fauna followed by severe environmental degradation. Therefore these zones are assigned a higher vulnerable weight of 5. Lowering the groundwater fluctuation (<1.5 m) denotes the stability of the groundwater table in various parts of the coastal stretch, whereas maintaining sediment moisture status decreases the erosion process, eventually protecting seawater intrusion; therefore, these areas are assigned a lower vulnerability weighted value of 1. Meanwhile, the groundwater table is seasonally fluctuating in the range of 2.0–4.0 m due to monsoonal rainfall; these areas are assigned to moderately vulnerability weighted values of 3–4. In various parts of the coastal stretch, environmental ecosystems are changing in growth and distribution, which depends on altering the capillary fringe of groundwater flow, particularly as it underlies nearshore areas (beaches, dunes, and wetlands) that significantly control erosion/accretion activities.

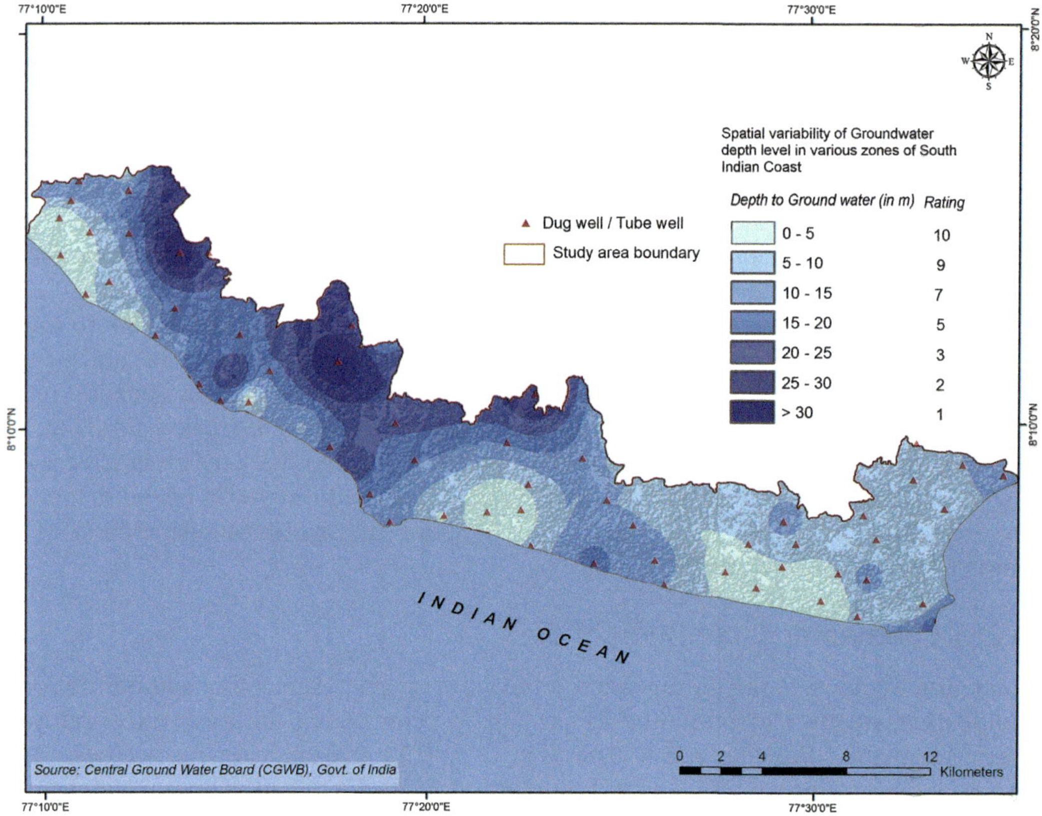

FIGURE 14.3 Spatial variability of groundwater table level at various parts of the South Indian coastal stretch.

4.3 Coastal sediment load change rate

Sediments supplied to various parts of the coastal zones depend on swashing and backwashing actions resulting from waves, currents, winds, and tides that determine the locations and distribution of environmental ecosystems. Changing the rate of sediment load influences the displacement of coastal habitats in beaches, dunes, tidal flats, lagoons, estuaries, sea-grass beds, coral reefs, mangroves, and other coastal wetlands (Villa and McLeod, 2002; Skondras et al., 2011). Fig. 14.4 shows the geomorphic landforms and their sediment load variability in the South Indian coastal stretch. The larger volume of sediment losses ($>-10.0\ \mathrm{m/km^2}$) in various parts of the coastal zones cause removal of vegetative covers and shifting of habitats, and therefore these areas are assigned higher vulnerability weighted value of 5. Many locations of the coastal stretch have noted a decreasing rate of sediment loads (-10.0 and $-5.0\ \mathrm{m/km^2}$) due to erosion processes; these zones are assigned a vulnerability weighted value of 4. Some parts of the coastal environment are seasonally influenced in natural structures depending on loss or gain of sediment from the marine and land sources; hence these zones are assigned moderate

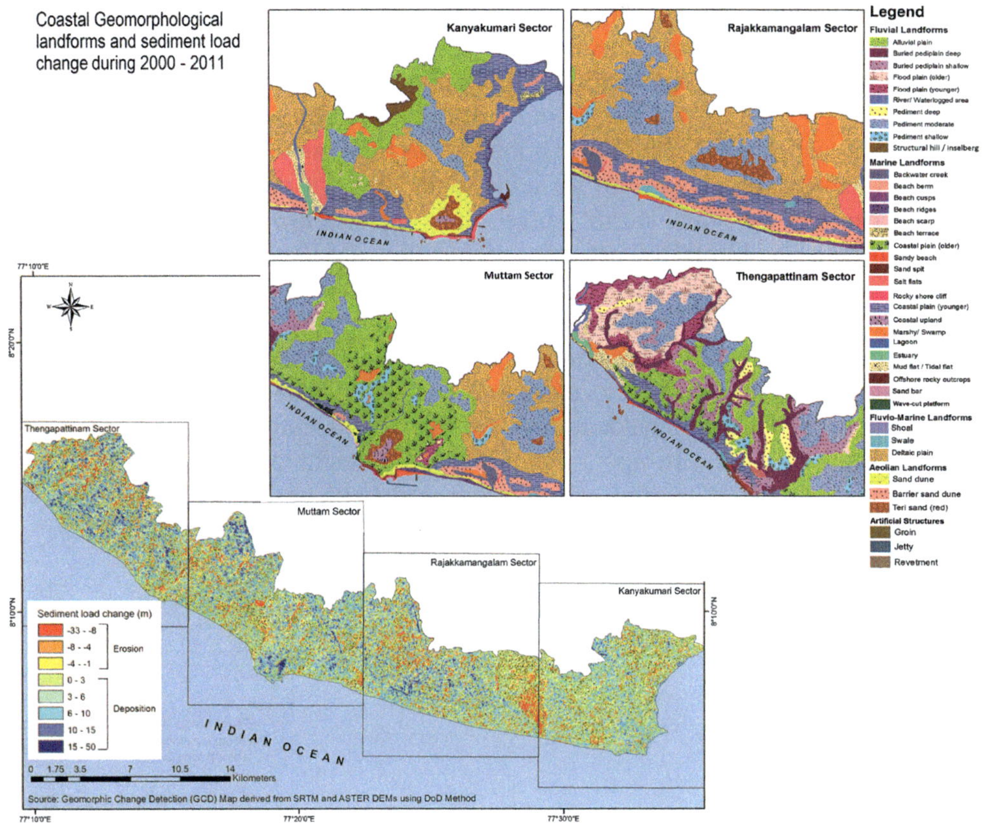

FIGURE 14.4 The geomorphic landforms and its sediment load variability in the South Indian coastal stretch.

vulnerability weighted values of 3. The coastal environmental systems are seasonally stripping and searing of living organisms. In the coastal stretch, the depositional zones include sandy beaches, dunes, ridges, and spits, which are assigned lower vulnerable weighted values of 1–2. These areas serve as natural habitats for a variety of ecosystems and protect their fragile environments from hazards, whereas the status of sediment budget is maintaining the equilibrium of environment resulting from littoral currents and river discharge.

4.4 Storm surge or wave run-up

The coastal environment is often modifying structures due to storm surge or wave run-up that massively displaces physical and ecological settings. Coastal zones experiencing storm surges or higher wave run-up produce adverse impacts like erosion and beach subsidence making it more feasible for intruding seawater into environmental ecosystems (Parris and Kates, 2003; Kasperson et al., 2005; Kaliraj et al., 2014). Fig. 14.5 shows potential zones of erosion impact during storm surges or wave run-ups in the South

Indian coastal stretch. Storm surges and higher wave run-up releasing at various locations of the South Indian coast causing physical damage to environmental ecosystems, for example, higher wave run-up resulting from cyclones and tsunami (i.e., tsunami waves on December 26, 2004), have at a height of 5 m or more in various parts. Site-specific wave run-up is derived using tide gauge and buoy data as integrated sources from SOI, INCOIS, NIOT, and PSMSL. The long-term observed data shows a mean value of 4.5 m during higher wave run-up and storm surge that affects low-lying landforms in the Thengapattinam and Manakudi estuaries, hence they are assigned a higher vulnerability weighted value of 5. In the coastal zones of Kanyakumari and Inayamputhenthurai, the wave run-up during storm surges extend to backshore up to 3.5–4.5 m, and are more susceptible for inundating nearshore landforms and intruding saltwater into surface water bodies and aquifers; therefore, these areas are assigned a vulnerability weighted value of 4. In the Pallam, Rajakkamangalamthurai, and Periyakadu coastal zones, the seasonal rising of wave run-up is estimated at a height of 2.5–3.5 m and is more vulnerable to damage of coastal environment and habitats during monsoonal surges;

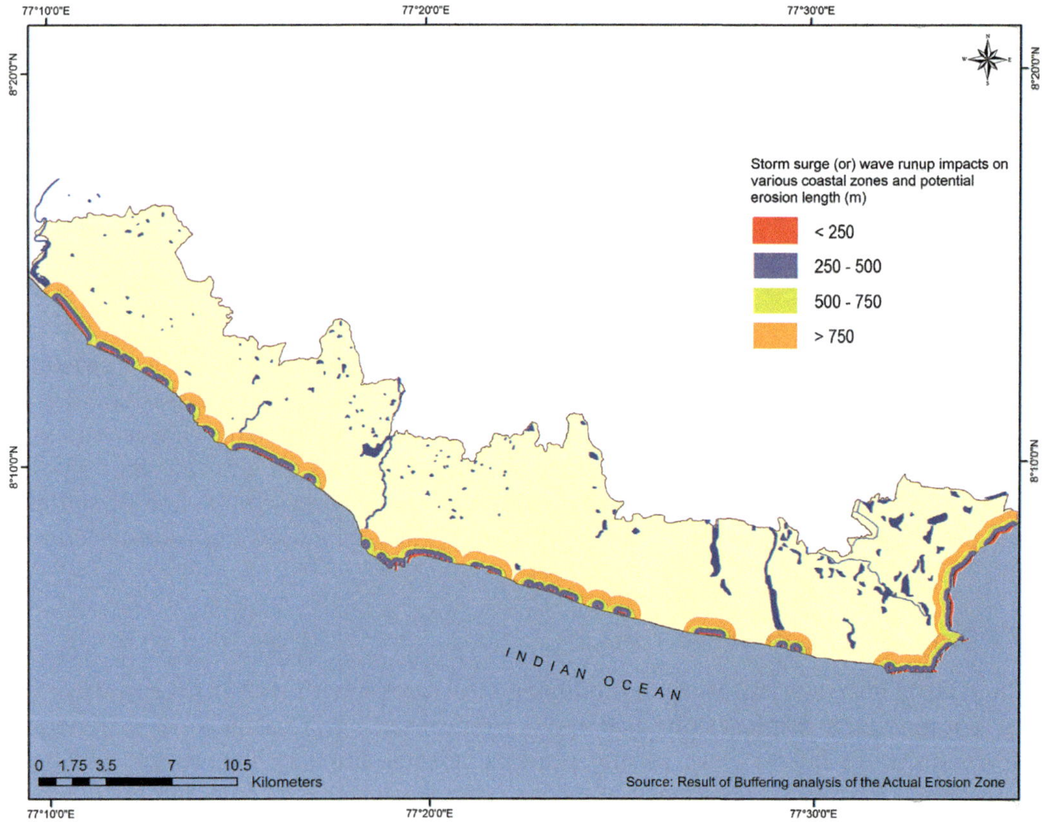

FIGURE 14.5 Potential zones of erosion impact during storm surges or wave run-ups in the South Indian coastal stretch.

therefore these areas are noted as moderate vulnerability zones with an assigned weighted value of 3. The uplands composed of lateritic formations found in the Colachel, Muttam, and Kanyakumari are potentially reducing impacts of storm surges and flooding due to geographical structures, so these areas are assigned lower vulnerability weighted values of 1–2. On the southern coast, the higher wave run-up resulting from storm surges affects the spatial distribution of environmental ecosystems at a certain level in the various coastal zones.

4.5 Suspended sediments or river discharge rate

Along the coastal stretch, transporting suspended sediments in coastal water is a major source for coastal landform formation and supplying nutrients to the ecosystems (Bossel, 1999; Cutter, 2003; Gallopin, 2006). The quantity of suspended sediments reaching the shoreline influences coastal environmental balancing activities (Pethick and Crooks, 2000; Kaliraj et al., 2013); for example, the suspended sediment mixing of toxicants affects the lifespan of the ecosystems in a particular region. In the Southern Indian coast, the amount of suspended sediments in coastal water is estimated quantitatively using Landsat ETM+ and OLI images. It is noted that coastal zones like Kanyakumari east, Manakudi, and Thengapattinam have supplied a higher rate of suspended sediments (>122 mg/L) from river discharge and eroded rocky shore, wherein the coastal habitats include sandy beaches, estuaries, mudflats, and sand barriers have been assigned a lower vulnerability weighted value of 1. In these areas, the swashing of suspended sediments toward the shoreline reduces vulnerability to environmental damage. Meanwhile, plenty of suspended sediments stirring in coastal water produce adverse impacts on living organisms; that is, reducing sunlight and dissolved oxygen (Barnett et al., 2008; Kaliraj et al., 2016). Some parts of the coastal zones, namely Pallam, Periyakadu, and Midalam, have estimated a lower rate of suspended sediment (<32 mg/L) in coastal water due to intervening of natural and manmade coastal structures. In these areas, the decreasing sediment supply leads to landform physical damages followed by ecosystem degradation. Waves and currents are sending sediments along with a variety of minerals and organic matter toward the up-drift side, causing vulnerability to coastal and marine ecosystems and human settlements; hence these areas are assigned a higher vulnerability weighted value of 5. Significantly, coastal structures (i.e., groins, seawalls, and jetties) constructed in different locations have to divert a larger quantity of suspended sediments, intervening food and nutrient supply to wetlands habitats (Pethick and Crooks, 2000; Lambert, 2001; Kaliraj et al., 2013), and therefore, these areas are assigned a vulnerability weighted value of 4. The coastal areas of Kovalam, Manavalakurichi, and Mandaikadu seasonally influence the suspended sediment concentration at a rate of 32–122 mg/L, which affects the balancing potential of marine and coastal ecosystems, and therefore, these areas are assigned to moderate vulnerability weighted values of 2–3. The availability of suspended sediments along the South Indian coastal stretch is one of the major factors for determining the landform stability influencing the life span of coastal ecosystems.

4.6 Coastal vegetation change rate

Coastal vegetative cover is an integrated ecosystem of marine and terrestrial biodiversity that provides food and shelter for a variety of fauna (United Nations, 1994; Pratt et al., 2002; Skondras et al., 2011). The vegetative cover is distributed mainly over dunes and mudflats along the shoreline. The change in the area of coastal vegetation is estimated from the Landsat ETM+ and TM images between 2000 and 2011 using ERDAS Imagine software. Fig. 14.6 shows the LULC features of the South Indian coastal stretch. In the coastal stretch, beaches and sand dunes are found with a variety of vegetative covers including shrubs, bushes, grasses, and woody species like casuarinas, coconuts, and other species in the coastal area. In the various locations, the dune vegetative cover is reduced in areas from 4.32 to 3.36 km^2 during 2000–2011 and the total loss is estimated at 0.96 km^2. The embryo dunes of grasses and bushes have disappeared in different parts in the Chothavilai, Sanguthurai, Puthenthurai, and Rajakkamangalamthurai coastal areas due to wave erosion, wind-induced salt spraying, and human encroachment activities (Pethick and Crooks, 2000; Kaliraj and Chandrasekar, 2012; Kaliraj et al., 2013).

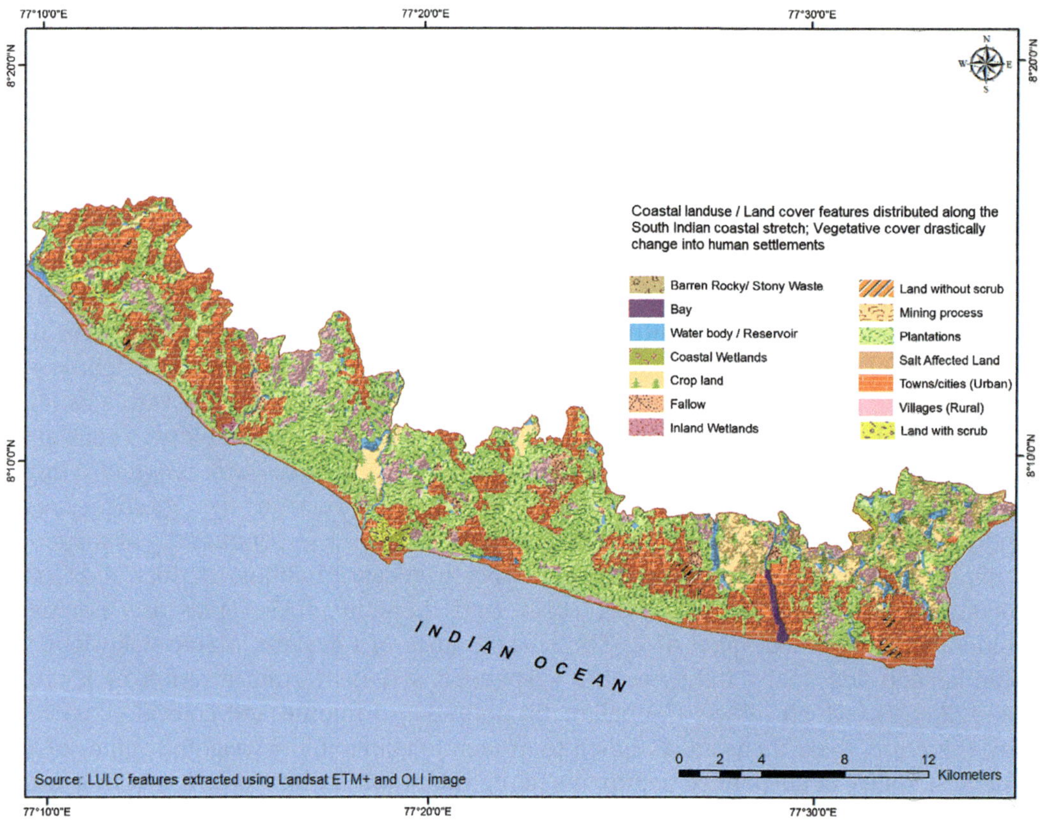

FIGURE 14.6 LULC features of the South Indian coastal stretch.

Meanwhile, the casuarinas tree covers known as woody fences are deforested or uprooted along the foredunes and backshore dunes in the Rajakkamangalamthurai, Pallam, Manavalakurichi, Mandaikadu, and Midalam coastal areas due to placer mining and expansion of settlements and built-ups. Decreasing the vegetative cover produces direct impacts on biodiversity that cause habitat deterioration and degradation along the coastal area. Therefore, these areas are assigned a higher vulnerability weighted value of 5. Decreasing the area of dune vegetative cover (1.5–2.0 km^2) in the South Thamaraikulam, Murungavilai, Midalam, and Inayamputhenthurai coastal zones causes physical damage of coastal habitats due to erosion resulting from storm surges and human encroachment activities; hence these zones are assigned a weighted value of 4. In the middle-eastern sections, vegetative cover like grass and weeds are seasonally changed with an area of 1.0–1.5 km^2 due to monsoonal effects; hence these areas are assigned a weighted value of 3. Some parts of the coastal tracts are spatially covered woody vegetation along the frontal dunes, providing sustained habitats to marine and coastal ecosystems and acting as natural barriers to beach erosion due to storms and tsunami waves. Therefore, these areas are assigned lower vulnerability weighted values of 1 and 2. It is noted that the changing vegetative cover is eroding sediments from various parts of the coastal stretch, which causes physical damage to coastal habitats.

4.7 Coastal mining and quarrying

Coastal mining and quarrying are some of the major activities for removing sediments that cause an adverse impact on the diversity of coastal ecosystems (Kasperson and Kasperson, 2001; Kaliraj et al., 2014). Landsat ETM+ and OLI images were used to estimate the areal extent of sites encroached for mining and quarrying in the South Indian coastline.Fig. 14.7 shows the different locations of mining and quarrying in the South Indian coastal stretch. Mining and quarries are found in various parts of the coastal stretch, and the total area is estimated as 1.30 km^2 in the Manavalakurichi, Mandaikadu, and Midalam coastal zones. From these areas, the larger sediments have frequently been removed for placer mining that affects growth and distribution of coastal habitats (Kaliraj et al., 2016) and therefore, these areas are assigned a higher vulnerability weighted value of 5. The various pits of abandoned placer mining are soaked with salt water during high tide, which intrudes salt matter into freshwater bodies and coastal aquifers; and therefore these areas are assigned a weighted value of 4. The terms and deposits found in coastal areas of Kovalam, Muttam, and Inayamputhenthurai are excavated due to anthropogenic activities and trigger physical damages to the coastal environment, so these areas are assigned a weighted value of 3. The placer mining sites in the Midalam, Puththalam, and Pillaithoppu are seasonally renovated by constructive wave actions, and these areas are assigned lower vulnerability weighted values of 1 and 2. However, in a few sites, the unregulated placer mining produces significant impact on coastal environmental ecosystems during certain periods.

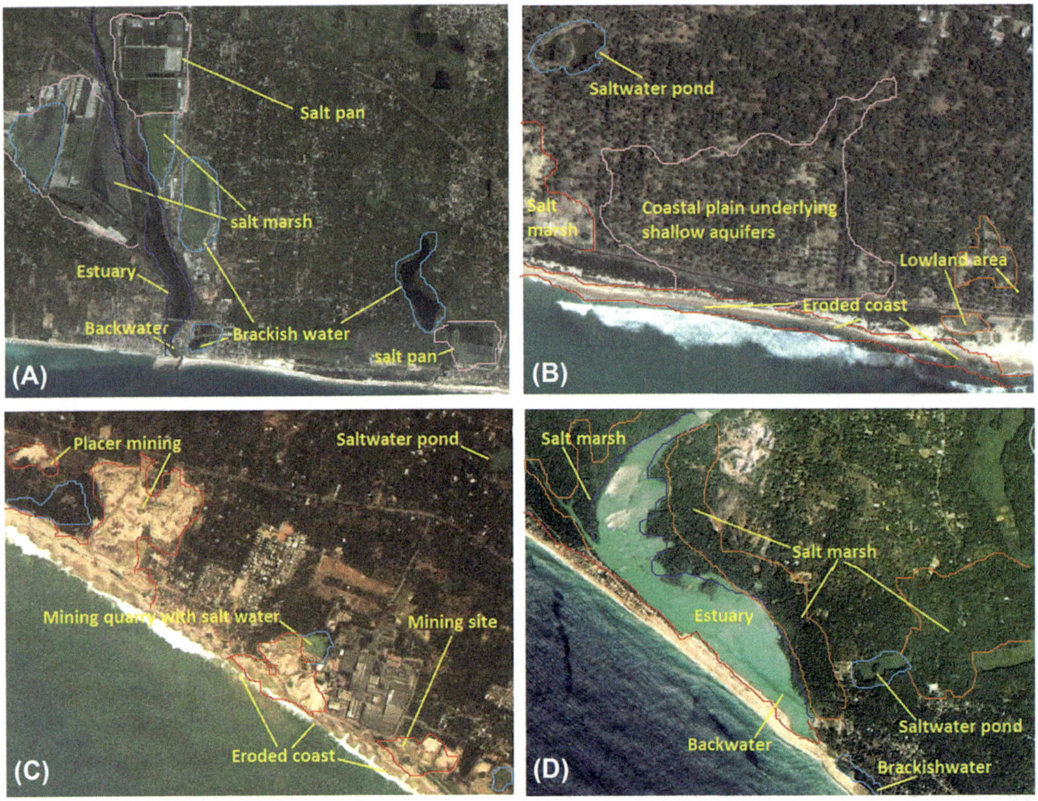

FIGURE 14.7 The different locations of mining and quarrying in the South Indian coastal stretch. *Courtesy: Google Earth.*

4.8 Seawater intrusion impacts (GALDIT score)

In the South Indian coastal stretch, water bodies and aquifers found more susceptibility to seawater intrusion vulnerability due to lowering groundwater table and recharge that gradually degrading the coastal environmental ecosystems (Lambert, 2001; Kasperson et al., 2005; Kaliraj et al., 2019). Fig. 14.8 shows the potential vulnerable zones to seawater intrusion in the South Indian coastal stretch based on GIS-based GALDIT model. The result reveals that the higher vulnerability to seawater intrusion (>7.45) is identified in the coastal zones of Kovalam, Manakudi, and Thengapattinam and the low-lying inland areas of South Thamaraikulam and Manavalakurichi, due to diffusing salt matter from existing saltwater bodies like salt ponds, brackish water, and saltpan. Therefore, these areas are assigned a higher vulnerability weighted value of 5. The coastal stretch between Manavalakurichi and Kadiyapattinam is noted as susceptible for seepage movement of seawater up to 2 km inland due to unregulated placer mining that removed sediments from beaches, dunes, and vegetative covers; these areas are assigned a weighted value of 4. In the western parts, the coastal zones Thengapattinam, Inayamputhenthurai,

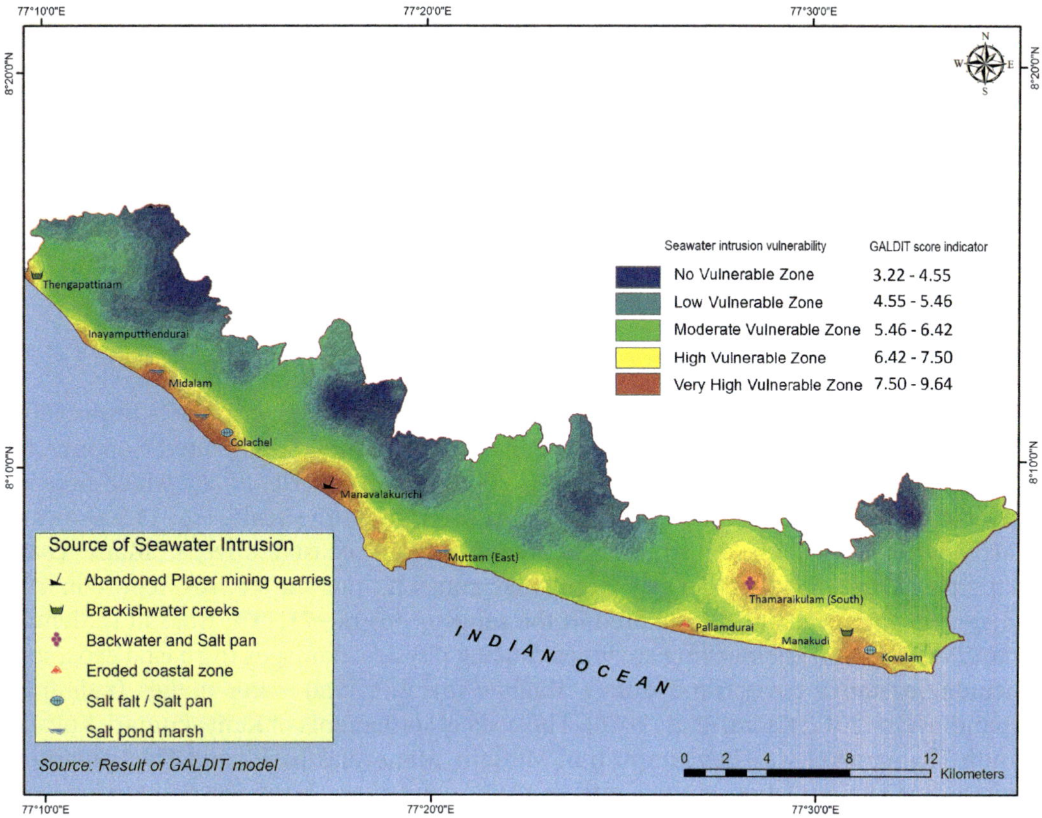

FIGURE 14.8 Seawater intrusion vulnerable zones in the South Indian coastal stretch based on GIS-based GALDIT model.

Colachel, and Midalam have experienced very high vulnerability to seawater intrusion due to the free movement of saline water from the estuary, brackish water, salt marsh, saltpan, salty marshes, salt pan, and other saltwater sources. Recently, coastal settlements are rapidly increasing seawater intrusion vulnerability (6.45–7.45) due to overpumping of groundwater; these areas are assigned a weighted value of 4.

Increasing salinity of freshwater bodies directly affects the growth and distribution of aquatic ecosystems in the coastal area. Moreover, the dissolved chemical constituents of seawater affect the geochemical budgets of aquatic organisms and play a role in the biological species composition and productivity of coastal ecosystems. In the middle-eastern parts, the coastal zones Puthenthurai, Sanguthurai, Rajakkamangalam, and Periyakadu have moderate vulnerability to seawater intrusion (i.e., estimated GALDIT score is 5.45–6.45) due to seasonally fluctuating groundwater table and recharge during monsoons, and are assigned weighted values of 2 and 3. Some parts of the coastal zones like Chothavilai, Ganapathipuram, and Pillaithoppu are found with static equilibrium of groundwater table due to accreted coastal landforms, wherein availability of larger sediments increase water hold capacity and prevent seawater intrusion at the interface zone. Likewise, the coastal uplands in Muttam, Kanyakumari, and Colachel are identified with low vulnerability conditions (<4.45) due to geological formation of garnet biotite gneiss and clay interbedded charnockite that stop seawater flowing further landward and therefore are assigned a lower vulnerability weighted value of 1. In the coastal stretch, the landward intruding seawater depends on the groundwater table and recharge rate is determined by surface flow, permeability, and landform composition, resulting in anthropogenic impact that affects the growth and spatial distribution of flora and fauna in the coastal environment.

4.9 The coastal population growth rate

Coastal environmental ecosystems experience a variety of adverse impacts, especially in the densely populated areas due to rapid growth of the population (United Nations, 1994; Pethick and Crooks, 2000; Parris and Kates, 2003; Skondras et al., 2011), which produces adverse impacts to environmental ecosystems for a long-time scale. Fig. 14.9 shows the village-wise population density in the South Indian coastal stretch. Many coastal settlements are found as densely populated zones as per the census report in 2011, whereas the growth rate is estimated in the range of 0.50%–71.39%. Coastal settlements found with human encroaching activities causing physical damages to coastal ecosystems include vegetative cover, sand dunes, shallow marshes, and water bodies (Kaliraj and Chandrasekar 2012; Kaliraj et al., 2016). The coastal settlements of Kanyakumari, Kovalam, South Thamaraikulam, Thengamputhur, Muttam, Manavalakurichi, and Thengapattinam have a higher population growth rate (>40.0%), and these areas are assigned a higher vulnerability weighted value of 5. Changing of LULC in and around populated zones leads to destructive impact on environmental features, such as eroding beaches and dunes, and polluting water bodies, hence these areas are assigned a vulnerability-weighted value of 4. In the middle-eastern parts, coastal settlements including Pallam, Puthenthurai, Rajakkamangalamthurai, and Ganapathipuram are recorded at population growth rate of 20.0%–30.0% during last decade, and the cultivable lands have become occupied for built-ups; these areas are assigned a weighted value of 3. The coastal settlements of Inayamputhenthurai, Thengamputhur, Simonkudiyiruppu, and Kesavanputhenthurai have a lower population growth rate (<10%) and are able to balance the environmental equilibrium status during favorable conditions, so these areas are assigned lower vulnerability weighted values of 1 and 2. It is noted that the increase in population growth in any coastal zones is deteriorating the environmental ecosystems due to human-induced activities.

4.10 Assessment and mapping of vulnerability to environmental damages

The South Indian coastline consists of a variety of ecosystems like beaches, dunes, cliffs, estuaries, salt marshes, and backwaters. During the past decades, various parts of the coastal stretch are facing potential vulnerability to environmental degradation due to natural and anthropogenic factors. The model explores site-specific vulnerability to

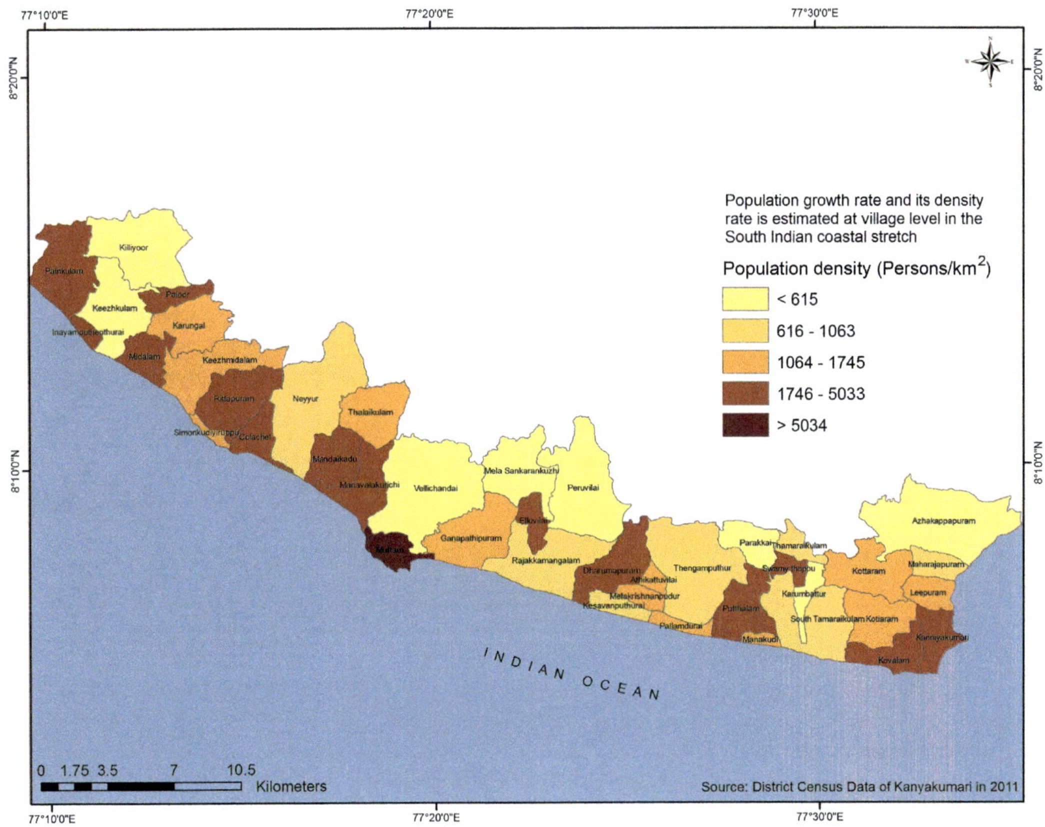

FIGURE 14.9 The population density village-wise of the South Indian coastal stretch.

environmental damage in the study area. Fig. 14.10 shows the result of the GIS-based EVI model, demarcating the potential vulnerable zones to environmental damage in the South Indian coastal stretch. This model analysis of multiple EVI parameters estimates site-specific vulnerability rate at raster grid (pixel size of 10 * 10 m) format, which indicates a pixel-wise potentiality to damage or resilience of environmental ecosystems. The EVI estimated the vulnerability index value in the range of 10–54 and mean vulnerability is 32. The South Indian coastal stretch is classified into five vulnerability classes: (1) resilience zone (coastal zones having ability to resisting or recovering from various environmental damages), (2) zone at risk level (coastal zones facing lower risk or vulnerability), (3) vulnerable zone (coastal zones having seasonal impacts on environmental systems), (4) high vulnerable zone (coastal zones having impacts on environmental systems while altering natural processes), and (5) extreme vulnerable zone (coastal zones having adverse impacts for damaging the environmental ecosystems). Table 14.2 shows the coastal village–wise EVI value of the South Indian coast and influencing factors. In the coastal stretch, 13% of the area has fallen under extreme

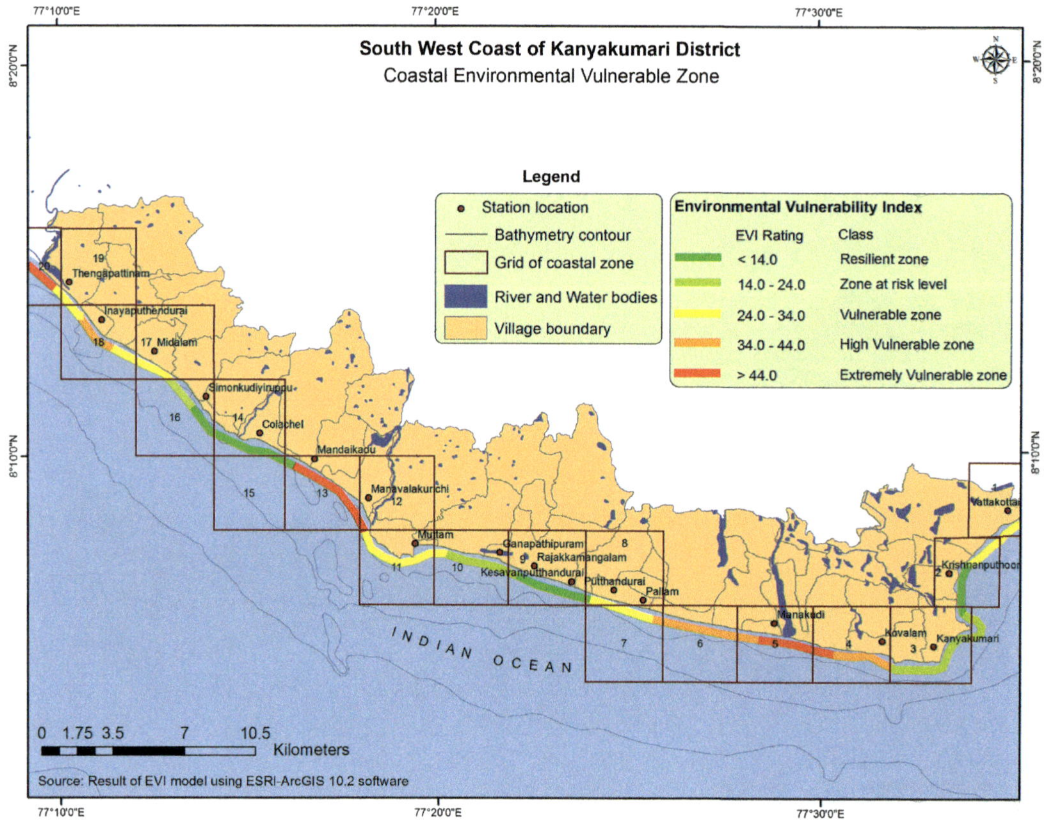

FIGURE 14.10 The result of GIS-based EVI model shows the potential environmental vulnerability in the South Indian coastal stretch.

vulnerability to damaging the environmental ecosystems due to adverse impact of natural and anthropogenic activities. The coastal zones of Manakudi, Thengapattinam, Manavalakurichi, and Mandaikadu are identified as extremely vulnerability conditions (the EVI value is estimated as > 44), whereas the altering of coastal landforms frequently affects environmental systems due to eroding sediments and intruding saltwater. Moreover, the human-induced activities (i.e., removal of sediments; encroaching beaches; constructing seawall, groins, and jetties; and placer mining and quarrying) are gradually degrading environmental ecosystems in various parts of the coastal stretch. The low-lying areas like estuaries, salt marshes, and sandy beaches are facing higher vulnerability to damaging the coastal ecosystems during energetic wave run-ups and storm surges (Pethick and Crooks, 2000; Kaliraj et al., 2016). In the shallow marshes (mudflat and tidal flat) of the Thengapattinam and Manakudi, the wetland ecosystems are facing higher vulnerability to inundating saltwater due to landward extending (up to 4.5 m) of storm surge and wave run-ups.

Table 14.2 The environmental vulnerability index (EVI) value at the coastal villages of the South Indian Coast and influencing factors.

Coastal sector	Grid no.	Coastal zone	EVI vulnerability Score	Class	Factors that influence coastal environmental vulnerability
Kanyakumari	1	Kanyakumari North east	30	Vulnerable level	Littoral current, high energy breaking wave action and narrow continental shelves cause cliff erosion that affects flora and fauna of rocky shore
	2	Kanyakumari east	12	Resilient zone	
	3	Kanyakumari	16	Risk level	
	4	Kovalam	34	High Vulnerable	
	5	Manakudi	54	Extreme level	
Rajakkamangalam	6	Pallam	42	High Vulnerable	Human-induced impact altering seasonal coastal processes that cause adverse impact on the coastal environment A few spots experience seasonal accretion of the coast due to littoral current and wave action.
	7	Puthenthurai	26	Vulnerable level	
	8	Kesavanputhenthurai	12	Resilient zone	
	9	Rajakkamangalam	10	Resilient zone	
Muttam	10	Ganapathipuram	22	Risk level	Coastal environment is highly vulnerability to placer mining and quarrying. Lack of sediment supply due to man-made construction of wave breakers.Offshore rocky outcrops acting like barriers to protect some spot of coastal environments and ecosystems
	11	Muttam	24	Vulnerable level	
	12	Manavalakurichi	46	Extreme level	
	13	Mandaikadu	44	Extreme level	
Colachel	14	Colachel	12	Resilient zone	Encroaching human settlements cause severe impact; i.e., seawater intrusion, beach erosion at down-drift that leads to environmental degradation in and around the coastal stretch.
	15	Simonkudiyiruppu	22	Risk level	
	16	Midalam	26	Vulnerable level	
Thengapattinam	17	Mel Midalam	32	Vulnerable level	Coastal environment consists of varsity of ecosystems; river discharge and suspended sediment supply have seasonally influenced flora and fauna growth and productivity.
	18	Inayamputhenthurai	38	High Vulnerable	
	19 & 20	Thengapattinam	52	Extreme level	

The placer mining activities in the coastal zones including Manavalakurichi, Mandaikadu, and Midalam are triggering vulnerability to damaging coastal habitats in the nearshore environment, whereas the removing of sediments is reducing nutrient supply to the species of flora and fauna and intruding seawater toward the backshore environment. In various parts of the coastal stretch, 18% of the area has high vulnerability to environmental damage, and the estimated EVI value is in the range of 34–44. The coastal zones Inayamputhenthurai, South Thamaraikulam, and Kovalam fall under high vulnerability conditions due to a decrease in the sediment and nutrient supply that is decreasing the availability of minerals, nutrients, and organic matter to marine and coastal ecosystems. In these areas, the construction of coastal structures like groins, seawalls, and revetments across the shore is directly intervening sediment movement along the coastline that causes shifting or damaging of the habitats of coastal and marine ecosystems on the down-drift side (Kaliraj et al., 2014). The coastal settlements experiencing higher population density and growth rates are severely deteriorating environmental quality due to human encroachment of beaches, dunes, vegetative cover, shallow marshes, and water bodies. Changing the sediment load due to seasonal coastal dynamics is directly influencing the environmental equilibrium status by controlling the availability of organic matter to marine and coastal organisms. Moreover, the low-lying landforms of various locations are highly susceptible to degrading environmental systems due to inundating seawater, accumulating pollutants and salt maters during high tides, wave run-ups, and storm surges.

In the coastal stretch, 26% of the area has moderate vulnerability to environmental damage, and the EVI value is estimated at 24–34. The coastal zones including Pallam, Muttam, Midalam, Puthenthurai, and the eastern parts of Kanyakumari are identified as moderately vulnerability conditions, whereas the coastal landforms are seasonally altering in morphological structures due to monsoonal coastal dynamic processes during north-east and south-west monsoons. Fourteen percent of the area is identified under risk for environmental damage (coastal zones having low vulnerability or initial risk stage), and the EVI value is estimated at 14–24. In coastal zones like Kanyakumari, Periyakadu and Midalam, the environmental ecosystems are undergoing risk due to damaging the coastal landforms; that is, high-energetic waves are eroding cliffs and sandy beaches. In some parts of the coastal stretch, environmental systems face certain risks while wave run-ups are reaching above-average heights (due to storm surge or tsunami) that damage the coastal ecosystem in beaches and wetlands. Twenty-nine percent of the total area is demarcated as resilient zones, whereas coastal landforms have the ability to tolerate environmental hazards, and the EVI value is estimated as (<14). In various parts of the South Indian coast, the depositional landforms that especially occur in Puthenthurai, Chothavilai, Sanguthurai, Ganapathipuram, Colachel, and Simonkudiyiruppu have resilient characteristics to environmental damages, whereas the coastal environmental systems are able to sustain themselves during the favorable seasons. In some places, sandy beaches are able to rebuild shapes and spatial patterns during constructive waves and littoral currents and sustain the environment from a variety of hazards. Coastal settlements of different locations facing rapid population growth are increasing the rate of vulnerability to environmental degradation through erosion, seawater intrusion, pollution, encroachment, and other anthropogenic activities.

5. Conclusions

The GIS-based EVI model is developed for assessment and mapping potential vulnerability to environmental damage or degradation in a variety of ecosystems of the South Indian coastal stretch including beaches, dunes, cliffs, estuaries, salt marshes, and backwaters. This model is believed to fulfill the limitations of the general EVI model by analyzing site-specific geoenvironmental parameters in the sense of diversity and rapidly changing conditions. An advantage of the GIS-based EVI model is that it is able to resolve the issues of heterogenic measurements (linear, areal, and volume) of EVI parameters and map environmental vulnerability at a user-defined scale. In the coastal stretch, changing of coastal landforms produces a variety of adverse impacts on environmental systems due to eroding sediments and intruding saltwater; removal of sediments; encroaching beaches; constructing seawalls, groins, and jetties; and placer mining and quarrying. The low-lying coastal zones—especially estuaries, salt marshes, and sandy beaches—are facing higher vulnerability during energetic wave run-ups and storm surges. The coastal environmental systems spread over the depositional landforms are sustaining themselves to environmental hazards during favorable seasons. However, in many places, the coastal settlements (urban and rural) having higher population growth are triggering vulnerability to damage environmental ecosystems through erosion, seawater intrusion, pollution, encroachment, and other anthropogenic activities. The GIS-based EVI model is very useful for mapping vulnerable zones of environmental degradation and provides insight for coastal sustainable development.

Acknowledgments

The corresponding author is thankful to the director, National Center for Earth Science Studies, Ministry of Earth Sciences, government of India, for providing support and encouragement to complete the research successfully. South Pacific Geoscience Commission (SOPAC) is highly acknowledged for the global assessment of EVI. The authors extend their gratitude to IMD, SOI, Census of India, NIO, INCOIS, USGS–Earth Explorer, and PSMSL–UK, for providing necessary data sources.

References

Adger, W.N., 1996. Approaches to Vulnerability to Climate Change. CSERGE (Centre for Social and Economic Research of the Global Environment, UK) Working Paper GEC 96-05, 63pp.

Alwang, J., Siegel, P., Jorgensen, S., 2001. Vulnerability: A View from Different Disciplines. Social protection discussion paper No. 0115. The World Bank, Washington, DC.

Barnett, J., Lambert, S., Fry, I., 2008. The hazards of indicators: insights from the environmental vulnerability index. Ann. Assoc. Am. Geogr. 98 (1), 102–119.

Bossel, H., 1999. Indicators for Sustainable Development: Theory, Method, Applications. Winnipeg, Canada: International Institute for Sustainable Development.

Report of the meeting of experts on the environmental vulnerability index. In: Briguglio, L., Kaly, U., Pratt, C. (Eds.), 1999. Suva, Fiji: South Pacific Applied Geoscience Commission.

Campbell, J., 1997. Examining Pacific Island vulnerability to natural hazards. In: Planitz, A., Chung, J. (Eds.), Proceedings, VIII Pacific Science Intercongress. United Nations Department for Humanitarian Affairs, South Pacific Programme Office, Suva, Fiji, pp. 53–62.

Cutter, S., 1996. Vulnerability to environmental hazards. Prog. Hum. Geogr. 20, 529–539.

Cutter, S., 2003. The vulnerability of science and the science of vulnerability. Ann. Assoc. Am. Geogr. 93, 1–12.

Esty, D., Levy, M., Srebotnjak, T., de Sherbinin, A., 2005. Environmental Sustainability Index: Benchmarking National Environmental Stewardship. Yale Center for Environmental Policy and Law, New Haven, CT.

Gallopin, G.C., 2006. Linkages between vulnerability, resilience and adaptive capacity. Glob. Environ. Change 16, 293–303.

IPCC, 1991. The Seven Steps to the Vulnerability Assessment of Coastal Areas to Sea-Level Rise - Guidelines for Case Studies. IPCC Report, p. 24pp.

Kaliraj, S., Chandrasekar, N., 2012. Geo-processing model on coastal vulnerability index to explore risk zone along the south west coast of Tamil nadu, India. Int. J. Earth Sci. Eng. 5 (5), 1138–1147.

Kaliraj, S., Chandrasekar, N., Magesh, N.S., 2013. Evaluation of coastal erosion and accretion process along the South West coast of Kanyakumari, Tamil nadu using Geospatial techniques. Arab. J. Geosci. https://doi.org/10.1007/s12517-013-1216-7.

Kaliraj, S., Chandrasekar, N., Magesh, N.S., 2014. Impacts of wave energy and littoral currents on shoreline erosion/accretion along the south west coast of Kanyakumari, Tamil Nadu using DSAS and geospatial technology. Environ. Earth Sci. 71 (10), 4523–4542.

Kaliraj, S., Chandrasekar, N., Simon Peter, T., Selvakumar, S., Magesh, N.S., 2015. Mapping of coastal aquifer vulnerable zone in the south west coast of Kanyakumari, South India, using GIS based DRASTIC model. Environ. Monit. Assess. https://doi.org/10.1007/s10661-014-4073-2.

Kaliraj, S., Chandrasekar, N., Ramachandran, K.K., 2016. Mapping of coastal landforms and volumetric change analysis in the south west coast of Kanyakumari, South India using remote sensing and GIS techniques. Egypt. J. Remote Sens. & Space Sci. https://doi.org/10.1016/j.ejrs.2016.12.006.

Kaliraj, S., Chandrasekar, N., Ramachandran, K.K., Srinivas, Y., Saravanan, S., 2017. Coastal landuse and land cover change and transformations of Kanyakumari coast, India using remote sensing and GIS. Egypt. J. Remote Sens. & Space Sci. https://doi.org/10.1016/j.ejrs.2017.04.003.

Kaliraj, S., Chandrasekar, N., Ramachandran, K.K., Selvakumar, S., 2019. Seawater intrusion vulnerability in the coastal aquifers of southern India—an appraisal of the GALDIT model, parameters' sensitivity, and hydrochemical indicators. Environ. Sci. Pollut. Control Ser 26 (10), 9755–9784. https://doi.org/10.1007/s11356-019-04401-0.

Kaly, U., Briguglio, L., McLeod, H., Schmall, S., Pratt, C., Pal, R., 1999. Environmental Vulnerability Index to Summarise National Environmental Vulnerability Profiles. SOPAC Technical Report 275. http://www.vulnerabilityindex.net/EVI Library.htm.

Kaly, U., Pratt, C., Howorth, R., 2002. A framework for managing environmental vulnerability in small island developing states. Dev. Bull. 58, 22–38.

Kaly, U., Pratt, C., Mitchell, J., 2004. The Environmental Vulnerability Index 2004. SOPAC Technical Report 384. http://www.vulnerabilityindex.net/EVI Library.htm.

Kasperson, J.X., Kasperson, R.E. (Eds.), 2001. Global Environmental Risk. United Nations University Press, Earthscan, London.

Kasperson, J., Kasperson, R., Turner, B., Hsieh, W., Schiller, A., 2005. Vulnerability to global environmental change. In: Kasperson, J., Kasperson, R. (Eds.), Social Contours of Risk, vol. II. Earthscan, London, pp. 245–285.

Lambert, S., 2001. The Assessment of Pacific Island Environmental Vulnerability: A Critical Study of an Environmental Vulnerability Index by the South Pacific Geoscience Commission. Master of Arts of Geography thesis, Department of Geography, The University of Canterbury.

National Research Council, 2000. Ecological Indicators for the Nation. National Academy Press, Washington, DC.

Pantin, D.A., 1997. Alternative Ecological Vulnerability Indices for Developing Countries with Special Reference to Small Island Developing States (SIDS). Report to UN Department of Economic and Social Affairs, 22pp.

Parris, T., Kates, R., 2003. Characterizing and measuring sustainable development. Annu. Rev. Environ. Resour. 28, 559–586.

Peterson, G.D., Allen, C.R., Holling, C.S., 1998. Ecological resilience, biodiversity, and scale. Ecosystems 1, 6–18.

Pethick, J., Crooks, S., 2000. Development of a coastal vulnerability index: a geomorphological perspective. Environ. Conserv. 27, 359–367.

Pratt, C., Kaly, U., Sale-Mario, E., Seddon, J., 2002. Towards a Global Environmental Vulnerability Index (EVI): Update on Progress March–June 2002 and Revised Funding Proposal 2002–2003. SOPAC Misc. Rep. 465. Suva, Fiji: South Pacific Applied Geoscience Commission.

Pratt, C., Kaly, U., Mitchell, J., 2004. Manual: How to Use the Environmental Vulnerability Index. SOPAC Technical Report 383. http://www.vulnerabilityindex.net/EVI Library.htm.

Skondras, N.A., Karavitis, C.A., Gkotsis, I.I., Scott, P.J.B., Kaly, U.L., Alexandris, S.G., 2011. Application and assessment of the environmental vulnerability index in Greece. Ecol. Indicat. 11, 1699–1706. https://doi.org/10.1016/j.ecolind.2011.04.010.

SOPAC, 2005. EVI Country Profiles. Last accessed 25 December 2018. http://www.vulnerabilityindex.net/EVI Country Profiles.htm.

Turner, B., Kasperson, R., Matson, P., McCarthy, J., Corell, R., Christensen, L., et al., 2003. A framework for vulnerability analysis in sustainability science. Proc. Natl. Acad. Sci. U. S. A. 100, 8074–8079.

United Nations, 1994. Report of the Global Conference on the Sustainable Development of Small Island Developing States. Bridgetown, Barbados, 25 April–6 May 1994. New York, USA.

Villa, F., McLeod, H., 2002. Environmental vulnerability indicators for environmental planning and decision making: guidelines and applications. Environ. Manag. 29, 335–348.

15

Evaluation of suspended sediment concentration and heavy metal distribution in Ashtamudi Lake, a Ramsar site in the southwest coast of India using remote sensing and GIS techniques

A. Krishnakumar, S.K. Aditya, Kaliraj Seenipandi, K. Anoop Krishnan, Jeenu Jose

NATIONAL CENTRE FOR EARTH SCIENCE STUDIES (NCESS), THIRUVANANTHAPURAM, KERALA, INDIA

Chapter outline

Remote Sensing of Ocean and Coastal Environments. https://doi.org/10.1016/B978-0-12-819604-5.00015-9

1. Introduction

In the Ashtamudi Lake environment, the amount of suspended sediments is derived from river discharge, tidal flow, shore erosion, rock weathering, and so on, which determines the physicochemical properties of water and sediments. Seasonal variability of suspended sediments controls nutrients and pollution status, which are source materials for physical, chemical, and biological inputs to ecosystems of coastal, estuarine, and lake environments. Water bodies associated with coastal, estuaries, and river flow are generally accumulated with larger suspended sediments of sand, silt, and clay matters, which are triggering the status of pollution by point sources and nonpoint sources (Nair et al., 1993; Jennerjahn et al., 2008; NarendraBabu et al., 2010). The suspended sediment concentration in the coastal water is an indicator of erosion and deposition of coastal landforms, transporting the suspended sediments to various places causing formation of constructing or destructing landforms (Kaliraj et al., 2016). Quantitative measurement of suspended sediment concentration (SSC) for larger water bodies is a cost-effective and time-consuming process using conventional in situ methods, and it is critical to identify the SSC spatiotemporal variability using laboratory-based analysis. The advanced development of remote sensing technology provides an effective platform for qualitative and quantitative estimation of SSC from terrestrial and marine water sources (Pavelsky and Smith, 2009; Kaliraj et al., 2014a,b; Chamberland et al., 2019). Satellite remote sensing provides images with synoptic coverage of Earth's objects that shows their distinct reflected spectral signatures for mapping and monitoring earth and environmental changes. Many remote sensing satellites such as IRS, Landsat, AVHRR/3, MODIS, Quick Bird, GOES, IKONOS, SPOT, Sentinel 2A, and ASTER have sensors with various capabilities that are used for acquiring Earth's features with different spectral and spatial resolutions, which are used worldwide for studying natural resources of marine, coastal, and inland environments (Colleen and Pavelsky, 2013; Wu et al., 2015; Dethier et al., 2019; Topp et al., 2020).

Many researchers have successfully performed SSC estimation using multispectral images acquired from Landsat MSS, TM and ETM+, and Sentinel 2A–MSI. Remote sensing applications involve various scientific studies including land, water, and atmosphere. The multispectral images are widely used for studying water quality properties like physicochemical parameters, suspended sediments, chlorophyll-a (chl-a), temperature, salinity, conductivity, turbidity, organic components, and colored dissolved matters (Olmanson et al., 2011; Palmer et al., 2015; Hansen et al., 2017; Kaliraj et al., 2019a,b). Multitemporal images are used for mapping and monitoring the amount of suspended sediment entering, residing, and moving in larger water bodies like lakes, reservoirs, estuaries, and nearshore (Kaliraj et al., 2014a,b; Larson et al., 2018). Satellite sensors acquired by multispectral or hyperspectral images are used for quantifying suspended sediments in any water bodies, based on the amount of reflected electromagnetic radiation that distinctly indicates spectral properties of clear water and other suspended matters and are executed for measuring the quantity of SSC, through mathematical algorithms showing potential results at a local and regional scale, with synoptic and multitemporal levels.

Recent developments of remote sensing (i.e., Sentinel 2A and GeoEye) provide images with higher spectral and spatial resolution for obtaining accurate spatiotemporal information of water quality (Colleen and Pavelsky, 2013; Wackerman et al., 2017). Moreover, many successful studies have been demonstrated for estimating SSC using visible (VIS) and near-infrared (NIR) bands of multispectral images, making remote sensing an essential tool for measuring SSC variability in coastal and fluvial systems, particularly in large water bodies (Kaliraj et al., 2014a,b). Many researchers have used remotely sensed images for mapping water quality parameters including suspended sediment concentrations, chl-a, and salinity (Carpenter and Carpenter, 1983; Mertes et al., 1993; Kaliraj et al., 2014a,b). The remotely sensed images available with a wide variety of spectral, spatial, and temporal resolutions have been used to evaluate the suspended sediments, chemical pollutants, and chlorophyll contents of the shallow water surface (Olmanson et al., 2011; Hansen et al., 2017). Seasonal measurement of SSC using multitemporal images provides vital information for sustainable development of coastal and lake environments. Quantitative measurement of SSC using a multivariate regression model is performed using the mathematical calculation between spectral reflectance of bands (VIS, NIR, mid-infrared (IR), and short-wave IR (SWIR)) and correlation coefficients derived from in situ data (Liqin, 2014; Wackerman et al., 2017; Larson et al., 2018; Chamberland et al., 2019). Many researchers have developed empirical models and linear regression algorithms for quantitative and qualitative estimation of SSC in lakes, estuarine, coastal, and inland waters using multispectral images; for example, Landsat TM and ETM + images provide SSC measurement with tolerable accuracy (Olmanson et al., 2011; Wu et al., 2015; Liu et al., 2017; Dethier et al., 2019; Topp et al., 2020). Multispectral images acquired from Landsat 8 OLI and Sentinel 2A satellites are potentially used for mapping and monitoring SSC variability, and show relatively accurate measurement as compared to in situ data (Palmer et al., 2015; Hansen et al., 2017). Overall observations have declared that remote sensing is considered a suitable tool and technology for SSC estimation in the shallow water sources inland and coastal environments.

GIS-based mapping and assessment of heavy metal contamination using in situ sediment sampling reveals risk and vulnerability to environmental ecosystems of lakes, estuaries, and coastal regions. Deteriorating water quality of lakes and estuaries is due mainly to anthropogenic activities like rapid growth of population and industrialization that are opening gateways for mixing heavy metals and other point source pollutants into water bodies through river discharges and oceanic dumping (Choi et al., 2007; Zhang et al., 2012; Yuan et al., 2012). Seasonal changes in geo-environmental factors via natural and anthropogenic forces lead surface and subsurface flow of pollutants into water sources, which ultimately affects their geochemical characteristics (Dassenakis et al., 1998; Akcay et al., 2003; Zhang et al., 2003; Abdel-Ghani and Elchaghaby, 2007; Nicolau et al., 2006; Nouri et al., 2011). In general, water bodies receive heavy metals and other nutrients from various sources like river discharge, surface runoff, leaking pesticide residue from agricultural lands, releasing industrial effluents, mixing urban sewage, and dumping solid waste. Suspended sediments act like a transporting source to carry a variety of materials and mix them into the water column of lakes, estuaries, and coastal environments, controlling growth and productivity of aquatic ecosystems (Sheela et al., 2014). Many researchers have stated heavy metals occurring in suspended sediments are significantly affecting water quality, causing adverse impacts to the estuarine and marine ecosystems (Marchand et al., 2006; Zhang et al., 2007; Christophoridis et al., 2009; Rahman and Ishiga, 2012). Water bodies' confluence suspended sediments that carry heavy metals create toxic and polluted environmental ecosystems and become injurious to health (Mandal and Suzuki, 2002; Jiang et al., 2014). Spatiotemporal variability of heavy metals in water sources is affecting aquatic resources and are in evidence for anthropogenic impact to water bodies (Luo et al., 2010; Zhan et al., 2010). In larger water bodies, the amount of suspended sediments discharging from various sources usually transporting heavy metals are considered an environmental indicator to aquatic systems (Unlu et al., 2008). In this study, the SSC monthly variability is estimated in the Ashtamudi lake water using Sentinel 2A–MSI images during 2018–19 based on empirical methods; outputs can be used as vital information for monitoring and management of the lake environmental ecosystems for long-term scale.

2. Study area

The Ashtamudi wetland system (AWS) is one of the wetlands of international importance as defined by the Ramsar convention for the conservation and sustainable utilization of wetlands. Located in the west coast of Kerala state, South India, it is one of the major estuarine wetlands and the second largest backwater system in Kerala (Fig. 15.1). The geographical extent of this lake lies between latitude 8°53′19.083″N to 9°1′11.5″N and longitude 76°32′6.157″E to 76°40′15.006″E, and the areal extend is about 61.42 km^2. Ashtamudi Lake possesses unique wetland ecosystems that are composed of fresh and saltwater habitats, and is a gateway to the backwater transports of Kerala state. Geologically, the lake is a formation of sedimentary rocks with quaternary sediments with marine and fluvial alluvium and the tertiary sediments with laterite, sandstones, and clays of Warkalai formation in the lake basin and environment (Nair et al., 1993; Jennerjahn et al., 2008; NarendraBabu et al., 2010). Topographically, the lake is formed as a large palm-shaped water body with multiple branches. The bunds around the lake have many tourist resorts and boat houses along with agricultural lands, coconut groves, and palm trees interspersed with towns and villages. Ashtamudi Lake is perennially discharging a large quantity of water and sediments from the Kallada River running from

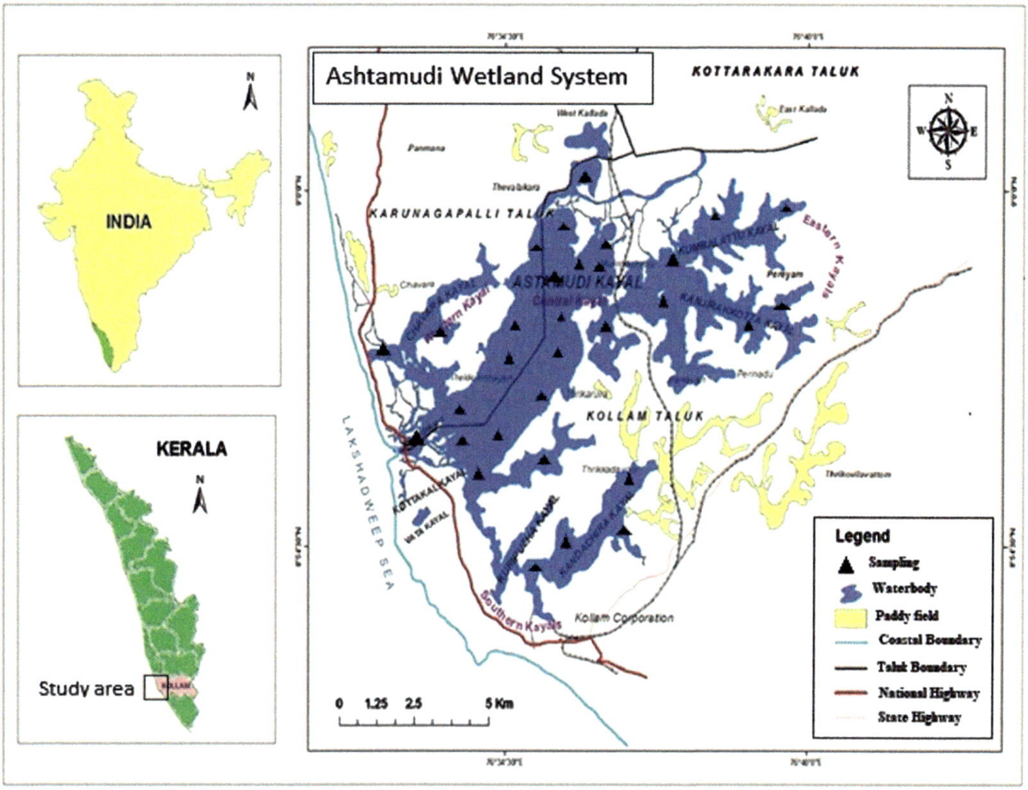

FIGURE 15.1 Location map of the study area.

the Western Ghats after traversing a distance of about 121 km through virgin forests. Ashtamudi Lake predominantly stores fresh water, while the estuary is mainly flowing salt water during pre- and postmonsoons. The average annual runoff from the river system is reported as 76 km^3, and is the major irrigation source for paddy and upland crops up to the areal extent of 61,630 ha (NCESS, 2015). Nowadays, the lake is susceptible to salinity ingress into the wetland due to seawater inflows during summer months. The lake environment's prevailing subtropical climatic condition has a temperature range of 25.5–27.5°C, and the annual rainfall is 24 cm. Economic evaluation of the AWS resources is assessed a total direct use value at Rs.66.8 million per annum, but it is presently showing a downward trend, particularly in fish resources in the estuary. The lake and backwater of the AWS environment is reported to be causing pollution due to various anthropogenic impacts like disposal of untreated sewage and solid wastes from human habitats; oil spills from thousands of fishing boats and tourist houseboats; and discharging industrial effluents from paper mills, mineral industries (aluminum, ceramics, seafood), as well as from coconut husk retting (NCESS, 2015). Hence, the AWS needs to be evaluated for sustainable management of environmental ecosystems and their economic importance for long-term perspectives.

3. Materials and methods

In the AWS, the evaluation of suspended sediment concentration and heavy metal distribution is performed using two different approaches: (1) estimating the SSC monthly variability using Sentinel 2A–MSI images and (2) in situ sediment samplings for heavy metal distribution during monsoon (M) and nonmonsoon (NM) seasons of the year 2018. Table 15.1 shows the sediment sampling locations selected for the detailed geochemical studies of the four sub-environment of the AWS.40 samples were collected from the entire areas of the AWS and for the assessment of heavy metal pollution studies, representative samples were used.

3.1 Estimation of SSC and its monthly variability

Sentinel 2A–MSI images are used for measurement of SSC monthly variability in the Ashtamudi lake water during 2018–19. The multispectral image is composed of 13 spectral bands; of which, bands including VIS, NIR, and SWIR are used for SSC estimation with a spatial resolution of 10 m. For this analysis, the Sentinel 2A–MSI images of orthorectified products (Level-1C and Level-2A) were download from ESA online portal of Copernicus Open Access Hub (https://scihub.copernicus.eu/dhus/#/home). Using MSI images, SSC is estimated using a linear regression algorithm based on spectral properties of water and suspended sediments that are distinctly measuring the quantity of SSC at local and regional levels with synoptic and multitemporal scales. Estimation of SSC is performed by following systematic procedures including (1) preprocessing Sentinel 2A–MSI images via calibration techniques like top-of-atmosphere and dark-object-subtraction for atmospheric and radiometric correction using Sen2Core v2.8 module of SNAP software; (2) resampling the multispectral bands of MSI image into

Table 15.1 Subenvironments of AWS with location and sample details.

S. No.	Name of subenvironment in AWS	Location		No. of samples collected
		Latitude	Longitude	
1	Ashtamudi main Lake (AML)	8°56′9″ to 8°59′25″N	76°32′1″ to 76°35′46.7″E	5 (M2,M5,M7, M9, M11)
2	ChavaraKayal (Ck)	8°55′7″ to 8°57′47″N	76°32′ 2″ to 76°35′59″E	6 (C1,C2, C3, C5, C7, C9)
3	Kureepuzha–Kandachira Kayal (Ks)	8°54′13″ to 8°55′45″N	76°34′13″ to 76°36′57″E	4 (Ks 1–Ks 4)
4	Kumbalathu–Kanjirakottu Kayal (K_E)	8°57′1″ to 8°59′58″N	76°36′1″ to 76°39′17″E	5 (K_E1, K_E3, K_E5, K_E7, K_E8)

common spatial resolution with reference to band 2 (pixel size at 10 m * 10 m) using ENVI 5.3 software; (3) land and cloud masking analysis for removing lands, clouds, and intertidal zones that generally comprise single or groups of pixels of nonwater objects; and (4) executing the empirical model on specific bands of MSI images using the Map Algebra module of ESRI-ArcGIS software (i.e., mathematical algorithms or linear regression analysis). In this analysis, the estimation of SSC using the linear regression algorithm is executed on VIS, NIR, and SWIR bands of Sentinel 2A–MSI image, wherein the SSC-derived products from band 4 (wavelength 650–680 nm) shows strong correlation with in situ sampling of SSC data. The equation is expressed as

$$SSC = 0.0262 * B4_{Rrs} - 7.5206 \quad (15.1)$$

where *SSC* refers to suspended sediment concentration (mg/L) and $B4_{Rrs}$ is the reflectance value of band 4 (red) of the Sentinel 2A–MSI image. Quantitative measurement of SSC is estimated using Eq. (15.1), in which the mathematical calculation is executed on band 4 spectral reflectance ($B4_{Rrs}$) values and correlation coefficients (derived from in situ SSC data) at a pixel scale of 10 m * 10 m. The SSC-derived products are cross-validated using in situ samplings data collected from the 40 locations in the Ashtamudi Lake based on linear regression analysis and root-mean-square error (RMSE) calculation, whereas the SSC-derived products show positive correlation compared to in situ SSC data with an R^2 value of 0.83 and RMSE estimated at 0.192. It is noted that the SSC-derived products using Sentinel 2A–MSI provides relatively accurate measurement of the suspended sediments that can be used for evaluating heavy metal distribution in a lake water environment.

3.2 Heavy metal: spatial distribution analysis

The present study also investigates spatial distribution of heavy metals and pollution status in the AWS. In this analysis, spatial distribution of heavy metal is identified using 40 sediment samples collected from various parts of the study area. In situ sampling is performed using a van Veen grab sampler for monsoon and nonmonsoon seasons during 2018 (Table 15.1). Quantitative measurement of heavy metals from each sediment sample is derived using systematic laboratory analysis, in which sediment samples were dried and powdered using an agate mortar, which in turn is used for geochemical analysis. The advanced X-ray fluorescence method (Bruker Model S4 pioneer sequential wavelength dispersive X-ray spectrometer) is used to measure heavy metals such as Cr, Co, Ni, Cu, Zn, Al, and so on. GIS-based inverse distance weighted technique is applied on a database of heavy metal concentration for mapping spatial distribution using ArcGIS v10.6 software. The pollution status of AWS is assessed using various heavy metals such as Cr, Al, Fe, Co, Ni, Cu, and Zn. The database of heavy metals is analyzed using multiple statistical indices like contamination factor (CF), contamination degree (Cd), pollution load index (PLI), and modified contamination degree (mCd) to evaluate degree of heavy metal pollution. Using the quantitative measurement of heavy metal concentration, the pollution risk for AWS is estimated through multiple analysis include CF, Cd, and mCd for monsoon and nonmonsoon periods. Moreover, these statistical outcomes are further analyzed using the potential ecological risk index (PERI) model for finding environmental risk in the AWS environment.

4. Results and discussion

Remote sensing images are widely used for mapping quantitative variability of sediment transport over large rivers, seawater, lakes, and estuaries based on spectral reflectance properties. Spatiotemporal sediment variability can be measured using multitemporal images, which is compared with monthly water sampling to calibrate an SSC retrieval mathematical modeling. In this analysis, the regression model is used to estimate SSC based on surface reflectance of Sentinel 2A–MSI images, which produces the best performance with $R^2 = 0.92$ ($N = 42$) for monthly SSC variability during 2018–19. The method offers a simple new approach to estimate the SSC along the Ashtamudi Lake and demonstrates the feasibility and reliability of remote sensing images for SSC measurement. In the Ashtamudi Lake, the SSC is spatially varying from place to place every month, indicating morphological landforms and ecosystems control as a primary factor for spreading heavy metals in water and sediment columns. The SSC monthly variability and impacts of spatial distribution of heavy metals and pollution status in the Ashtamudi Lake environment is described in detail.

4.1 Estimation of suspended sediment concentration

In Ashtamudi lake water, the monthly variability of SSC is estimated using Sentinel 2A–MSI images. Fig. 15.2 shows quantitative measurement of SSC during an 8 month period of 2018–19. In the lake-water column, the monthly variability of SSC is estimated in the range of 1.34–64.63 mg/L, wherein the maximum SSC found in two major sites includes (1) lake–Arabian Sea meeting point, and (2) lake–Kallada River meeting point. Meanwhile, in the eastern parts of lake, the water column far-flung from river and sea confluences is noted with a minimum amount of SSC during all the seasons due to lower discharge from the external sources. During northeast monsoon (October–December 2018), the SSC in lake water was estimated at 6.08–62.83 mg/L in October, and gradually decreased to 1.34–43.44 mg/L in November 2018 and 3.59–39.96 in December 2018. The lower amount of SSC is estimated during postmonsoon in the range of 7.86–34.66 mg/L in January 2019, 3.25–49.96 mg/L in February 2019, and 1.68–38.39 mg/L in March 2019. However, during premonsoon, the amount of SSC was found at a moderate level in the range of 2.75–46.19 mg/L in April 2019 and 2.29–64.63 mg/L in May 2019. Higher SSC is found in the mouth of the lake (lake–sea connecting area) during all seasons due to the release of a larger quantity of sediments from the Arabian Sea by the actions of littoral current and tidal waves. However, the river mouth (lake–river connecting area)

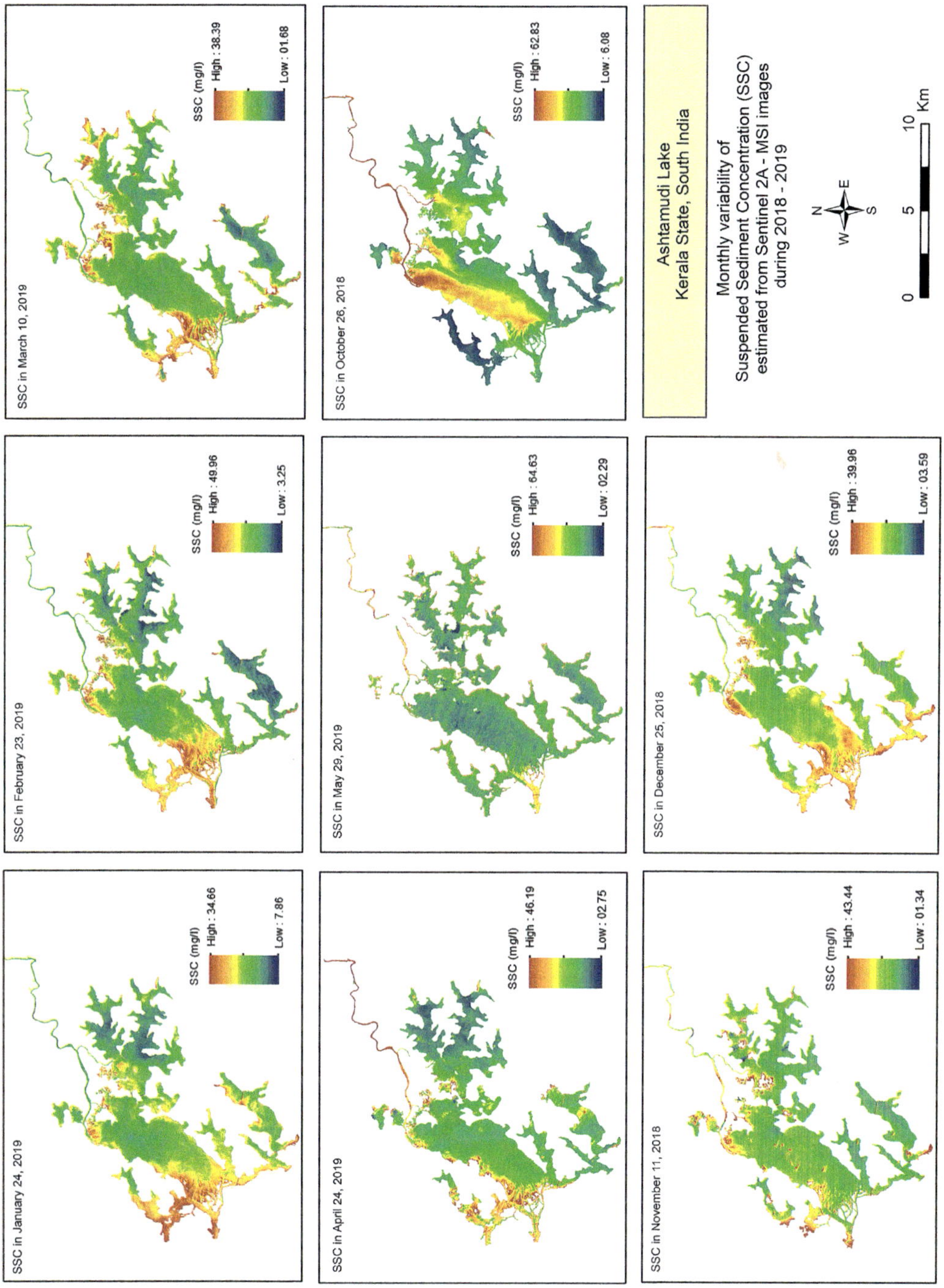

FIGURE 15.2 Suspended sediment concentration (SSC) and its monthly variability during 2018–19.

has higher SSC during the northeast monsoon due to heavy run-off from the Western Ghats catchment area. In the Ashtamudi lake environment, the larger quantity of suspended sediments settles in proximity of backwater and river mouth areas, whereas the lake's water physicochemical properties frequently change due to mixing of external materials (Nair et al., 1993; Jennerjahn et al., 2008; NarendraBabu et al., 2010). Moreover, in other parts of lake water, the SSC is found at a significant level due to shoaling effects and flowing water-current triggering the movement of suspended sediments toward different directions, especially the southern side. In the mid-lake's water column, the SSC variations during different months depend on velocity of inflow from sea and river discharge. The spatial distribution of SSC is found to gradually decrease with the increase of distance from the points of confluence and it is estimated in the range of 1.34–7.86 mg/L during all seasons. In the Ashtamudi lake water, the monthly variability of SSC is directly proportional to inflow of coastal water and river discharge during the study periods.

4.2 GIS mapping of heavy metal spatial distribution

In the AWS environment, spatial distribution of heavy metal concentration varies from place to place due to impacts of various geoenvironmental factors. The GIS-based mappings of spatial distribution of heavy metals are shown in Figs. 15.3–15.9. The results of heavy metal analysis shows that the Cr level ranges from 99 to 183 ppm

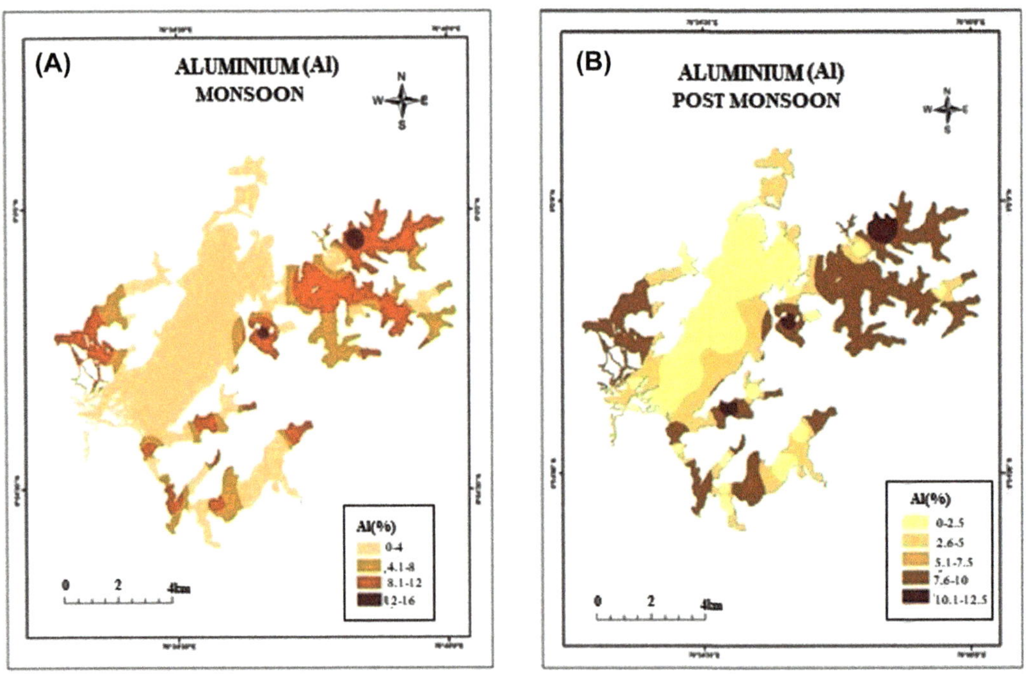

FIGURE 15.3 Spatial distribution of aluminum in the sediments of AWS.

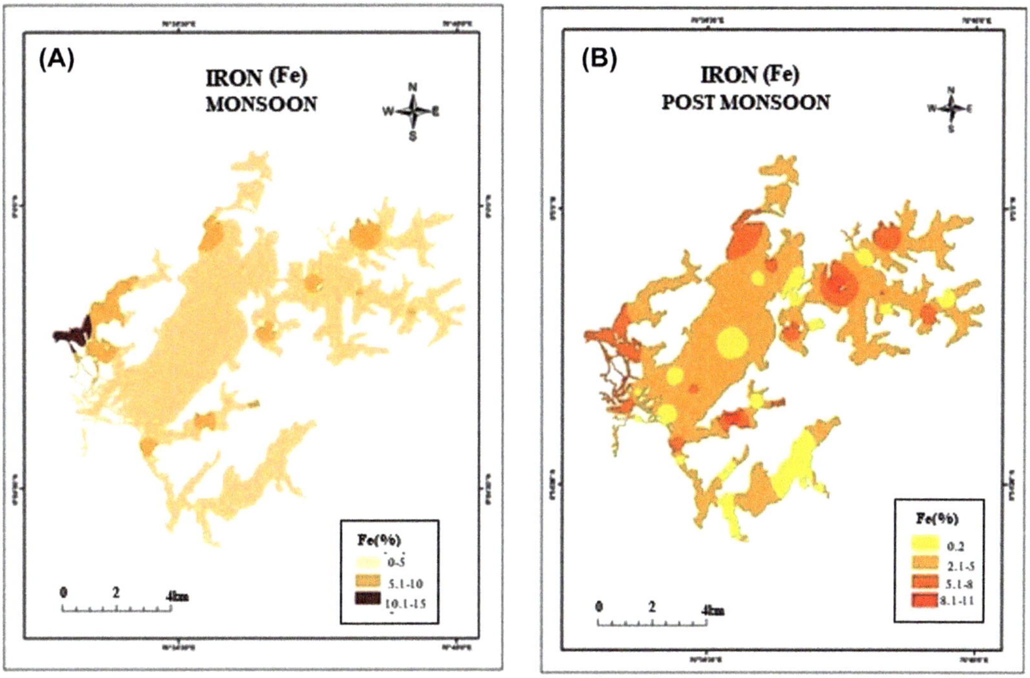

FIGURE 15.4 Spatial distribution of iron in sediments of AWS.

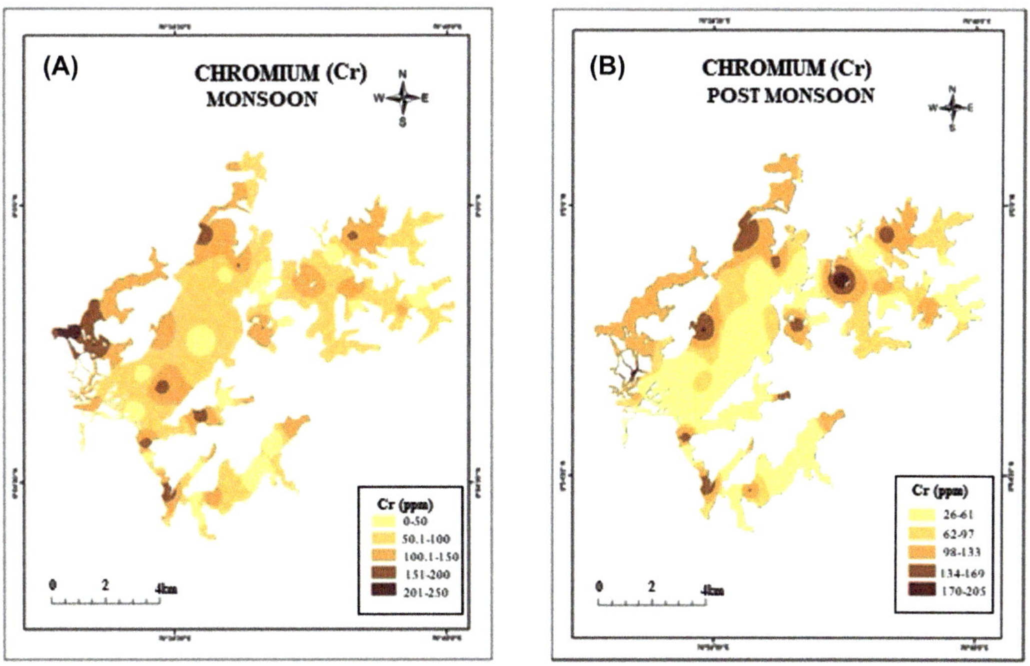

FIGURE 15.5 Spatial distribution of chromium in the sediments of AWS.

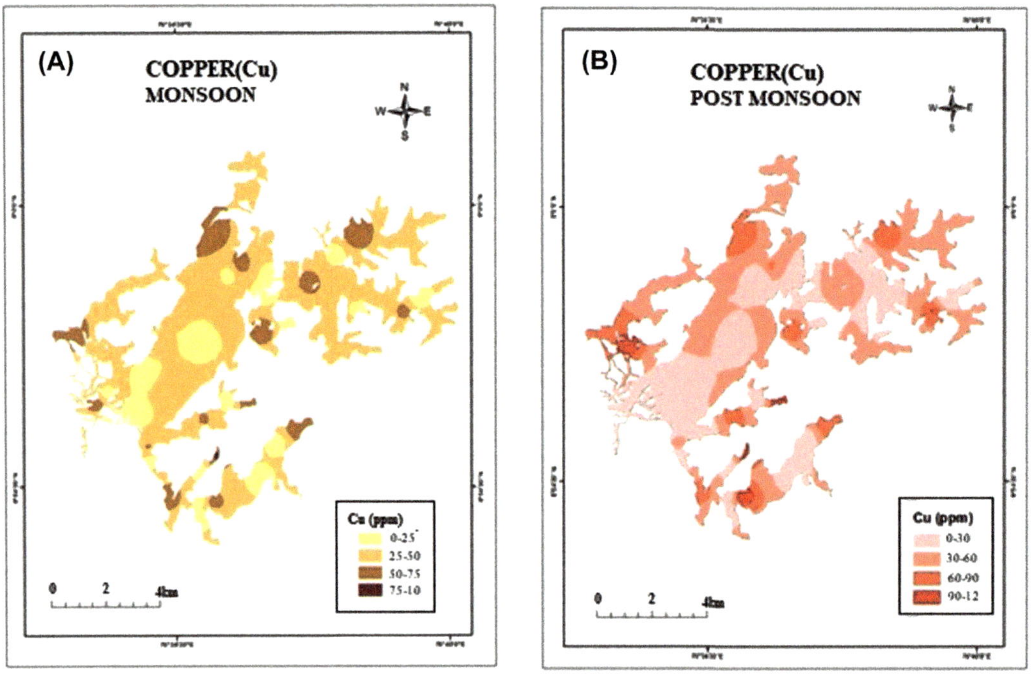

FIGURE 15.6 Spatial distribution of copper in the sediments of AWS.

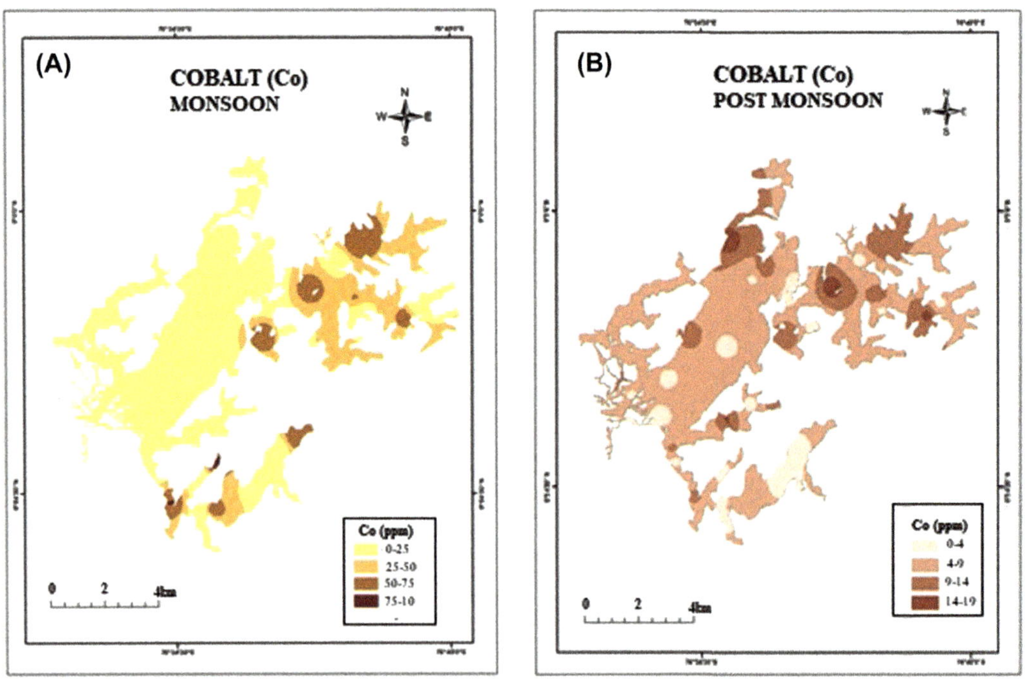

FIGURE 15.7 Spatial distribution of cobalt in the sediments of AWS.

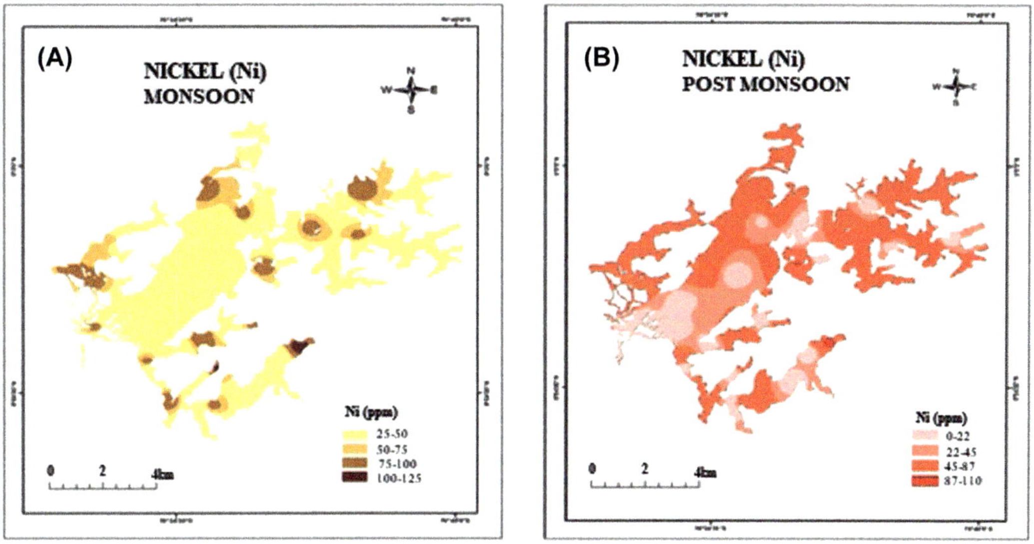

FIGURE 15.8 Spatial distribution of nickel in the sediments of AWS.

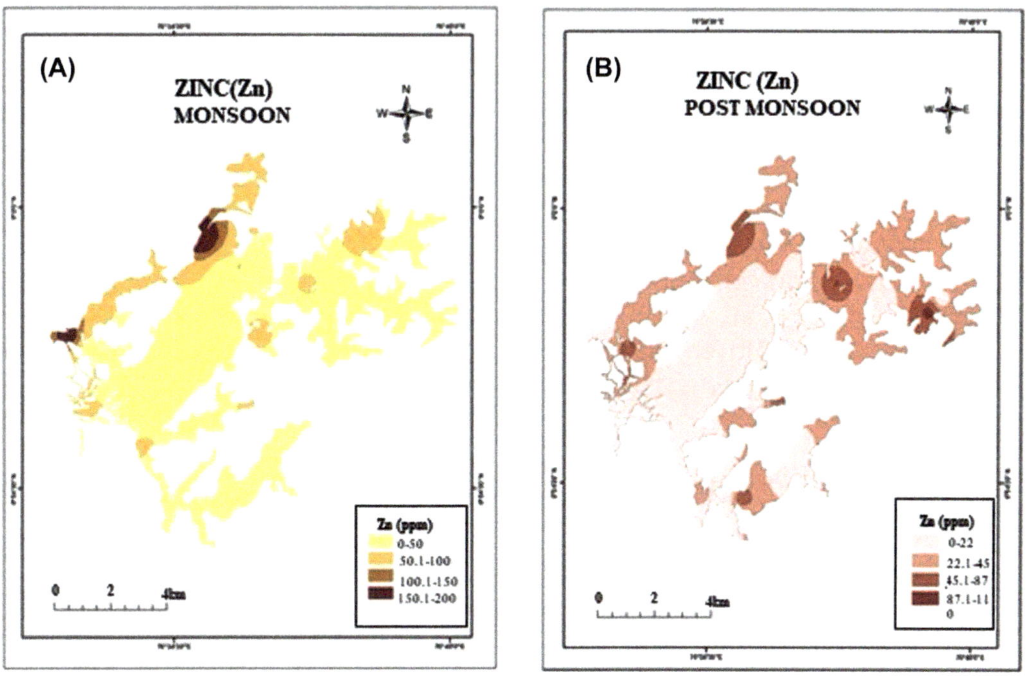

FIGURE 15.9 Spatial distribution of zinc in the sediments of AWS.

(average value is 141 ppm). Al spans between 3.08% and 12.62% (avg = 7.85%). The concentration of Fe ranges between 1.58% and 6.65%. The average of trace elements Co, Ni, Cu, and Zn were 12, 49, 38, and 48.5 ppm, respectively. The order of abundance of elements in Ashtamudi main kayal (AM) is denoted in sequences as Cr > Ni > Zn > Cu > Co > Al > Fe. In Kumbalath–Kanjirakottukayals, the Cu ranges between 25 and 74 ppm (avg = 49.5 ppm). The Zn concentration varies from 4 to 103 ppm (avg = 53.5 ppm). Ni ranges between 61 and 92 ppm (76.5 ppm). The annual average of Co was 14.5 ppm. Cr concentration is estimated as 103–197 ppm, whereas the annual average is 150 ppm. Fe ranges between 0.42% and 1.09% (0.75%), and Al concentration varies between 8.36% and 12.84% (avg = 10.6%). The order of abundance of elements is noted as Cr > Ni > Zn > Cu > Co > Al > Fe. In Kureepuzha–Kandachirakayals, Cu ranges between 63 and 110 ppm (avg = 86.5 ppm). The Zn concentration varies from 17 to 71 ppm (avg = 44 ppm). Ni ranges between 76 and 115 ppm (avg = 95.5 ppm). The annual average of Co is 9 ppm. Meanwhile, Cr concentration is estimated in the range of 95–196 ppm (avg = 145.5 ppm). Fe ranges between 3.09% and 4.62% (3.85%) and Al concentration varies between 7.94% and 9.43% (mean value of 8.68%). The order of abundance of elements is noted as Cr > Ni > Cu > Zn > Co > Al > Fe. In Chavara, Cu ranges between 33 and 112 ppm (avg = 72.5 ppm). Zn concentration varies from 2 to 155 ppm (avg = 78.5 ppm), and Ni ranges between 46 and 98 ppm (avg = 72 ppm). In this area, the annual average of Co is estimated at 10 ppm, and the concentration of Cr is estimated as 90–219 ppm (avg = 154.5 ppm). Fe ranges between 5.25% and 14.46% (9.85%) and Al concentration varies between 6.72% and 11.54% (avg = 9.13). The order of abundance of elements was Cr > Zn > Cu > Ni > Co > Fe > Al.

It is seen from the spatiotemporal results that Fe concentration in the lake is considerably less and with little or no significant role in controlling other metal ions. This might be due to the dredging activities widely practiced in the estuary, which have resulted in the mixing of sediments throughout the wetland system, which in turn has resulted in the dilution and dispersion of heavy metals (Krishnakumar et al., 2015). Spatial distribution of heavy metal concentration varies from place to place depending on amount of SSC discharge deposited in and around the lake environment (Nair et al., 1993; Jennerjahn et al., 2008; NarendraBabu et al., 2010). It is noted that human-driven location-specific issues are the major reason for elevated and altered concentration values in the lake environment (Krishnakumar et al., 2015).

4.3 Assessment of heavy metal pollution status

In the AWS, the assessment of heavy metal pollution is performed using statistical indices such as PLI (Tomlinson et al., 1980) and PERI (Hakanson, 1980). These models are used to analyze heavy metals in order to identify potential vulnerability to pollution risk in the Ashtamudi Lake environment.

4.3.1 Pollution load index

Modeling of PLI integrates multiple heavy metal factors to calculate the contamination potentiality using the following equation, expressed as

$$PLI = (CF_1 * CF_2 * CF_3 * \ldots * CFn)1/n \quad (15.2)$$

in which $CF = C_{\text{metal}}/C_{\text{background}})$.

PLI refers to pollution load index value, *CF* refers to contamination factor of each element, and *n* is the total number of parameters. In this analysis, the result shows that the value CF < 1 refers to low contamination, $1 =$ CF < 3 implies moderate contamination, $3 =$ CF ≤ 6 indicates considerable contamination, and CF > 6 indicates severe (very high) contamination.

The estimated values of PLI indicate the degree of pollution based on potential contribution of all elements in a particular location. Many researchers used CF as a versatile tool to identify contamination level in the environmental matrix analysis (Devanesan et al., 2017). In this analysis, for computing CF values from sediment samplings, the world shale average concentrations have been taken as background values (Turekian and Wedephol, 1961). The CF values of Cr, Al, Fe, Co, Ni, Cu, and Zn indicate that the four subenvironments of AWS have appeared to be in the low to moderate contamination categories.

Fig. 15.10 shows the spatial variability of PLI in various sampling locations in an AWS environment. The result shows that the estimated PLI values vary from place to place in the AWS environment depending on heavy metal concentrations whereas site-specific PLI values are estimated at various locations. The ranges of PLI in Ashtamudi Main Kayal is estimated to be 0.24–1.44, Chavara Kayal is noted as 0.59 to 1.53, Kumbalath–Kanchirakottukayal is measured as 0.63–1.27, and the Kureepuzha–Kandachirakayal is found with the PLI value of 0.81–1.18. The result shows that 47% of samples are noted as protected zones from heavy metal pollution, with PLI value (PLI < 1) estimated below 1. Meanwhile, 53% of samples are found with PLI > 1, indicating severe heavy metal pollution status. This is due to the inflow of discharge from various sources that may carry and mix pollutants from point and nonpoint sources and their subsequent attachment to the finer particles of the sediments (Arunkumar et al., 2010). PLI value for the four subenvironments depends mainly on various external factors such as fish processing units located in the western side (Chavara kayal) of the estuary, which discharge organic wastes in enormous quantities into the system. Retting of coconut husks for the coir industry is prevalent at several regions in the eastern (Kumbalath–Kanjirakottu kayal) and northern branches of the estuary. In the southern part of AWS (Kureepuzha–Kandachira kayal) a canal running through the city brings considerable municipal waste). Earlier studies have also confirmed the deterioration of water and sediment quality of AWS is due mainly to the inputs of contaminants from industrial effluents, municipal sewage, waste disposal, and coconut husk retting (Sajan and Damodaran, 1981; NarendraBabu et al., 2009; Krishnakumar et al., 2017).

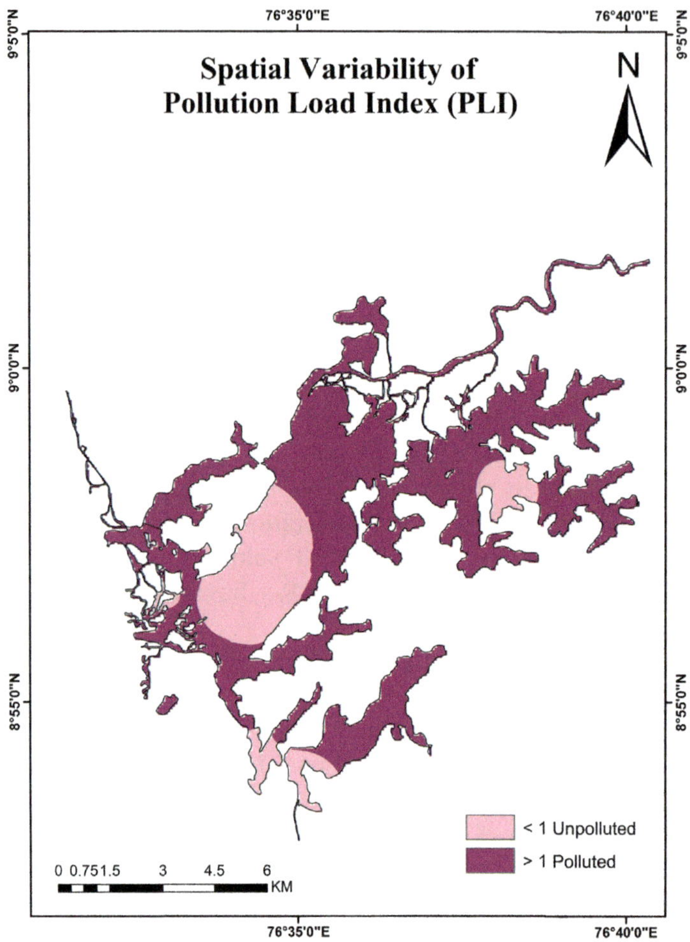

FIGURE 15.10 Spatial variability of PLI in sampling locations of AWS.

4.3.2 Contamination degree

Estimation of Cd suggested by Hakanson (1980) is used as a diagnostic tool to simplify contamination control. Cd was calculated as the sum of the CF for each sample and is expressed as

$$Cd = \Sigma Cf \tag{15.3}$$

where *Cd* refers to contamination degree derived from the total value of the contamination factor (*Cf*). According to Man et al. (2004), in the classification for Cd, the contamination classes are categorized as follows: Cd $<$ 6 indicates a low degree of contamination, 6 $<$ Cd $<$ 12 is a moderate degree of contamination, 12 $<$ Cd $<$ 24 is a considerable degree of contamination, and Cd $>$ 24 is a high degree of contamination. In the AWS, the calculated values of Cd are given in Tables 15.2–15.5.

The result showed that the *Cd* value in the Ashtamudi Main Kayal was found in the low to moderate category at a range of 1.1–8.6. The Cd value in the Chavara Kayal is estimated at 2.36–11.47, which reflects the area under moderate degree of

Table 15.2 Statistical measurement of Cf, Cd, and mCd in the AML during monsoon (M) and nonmonsoon (Nm) periods.

		Contamination factor (CF) in the AML sector													
		Cr		Al		Fe		Co		Ni		Cu		Zn	
SL. NO	Sample ID	M	Nm	M	Nm	M	Nm	M	Nm	M	Nm	M	Nm	M	Nm
1	M2	1.74	1.1	1.43	0.35	0.34	1.51	0.84	0.36	1.72	0.08	1.51	0.06	0.97	0.05
2	M5	2.03	0.9	0.96	0.54	0.37	1.04	0.73	0.36	1.5	0.38	0.64	0.48	0.18	0.04
3	M7	1.24	1.85	0.70	0.981	0.86	0.5	0.52	0.57	0.66	1.54	0.35	0.95	0.31	Negligible
4	M9	1.81	1.74	1.41	1.34	1.32	1.45	0.89	0.84	1.52	1.42	1.51	1.622	1.91	0.75
5	M11	1.78	1.68	1.25	1.17	1.12	1.25	0.73	0.68	1.88	1.72	1.55	1.24	0.48	0.26
Contamination degree (Cd)		8.6	7.27	5.75	4.38	4.01	5.75	3.71	2.81	7.28	5.14	5.56	4.35	3.85	1.1
Modified contamination degree (mCd)		1.22	1.03	0.82	0.62	0.57	0.82	0.53	0.40	1.04	0.73	0.79	0.72	0.55	0.62

Table 15.3 Statistical measurement of Cf, Cd, and mCd in the Chavara kayal during monsoon (M) and nonmonsoon (Nm) periods.

		Contamination factor (CF) in the Ck sector													
		Cr		Al		Fe		Co		Ni		Cu		Zn	
SL. NO	Sample ID	M	Nm	M	Nm	M	Nm	M	Nm	M	Nm	M	Nm	M	Nm
1	C1	2.43	1.26	1	0.88	3.14	1.16	0.94	0.26	1.38	1.42	1.24	1.11	1.63	0.05
2	C2	2.22	1	0.94	0.84	1.7	1.03	0.73	0.10	1.6	1.28	1	2.06	Negligible	0.54
3	C3	1.4	2.2	0.76	0.95	1.14	1.74	0.68	0.68	0.92	1.64	1	0.97	0.71	0.70
4	C5	2.26	1.83	1	0.95	2	1.56	0.84	0.57	1.72	1.44	1.23	0.73	1	0.02
5	C7	2.02	1.57	0.96	1.31	1.64	2.05	0.52	0.78	1.96	1.54	1.11	1.46	0.21	0.33
6	C9	1.14	1.7	0.84	0.87	1.18	1.41	0.21	0.63	1.4	1.22	1.77	2.48	0.75	0.72
Contamination degree (Cd)		11.47	5.5	10.8	5.8	3.92	8.95	8.68	3.02	7.35	8.54	4.3	8.81	5.5	2.36
Modified contamination degree (mCd)		1.63	0.61	0.78	0.82	1.54	1.27	0.56	0.43	1.24	1.22	1.05	1.25	0.61	0.33

contamination. In the Kumbalath and Kureepuzha Kayals, the Cd value shows a lower contamination category and is estimated at 2.26–8.06 and 0.87–7.98, respectively. The sediments recovered from the frequent dredging activities in Ashtamudi Lake are normally dumped on its banks. Ultimately these sediments are again brought into the lake system through rains and other human activities. Hence in AWS, dredging activities are caused by the transportation and mixing of sediments throughout the water column, dispersing heavy metals into various parts. The sediments deposited in the bottom of shelf and nearshore due to tidal waves and littoral currents contribute to stratification of sediments. Moreover, AWS varies spatially in contamination degree level, which depends on various anthropogenic activities.

Table 15.4 Statistical measurement of Cf, Cd, and mCd in the Kureepuzha–Kandachirakayal during monsoon (M) and nonmonsoon (Nm) periods.

		Contamination factor (CF) in the Ks sector													
		Cr		Al		Fe		Co		Ni		Cu		Zn	
SL. NO	**Sample ID**	**M**	**Nm**	**M**	**Nm**	**M**	**Nm**	**M**	**Nm**	**M**	**Nm**	**M**	**Nm**	**M**	**Nm**
1	Ks1	2.17	1.81	0.99	0.92	1	0.84	0.68	0.52	1.78	1.52	1.55	1.53	0.17	0.21
2	Ks2	1.52	1.05	1.07	0.90	0.91	0.70	0.47	0.26	2.14	1.62	1.82	2.02	–	0.28
3	Ks3	1.56	1.48	0.921	0.93	0.92	0.83	0.47	0.36	1.76	1.66	1.46	2.44	0.42	0.74
4	Ks4	1.72	1.27	0.98	0.92	0.83	0.67	0.47	0.26	2.3	1.88	1.4	1.4	0.28	0.44
Contamination degree (Cd)		6.97	5.61	3.961	3.67	3.66	3.04	2.09	1.4	7.98	6.68	6.23	7.39	0.87	1.67
Modified contamination degree (mCd)		0.99	0.80	0.56	0.52	0.52	0.43	0.29	0.2	1.14	0.95	0.89	1.05	Negligible	Negligible

Table 15.5 Statistical measurement of Cf, Cd, and mCd in the Kumbalathu–Kanjirakottukayal during monsoon (M) and nonmonsoon (Nm) periods.

		Contamination factor (CF) Ke													
		Cr		Al		Fe		Co		Ni		Cu		Zn	
SL. NO	**Sample ID**	**M**	**Nm**	**M**	**Nm**	**M**	**Nm**	**M**	**Nm**	**M**	**Nm**	**M**	**Nm**	**M**	**Nm**
1	K_E1	1.68	1.6	1.45	1.18	1.43	1.21	0.68	0.68	1.8	1.84	1.51	1.51	0.73	Negligible
2	K_E3	1.55	2.18	1.26	0.95	1.27	2.02	0.73	0.94	1.6	1.24	1.33	1.04	0.56	0.95
3	K_E5	1.74	1.53	1.48	1.31	1.51	1.22	0.84	0.73	1.8	1.64	1.44	1.51	0.97	0.26
4	K_E7	1.61	1.28	1.30	1.15	1.20	1.03	0.63	0.63	1.74	1.22	1	0.55	Negligible	0.12
5	K_E8	1.14	1.33	1.07	1.20	1.13	1.24	0.57	0.84	1.12	1.46	1.22	1.64	Negligible	1.08
Contamination degree (Cd)		7.72	7.92	6.56	5.79	6.54	6.72	3.45	3.82	8.06	7.4	6.5	6.25	2.26	2.41
Modified contamination degree (mCd)		1.28	1.32	1.09	0.96	1.09	1.12	0.57	0.63	1.34	1.23	1.08	1.04	0.37	0.40

4.3.3 Modified contamination degree

The modified contamination degree (mCd) developed by Abrahim and Parker (2008) is the updated model of Cd that is used as a diagnostic tool to simplify contamination control based on total sum of contamination factors (CF) and number of parameter (n). This is introduced to estimate the overall degree of contamination at a given site using the following equation:

$$mCd = (\Sigma CF)/n \tag{15.4}$$

where n refers to number of analyzed elements and CF is the contamination factor. The output of mCd is classified into different classes to reflect degree of contamination: $mCd < 1.5$ is very low degree of contamination; $1.5 < mCd < 2$ is low degree of contamination;

$2 < mCd < 4$ is moderate degree of contamination; $4 < mCd < 8$ is high degree of contamination; $8 < mCd < 16$ is very high degree of contamination; $16 < mCd < 32$ is extreme degree of contamination; and $mCd < 32$ is ultrahigh degree of contamination.

In the AWS, the estimated mCd value shows a lower degree of contamination for all heavy metal elements except Cr and Fe in the Chavara kayal (monsoon). According to Krishnakumar et al. (2015), the elevated Cr concentration in the study area is attributed to the industrial effluents, mainly metallurgical industries adjacent to AWS.

4.3.4 Potential ecological risk index

The PERI model developed by Hakanson (1980) is used to assess the potential risk and vulnerability to ecosystems from impacts of pollution or contamination based on sedimentological perspectives. In the analysis, the PERI for various locations is calculated based on toxic response factor of individual heavy metal elements collected from sediment samplings. In addition, this model indicates potential ecological risk based on impacts and toxicity of heavy metal elements to biological responses. The PERI model is applied worldwide for assessment of potential risk and vulnerability to ecosystems using the following equation:

$$RI = \Sigma Er_i \tag{15.5}$$

wherein

$$C_F = Cd/Cb$$
$$Er_i = Tr_i{}^*CF$$

where RI is potential risk index value derived from sum of Er_i of individual elements, Er_i refers ecological risk index of individual heavy metal element, Tr_i refers toxic response factor of heavy metal element, CF is contamination factor of heavy metal element, Cd is heavy metal concentration, and Cb is background of heavy metal concentration of sediment samplings. According to Hakanson (1980), the toxic response factor values for Cr, Cu, Zn, Mn, Pb, and Ni are assigned as 2, 5, 1, 1, 5, and 5, respectively. The crustal average value is considered as background value (*Cb*) for executing the contamination factor (*CF*) of heavy metal elements (Taylor, 1964; Li et al., 1986).

Fig. 15.11 shows spatial variability of PERI in an AWS environment. From the result, it is noted that the estimated Er_i values for all heavy metal elements indicate lower ecological risk category; similarly, the estimated *RI* value suggests that the ecosystems have categorized under low risk zones subject to impacts of heavy metal elements in the AWS environment. Table 15.6 shows the statistical measurement of PERI ($\mathrm{Er_i}$) and Risk index (RI) for four subenvironment of the AWS. It is observed that the Er_i values are estimated as <40 for heavy metal elements including Cr, Cu, Ni, and Zn, which indicate lower risk to ecosystems in the AWS environment. In the Ashtamudi main kayal, the *RI* value estimated is 85.25 during monsoon and 63.29 in nonmonsoon periods. Similarly, the *RI* values for Kumbalath kayal is estimated as 82.22 in monsoon and 86.5 in nonmonsoon. Moreover, the *RI* value is estimated to be 85.86 in the Kureepuzha kayal during monsoon. These zones are noted in the low-grade level of potential ecological

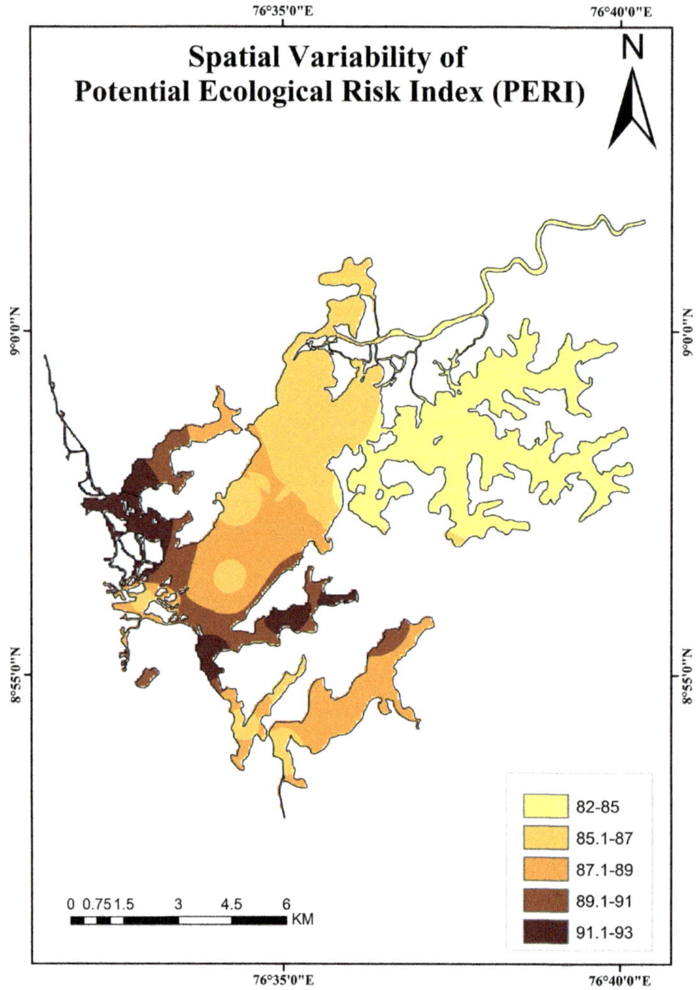

FIGURE 15.11 Spatial variability of potential ecological risk index (PERI) AWS.

risk. Meanwhile, in the Chavara kayal, the *RI* is estimated to be 92.29 during monsoon and it increases to 108.23 in nonmonsoon periods, which indicates moderate level of potential ecological risk. The difference in RI between seasons can be attributed to the accumulation of sediments at different regions of AWS, which has altered the sediment water dynamics of AWS. It is also observed that the AWS environment denotes the spatiotemporal variability in potential ecological risk (*RI*) level, which depends on accumulation of sediments into the different parts of AWS, affecting growth and production of ecosystem in the AWS environment. Moreover, sediment dynamics and water quality in the lake-water column have been frequently affected by human-induced activities like dredging of river sand and deepening of harbors and fishing ports in the region.

Table 15.6 Statistical measurement of potential ecological risk index (Er_i) and risk index (RI) for the four subenvironments of AWS.

Potential ecological risk index value ($Er_i = Tr * CF$)									
Ashtamudi main kayal (AML)	**Sample ID**	**Cr**		**Cu**		**Ni**		**Zn**	
RI = 85.25 (M) RI = 63.29 (Nm)		**M**	**Nm**	**M**	**Nm**	**M**	**Nm**	**M**	**Nm**
	M2	3.48	2.2	7.55	0.3	8.6	0.4	0.97	0.05
	M5	4.06	1.8	3.2	2.4	7.5	1.9	0.18	0.04
	M7	2.48	3.7	1.75	4.75	3.3	7.7	0.31	–
	M9	3.62	3.48	7.55	8.1	7.6	7.3	1.91	0.75
	M11	3.56	3.36	7.75	6.2	9.4	8.6	0.48	0.26
Chavara kayal	C1	4.86	2.52	6.2	5.55	6.9	7.1	1.63	0.05
RI = 92.29(M)	C2	4.44	2	5	10.3	8	6.4	–	0.54
RI = 108.2(Nm)	C3	2.8	4.4	5	4.85	4.6	8.2	0.71	0.70
	C5	4.52	3.66	6.15	3.65	8.6	7.2	1	0.02
	C7	4.04	3.14	5.55	7.3	9.8	7.7	0.21	0.33
	C9	2.28	3.4	8.85	12.4	7	6.1	0.75	0.72
Kureepuzha–Kandachira kayal	Ks1	4.34	3.62	7.75	7.65	8.9	7.6	0.17	0.21
RI = 85.86(M)	Ks2	3.04	2.1	9.1	10.1	10.7	8.1	–	0.28
RI = 152.37(nm)	Ks3	3.12	0.74	7.3	12.2	8.8	8.3	0.42	74
	Ks4	3.44	0.63	7	7	11.5	9.4	0.28	0.44
Kumbalath–Kanjirakottu kayal	K_E 1	3.36	3.2	7.55	7.55	7.55	9.2	0.73	–
RI = 82.22(M)	K_E3	3.1	4.36	6.65	5.2	8	6.2	0.56	0.95
RI = 86.5(Nm)	K_E5	3.48	3.06	7.2	7.55	7.55	8.2	0.97	0.26
	K_E7	3.22	2.56	5	2.75	8.7	6.1	–	0.12
	K_E8	2.28	2.66	6.1	8.2	0.22	7.3	–	1.08

5. Conclusions

In the AWS environment, the seasonal variability of SSC has significant influence on physicochemical and biological properties. Remote sensing technology performs a vital role for measuring and monitoring SSC in water columns of large water bodies including coastal regions. Satellite images with multispectral characteristics used for SSC quantitative measurement can be used as primary information to researchers and management authorities for understanding sediment transportation and coastal dynamic processes. In the Ashtamudi lake water, a higher rate of SSC was found in the mouth of the lake (lake–sea connecting area) during all seasons resulting from actions of littoral current and tidal waves. Meanwhile, the river mouth (lake–river connecting area) also received larger SSC during northeast monsoon through river discharge from the Western Ghats catchment area. It is noted that the larger quantity of suspended sediments settled within proximity of backwater and river mouth areas, whereas the lake's water physicochemical properties are frequently changing due to mixing of external materials. Using sediment-sampling analysis, the assessment of pollution status is performed based on heavy metal concentration and shows the AWS environment is in a lower category for ecological risk. The CF, Cd, and mCd, analyzed for monsoon and nonmonsoon seasons suggested low to moderate contamination in all four subdivided locations, whereas PLI suggested 53% of samples as polluted (>1). The Hakanson PERI indicated that most areas of AWS during monsoon and nonmonsoon seasons belong to the low risk category. This study proves that the integrated remote sensing and GIS techniques become essential for studying the environmental ecosystems in coastal and inland water bodies. Moreover this study emphasizes the need for sustainable management measures and continuous monitoring of location-specific action plans for effective conservation and utilization of an important RAMSAR wetland system.

Acknowledgments

The corresponding author is thankful to the Director, National Center for Earth Science Studies (NCESS), Ministry of Earth Sciences, Government of India, for his constant encouragement and support. The Dept of Port and Fisheries, Government of Kerala is acknowledged for support to study AWS. The authors extend their gratitude to ESA and USGS–Earth Explorer for providing necessary satellite images.

References

Abel-Ghani, N.T., Elchaghaby, G.A., 2007. Influence of operating conditions on the removal of Cu, Zn, Cd and Pb ions from wastewater by adsorption. Int. J. Environ. Sci. & Technol. 4, 451–456.

Abrahim, G.M.S., Parker, R.J., 2008. Assessment of heavy metal enrichment factors and the degree of contamination in marine sediments from Tamaki Estuary, Auckland, New Zealand. Estuar. Coast Shelf Sci. 136, 227–238.

Akcay, H., Oguz, A., Karapire, C., 2003. Study of heavy metal pollution and speciation in Buyak Menderes and Gediz river sediments. Water Res. 37, 813–822.

Arunkumar, K.S., Joseph, S., Thomas, J., 2010. Heavy metal concentration in the coastal wetlands of Thiruvananthapuram district, southern India. J. Environ. Sci. Eng. 52 (2), 131–136.

Carpenter, D., Carpenter, S., 1983. Modeling inland water quality using Landsat data. Remote Sens. Environ. 13, 345–352.

Chamberland, J., Bennett, J., Yue, B., 2019. Water surface temperature and suspended sediment concentration (SSC) retrieval from Landsat imagery, Northwest Territories, Canada. Can. J. Rem. Sens. 1–14. https://doi.org/10.1080/07038992.2019.1643706.

Choi, M.S., Yi, H.I., Yang, S.Y., Lee, C.B., Cha, H.J., 2007. Identification of Pb sources in Yellow Sea sediments using stable Pb isotope ratios. Mar. Chem. 107, 255–274.

Christophoridis, C., Dedepsidis, D., Fytiannos, K., 2009. Occurrence and distribution of selected heavy metals in the surface sediments of Thermaikos Gulf, N.Greece -Assessment using pollution indicators. J. Hazard Mater. 168, 1082–1091.

Colleen, M.L., Pavelsky, T.M., 2013. Remote sensing of suspended sediment concentration and hydrologic connectivity in a complex Wetland environment. Remote Sens. Environ. 129, 197–209. https://doi.org/10.1016/j.rse.2012.10.019.

Dassenakis, M., Scoullos, M., Foufa, E., Krasakopoulou, E., Pavlidou, A., Kloukiniotou, M., 1998. Effects of multiple source pollution on a small Mediterranean river. Appl. Geochem. 13, 197–211.

Dethier, E.N., Sartain, S.L., David, A., Lutz, D.A., 2019. Heightened levels and seasonal inversion of riverine suspended sediment in a tropical biodiversity hot spot due to artisanal gold mining. Proc. Natl. Acad. Sci. Unit. States Am. https://doi.org/10.1073/pnas.1907842116.

Devanesan, E., Gandhi, M.S., Selvapandiyan, M., Senthilkumar, G., Ravishankar, R., 2017. Heavy metal and potential ecological risk assessment in sediments collected from Poombuhar to Karaikal Coast of Tamilnadu using Energy dispersive X-ray fluorescence (EDXRF) technique. J. Basic Appl. Sci. 6, 285–292.

Hakanson, L., 1980. An ecological risk index for aquatic pollution control. A sedimentological approach. Water Res. 14, 975–1001.

Hansen, C.H., Burian, S.J., Dennison, P.E., Williams, G.P., 2017. Spatiotemporal variability of lake water quality in the context of remote sensing models. Rem. Sens. 9, 409.

Jennerjahn, T.C., Soman, K., Ittekkot, V., Nordhaus, I., Sooraj, S., Priya, R.S., Lahajnar, N., 2008. Effect of land use on the biogeochemistry of dissolved nutrients and suspended and sedimentary organic matter in the tropical Kallada River and Ashtamudi estuary, Kerala, India. Biogeochemistry 90 (1), 29–47. https://doi.org/10.1007/s10533-008-9228-1.

Jiang, X., Teng, A., Xu, W., Liu, X., 2014. Distribution and pollution assessment of heavy metals in surface sediments in the yellow Sea. Mar. Pollut. Bull. 83, 366–375.

Kaliraj, S., Chandrasekar, N., Magesh, N.S., 2014a. Multispectral image analysis of suspended sediment concentration along the Southern coast of Kanyakumari, Tamil Nadu, India. J. Coast. Sci. 1 (1), 63–71.

Kaliraj, S., Chandrasekar, N., Magesh, N.S., 2014b. Impacts of wave energy and littoral currents on shoreline erosion/accretion along the South-west coast of Kanyakumari, Tamil nadu using DSAS and geospatial technology. Environ. Earth Sci. https://doi.org/10.1007/s12665-013-2845-6.

Kaliraj, S., Chandrasekar, N., Ramachandran, K.K., 2016. Mapping of coastal landforms and volumetric change analysis in the south west coast of Kanyakumari, South India using remote sensing and GIS techniques, Egypt. J. Remote Sens. Space Sci.

Kaliraj, S., Chandrasekar, N., Ramachandran, K.K., 2019a. Phytoplankton (chl-a) biomass seasonal variability in the gulf of Mannar, South India: a remote sensing perspective. In: Kumar, P., et al. (Eds.), Applications and Challenges of Geospatial Technology. https://doi.org/10.1007/978-3-319-99882-4_5.

Kaliraj, S., Chandrasekar, N., Ramachandran, K.K., Selvakumar, S., 2019b. Seawater intrusion vulnerability in the coastal aquifers of southern India—an appraisal of the GALDIT model, parameters' sensitivity, and hydrochemical indicators. Environ. Sci. Pollut. Control Ser. 26, 9755–9784. https://doi.org/10.1007/s11356-019-04401-0.

Krishnakumar, A., Saranya, P., Geethu Prasad, P., Rakhi, C., 2015. Holocene records of human driven geological impacts in a Ramsar wetland of India. J. Aquat. Procedia 373–380. Elsevier.

Krishnakumar, A., Saranya, P., RevathyDas, 2017. Hydrogeochemistry and environmental issues of the Wetlands of Kerala, Southwestern India. Wetl. Sci. 145–161. Springer.

Larson, M.D., Milas, A.S., Vincent, R.K., Evans, J.E., 2018. Multi-depth suspended sediment estimation using high-resolution remote-sensing UAV in Maumee River, Ohio. Int. J. Rem. Sens. 39 (15–16), 5472–5489. https://doi.org/10.1080/01431161.2018.1465616.

Li, J., Zeng, B.W., Yao, Y.Y., Zhang, L.C., Qui, C.Q., Qian, X.Z., 1986. Studies on environmental background levels in waters of Dongting lake system. Environ. Sci. J. Integr. Environ. Res. 7, 62–68.

Liqin, Q., 2014. Remote Sensing Suspended Sediment Concentration in the Yellow River. Doctoral Dissertations, 383. https://opencommons.uconn.edu/dissertations/383.

Liu, H., Li, Q., Shi, T., Hu, S., Wu, G., Zhou, Q., 2017. Application of Sentinel 2 MSI images to retrieve suspended particulate matter concentrations in Poyang lake. Rem. Sens. 9, 761. https://doi.org/10.3390/rs9070761.

Luo, W., Lu, Y., Wang, T., Hu, W., Jiao, W., Naile, J.E., Khim, J.S., Giesy, J.P., 2010. Ecological risk assessment of arsenic and metals in sediments of coastal areas of northern Bohai and Yellow seas, China. Ambio 39, 367–375.

Man, K.W., Zheng, J., Leung, A.P.K., Lam, P.K.S., Lam, M.H.W., Yen, Y.F., 2004. Distribution and behavior of trace metals in the sediment and pore water of a tropical coastal Wetland. Sci. Total Environ. 327, 295–314.

Mandal, B.K., Suzuki, K.T., 2002. Arsenic round the world: a review: review article. Alanta 58 (1), 201–235.

Marchand, C., Lallier-Verges, E., Baltzer, F., Alberica, P., Cossac, D., Baillif, P., 2006. Heavy metal distribution in mangrove sediments along the mobile coastline of French Guiana. Mar. Chem. 98, 1–17.

Mertes, L.A.K., Smith, M.O., Adams, J.B., 1993. Estimating suspended sediment concentration in surface waters of the Amazon river Wetlands from Landsat images. Remote Sens. Environ. 43, 281–301.

Nair, N.B., Aziz, P.K., Dharmaraj, K., Arunachalam, M., Krishnakumar, K., Balasubramanian, N.K., 1993. Ecology of Indian estuaries: Part I—physico-chemical features of water and sediment nutrients of Ashtamudi estuary. Indian J. Mar. Sci. 12, 143–150.

NarendraBabu, K., Omana, P.K., Mohan, M., 2009. Water and sediment quality of Ashtamudi estuary, a Ransar site,Southwest coast of India- a statistical appraisal. Environ. Monit. Assess. 165, 307–3019.

NarendraBabu, K., Omana, P.K., Mohan, M., 2010. Water and sediment quality of Ashtamudi estuary, a Ramsar site, southwest coast of India—a statistical appraisal. Environ. Monit. Assess. 165, 307–319. https://doi.org/10.1007/s10661-009-0947-0.

National Centre for Earth Science Studies, 2015. Environmental Studies of Kollam-Neendakara Wetlands and Associated Inlands. Project report submitted to the Department of Port and Fisheries, Govt. of Kerala, India.

Nicolau, R., Galera, C.A., Lucas, Y., 2006. Transfer of nutrients and labile metals from the content to the sea by a small Mediterranean river. Chemosphere 63, 469–476.

Nouri, J., Lorestani, B., Yousefi, N., Khorasani, N., Hasani, A.H., Seif, S., Cheraghi, M., 2011. Phytoremediation potential of native plants grown in the vicinity of Ahangaran lead-zinc mine (Hamedan, Iran). Environ. Earth Sci. 62, 639–644.

Olmanson, L.G., Brezonik, P.L., Bauer, M.E., 2011. Evaluation of medium to low resolution satellite imagery for regional lake water quality assessments. Water Resour. Res. 47, 1900–1904.

Palmer, S.C., Kutser, T., Hunter, P.D., 2015. Remote sensing of inland waters: challenges, progress and future directions. Remote Sens. Environ. 157, 1–8.

Pavelsky, T.M., Smith, L.C., 2009. Remote sensing of suspended sediment concentration, flow velocity, and lake recharge in the Peace-Athabasca Delta, Canada. Water Resour. Res. 45, W11417. https://doi.org/10.1029/2008WR007424.

Rahman, M.A., Ishiga, H., 2012. Trace metal concentrations in tidal flat coastal sediments, Yamaguchi Prefecture, Southwest Japan. Environ. Monit. Assess. 184, 5755–5771.

Sajan, K., Damodaran, K.T., 1981. Studies on the distribution of organic matter content in sediments of Ashtamudi lake, Kerala. Bull. Dept. Mar. Sci. Univ. Cochin 12, 155–160.

Sheela, A.M., Letha, J., Joseph, S., Joseph, J., Thomas, J., 2014. Multivariate analysis of the sediment quality of a tropical coastal lake system. Lake Reservoir Manag. 19, 306–317.

Taylor, S.R., 1964. Abundance of chemical elements in the continental crust: a new table. Geochem. Cosmochim. Acta 28, 1273–1285.

Tomlinson, D.C., Wilson, J.G., Harris, C.R., Jeffrey, D.W., 1980. Problems in the assessment of heavy metals in estuaries and the formation of pollution index. Helgol. Meeresunters. 33, 566–575.

Topp, S.N., Pavelsky, T.M., Jensen, D., Simard, M., Ross, M.R.V., 2020. Research trends in the use of remote sensing for inland water quality science: moving towards multidisciplinary applications. Water 12 (1), 169. https://doi.org/10.3390/w12010169.

Turekian, K.K., Wedepohl, K.H., 1961. Distribution of the elements in some major units of the earth's crust. Geol. Soc. Am. 72, 175–192.

Unlu, S., Topcuoglu, S., Alpar, B., Kirbasoglu, C., Yilmaz, Y.Z., 2008. Heavy metal pollution in surface sediment and mussel samples in the gulf of Gemlik. Environ. Monit. Assess. 144, 169–178.

Wackerman, C., Hayden, A., Jonik, J., 2017. Deriving spatial and temporal context for point measurements of suspended-sediment concentration using remote-sensing imagery in the Mekong Delta. Continent. Shelf Res. 147, 231–245. https://doi.org/10.1016/j.csr.2017.08.007.

Wu, G., Cui, L., Liu, L., Chen, F., Fei, T., Liu, Y., 2015. Statistical model development and estimation of suspended particulate matter concentrations with Landsat 8 OLI images of Dongting Lake, China. Int. J. Rem. Sens. 36, 343–360.

Yuan, H., Song, J., Li, X., Li, N., Duan, L., 2012. Distribution and contamination of heavy metals in surface sediments of the South Yellow Sea. Mar. Pollut. Bull. 64, 2151–2159.

Zhan, S., Peng, S., Liu, C., Chang, Q., Xu, J., 2010. Spatial and temporal variations of heavy metals in surface sediments in Bohai Bay, North China. Bull. Environ. Contam. Toxicol. 84, 482–487.

Zhang, L., Wang, G., Yao, D., Duan, G.Z., 2003. Environmental significance and research of heavy metals in offshore sediments. Mar. Ecol. Lett. 19, 6–9.

Zhang, L., Ye, X., Feng, H., Jing, Y., Ouyang, T., Yu, X., Liang, R., Gao, C., Chen, W., 2007. Heavy metal contamination in western Xiamen Bay sediments and its vicinity, China. Mar. Pollut. Bull. 54, 974–982.

Zhang, G., Liu, D., Wu, H., Chen, L., Han, Q., 2012. Heavy metal contamination in the marine organisms in Yantai coast, northern Yellow Sea of China. Ecotoxicology 21, 1726–1733.

16

Seasonal variability of sea surface temperature in Southern Indian coastal water using Landsat 8 OLI/TIRS images

Kaliraj Seenipandi[1], K.K. Ramachandran[1], Prashant Ghadei[2], Sulochana Shekhar[2]

[1]NATIONAL CENTRE FOR EARTH SCIENCE STUDIES (NCESS), THIRUVANANTHAPURAM, KERALA, INDIA; [2]DEPARTMENT OF GEOGRAPHY, CENTRAL UNIVERSITY OF TAMIL NADU, THIRUVARUR, TAMIL NADU, INDIA

Chapter outline

1. Introduction

Coastal water is dynamically changing its physicochemical properties subject to various geo-environmental and climatic factors that influence growth and development of environmental ecosystems (Edwards et al., 2010; Gholizadeh et al., 2016). The coastal water column receives various input matter from waves, tides, littoral currents,

Remote Sensing of Ocean and Coastal Environments. https://doi.org/10.1016/B978-0-12-819604-5.00016-0

river discharges, and anthropogenic activities that control formation of coastal landforms and provide sources of physical, chemical, and biological input to the offshore (Meijun et al., 2015; Liu et al., 2017). Land and water surface temperature (L/WST) are essential variables for understanding Earth's climate and are important driving factors in ecology, biodiversity, and species distribution (Rongali et al., 2018). Sea surface temperature (SST) is a dynamic factor in coastal water that controlling various biogeochemical processes of coastal and marine ecosystems (Schott et al., 2009; Geetha et al., 2011; Kaliraj et al., 2019a,b). The coastal and ocean water column consists of different physicochemical properties such as salinity, SST, total dissolved solids (TDS), turbidity, conductivity, chlorophyll (chl-a), biological oxygen demand (BOD), total phosphate and organic matters, and so on (Miyazawa et al., 2013; Skakala and Timothy, 2016; Kaliraj, 2016). In situ survey for oceanographic parameters are difficult to produce within user required time and finding spatiotemporal variability within or near real time. Remote sensing is the art, science, and technology and is being used as an effective platform for resolving Earth's issues at a spatiotemporal scale, especially in the oceans and coastal environmental issues (Park et al., 2014; Casal and Lavender, 2017; Wahyudi et al., 2018).

Multiple satellite images are available for the insight of ocean and coastal environments, such as Landsat series, IRS series, SPOT, ESA, MODIS, Oceansat, Sentinel, and SAR microwave. MODIS and AVHRR satellite images are often used for SST retrieval for larger areas but with lower scale; this may perform poorly along the coastline due to mixed pixels at the border of land and water (Hosoda, 2011; Liu et al., 2017). Landsat 8 OLI/TIRS images (spatial resolution is 30 m) have been used for estimating SST at higher resolution from coastal, estuaries, lakes, reservoirs, and so on (Schaeffer et al., 2018). Several researchers have used thermal infrared sensor (TIRS) images for estimating SST, which are used for producing higher performance over water property studies like coastal and estuarine plumes (Chen et al., 2014; Cahyono et al., 2017; LSDS-1574, 2019). Satellite images provide synaptic coverage of ocean and coastal environments that are very helpful for understanding physicochemical properties at a user-defined scale (Gholizadeh et al., 2016). In the ocean and coastal environment, SST is one of major parameters controlling physical and biogeochemical processes, especially salinity, accumulating, and chl-a. Estimating the SST using total internal reflection (TIR) bands are used for primary information for understanding oceanic and climate dynamics in the field of weather forecasting (Chen et al., 2014; Wahyudi et al., 2018). The SST of coastal water is a vital source for coastal ecosystem management like that of coral reefs, estuaries, swamps, and tidal flats toward climate change impact (Kaliraj et al., 2013, 2017). Exploring the spatiotemporal variability of SST using thermal infrared images is very useful for understanding seawater currents, upwelling, front, mixing, and hydrodynamic processes. Monitoring SST seasonal variability in and around the coastal water and coral islands is a vital source for studying coral bleaching events for long-term scale (Geetha et al., 2011; Gholizadeh et al., 2016).

Hyperspectral and multispectral satellite images are used worldwide for measuring and monitoring water quality (major physical and chemical properties) in surface water sources like coastal water, sea surface, river, lake, and reservoirs (Miyazawa et al., 2013; Kaliraj et al., 2019a,b). Remote sensing provides synoptic coverage of Earth's surface in images with unique spatial and spectral resolutions that are useful for observing ocean and coastal environments during past and present scenarios (Sobrino et al., 2001; Kaliraj et al., 2013; Meijun et al., 2015). The multitemporal satellite images like Landsat, Sentinel, and IRS are used for exploring coastal water properties such as salinity, temperature, turbidity, dissolved organic matter, suspended sediments, biological and chemical oxygen demand, conductivity, depth, chlorophyll, organic carbon, and total phosphorus at a regional or global scale (Kaliraj et al., 2014a,b; Skakala and Timothy, 2016). Satellite remote sensing is an effective platform for observing the physicochemical properties of water bodies like suspended sediments, organic components, colored dissolved matters, and so on based on interaction between the electromagnetic waves and the molecules of water and its components (Casal and Lavender, 2017). TIRS images reflect the characteristics of emissivity of objects of Earth surfaces. Currently physicochemical properties in the North Indian Ocean are frequently changing due to climate change and related processes (Wahyudi et al., 2018). The spatial variability of salinity and SST are acting as major factors that determine primary productivity (chlorophyll) and other nutrients to the coastal ecosystems along the Southern Indian coastal water environment (Kaliraj et al., 2019a,b).

The SST along coastal water is one of the important oceanic factors for understanding coastal and marine environmental ecosystem activities. The TIR images acquired by the satellites like MODIS, Oceansat, and Landsat are widely used for measuring SST and sea surface salinity (Cahyono et al., 2017). The Landsat 8 TIRS images are characterized as thermal emissivity properties (TIR bands 10 and 11) for measuring L/WST at 30 * 30 m pixel scale, and the derived SST output is strongly correlated with instruments based on on-site measurements (Rongali et al., 2018; LSDS-1574, 2019). Landsat 8 OLI/TIRS images are widely used for a variety of coastal and oceanic applications such as measuring and monitoring of salinity, temperature, TDS, electrical conductivity (EC), BOD, chemical oxygen demand (COD), and sediment concentration. Landsat 8 OLI/TIRS—derived SST products are used for studies of water properties due to the known spectral emissivity and the generally homogeneous temperature of water bodies within the image pixel (Miyazawa et al., 2013; Park et al., 2014; Edwards et al., 2010). It is vitally necessary for exploring seasonal variability of coastal water properties using multitemporal satellite images. Presently, the derived land surface temperature/SST products using Landsat 8 TIRS images are generally not yet available for worldwide studies; however, the present study deals with the systematic analysis of Landsat 8 OLI/TIRS images for estimating the SST using the empirical algorithms based on in situ inputs for monitoring spatiotemporal status of ocean and coastal water environments.

2. Study area

The Southern Indian coastal water presents dynamic changing of major physical and chemical properties such as salinity, temperature, chlorophyll, and total suspended sediments due to monsoonal variability of oceanic and coastal processes (Kaliraj and Chandrasekar, 2012; Kaliraj et al., 2016). Geographically, the study area is located within latitude 7°38′ N to 9°20′ N and longitude 76°50′ E to 78°53′ E and the coastal water column is about 8440.22 km^2 that stretches parallel to the Southern Indian coastline at a length of 410.72 km. The coastal water constitutes the coastal water of the Northern Indian Ocean with the flows of the Arabian Sea and Bay of Bengal (Fig. 16.1). The coastal slope tends to seaward comprising the laterite uplands, estuaries, sand dunes, and outlets of the nonperennial river. The coastal water receives a larger amount of discharge and runoff materials from drainage networks like Thamirabarani, Valliyar, Pazhayar, and Vaippar during northeast and southwest monsoons (Kaliraj et al., 2014a,b, 2015, 2017). The study area has characteristics of shallow depth of nearshore and surf zones at the depth ranges of 1–10 m, and it is gradually increasing to 20–30 m with the increment of seaward distance of 5–6 km. The Southern Indian coast presents subtropical climatic conditions with temperature ranges from 23.78 to 33.95°C and annual average rainfall between 826 and 1456 mm. The average value of wave energy is estimated at 0.5–8.5 kJ/km^2 and the mean current velocity is measured as 0.14 ms^{-1}. The physicochemical

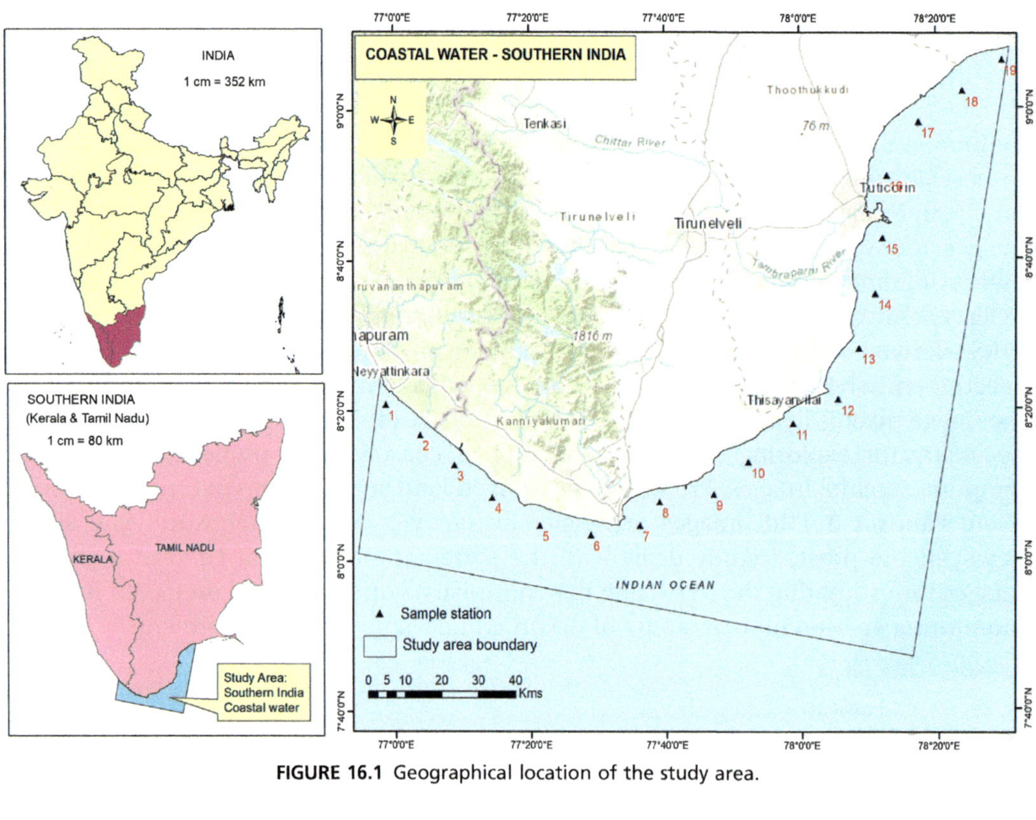

FIGURE 16.1 Geographical location of the study area.

properties of coastal water like pH, EC, TDS, colored dissolved organic matter (CDOM), BOD, COD, salinity, and temperature seasonally varying from place to place depend on hydrodynamics like waves, wind and currents, and so on (Kaliraj et al., 2019a,b). In this study, the coastal water region of the Southern Indian coastal stretches (Keelakarai–Thiruvananthapuram) is explored using multitemporal Landsat 8 TIR images and the outcome provides vital information for understanding ocean and coastal hydrodynamic characteristics.

3. Materials and methods

Landsat 8 OLI/TIRS images of thermal infrared bands used for estimating SST in the Southern Indian Coastal water column for the three seasons (pre-, post-, and during monsoons) during the periods of 2017–2018. The TIR bands (B10 and B11) are characterized with thermal infrared spectrum at the wavelength (λ) (TIR1 = 10.60–11.19 μm) and (TIR2 = 11.50–12.51 μm); the spatial resolution is 30 m (after the resampling process). NASA–Goddard Space Flight Center, Greenbelt, USA, fabricates Landsat 8 TIRS sensor with quantum well infrared photodetectors (QWIPs). The Landsat 8 OLI/TIRS images are obtained from the USGS/Earth Explorer online portal for the periods of 2017 and 2018. These Level 1T (terraincorrected) images available in GeoTiFF format are calibrated via top-of-atmosphere and dark-object-subtraction techniques for atmospheric and radiometric correction using ENVI 5.3 software (LSDS-1574, 2019). The TIRS QWIPs use two TIR bands that help separate the emitted thermal energy from Earth's surface objects and record to create a digital image. The Landsat 8 TIRS image reflects the thermal emissivity from the objects of Earth's surface is recorded as digital number (DN) values to represent L/WST at pixel scale. The split-window algorithm (SWA) is used to calibrate the TIR bands (B10 and B11) for estimating the SST at a pixel level (30 * 30 m). In the TIR bands (bands 10 and 11) of Landsat 8 OLI/TIRS images, the pixel size (spatial resolution) of 60 m becomes resampled into 30 m using data fusion techniques to bring better output on a large scale (Chen et al., 2014). According to USGS – Earth Explorer recommendation, in this analysis, Landsat 8 OLI/TIRS images was updated for full stray light correction based on metadata of Landsat Collection 1 dataset distributed by USGS. After the development of a stray light correction, the TIRS images reportedly produced high fidelity in SST using a SWA model comparable to that of MODIS images (Hosoda, 2011; Liu et al., 2017). Second, the TIR bands are analyzed via land masking techniques for removing land, clouds, and intertidal zones that generally restricted the pixels between water and nonwater regions. In situ data of SST and other coastal water properties obtained from buoys of the selected locations are used for validating the output of SST derived from TIRS images. The SST of TIRS-derived values is reported with spatial variability of 0.2°C from buoy measurement. In this analysis, the SWA model is used for estimating SST using Landsat 8 TIRS images and the result is available in raster format at 30 * 30 m pixel scale. The SWA is used for Landsat 8 TIRS bands (B10 and B11) for retrieving relatively accurate SST with the root-mean-squared deference (RMSD) value of 0.7, which indicates low bias in the derived SST. However, the SST is scattered at the proximity of pixels associated with land, turbid water, and clouds whereas the RMSD value increases to K1 and K2, respectively, with an approximately 0.5 K cold bias. The Landsat 8 TIRS image is calibrated using USGS-Collection 1 data (Landsat 8 – Metadata) that allows for retrieving absolute SST from the coastal water region. Fig. 16.2 shows the

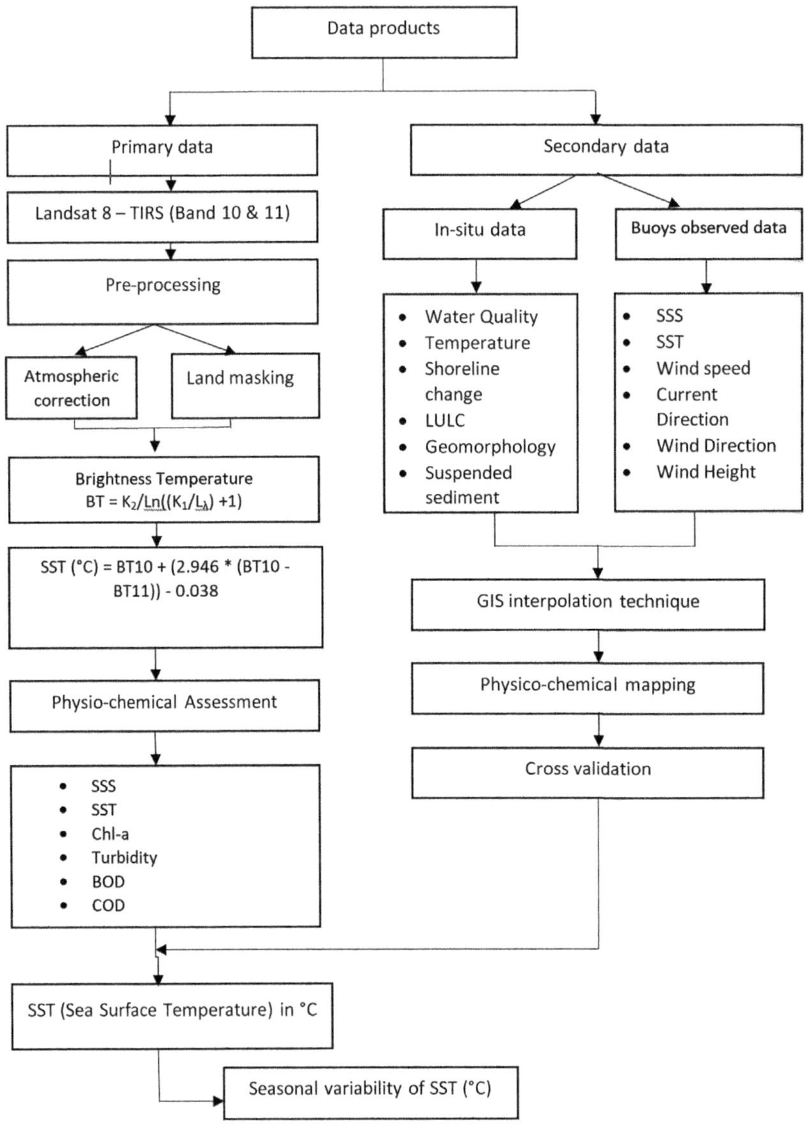

FIGURE 16.2 Methodology flowchart for estimating SST seasonal variability.

systematic methodology flowchart for estimating seasonal variability of SST using Landsat TIRS images.

The systematic analysis of SST estimation is starting from steps like (1) image preprocessing atmospheric and radiometric correction using SNAP–Sen2Core module, (2) layer stacking and land masking using ENVI software, (3) extraction of coastal water column, and (4) SWA analysis for estimating SST using ArcGIS–Raster Calculator module.

In this analysis, Landsat 8 TIRS images are used for estimating SST along the Southern Indian coastal water regions for the three seasons (premonsoon, postmonsoon, and monsoon) during 2017–2018. Table 16.1 shows the Landsat 8 TIRS metadata parameters such as thermal constant and rescaling factor values. The SWA model is executed using Landsat 8 TIRS image (bands 10 and 11) for estimating calculated brightness temperature (BT) for bands 10 and band 11; this becomes input for estimating SST and the equation is represented as

$$Ts = BT10 + (2.946 * (BT10 - BT11)) - 0.038 \tag{16.1}$$

where *Ts* refers to SST (°C), *BT10* is the brightness temperature (°C) of band 10, and *BT11* is the brightness temperature (°C) of band 11. The root-mean-square error (RMSE) value is estimated as (<0.40), which indicates good performance of SST, and as the result of SST, it is calculated as 0.19–0.22 using INCOIS-buoy-observed SST data.

First, the DN value of Landsat 8 TIRS images (bands 10 and 11) are converted into radiance (emissary) value using the equation

$$L\lambda = M_L * Q_{cal} + A_L \tag{16.2}$$

where L_λ = spectral reflectance (W/m^2 * sr * μm) (i.e., emissivity for Landsat 8 TIR bands10 and 11), M_L = radiance multiplicative scaling factor (Landsat 8 OLI/TIRS metadata), A_L = radiance additive scaling factor (Landsat 8 OLI/TIRS metadata), Q_{cal} refers to the DN value of pixel at Landsat 8 TIRS image of level-1 product.

Table 16.1 Landsat 8 TIRS bands: thermal constant and rescaling factor values.

Landsat 8 TIRS statistical data				
Thermal constant	Band 10	Band 11	Path/Row	Acquisition date
K1	774.89	480.89	143/54	December 29, 2017
K2	1321.08	1201.14		March 3, 2018
				September 11, 2018
Landsat 8 OLI/TIRS rescaling factors (Landsat 8 Metadata)				
Thermal constant	Band 10	Band 10	Path/Row	Acquisition date
Multiplicative scaling factor (ML)	3.3420E-04	3.3420E-04	143/54	December 29, 2017
Additive scaling factor (AL)	0.10000	0.10000		March 3, 2018
				September 11, 2018

Landsat 8-Metadata.

Second, the radiance value of Landsat 8 TIRS image is converted into effective temperature value (°C) as the BT value for bands 10 and 11. The spectral radiance of Landsat 8 TIRS image (bands 10 and 11) is converted into the effective temperature value (°K) using the equation

$$T = K_2 / Ln((K_1 / L_\lambda) + 1) \tag{16.3}$$

where T refers to BT (°K), L is the spectral radiance (Watts/(m2 * sr * μm), K_1 is the temperature constant (°K) for band 10, K_2 refers to temperature constant (°K) for band 11.

Third, the effective temperature value in Kelvin (°K) is converted to effective temperature in (°C) using the equation

$$T\ (^\circ C) = T\ (^\circ K) - 273 \tag{16.4}$$

where T (°C) = BT value in (°C), T (°K) = BT value (°K).

Using the SWA model algorithm, the SST is estimated using a Landsat 8 TIRS image that produces good performance for retrieval of SST in the coastal water column at high-resolution scale. This derived SST can be used for monitoring salinity variation, chl-a accumulation, coral bleaching, coastal plumes, thermal pollution by factories and power plants, and so on. It is noticed that the Landsat 8 TIRS image with the combination of the visible and near-infrared (VNIR) and short-wave infrared bands are required for a robust automated land and cloud masking from TIRS bands for quality estimation of the thermal emissivity properties.

4. Results and discussion

Satellite remote sensing provides vital sources for estimating major physicochemical properties of coastal water like salinity, temperature, conductivity, turbidity, dissolved organic matter, suspended sediments, oxygen demand, total nitrate, total phosphate, and other major variables and overcomes measurement difficulties in traditional methods. Measuring and monitoring these parameters is essential for understanding coastal and marine hydrodynamics and environmental ecosystems. Satellite remote sensing for assessing coastal and ocean water properties provide vital information for many applications including marine physicochemical, climatic, and aquatic ecology studies. TIR sensing is used for assessment of L/WST, and is used as vital information for understanding hydrothermodynamics, ocean and coastal currents, primary productivity, heat budget, evaporation, and related biogeochemical processes in the coastal water regions. Measuring and monitoring of SST using multitemporal Landsat 8 TIRS images provides evidence for long-term changes in coastal and ocean productivity, habitats, and biological processes. An SST database is a vital source for finding frequency and intensity of coral bleaching that is becoming evident for climate change vulnerability. In the present study, the performance of the Landsat 8 TIRS images are used for estimating the seasonal variability of SST from the Southern Indian coastal water region that significantly provides insight toward understanding of spatiotemporal variability of hydrothermodynamics in nature.

4.1 SST spatial variability in monsoon (December 2017)

Remote sensing-based estimation of SST in the coastal water at seasonal scale represents the annual trending of water surface temperature and is used for studying coastal, ocean water properties, and their environmental ecosystems. In the Southern Indian coastal water region, the SST shows the annual average in the range of 25.10°C –33.08°C and the mean SST is 29.52°C. The SST variability is noted as standard deviation (SD) value of 0.3–1.40 among all three seasons. During monsoon, the spatial variability of SST is estimated at 27.32°C –30.48°C (Fig. 16.3) and it is significantly decreased to 27.32–28.90°C along the shoreline. Major factors are strongly influencing SST in coastal water during monsoon (October–December) such as (1) prevailing atmospheric lower temperature, (2) movement of coastal water current, (3) seaward flow of terrestrial chilly wind, and (4) discharging freshwater from river outlets at various locations. Tables 16.2 and 16.3 shows the SST seasonal variability for the 19 stations located near major coastal settlements (Fig. 16.1) for interpreting the SST characteristics in the coastal water region., whereas the SST is estimated in the range of 28.03–28.83°C during monsoon,

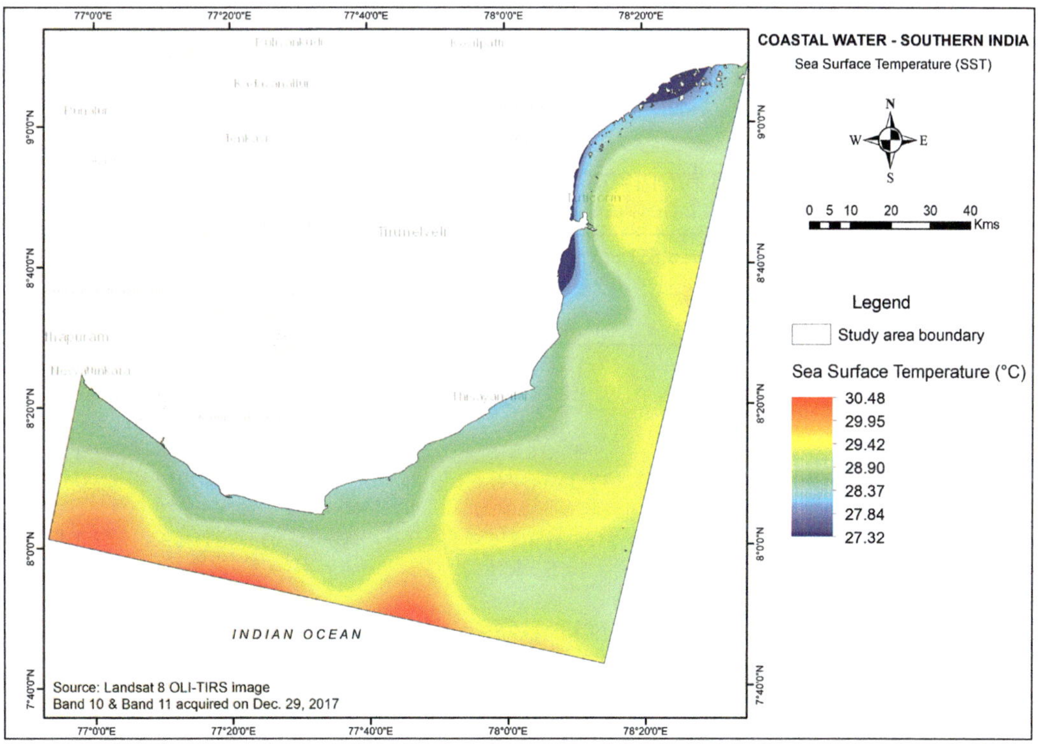

FIGURE 16.3 SST spatial variability during monsoon (December 2017).

Table 16.2 SST seasonal variability in selected stations along the Southern Indian coast (SST mean refers average of three seasons; *SD*, standard deviation).

	SST seasonal variability						
Stations	Dec 2017 (monsoon)	Mar 2018 (postmonsoon)	Sep 2018 (premonsoon)	SSTmaximum (°C)	SSTminimum (°C)	SSTmean (°C)	SD
1	28.65	26.08	27.36	28.65	26.08	27.36	1.28
2	28.52	31.02	29.77	31.02	28.52	29.77	1.25
3	28.65	31.08	29.86	31.08	28.65	29.86	1.21
4	28.45	29.55	29.00	29.55	28.45	29.00	0.55
5	28.62	29.42	29.02	29.42	28.62	29.02	0.40
6	28.70	28.77	28.74	28.77	28.70	28.74	0.03
7	28.69	29.58	28.07	29.58	28.07	28.78	0.76
8	28.47	29.50	30.33	30.33	28.47	29.44	0.93
9	28.73	29.50	30.24	30.24	28.73	29.49	0.75
10	28.83	31.63	30.03	31.63	28.83	30.16	1.40
11	28.39	31.16	30.05	31.16	28.39	29.87	1.40
12	28.57	30.07	30.13	30.13	28.57	29.59	0.89
13	28.77	29.26	30.08	30.08	28.77	29.37	0.66
14	28.30	29.10	30.05	30.05	28.30	29.15	0.88
15	28.52	29.78	30.12	30.12	28.52	29.47	0.84
16	28.68	30.19	30.38	30.38	28.68	29.75	0.93
17	28.59	30.23	30.23	30.23	28.59	29.68	0.94
18	28.07	28.93	30.35	30.35	28.07	29.12	1.15
19	28.03	29.10	30.47	30.47	28.03	29.20	1.22

Table 16.3 SST validation for premonsoon (September 2018) using SST observed (INCOIS buoys data) and SST derived (Landsat 8 TIRS).

	INCOIS-buoys observed SST			Landsat 8 TIRS derived SST			
Buoys station	SST observed (°C)	Date/ time	Location	Station	SST derived (°C)	Date/ time	Location
CB02	28.25	11-09-2018 05:55:03	72°12′32.4″E 1052′26.4″N	1	28.10	11-09-2018 05:05:29	77°1′26.82″E 8°11′33.505″N
CB04	27.11	11-09-2018 05:55:03	73°46′4.8″E 15°24′14.4″N	2	26.84	11-09-2018 05:05:29	77°1′55.435″E 8°20′18.605″N
CB06	27.91	12-09-2018 05:55:53	80°19′1.2″E 13°6′3.6″N	3	27.93	11-09-2018 05:05:29	77°1′12.253″E 8°20′27.436″N
09 CE	28.83	11-09-2018 05:00:00	73°18′14.5116″E 8°15′3.6792″N	4	28.61	11-09-2018 05:05:29	76°58′50.072″E 8°7′33.178″N
BD11	29.97	11-09-2018 06:00:00	84°6′32.2128″E 13°28′41.3364″N	5	30.12	11-09-2018 05:05:29	78°22′3.498″E 8°27′4.544″N

and the maximum SST is noted in stations 10 and 13 (Table 16.2). The SST is estimated with the average value of 28.54°C, in December 2017. The SST lowering value found in the nearshore versus that of offshore due to interacting monsoonal wind (northeast monsoon) between land and water. The spatial variation of SST is very limited in different places, maintaining the physicochemical properties and ecosystem's productivity during monsoon time (Tables 16.4 and 16.5).

4.2 SST spatial variability in premonsoon (September 2018)

During premonsoon (July–September), the spatial variation of SST shows higher values (29.89–30.85°C) in the east and northeast coastal water column due to (1) transforming the heat waves through coastal water currents with high temperature, and (2) interacting high temperature atmospheric wind between land and water. Fig. 16.4 shows the spatial variability of SST during premonsoon. The SST is estimated in the range of 25.10–30.85°C in September 2018, and the average value is 29.70°C (Table 16.3). However, the SST is found with lower values (25.10–27.98°C) along the southern coastal water column (Kanyakumari–Thiruvananthapuram) due to (1) eastward movement of coastal water current with low temperature, and (2) on-setting the southwest monsoon in the Indian Ocean mixing precipitation and river discharge during premonsoon periods.

Table 16.4 SST validation for monsoon (December 2017) using SST observed (INCOIS buoys data) and SST derived (Landsat 8 TIRS).

	INCOIS-Buoys observed SST			Landsat 8 TIRS derived SST			
Buoys Station	**SST observed (°C)**	**Date/time**	**Location**	**Station**	**SST derived (°C)**	**Date/time**	**Location**
CB02	28.33	03-01-2018 05:55:05	72°12′32.4"E 10°52′26.4"N	1	28.38	29-12-2017 05:06:03	77°1′26.82"E 8°11′33.505"N
CB04	28.43	03-01-2018 05:55:04	73°46′4.8"E 15°24′14.4"N	2	28.41	29-12-2017 05:06:03	77°1′55.435"E 8°20′18.605"N
CB06	26.9	03-01-2018 05:55:04	80°19′1.2"E 13°6′3.6"N	3	27.32	29-12-2017 05:06:03	77°1′12.253"E 8°20′27.436"N
AD09	28.31	03-01-2018 05:00:00	73°18′14.5116"E 8°15′3.6792"N	4	28.37	29-12-2017 05:06:03	76°58′50.072"E 8°7′33.178"N
BD11	27.36	03-01-2018 05:00:00	84°6′32.2128"E 13°28′41.3364"N	5	27.35	29-12-2017 05:06:03	78°22′3.498"E 8°27′4.544"N

Table 16.5 SST validation for postmonsoon (March 2018) using SST observed (INCOIS buoys data) and SST derived (Landsat 8 TIRS).

INCOIS-buoys observed SST				Landsat 8 TIRS derived SST			
Buoys station	SST observed (°C)	Date/ time	Location	Station	SST derived (°C)	Date/ time	Location
CB02	28.95	03-03-2018 05:55:03	72°12′32.4″E 10°52′26.4″N	1	28.97	03-03-2018 05:05:34	77°1′26.82″E 8°11′33.505″N
CB04	28.42	03-03-2018 05:55:03	73°46′4.8″E 15°24′14.4″N	2	28.71	03-03-2018 05:05:34	77°1′55.435″E 8°20′18.605″N
CB06	27.64	03-03-2018 05:55:04	80°19′1.2″E 13°6′3.6″N	3	27.84	03-03-2018 05:05:34	77°1′12.253″E 8°20′27.436″N
09 CE	29.17	03-03-2018 05:00	73°18′14.5116″E 8°15′3.6792″N	4	29.07	03-03-2018 05:05:34	76°58′50.072″E 8°7′33.178″N
BD11	27.86	03-03-2018 05:00:00	84°6′32.2128″E 13°28′41.3364″N	5	27.88	03-03-2018 05:05:34	78°22′3.498″E 8°27′4.544″N

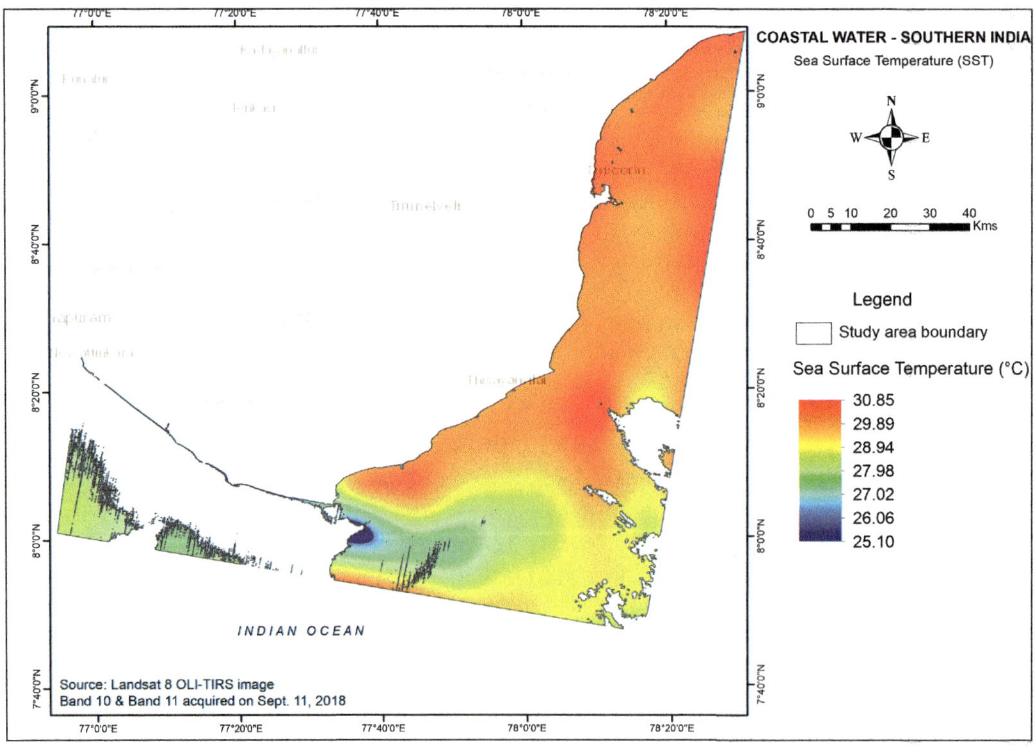

FIGURE 16.4 SST spatial variability in premonsoon (September 2018).

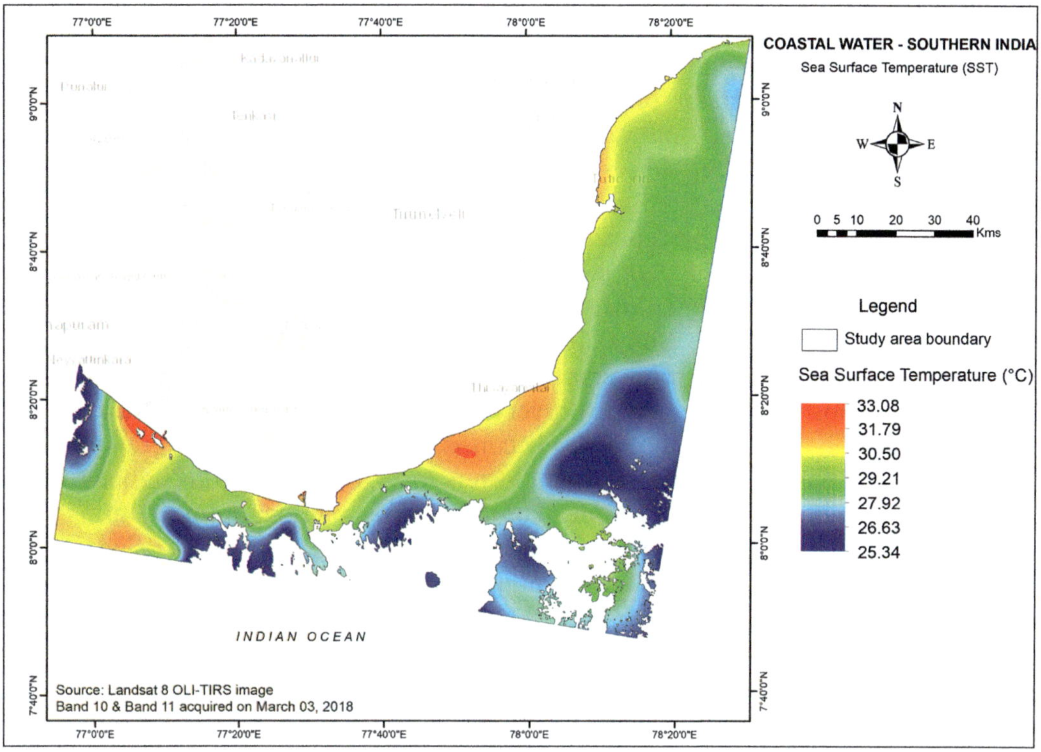

FIGURE 16.5 SST spatial variability in postmonsoon (March 2018).

4.3 SST spatial variability in postmonsoon (March 2018)

During postmonsoon (March–May), the SST is spatially varying in different locations and it is estimated in the range of 25.34–33.08°C. Fig. 16.5 shows the SST spatial variability along the Southern Indian nearshore and offshore water column during postmonsoon (March 2018). Higher SST values are noticed in some isolated places in the nearshore. This is due to anthropogenic activities like (1) hot water releasing from an atomic power plant and (2) discharging of lower quantity of Thamirabarani river flow through terrain surface increases the coastal water SST near urban Thengapattinam. For the 19 sampling stations, the average SST is measured at 29.67°C, and it indicates that the coastal water prevails higher SST along the nearshore than that of offshore, due to transforming terrestrial-facing high-temperature wind flow toward the coastal water surface unable to reach to offshore. In the east and northeast offshore area, the spatial variation of SST is estimated in a lower range (25.34–29.21°C) due to circulation of cold-water current flowing from the North Indian Ocean and Bay of Bengal, causing the lower SST in vast spatial coverage during these periods.

4.4 SST seasonal variability during 2017–2018

The SST seasonal variability in the Southern Indian coastal water is characterized from three seasonal measurement of SST using Landsat 8 TIRS images for 2017–2018. Fig. 16.6 shows the SST seasonal variability for selected stations along the South Indian coastal water region. The SST observed from the 19 stations nearby coastal settlements (Fig. 16.1) obviously represented spatiotemporal variability of SST during the monsoon season (December 2017), premonsoon (September 2018), and postmonsoon (March 2018). Table 16.3 shows the SST seasonal variability for selected stations along the Southern Indian coastal water, whereas the overall average of SST is 29.50°C, the maximum SST value is 30.16°C in station 10, and the minimum SST value is 27.36°C in station 1. The SST is noted as a lower rate (28.0–28.50°C) during monsoon and relatively constant in almost all the stations with very little variation at stations 18 and 19. It is noted that the SST maintains the relatively constant temperature during premonsoon at 30.0–30.50°C along the stations 8–19, but lowers to 28.0–29.50°C from stations 1 and 7 (i.e., coastal water column of Kanyakumari–Thiruvananthapuram). During post-monsoon, the SST values highly fluctuate in almost all the stations, while rising to higher values (31.0–31.50°C) at the selected stations of 3 and 10–11 due to anthropogenic activities that discharge warm water into the sea. The SST seasonal variability influences temperature, salinity, chl-a, organic matter, and other water properties based on upwelling–downwelling processes and can be associated with mixing of surface and deep water properties.

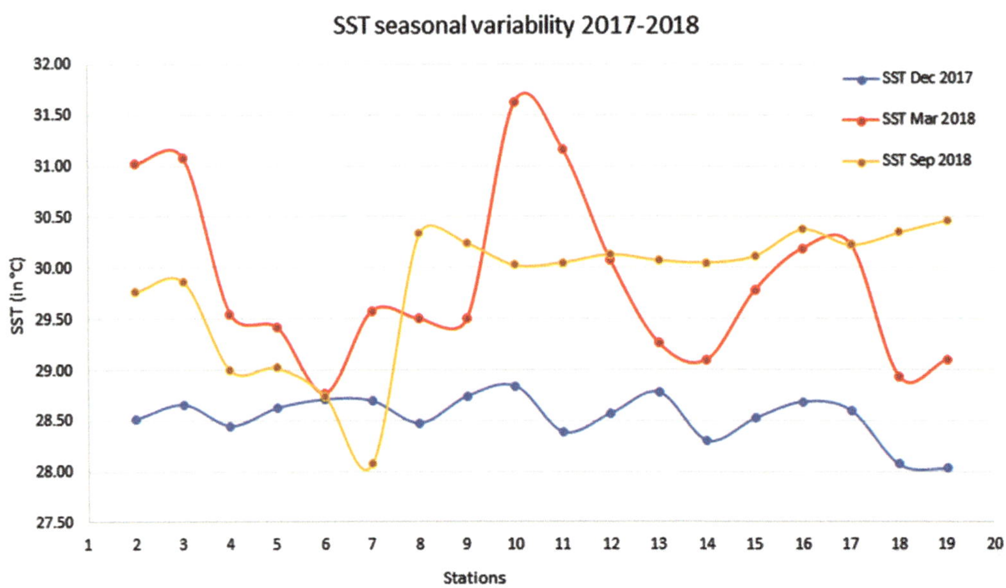

FIGURE 16.6 Seasonal variability of SST during 2017–2018.

Table 16.6 Linear regression and RMSE estimation using SST observed (INCOIS buoys data) and SST derived (Landsat 8 TIRS) for 2017–2018.

Season	Date/Month/Year	R^2 value for mean SST	RMSE for mean SST
Premonsoon	Sep 11, 2018	0.9882	0.210
Monsoon	Dec 29, 2017	0.9507	0.228
Postmonsoon	Mar 03, 2018	0.9523	0.192

4.5 SST validation using linear regression analysis and RMSE estimation

The SST is estimated for all three seasons using Landsat 8 TIRS images for the South Indian coastal water column in 10 * 10 m pixel scale. Accuracy assessment of SST-derived products for the three seasons is executed by the two statistical methods such as (1) linear regression analysis (LRA; coefficient of determination) (R^2) and (2) RMSE estimation. The SST-derived products using Landsat 8 TIRS images are cross-validated using INCOIS-buoys SST observed data (in situ) of the corresponding date and time. Table 16.6 shows the statistical measurement of linear regression and RMSE between SST observed (INCOIS buoys data) and SST derived (Landsat 8 TIRS) for the period of 2017–2018. In this analysis, the LRA is applied to evaluate the linear relationships between SST derived and SST observed, and the result shows the R^2 value in the range of 0.95–0.98, which indicates a higher relationship of SST-derived products for all three seasons (Table 16.5). Fig. 16.7 shows SST cross-validation using linear regression relationship and RMSE using SST derived (Landsat 8 TIRS) and SST observed (INCOIS buoys data) during 2017–2018.

The RMSE is executed for accuracy validation of SST-derived products using INCOIS-buoys observed SST data using the equation as follows:

$$\text{RMSE} = \sqrt{\sum_{i=1}^{n} (Xd - Xo)^2 / N} \tag{16.5}$$

where X_d refers to the SST-derived products, X_o refers to the INCOIS-buoys observed SST data (in situ data), and N refers the total number of stations for observed and derived SST. To assess the accuracy of Landsat 8 TIRS–derived SST products, the INCOIS-buoys observed SST values for five points were applied for RMSE estimation, and the estimated RMSE value below 1.000 shows a good performance (i.e., output with high accuracy). It means that the lower RMSE (<1.0) indicates higher accuracy of output products. In this study, the RMSE is estimated for three seasons in the range of 0.192–0.228 and indicates high accuracy of the SST-derived products using Landsat 8 TIRS images.

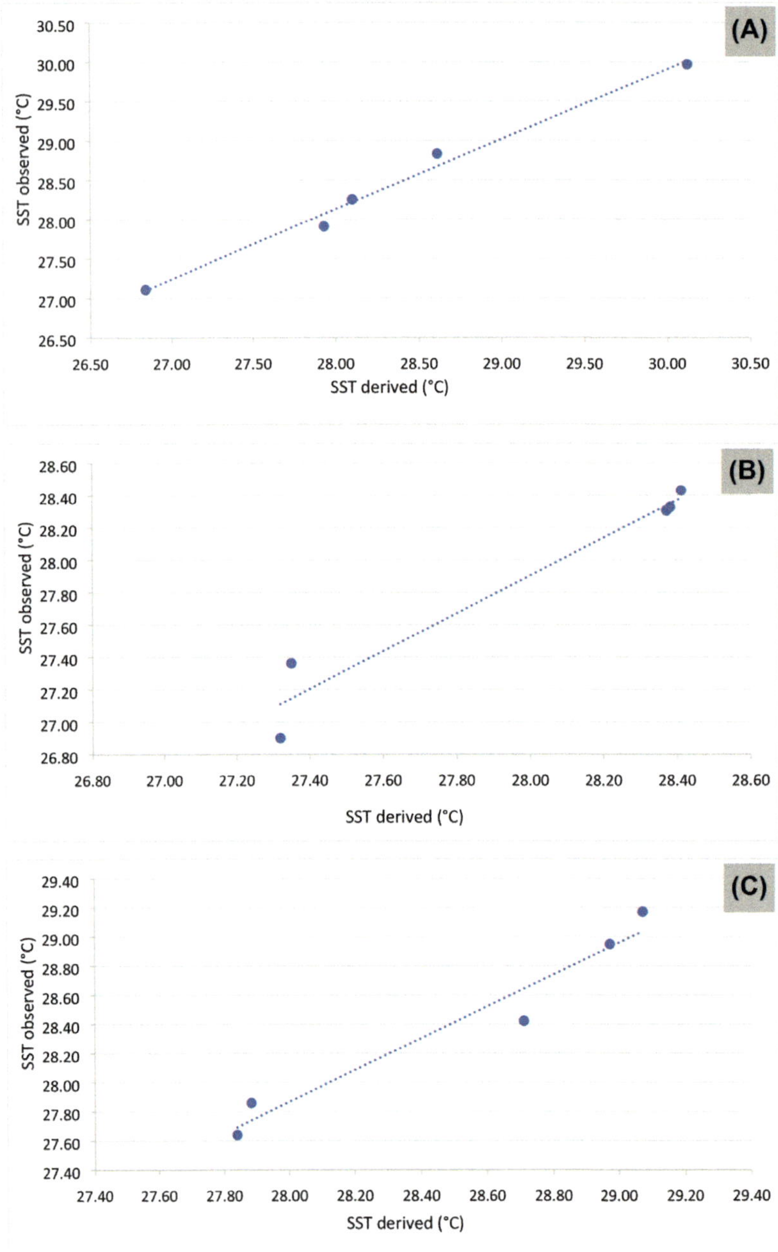

FIGURE 16.7 Linear regression and RMSE estimation using SST observed (INCOIS buoys data) and SST derived (Landsat 8 TIRS) for 2017–2018; premonsoon (A), monsoon (B), and postmonsoon (C).

5. Conclusions

Remote sensing applications cover various aspects of ocean and coastal environment studies. SST monitoring and mapping provide primary information for oceanographers and earth and environmental scientists for exploring ocean and coastal environment and ecosystems. Estimating SST in the coastal water using the SWA is used for understanding seasonal and spatial variability of biomass, phytoplankton (chl-a) assemblage, and potential fishing zones at a site-specific level. The SWA model is highly automated, globally applicable, and compatible to GIS software processing and suited for temperature retrieval over water and nonwater targets like ocean, coastal, urban, agricultural, and forested areas based on pixel emissivity rate. The SST products derived from Landsat 8 TIRS images using the SWA model shows better match-up performance with the corresponding in situ measurements (INCOIS-buoys observed SST data). Landsat 8 TIRS image–derived SST indicates spatiotemporal characteristics of temperature, salinity, BOD, COD, TDS, total nitrate, total phosphate, pH, conductivity, chlorophyll (chl-a), and other organic matter in the coastal water column.

SST seasonal variability is used for understanding the spatial and temporal patterns of changing of coastal and ocean water properties essential for ensuring sustainable development and conservation of environmental ecosystems for long-term scale.

Acknowledgments

The corresponding author is thankful to the Director, National Center for Earth Science Studies (NCESS), Ministry of Earth Sciences, Government of India, for providing continued support and encouragement. The authors extend their gratitude to IMD, SOI, NIOT, INCOIS, NOAA, NASA, USGS–Earth Explorer, and PSMSL-UK, for providing necessary data sources.

References

Cahyono, A.B., Saptarini, D., Pribadi, C.B., Armono, H.D., 2017. Estimation of sea surface temperature (SST) using split window methods for monitoring. Ind. Act. Coast. Area 862, 90–95, 1662-7482.

Casal, G., Lavender, S., 2017. Spatio-temporal variability of sea surface temperature in Irish waters (1982–2015) using AVHRR sensor. J. Sea Res. 129, 89–104.

Chen, D., Huazhong, R., Qiming, Q., Jinjie, M., Jing, L., 2014. Split-Window Algorithm for estimating land surface temperature from Landsat 8 TIRS data, IEEE Geoscience and Remote Sensing Symposium, pp. 3578–3581. Quebec City, QC. https://doi: 10.1109/IGARSS.2014.6947256.

Edwards, M., Gregory, B., Graeme, C.H., Koslow, J.A., Anthony, J.R., 2010. Multi-decadal oceanic ecological datasets and their application in marine policy and management. Trends Ecol. & Evol. 25, 602.

Geetha, G., Bhaskar, T.V.S.U., Rao, E.P.R., 2011. Argo data and product of Indian ocean for low bandwith users. Int. J. Oceans Oceanogr. 5, 1.

Gholizadeh, M.H., Melesse, A.M., Reddi, L., 2016. A comprehensive review on water quality parameters estimation using remote sensing techniques. Sensors 16 (8), 1298. https://doi.org/10.3390/s16081298.

Hosoda, K., 2011. Algorithm for estimating sea surface temperature based on aqua/MODIS global ocean data. 2. Automated quality check process for eliminating cloud contamination. J. Oceanogr. 67, 791.

Kaliraj, S., 2016. Geospatial Analysis of Coastal Geomorphological Vulnerability on South West Coast of Kanyakumari Using Remote Sensing Technology. Unpublished Ph.D. Thesis. Manonmaniam Sundaranar University, Tirunelveli, pp. 83–110.

Kaliraj, S., Chandrasekar, N., 2012. Geo-processing model on coastal vulnerability index to explore risk zone along the south west coast of Tamilnadu, India. Int. J. Earth Sci. Eng. 5 (5), 1138–1147.

Kaliraj, S., Chandrasekar, N., Magesh, N.S., 2013. Evaluation of coastal erosion and accretion processes along the southwest coast of Kanyakumari, Tamil Nadu using geospatial techniques. Arab. J. Geosci.. https://doi.org/10.1007/s12517-013-1216-7.

Kaliraj, S., Chandrasekar, N., Magesh, N.S., 2014a. Impacts of wave energy and littoral currents on shoreline erosion/accretion along the south-west coast of Kanyakumari, Tamil nadu using DSAS and geospatial technology. Environ. Earth Sci.. https://doi.org/10.1007/s12665-013-2845-6.

Kaliraj, S., Chandrasekar, N., Magesh, N.S., 2014b. Multispectral image analysis of suspended sediment concentration along the Southern coast of Kanyakumari, Tamil Nadu, India. J. Coast. Sci. 1 (1), 63–71.

Kaliraj, S., Chandrasekar, N., Simon Peter, T., Selvakumar, S., Magesh, N.S., 2015. Mapping of coastal aquifer vulnerable zone in the south west coast of Kanyakumari, South India, using GIS-based DRASTIC model. Environ. Monit. Assess.. https://doi.org/10.1007/s10661-014-4073-2.

Kaliraj, S., Chandrasekar, N., Ramachandran, K.K., 2016. Mapping of coastal landforms and volumetric change analysis in the south west coast of Kanyakumari, South India using remote sensing and GIS techniques. Egypt. J. Remote Sens. & Space Sci.. https://doi.org/10.1016/j.ejrs.2016.12.006.

Kaliraj, S., Chandrasekar, N., Ramachandran, K.K., Srinivas, Y., Saravanan, S., 2017. Coastal landuse and land cover change and transformations of Kanyakumari coast, India using remote sensing and GIS. Egypt. J. Remote Sens. & Space Sci.. https://doi.org/10.1016/j.ejrs.2017.04.003.

Kaliraj, S., Chandrasekar, N., Ramachandran, K.K., 2019a. Phytoplankton (chl-a) biomass seasonal variability in the gulf of Mannar, south India: a remote sensing perspective. In: Kumar, P., et al. (Eds.), Applications and Challenges of Geospatial Technology. https://doi.org/10.1007/978-3-319-99882-4_5.

Kaliraj, S., Chandrasekar, N., Ramachandran, K.K., Selvakumar, S., 2019b. Seawater intrusion vulnerability in the coastal aquifers of southern India—an appraisal of the GALDIT model, parameters' sensitivity, and hydrochemical indicators. Environ. Sci. Pollut. Res. 26, 9755–9784. https://doi.org/10.1007/s11356-019-04401-0.

Liu, C., Sun, Q., Xing, Q., Liang, Z., Deng, Y., Zhu, L., 2017. Spatio-temporal variability in sea surface temperatures for the Yellow Sea based on MODIS dataset. Ocean Sci. J 52, 1–10. https://doi.org/10.1007/s12601-017-0006-7.

LSDS-1574, 2019. Landsat 8 - Data Users Handbook. https://prd-wret.s3-us-west-2.amazonaws.com/assets/palladium/production/atoms/files/LSDS-1574_L8_Data_Users_Handbook-v5.0.pdf.

Meijun, J., Li, J., Wang, C., Shang, R., 2015. A practical split-window algorithm for retrieving land surface temperature from Landsat-8 data and a case study of an urban area in China. Open Access J. Remote Sens. 7 (4), 4371–4390. https://doi.org/10.3390/rs70404371. http:www.mdpi.com/2072-4292/7/4/4371.

Miyazawa, Y., Hiroshi, M., Toru, M., Sergey, M.V., Xinyu, G., Takuji, W., Sourav, S., 2013. Data assimilation of the high-resolution sea surface temperature obtained from the AquaTerra satellites (MODIS-SST) using an ensemble Kalman Filter. Rem. Sens. 5, 3123.

Park, K.A., Lee, E.Y., Chang, E., Hong, S., 2014. Spatial and temporal variability of sea surface temperature andWarming Trends in the Yellow sea. J. Mar. Syst. 143. https://doi.org/10.1016/j.jmarsys.2014.10.013.

Rongali, G., Keshari, A.K., Gosain, A.K., Khosa, R., 2018. A Mono-window algorithm for land surface temperature estimation from Landsat 8 thermal infrared sensor data: a case study of the beas river basin, India. Pertanika J. Sci. & Technol. 26 (2), 829–840.

Schaeffer, B.A., Bailey, S.W., Conmy, R.N., Galvin, M., Ignatius, A.R., Johnston, J.M., Keith, D.J., Lunetta, R.S., Parmar, R., Stumpf, R.P., Urquhart, E.A., Werdell, P.J., Wolfe, K., 2018. Mobile device application for monitoring cyanobacteria harmful algal blooms using Sentinel-3 satellite Ocean and Land Colour Instruments. Environ. Modell. Softw. 109, 93–103. https://doi.org/10.1016/j.envsoft.2018.08.015.

Schott, F.A., Shang, P.X., Julian, P.M., 2009. I'ndian ocean circulation and climate variability. Rev. Geophys. 47, 1.

Skakala, J., Timothy, J.S., 2016. Simple Heterogeneity parametrization for sea surface temperature and chlorophyll. J. JMARSYS. 158, 52.

Sobrino, J., Jiménez, J.C., Laporta, S., Nerry, F., 2001. Split-Window Methods for Surface Temperature Estimation From DAIS Data, the Digital Airborne Spectrometer Experiment (DAISEX). Proceedings of the Workshop held July, 2001.

Wahyudi, A., Susilo, S.B., Panjaitan, J.P., 2018. Seasonal-temporal variability of sea surface temperature derived from satellite and in situ data: case study of Indian ocean western part of sumatera island. Earth & Environ. Sci. 135, 012016. https://doi.org/10.1088/1755-1315/135/1/012016.

17

Ocean remote sensing of suspended sediment variability in Southern Indian coastal water region using Landsat 8 OLI images

Kaliraj Seenipandi[1], K.K. Ramachandran[1], Prashant Ghadei[2], Sulochana Shekhar[2]

[1]NATIONAL CENTRE FOR EARTH SCIENCE STUDIES (NCESS), THIRUVANANTHAPURAM, KERALA, INDIA; [2]DEPARTMENT OF GEOGRAPHY, CENTRAL UNIVERSITY OF TAMIL NADU, THIRUVARUR, TAMIL NADU, INDIA

Chapter outline

1. Introduction

In the coastal water column, the spatiotemporal variability of suspended sediments depends on littoral currents, beach erosion, rocky shore weathering, and river discharges. Suspended sediment is considered a portion of sediment matter that includes sand, silt, clay, and such from discharges of coastal and inland sources that are

Remote Sensing of Ocean and Coastal Environments. https://doi.org/10.1016/B978-0-12-819604-5.00017-2

mixed and transported to various parts of the shoreline through littoral currents, waves, tides, winds, and other hydrodynamic processes (Nair et al., 1993; Jennerjahn et al., 2008; Narendra Babu et al., 2010; Tripathy et al., 2018). In general, the large amount of suspended sediments accumulates more within the shallow depth of nearshore than offshore region and influences morphological changes in the coastal and marine environment (Carpenter and Carpenter, 1983; Mertes et al., 1993; Qu, 2014; Larson et al., 2018; Kaliraj et al., 2019a,b). In the coastal systems, suspended sediments transporting and dispersing is regulated mainly through hydrodynamic processes including littoral current, tidal fluctuation, wave height, and wind direction, and it tends to create erosion and accretion processes (Forget and Ouillon, 1998; Doxaran et al., 2002; Wackerman et al., 2017). Suspended sediments provide input sources for physical, chemical, and biological activities, and for regulating nutrient availability and formation of coastal landforms.

Monitoring and measurement of suspended sediment concentration (SSC) in coastal waters is very important for the assessment and monitoring of coastal settings and their ecology. A large number of noteworthy studies have been conducted regarding the spatiotemporal deviation of SSCs in systems all over the world (Kaliraj and Chandrasekar, 2012). Sediment transportation in coastal water is seasonally varying due to the action of various factors like tides, waves, river discharges, wind, and currents; temporal capacity of remotely sensed image provides the feasibility to monitor changes in suspended sediment concentration (Chen et al., 1992; Tassan, 1997). In the South Indian coastal water region, the quantity of suspended sediments differs from place to place during different seasons based on river discharge, tidal flow, shore erosion, and rocky weathering, which frequently controls physicochemical properties of water and sediments. In the coastal water column, suspended sediments are predominantly accumulated with sand, silt, and clay matters, wherein suspended sediments transporting to various places cause formation of destructive landforms (Kaliraj, 2016) and act like indicators for environmental degradation from point sources and nonpoint sources pollutants (Nair et al., 1993; Jennerjahn et al., 2008; Narendra Babu et al., 2010).

The seasonal variability of SSC in coastal water columns determines an availability of sources for physical, chemical, and biological inputs to coastal and marine living organisms and determines nutrient supply and pollution status in coastal, estuarine, and lake environments. Quantitative measurement of SSC in larger water bodies like ocean, sea, lake, reservoir, and estuarine is a very difficult task, and is cost effective but time consuming using conventional methods, and it is critical to identify the SSC spatiotemporal variability using in situ and laboratory analysis. Monitoring SSC seasonal variability in a coastal water column is extremely difficult using field sampling analysis due to hydrodynamic actions.

Remote sensing technology provides an effective platform for qualitative and quantitative estimation of SSC from terrestrial and marine water bodies at a regional and site-specific scale (Pavelsky and Smith, 2009; Kaliraj et al., 2014a,b; Chamberland et al., 2019). Consequently, the development of remote sensing over the last few decades provides multispectral satellite images with higher spectral and spatial resolutions that increase researchers' abilities for measuring and monitoring SSC in coastal and ocean water environments (Curran and Novo, 1988; Nechad et al., 2010). Satellite images cover a synoptic view of Earth's surface based on reflectance and used for assessment of water quality and SSC. Geospatial technology has also been widely used by many hydrologists, as it has the ability to answer complex spatial and temporal questions (Marcus and Fonstad, 2010; Kilham et al., 2012; Colleen and Pavelsky, 2013). The advanced space technology of remote sensing provides synoptic coverage of Earth's surface image by continuous observation for mapping and monitoring the coastal changes and helps us to understand how the changes happened in various parts of the environment including coastal waters (Xia, 1993).

Landsat satellite multispectral images of MSS, TM, and ETM+have an ability to detect and map the amount of suspended sediment entering, residing, and moving in the nearshore region (Oestlund et al., 2001; Chamberland et al., 2019). The advantage of satellite images with high spectral properties provide better information on movement and concentration of suspended sediments compared to point data collected by on-site instruments during in situ survey (Stumpf et al., 1992; Tassan, 1994). The coastal water surface of an optical image has high spectral reflectance in blue bands and it increases in green and red bands with high suspended sediment concentrations and low reflectance with increased salinity and depth (Sridhar et al., 2008; Kaliraj and Chandrasekar, 2012). Landsat 8 Operational Land Imager (OLI)/Thermal Infrared Sensor (TIRS) images (spatial resolution is 30 m) were used for estimating SSC at higher resolution from coastal, estuaries, lake, and reservoir (Seelanki et al., 2018). Nowadays, several researchers have used Landsat OLI images for estimating SSC and are used for producing higher performance over water property studies like coastal and estuarine plume (Chen et al., 2014; LSDS-1574, 2019). Satellite images provide synoptic coverage of the ocean and coastal environment that are very helpful for understanding physicochemical properties at a user-defined scale (Gholizadeh et al., 2016). Satellite remote sensing provides images with synoptic coverage of Earth's objects that shows their distinct reflected spectral signatures for mapping and monitoring major physicochemical parameters of water bodies.

Many remote sensing satellites, namely IRS, Landsat, AVHRR/3, MODIS, Quick Bird, GOES, IKONOS, SPOT, Sentinel 2A, and ASTER, have sensors with various capabilities and are used for acquiring Earth's features with different spectral and spatial resolutions that are used worldwide for studying coastal and marine environments and their dynamic changes (Colleen and Pavelsky, 2013; Wu et al., 2015; Dethier et al., 2019; Topp et al., 2020). SSC estimation is successfully performed using multispectral images acquired from Landsat MSS, TM, and ETM+ and Sentinel 2A–MSI for many places worldwide (Olmanson et al., 2011; Palmer et al., 2015; Hansen et al., 2017; Kaliraj et al., 2019a,b).

Measuring SSC seasonal variability using multitemporal images provides vital information for sustainable development of coastal and lake environments. Multitemporal images are used for mapping and monitoring the amount of suspended sediment entering, residing, and moving in larger water bodies like lake, reservoir, estuaries, and nearshore (Kaliraj et al., 2014a,b; Larson et al., 2018). Remote sensing images with higher spectral and spatial resolution derived from Sentinel 2A and GeoEye are used for obtaining water quality parameters on a large scale (Colleen and Pavelsky, 2013; Wackerman et al., 2017). Many researchers have used multispectral images acquired from Landsat TM, ETM+, OLI, and Sentinel 2A–MSI for mapping water quality parameters including suspended sediment concentrations, chlorophyll-a, and salinity (Carpenter and Carpenter, 1983; Mertes et al., 1993; Kaliraj et al., 2014a,b). Landsat 8 OLI and Sentinel 2A satellites acquired multispectral images that are potentially used for mapping and monitoring SSC variability, showing relatively accurate measurement as compared to in situ data (Palmer et al., 2015; Hansen et al., 2017). Quantitative measurement of SSC using a multivariate regression model performed the mathematical calculation between spectral reflectance of bands (visible (VIS), near-infrared (NIR), mid-infrared, and short-wave infrared (SWIR)) and correlation coefficients derived from in situ data (Qu, 2014; Wackerman et al., 2017; Larson et al., 2018; Chamberland et al., 2019). Nowadays, many researchers have developed empirical models and linear regression algorithms for quantitative and qualitative estimation of SSC in lakes, estuarine, coastal, and inland waters using multispectral images of Landsat TM, ETM+, OLI, and Sentinel 2A–MSI with tolerable accuracy (Olmanson et al., 2011; Wu et al., 2015; Liu et al., 2017; Dethier et al., 2019; Topp et al., 2020). Satellite images are widely used for measurement of SSC seasonal variability in coastal and ocean water that can deliver primary information for sustainable development of environmental ecosystems.

2. Profile of the study area

In the Southern Indian coastal water region, seasonal variability of SSC plays a significant role in formation of landforms by controlling erosion and accretion processes. In the present study, SSC seasonal variability is estimated in the water column, which lies within latitude 7°38′ N to 9°20′ N and longitude 76°50′ E to 78°53′ E. The areal extent of this area is about 8440.22 km^2 and the shoreline length is 410.72 km, parallel to the Keelakarai–Thiruvananthapuram coastline. The coastal water constitutes mixing of sediments from river discharge and littoral currents from inflow of the Arabian Sea and Bay of Bengal (BoB) (Fig. 17.1). Moreover, larger amount of sediments release into coastal water from a variety of landforms like laterite uplands, estuaries, sand dunes, and outlets of the nonperennial river. The coastal water receives river discharges along with runoff matter from different drainage networks like Thamirabarani, Valliyar, Pazhayar, and Vaippar during northeast and southwest monsoons (Kaliraj and Chandrasekar, 2012; Kaliraj et al., 2015, 2016). The water column has a shallow depth (1–10 m) between nearshore and surf zones, and gradually increases to 20–30 m toward a seaward distance of 5–6 km (Kaliraj et al., 2013, 2014a,b, 2017). The water column is seasonally varying in hydrodynamics like waves, wind, and currents, and the average value of wave energy is about 0.5–8.5 kJ/km^2 and the mean current velocity is measured as 0.14 ms^{-1}.

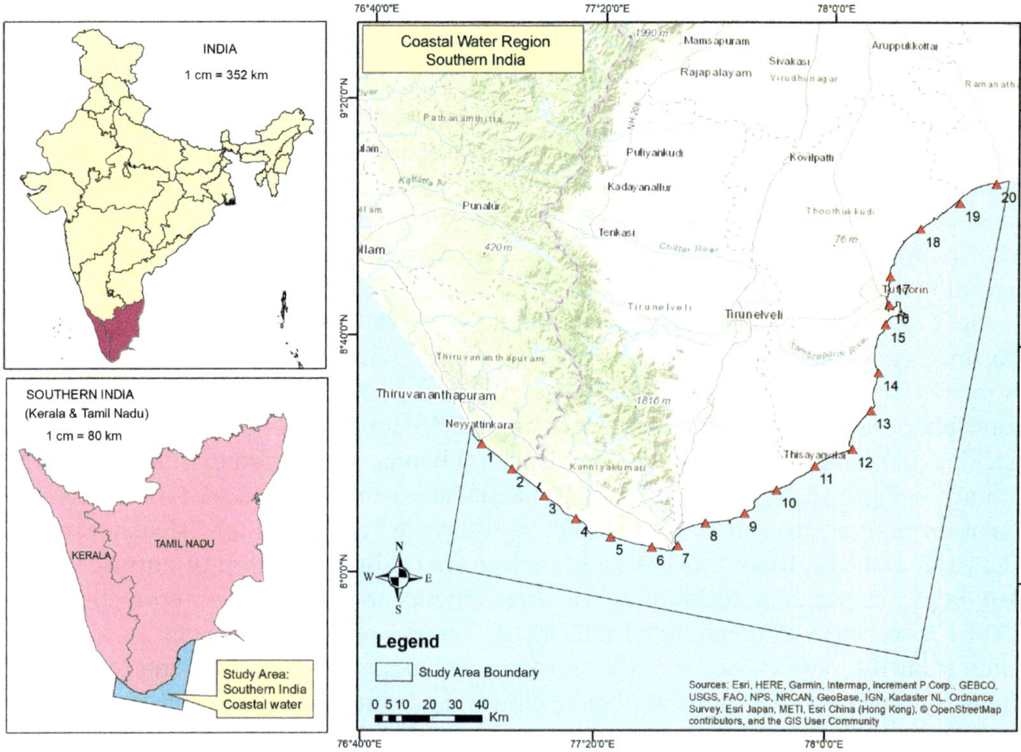

FIGURE 17.1 Geographical location of the study area.

Quantitative distribution of sediments in site-specific levels depends on the amount of materials released from erosion and accretion and magnitudes of littoral drift. The study area has subtropical climatic conditions where temperature ranges from 23.78 to 33.95°C and annual average rainfall between 826 and 1456 mm. Monitoring and assessment of SSC and its seasonal variability using multitemporal Landsat 8 OLI images is vital for understanding coastal landform formation and marine-coastal ecology in the study area.

3. Materials and methods

In this study, the SSC of three seasons of the year 2017–2018 and its spatial distribution was estimated along the southern India coastal water region using Landsat 8 OLI images acquired during premonsoon, monsoon, and postmonsoon. Landsat 8 OLI images (level-1 GeoTIFF data product) are collected from USGS–Earth Explorer online portal. Landsat 8 satellite has two sensors, OLI and TIRS. Landsat 8 OLI images have improved radiometric precision over a 12-bit dynamic range that improves overall signal-to-noise ratio and enables radiometric characterization for exploring land and water conditions (LSDS-1574, 2019). The OLI sensor collects image data for 9 spectral bands (VIS, NIR, SWIR) with a 30 m spatial resolution except Panchromatic band (15 m). The two new bands in addition to the legacy Landsat bands (1–5, 7, and Pan) are (1) coastal/aerosol band (band 1; wavelength = 0.435–0.451 μm), principally used for ocean color observations, similar to ALI's band 1; and (2) cirrus band (band 9; wavelength = 1.36–1.38 μm), which aids in the detection of cirrus clouds. Landsat 8 OLI image with multispectral bands (spatial resolution = 30 m) covers wide areas of Earth's landscape while providing sufficient resolution to distinguish land and water features within 16-day repetitive cycle. Landsat 8 OLI images are analyzed using linear regression equation for mapping SSC with a comprehensive and accurate pattern of spatial and seasonal variability in the coastal water region.

Fig. 17.2 shows the flowchart of systematic methodology for estimating the SSC and seasonal variability using Landsat 8 OLI images. The systematic analysis for SSC estimation includes the following steps: (1) Landsat 8 OLI image preprocessing for atmospheric and radiometric correction using FLAASH module of ENVI software, (2) layer stacking and land masking for OLI multispectral bands, (3) extraction of land and water column using isodata, an unsupervised classification technique, and (4) executing the linear regression equation for estimating SSC using the ArcGIS–Raster Calculator module. The Landsat 8 OLI (Level 1T - GeoTiFF) images are calibrated via top-of-atmosphere and dark-object-subtraction techniques for atmospheric and radiometric correction using ENVI 5.3 software. The calibrated OLI image shows spectral reflectance as a reflected value from the objects. Second, the masking technique is applied on the OLI image–multispectral bands for removing lands, clouds, and intertidal zones and distinguishing the boundary between water and nonwater regions. Finally, the OLI images (FCC format) are analyzed for estimating the SSC at 30 m × 30 m pixel scale using the empirical algorithm executed via the Map Algebra tool, ArcGIS 10.6 software, and the equation is expressed as

$$\begin{aligned} \text{SSC (mg/L)} = [&-4.281 + 23.628 * \text{Band 4} - 15.675 * \text{Band 5} - 14.653 * \text{Band 2} + 79.251 \\ &* \text{Band 3} + 21.303 * \text{Band 9} + 9.709 * \text{Band 6} + 10.963 * \text{Band 7}] \end{aligned} \tag{17.1}$$

where SSC is suspended sediment concentration (mg/L), Band 2, 3, ... 9 refers to the multispectral bands of the Landsat 8 OLI image, and numerical value refers to sediment coefficient values derived from in situ measurements. In this empirical equation, the SSC is estimated using the OLI image based on reflectance variance between suspended sediment and clear water; sediment concentration reflects more green, red, and NIR than the other bands, and this scenario is attributed in the empirical algorithm (Eq. 17.1) to estimate SSC in coastal water columns.

4. Results and discussion

In the Southern Indian coastal water, seasonal variability of SSC is estimated using multitemporal Landsat 8 OLI images. The results show SSC spatiotemporal variability in the range of 0.61–9.70 mg/L during three seasons of 2017–2018. The derived SSC

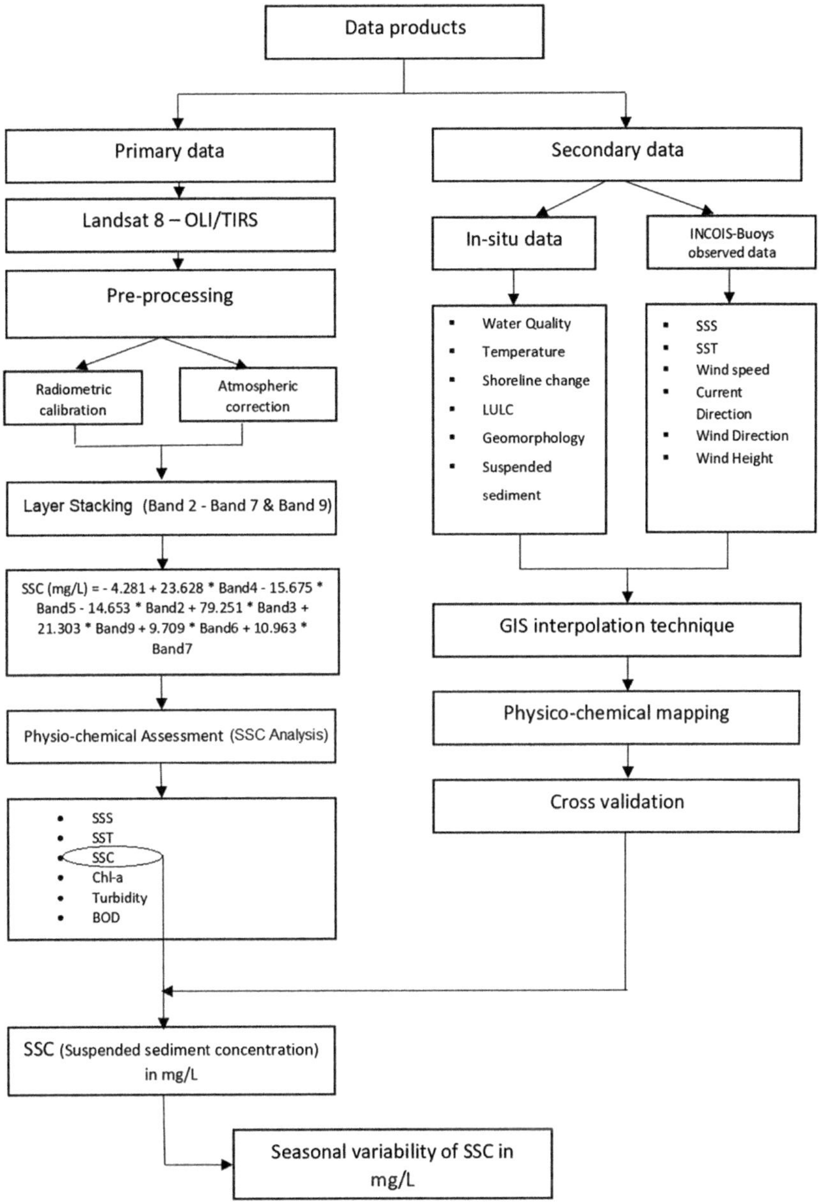

FIGURE 17.2 Methodology flowchart for estimating SSC seasonal variability.

product is estimated with R^2 value (0.92) and root-mean-square error (0.21), which indicates reliability and best performance compared to in situ data. In a coastal water column, SSC varies in location during different seasons, which controls activities of coastal environmental ecosystems. The site-specific distribution of sediments depends on the amount of materials released from erosion and accretion and magnitudes of littoral drift. It is observed that the nearshore region is estimated with higher sediment concentration than other places during all three seasons of 2017–2018. The results reveal that the SSC is seasonally changing in location based on different factors like precipitation and river discharges. During premonsoon (September 2018) and postmonsoon (March–May 2018), the SSC spatial variability is estimated at a higher range of more than 9 mg/L, this is due to movement of littoral current flowing eastward from the Arabian Sea carrying larger sediments by receiving discharge during southwest monsoon. The lower concentration of SSC is estimated in majority of areas at a range of less than 1 to 7.29 mg/L, during monsoon (December 2017) because of mixing of fresh water through rainfall and falling of inland stream channels to the sea. The SSC measurement for three seasons shows higher sediments in nearshore water (shallow depth region) and lower sediments in offshore water (deep depth region) due to the influences of waves, currents, winds, and river discharges that mix sediments in coastal water at various ranges depending on spatiotemporal variability of ocean hydrothermodynamics (Vinayachandran and Nanjundiah, 2009). The seasonal variability of SSC influences major physicochemical and biological properties of seawater that affect marine and coastal ecosystems due to monsoonal variability of oceanic and coastal processes (Seelanki et al., 2018; Kaliraj et al., 2019a,b).

4.1 SSC spatial variability in premonsoon (September 2018)

During premonsoon (September 2018), SSC is spatially varying in the range of 0.99–9.28 mg/L, whereas higher SSC is found in coastal water columns than in offshore water. Fig. 17.3 shows SSC spatial variability in premonsoon. The widespreading water column of offshore region (deep sea) shows a lower range of SSC at the rateof 0.99–5.26 mg/L. However, in the nearshore water column of Thiruvananthapuram–Kanyakumari–Tuticorin–Keelakarai, the SSC is estimated at a very high value of 5.27–9.28 mg/L as compared to other regions. This is due to adding a large amount of sediment through littoral current from the BoB and Arabian Sea. Moreover, inflowing SSC toward the nearshore water column is induced by prevailing hydrodynamic processes over the continental shelf of the South Indian coastal stretch.

4.2 SSC spatial variability in monsoon (December 2017)

SSC variability is estimated at a lower value during monsoon in the range of 0.61–7.29 mg/L due to heavy precipitation over the BoB region and it is significantly diluting seawater in both nearshore and offshore. However, higher SSC is found in the nearshore water column than the offshore region and it is estimated at 4.04–7.29 mg/L. This is due to discharge of river sediments into coastal water. Fig. 17.4 shows SSC spatial variability in coastal water during monsoon. The result reveals that the SSC found higher values in nearshore water than that of offshore (deep-sea water). The higher SSC is found in the coastal water column, especially eastern nearshore with an average value of 5.22 mg/L. However, in a few locations, the coastal water is recorded with lower sediments due to wave action and ocean current flow. The higher SSC is estimated in coastal water within proximity of 1 km from the shoreline due to discharging of sediments from river

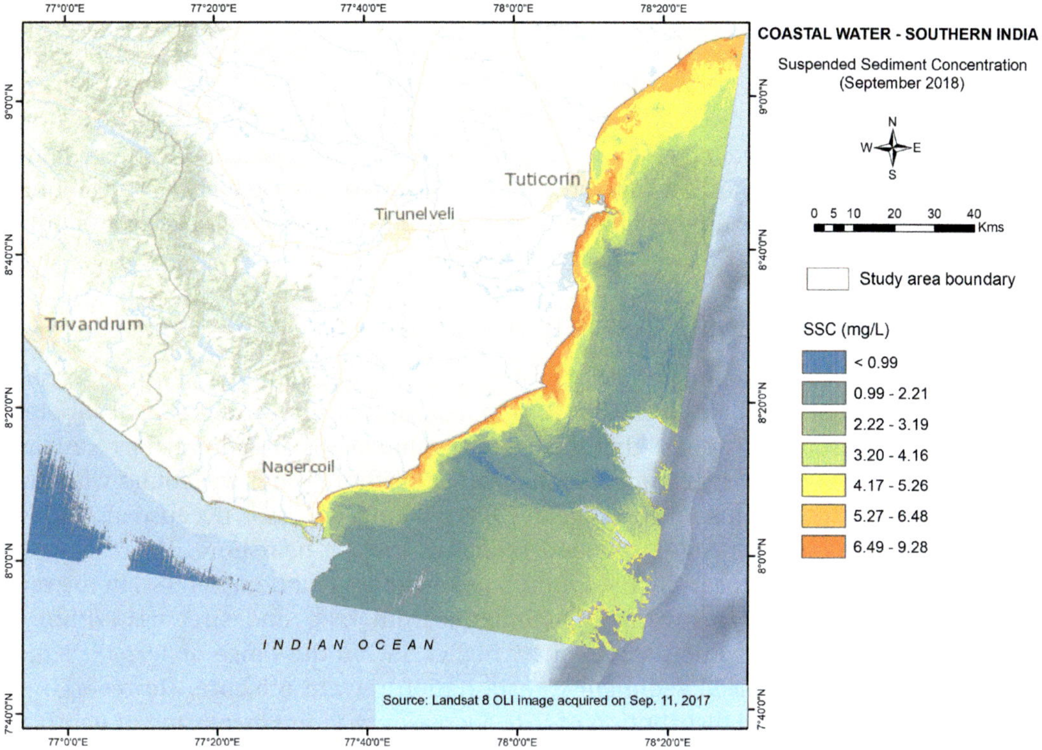

FIGURE 17.3 SSC spatial variability in premonsoon (September 2018).

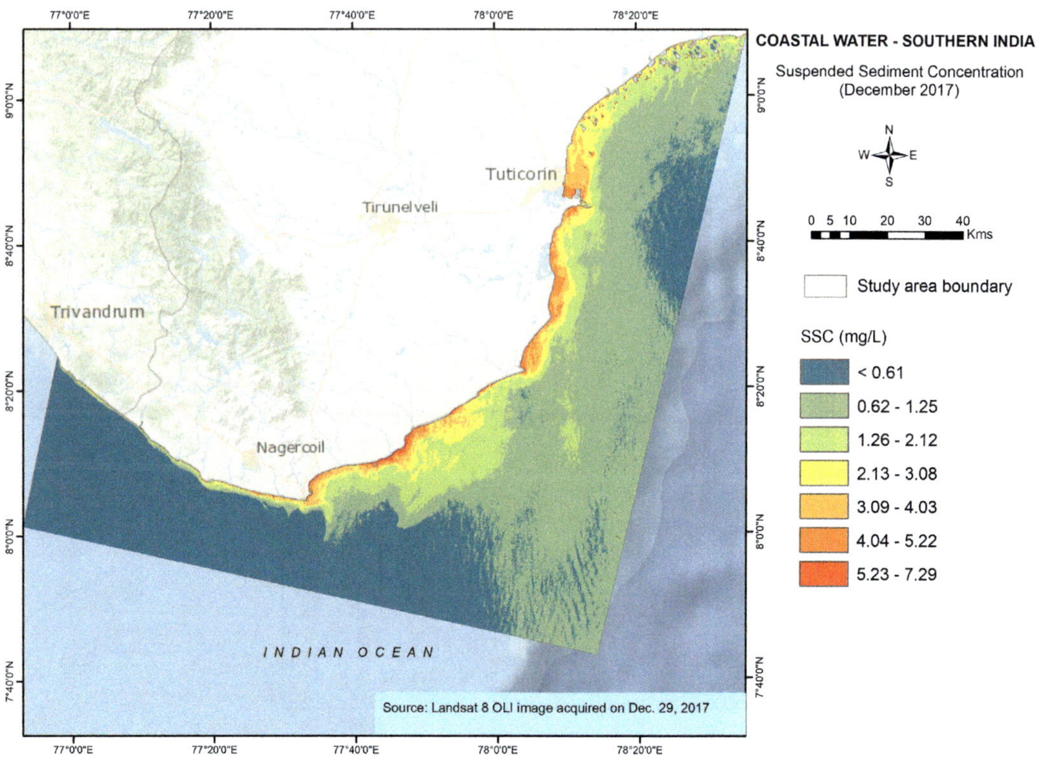

FIGURE 17.4 SSC spatial variability during monsoon (December 2017).

and streams. Moreover, the major part of offshore water experiences higher SSC during monsoon due to the shoaling effect and rip currents that are attributed to movement of the near-bottom SSC toward the offshore region. It is noted that the higher SSC is found in coastal water within a proximity of backwater and river mouth whereas a larger quantity of sediments discharge from river and streams during monsoon.

4.3 SSC spatial variability in postmonsoon (March 2018)

During postmonsoon, the higher SSC is estimated in water columns of both nearshore and offshore regions at the rate of 0.82–9.70 mg/L. Fig. 17.5 shows SSC spatial variability during postmonsoon. It is noted that the average value of SSC is estimated as 5.26 mg/L within nearshore water. Meanwhile, the majority of offshore water was found with lower concentration of sediments in the range of 0.83–2.06 mg/L. This is because of inflowing water currents that diverting suspended sediment toward nearshore. However, a few pockets of coastal water near the river mouth are estimated with higher SSC in the range of 6.00–9.70 mg/L due to mixing of sediments from river and streams. Within the offshore water column, some patches show higher SSC in the range of 3.52–5.99 mg/L due to rip current–induced sediment transportation toward offshore. However, in the coastal water, the spatial distribution of SSC decreases with the increasing of depth and distance from shoreline during postmonsoon.

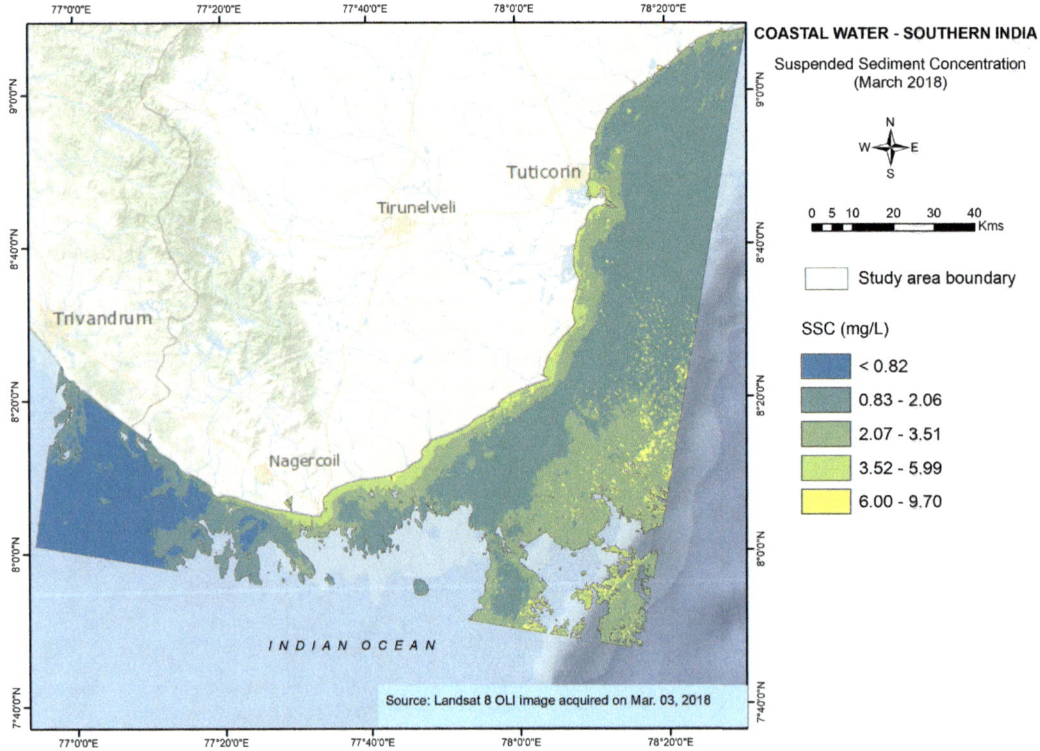

FIGURE 17.5 SSC spatial variability in postmonsoon (March 2018).

4.4 SSC seasonal variability in 2017–2018

Seasonal variability of SSC in the Southern Indian coastal water is characterized from SSC measurements using Landsat 8 OLI images for three seasons during 2017–2018. Fig. 17.6 shows seasonal variability of SSC in the Southern Indian coastal water. The SSC measured values at the 20 stations (Fig. 17.1 shows geographical locations of selected stations in the study area) obviously represent spatiotemporal variability of SSC during monsoon (October–December 2017), premonsoon (June–September 2018), and postmonsoon (March–May 2018). Table 17.1 shows SSC seasonal variability for the 20 selected stations along the Southern Indian coastal water. The SSC mean value for three seasons is noted as 4.32 mg/L, wherein the overall maximum SSC value is 8.14 mg/L in station 15, and the overall minimum SSC value is 1.11 mg/L in station 2. In a wide-spreading water column of offshore (deep sea), the SSC seasonal variability is estimated at 2.22–3.19 mg/L and it is relatively constant with a very little spatiotemporal variation during pre- and postmonsoons. The SSC seasonal variability shows the standard deviation (SD) value of 0.30–0.98 for majority of stations that indicates that SSC in the water column of the nearshore (shallow depth) is relatively constant during all three seasons, but changes in the offshore water (deep sea) column during different seasons. SSC seasonal variability shows the higher SD as 1.11–2.58 in a few coastal water packets found over stations 7, 11–15, and 18–20, whereas SSC is strongly affected from inland rivers and streams discharged during monsoon. In the offshore water column, the SSC found lower values mainly due to wave action and ocean current flow from BoB and TIO (Tropical Indian Ocean) that may cause dilution over wide-spreading areas (Tripathy et al., 2018; Kaliraj et al., 2019a,b). The lower SSC found in offshore water also is due to mixing of freshwater by precipitation to BoB and Arabian Sea during northeast southwest monsoons, respectively. The SSC found lower values in the western side than that of the eastern side of nearshore.

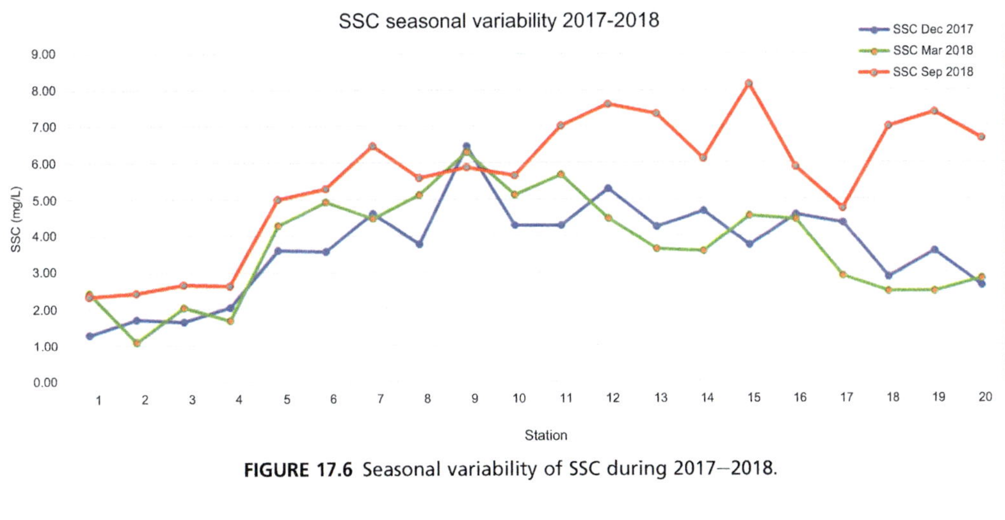

FIGURE 17.6 Seasonal variability of SSC during 2017–2018.

Table 17.1 SSC seasonal variability in selected stations along the Southern Indian coast.

	SSC seasonal variability (mg/L)						
Station	December 2017 (monsoon)	March 2018 (postmonsoon)	September 2018 (premonsoon)	SSC maximum (mg/L)	SSC minimum (mg/L)	SSC mean (mg/L)	SD
1	1.29	2.43	2.34	2.43	1.29	2.02	0.63
2	1.71	1.11	2.44	2.44	1.11	1.75	0.67
3	1.65	2.03	2.67	2.67	1.65	2.12	0.52
4	2.05	1.67	2.64	2.64	1.67	2.12	0.49
5	3.59	4.29	5.00	5.00	3.59	4.29	0.70
6	3.58	4.92	5.28	5.28	3.58	4.59	0.90
7	4.61	4.48	6.47	6.47	4.48	5.19	1.11
8	3.77	5.12	5.57	5.57	3.77	4.82	0.94
9	6.46	6.31	5.88	6.46	5.88	6.22	0.30
10	4.27	5.12	5.64	5.64	4.27	5.01	0.69
11	4.28	5.68	7.02	7.02	4.28	5.66	1.37
12	5.28	4.49	7.60	7.60	4.49	5.79	1.61
13	4.26	3.63	7.35	7.35	3.63	5.08	1.99
14	4.67	3.56	6.11	6.11	3.56	4.78	1.28
15	3.73	4.53	8.14	8.14	3.73	5.47	2.35
16	4.56	4.46	5.87	5.87	4.46	4.96	0.79
17	4.34	2.87	4.73	4.73	2.87	3.98	0.98
18	2.87	2.47	6.97	6.97	2.47	4.10	2.50
19	3.57	2.48	7.38	7.38	2.48	4.48	2.58
20	2.64	2.82	6.65	6.65	2.64	4.04	2.26

SSC mean refers to average of three seasons; SD, standard deviation.

During 2017–2018, the SSC seasonal variability within coastal water (shallow depth) was maximum during premonsoon and postmonsoon, up to values of more than 9 mg/L, but reached minimum value (<7.29 mg/L) during northeast monsoon (December). This is because of adding fresh water by rainfall and other forms of water discharge to the open sea. However, in offshore water, the SSC is lower at a value of (<2.12 mg/L) and has a relatively homogenous pattern during all three seasons. Hydrodynamic factors change in different seasons and affect SSC in coastal and ocean water columns particularly in the BoB, whereas the SSC is influencing on a site-specific scale due to other factors like precipitation and freshwater fluxes (Vinayachandran and Nanjundiah 2009; Seelanki et al., 2018). In the Southern Indian coastal water, the SSC varies at spatiotemporal scale depending on seasonally influencing factors like during monsoon, whereas other monsoonal factors control SSC such as river discharge, freshwater flux, and precipitation, especially in the coastal water column. Meanwhile, offshore water is affected by various wave actions and ocean current flow. In the study area, the higher SSC is distributed in shallow depth (<5 m) of the coastal water column, and within a distance 1 km from the shoreline. Moreover, in offshore water, the SSC decreases rapidly with the increase of distance to the shoreline and depth to the seabed due to the shoaling process in the surf zone due to frequent wind-generated wave action and rip currents. As the depth increases, the lower amount of sediment are available to move toward the shore, so lower concentrations are seen in the surface water. Meanwhile, the water column far-flung from points of river and stream confluences is noted with a minimum amount of SSC during all seasons due to lower discharge from external sources. During all three seasons, higher SSC is found in coastal water within proximity of nearshore than that of offshore due to a larger quantity of sediments released through river discharge and littoral currents. However, the larger quantity of suspended sediments settles within proximity of river mouth areas during monsoon. It is noted that the spatial distribution of SSC gradually decreases with the increasing of depth and distance from the shoreline. Hence, the SSC seasonal variability within coastal water is directly proportional to depth of seabed (bathymetry), distance to the shoreline, inflow of littoral current, and river discharge; and it has positive correlation with wave direction, shoaling action, and water currents in the offshore water column. In the South Indian coastal water, the higher SSC is found within proximity of coastal water (nearshore area) than that of offshore during all three seasons.

5. Conclusions

Satellite remote sensing provides multitemporal images with a wide variety of spectral and spatial resolutions for evaluating major physicochemical properties of water bodies. Landsat 8 OLI images are executed for measuring SSC and its seasonal variability using mathematical algorithms. In the Southern Indian coastal water column, during all three seasons, the SSC spatiotemporal variability shows a higher rate within proximity of the coastal water column than that of offshore water. The wide-spreading water column of offshore region (deep sea) has lower SSC and is relatively constant with very little spatiotemporal variability during monsoon. It is noted that the SSC seasonal variability is mainly due to changing physical and climatic factors like precipitation, waves, currents, and winds. On the other hand, in the coastal water column, the spatiotemporal variability of SSC depends on river discharges, runoff, and freshwater flux. Mapping and monitoring of SSC seasonal variability is necessary for understanding coastal ecology, landform formation, and so on, and is helpful for the development of sustainable management plans for environmental ecosystems on a long-term scale. Overall observations have declared that remote sensing is considered a suitable tool and technology for SSC estimation in coastal and ocean environments.

Acknowledgments

The corresponding author is thankful to the Director, National Center for Earth Science Studies (NCESS), Ministry of Earth Sciences, government of India, for providing continued support and encouragement. The authors extend their gratitude to IMD, SOI, NIOT, INCOIS, NOAA, NASA, USGS—Earth Explorer, and PSMSL—UK, for providing necessary data sources.

References

Carpenter, D., Carpenter, S., 1983. Modeling inland water quality using Landsat data. Remote Sens. Environ. 13, 345–352.

Chamberland, J., Bennett, J., Yue, B., 2019. Water surface temperature and suspended sediment concentration (SSC) retrieval from Landsat imagery, northwest territories, Canada. Can. J. Remote Sens. 1–14. https://doi.org/10.1080/07038992.2019.1643706.

Colleen, M.L., Pavelsky, T.M., 2013. Remote sensing of suspended sediment concentration and hydrologic connectivity in a complex wetland environment. Remote Sens. Environ. 129, 197–209. https://doi.org/10.1016/j.rse.2012.10.019.

Chen, J., Cui, T., Qiu, Z., Lin, C., 2014. A three-band semi-analytical model for deriving total suspended sediment concentration from HJ-1A/CCD data in turbid coastal waters. ISPRS J. Photogram. Remote Sens. 93, 1–13. https://doi.org/10.1016/j.isprsjprs.2014.02.011.

Chen, Z., Curran, P.J., Hansom, J.D., 1992. Derivative reflectance spectroscopy to estimate suspended sediment concentration. Remote Sens. Environ. 40, 67–77. https://doi.org/10.1016/0034-4257(92)90127-6.

Curran, P.J., Novo, E.M.M., 1988. The relationship between suspended sediment concentration and remotely sensed spectral radiance: a review. J. Coast Res. 4, 351–368.

Dethier, E.N., Sartain, S.L., David, A., Lutz, D.A., 2019. Heightened levels and seasonal inversion of riverine suspended sediment in a tropical biodiversity hot spot due to artisanal gold mining. Proc. Natl. Acad. Sci. U. S. A. https://doi.org/10.1073/pnas.1907842116.

Doxaran, D., Froidefond, J.M., Lavender, S., Castaing, P., 2002. Spectral signature of highly turbid waters application with SPOT data to quantify suspended particulate matter, concentrations. Remote Sens. Environ. 81, 149–161.

Forget, P., Ouillon, S., 1998. Surface suspended matter off the Rhone river mouth from visible satellite imagery. Oceanologica Acta 21 (6), 739–749. https://doi.org/10.1016/s0399-1784(99)80003-6.

Gholizadeh, M., Melesse, A., Reddi, L., 2016. A comprehensive review on water quality parameters estimation using remote sensing techniques. Sensors 16 (8), 1298. https://doi.org/10.3390/s16081298.

Hansen, C.H., Burian, S.J., Dennison, P.E., Williams, G.P., 2017. Spatiotemporal variability of lake water quality in the context of remote sensing models. Remote Sens. 9, 409.

Jennerjahn, T.C., Soman, K., Ittekkot, V., Nordhaus, I., Sooraj, S., Priya, R.S., Lahajnar, N., 2008. Effect of land use on the biogeochemistry of dissolved nutrients and suspended and sedimentary organic matter in the tropical Kallada river and Ashtamudi estuary, Kerala, India. Biogeochemistry 90 (1), 29–47. https://doi.org/10.1007/s10533-008-9228-1.

Kaliraj, S., 2016. Geospatial Analysis of Coastal Geomorphological Vulnerability on South West Coast of Kanyakumari Using Remote Sensing Technology. Unpublished Ph.D. Thesis. Manonmaniam-Sundaranar University, Tirunelveli, pp. 83–110.

Kaliraj, S., Chandrasekar, N., 2012. Geo-processing model on coastal vulnerability index to explore risk zone along the south west coast of Tamilnadu, India. Int. J. Earth Sci. Eng. 5 (5), 1138–1147.

Kaliraj, S., Chandrasekar, N., Magesh, N.S., 2013. Evaluation of coastal erosion and accretion processes along the southwest coast of Kanyakumari, Tamil Nadu using geospatial techniques. Arab. J. Geosci. https://doi.org/10.1007/s12517-013-1216-7.

Kaliraj, S., Chandrasekar, N., Magesh, N.S., 2014a. Multispectral image analysis of suspended sediment concentration along the southern coast of Kanyakumari, Tamil Nadu, India. J. Coast. Sci. 1 (1), 63–71. ISSN: 2348 – 6740.

Kaliraj, S., Chandrasekar, N., Magesh, N.S., 2014b. Impacts of wave energy and littoral currents on shoreline erosion/accretion along the south-west coast of Kanyakumari, Tamil nadu using DSAS and geospatial technology. Environ. Earth Sci. https://doi.org/10.1007/s12665-013-2845-6.

Kaliraj, S., Chandrasekar, N., Simon Peter, T., Selvakumar, S., Magesh, N.S., 2015. Mapping of coastal aquifer vulnerable zone in the south west coast of Kanyakumari, South India, using GIS-based DRASTIC model. Environ. Monit. Assess. https://doi.org/10.1007/s10661-014-4073-2.

Kaliraj, S., Chandrasekar, N., Ramachandran, K.K., 2016. Mapping of coastal landforms and volumetric change analysis in the south west coast of Kanyakumari, South India using remote sensing and GIS techniques. Egypt. J. Remote Sens. & Space Sci. https://doi.org/10.1016/j.ejrs.2016.12.006.

Kaliraj, S., Chandrasekar, N., Ramachandran, K.K., Srinivas, Y., Saravanan, S., 2017. Coastal landuse and land cover change and transformations of Kanyakumari coast, India using remote sensing and GIS. Egypt. J. Remote Sens. & Space Sci. https://doi.org/10.1016/j.ejrs.2017.04.003.

Kaliraj, S., Chandrasekar, N., Ramachandran, K.K., 2019a. Phytoplankton (chl-a) Biomass seasonal variability in the Gulf of Mannar, South India: a remote sensing perspective. In: Kumar, P., et al. (Eds.), Applications and Challenges of Geospatial Technology. https://doi.org/10.1007/978-3-319-99882-4_5.

Kaliraj, S., Chandrasekar, N., Ramachandran, K.K., Selvakumar, S., 2019b. Seawater intrusion vulnerability in the coastal aquifers of Southern India—an appraisal of the GALDIT model, parameters' sensitivity, and hydrochemical indicators. Environ. Sci. Pollut. Res. 26, 9755–9784. https://doi.org/10.1007/s11356-019-04401-0.

Kilham, N.E., Roberts, D., Singer, M.B., 2012. Remote sensing of suspended sediment concentration during turbid flood conditions on the Feather River, California—a modeling approach. Water Resour. Res. 48, 1.

Larson, M.D., Milas, A.S., Vincent, R.K., Evans, J.E., 2018. Multi-depth suspended sediment estimation using high-resolution remote-sensing UAV in Maumee River, Ohio. Int. J. Remote Sens. 39 (15–16), 5472–5489. https://doi.org/10.1080/01431161.2018.1465616.

Liu, H., Li, Q., Shi, T., Hu, S., Wu, G., Zhou, Q., 2017. Application of Sentinel 2 MSI images to retrieve suspended particulate matter concentrations in poyang lake. Remote Sens. 9, 761. https://doi.org/10.3390/rs9070761.

LSDS-1574, 2019. Landsat 8 (L8) Data Users Handbook. https://prd-wret.s3-us-west-2.amazonaws.com/assets/palladium/production/atoms/files/LSDS1574_L8_Data_Users_Handbook-v5.0.pdf.

Marcus, W.A., Fonstad, M.A., 2010. Remote sensing of rivers: the emergence of a subdiscipline in the river sciences. Earth Surf. Proc. Landf. 35 (15), 1867–1872. https://doi.org/10.1002/esp.2094.

Mertes, L.A.K., Smith, M.O., Adams, J.B., 1993. Estimating suspended sediment concentration in surface waters of the Amazon river wetlands from Landsat images remote sens. Environ. Times 43, 281–301.

Nair, N.B., Aziz, P.K., Dharmaraj, K., Arunachalam, M., Krishnakumar, K., Balasubramanian, N.K., 1993. Ecology of Indian estuaries: part I—physico-chemical features of water and sediment nutrients of Ashtamudi estuary. Indian J. Mar. Sci. 12, 143–150.

Narendra Babu, K., Omana, P.K., Mohan, M., 2010. Water and sediment quality of Ashtamudi estuary, a Ramsar site, southwest coast of India—a statistical appraisal. Environ. Monit. Assess. 165, 307–319. https://doi.org/10.1007/s10661-009-0947-0.

Nechad, B., Ruddick, K.G., Park, Y., 2010. Calibration and validation of a generic multi sensor algorithm for mapping of total suspended matter in turbid waters. Remote Sens. Environ. 114, 854–866.

Oestlund, C., Flink, P., Stroembeck, N., Pierson, D., Lindell, T., 2001. Mapping of the water quality of Lake Erken, Sweden, from imaging spectrometry and Landsat Thematic Mapper. Sci. Tot. Environ. 268, 139–154.

Olmanson, L.G., Brezonik, P.L., Bauer, M.E., 2011. Evaluation of medium to low resolution satellite imagery for regional lake water quality assessments. Water Resour. Res. 47, 1900–1904.

Palmer, S.C., Kutser, T., Hunter, P.D., 2015. Remote sensing of inland waters: challenges, progress and future directions. Remote Sens. Environ. 157, 1–8.

Pavelsky, T.M., Smith, L.C., 2009. Remote sensing of suspended sediment concentration, flow velocity, and lake recharge in the Peace-Athabasca Delta, Canada. Water Resour. Res. 45, W11417. https://doi.org/10.1029/2008WR007424.

Qu, L., 2014. Doctoral Dissertations. Remote Sensing Suspended Sediment Concentration in the Yellow River, 383. https://opencommons.uconn.edu/dissertations/383.

Seelanki, V., Sreenivas, P., Prasad, K.V.S.R., 2018. Impact of aquarius sea-surface salinity assimilation in improving the ocean analysis over Indian ocean. Marine Geodesy 41 (2), 144–158. https://doi.org/10.1080/01490419.2017.1422817.

Sridhar, P.N., Ali, M.M., Vethamony, P., Babu, M.T., Ramana, I.V., Jayakumar, S., 2008. Seasonal occurrence of unique sediment plume in the Bay of Bengal. EOS Trans. American Geophys Union 89 (3), 22–23.

Stumpf, R.P., Goldschmidt, P.M., 1992. Remote sensing of suspended sediment discharge into the western Gulf of Maine during the April 1987 100-year flood. J. Coast. Res. 8, 218–225.

Tassan, S., 1994. Local algorithm using Sea WiFS data for retrieval of phytoplankton pigment, suspended sediments and yellow substances in coastal waters. Applied Optics 12, 2369–2378.

Tassan, S., 1997. A numerical model for the detection of sediment concentration in stratified river plumes using Thematic Mapper data. Int. J. Remote Sens. 18 (12), 2699–2705. https://doi.org/10.1080/014311697217567.

Topp, S.N., Pavelsky, T.M., Jensen, D., Simard, M., Ross, M.R.V., 2020. Research trends in the use of remote sensing for inland water quality science: moving towards multidisciplinary applications. Water 12 (1), 169. https://doi.org/10.3390/w12010169.

Vinayachandran, P.N., Nanjundiah, R.S., 2009. Indian Ocean sea surface salinity variations in a coupled model. Clim. Dyn. 33, 245–263.

Wackerman, C., Hayden, A., Jonik, J., 2017. Deriving spatial and temporal context for point measurements of suspended-sediment concentration using remote-sensing imagery in the Mekong Delta. Continent. Shelf Res. 147, 231–245. https://doi.org/10.1016/j.csr.2017.08.007.

Wu, G., Cui, L., Liu, L., Chen, F., Fei, T., Liu, Y., 2015. Statistical model development and estimation of suspended particulate matter concentrations with Landsat 8 OLI images of Dongting Lake, China. Int. J. Remote Sens. 36, 343–360.

Xia, L., 1993. A united model for quantitative remote sensing of suspended sediment concentration. Int. J. Rem. Sens. 14 (14), 2665–2676. https://doi.org/10.1080/01431169308904300.

Further reading

Baban, S.M.J., 1993. Detecting water quality parameters in Norfolk Broads, UK., using Landsat imagery. Int. J. Remote Sens. 14, 1247–1267.

Baban, S.M.J., 1995. The use of Landsat imagery to map fluvial sediment discharge into coastal waters. Mar. Geol. 123, 263–270.

Baban, S.M.J., 1997. Environmental monitoring of estuaries: estimating and mapping various environmental indicators in Breydon Water Estuary, U.K., using Landsat TM imagery. Estuar. Coast Shelf Sci. 44, 589–598.

Bhargava, D.S., Mariam, D.W., 1991. Effects of suspended particle size and concentration on reflectance measurement. Photogramm. Eng. Remote Sens. 57, 519–529.

Byers, A.C., 1992. Soil loss and sediment transport during the storms and landslides of May 1988 in Ruhengeri prefecture, Rwanda. Nat. Hazards 5, 279–292.

Chauhan, P., Nayak, S., Ramesh, R., Krisnamoorthy, R., Ramachandran, S., 1996. Remote sensing of suspended sediments along the Tamil Nadu coastal waters. Photonirvachak 24 (2), 105–114.

European Centre, 2018. For Medium-Range Weather Forecasts: ERA Interim. Available at. https://www.ecmwf.int/. Lastaccess:February 2017.

Gao, P., 2008. Understanding watershed suspended sediment transport. Prog. Phys. Geogr. 32, 243–263.

Gordon, H., Wang, M., 1994. Retrieval of water-leaving radiance and aerosol optical thickness over the oceans with SeaWiFS: a preliminary algorithm. Appl. Optic. 33, 443–452.

Islam, M.R., Yamaguchi, Y., Ogawa, K., 2001. Suspended sediment in the Ganges and Brahmaputra river in Bangladesh: observation from TM and AVHRR data. Hydrol. Process. 15, 493–509.

Wong, M.S., Lee, K.H., Kim, Y.J., Nichol, J.E., Li, Z., Emerson, N., 2007. Modeling of suspended solids and sea surface salinity in Hong Kong using Aqua/MODIS satellite images. Korean J. Remote Sens. 23 (3), 161–169.

Tripathy, B.R., Kaliraj, S., Sajjad, H., Joshi, P.K., Chaudhary, B.S., Kurmar, P., 2018. Monitoring of seasonal variability and movement of suspended sediment concentration salong the Thiruvananthapuram coast, southern India, using the Landsat OLI sensor. Ocean Sci. 14, 1085–1092. https://doi.org/10.5194/os-14-1085-2018.

18

Modeling of coastal vulnerability to sea-level rise and shoreline erosion using modified CVI model

Kaliraj Seenipandi[1], K.K. Ramachandran[1], N. Chandrasekar[2]

[1]NATIONAL CENTRE FOR EARTH SCIENCE STUDIES (NCESS), THIRUVANANTHAPURAM, KERALA, INDIA; [2]CENTRE FOR GEOTECHNOLOGY, MANONMANIAM SUNDARANAR UNIVERSITY & FRANCIS XAVIER ENGINEERING COLLEGE, TIRUNELVELI, TAMIL NADU, INDIA

Chapter outline

Remote Sensing of Ocean and Coastal Environments. https://doi.org/10.1016/B978-0-12-819604-5.00018-4

1. Introduction

The coastal area is a transitional zone between land and water that influences dynamically in spatial and temporal patterns by terrestrial and ocean forces. The processes of oceanographic, climatic, and morphological factors characterize coastal landforms and habitats (Ahnert, 1998). The coastal areas are highly prone to various hazards and disasters as exposed to a variety of natural and human-induced impacts such as erosion, storm surge, beach subsidence, flooding, seawater inundation, cyclones, and tsunamis (Blanchard et al., 2010). Coastal vulnerability due to the occurrence of a natural or man-made phenomena have the potential for causing damage to landforms and ecosystems, and problems in building a relationship between natural ecosystems and man-made infrastructure along the coastal area. According to the Intergovernmental Panel on Climate Change (IPCC, 2014) report, the eustatic sea-level rise due to global warming is estimated at 18–59 cm by 2100, and hence, the major coastal ecosystem and habitats are identified as being under higher vulnerability along with coastal areas, throughout the world. The coastline of different countries, especially in the Asian continent, are highly vulnerable to physical, environmental, and habitat damage due to natural and anthropogenic factors.

In the Indian coastline with a length of 7517 km, the landforms of various locations are directly exposed to various hazards to coastal habitats resulting from erosion, flooding, seawater intrusion, storm surges, cyclones, and tsunamis. During the 21st century, the most threatening issues along the Indian coast are recorded as erosion, pollution, and sea inundation (Unnikrishnan et al., 2006; Kaliraj et al., 2014). The recent Indian Ocean tsunami on December 26, 2004, claimed many lives and produced major changes in the coastal landforms along the coastal area (Chandrasekar, 2005; Murthy et al., 2006; Hentry et al., 2012). Significantly, the southern coast of Tamil Nadu is highly prone to wave-induced erosion, resulting in seasonal altering of coastal waves and currents (Antony, 1976; Krishnamoorthy et al., 2002). During the past decades, the southwest coast has experienced higher pressure on habitats of human as well as coastal ecosystems due to anthropogenic activities (Kaliraj and Chandrasekar, 2012). The southern Tamil Nadu coast frequently alters the shoreline and landform due to seasonal changes in wave direction and littoral currents (Kaliraj et al., 2013). In recent decades, anthropogenic activities are increasing pressure on the coastal environment by changing the cyclic coastal dynamic processes causing damage of landforms and habitats along the coast (Loveson and Rajamanickam, 1998; Kaliraj et al., 2017a). Seawater extending landward is one of the major issues producing adverse effects on large-scale changes in the coastline, and in the configuration of continental margins and ocean floors (UNEP, 2007; Loveson et al., 1990). Coastal structures like groins, revetments, and seawalls are reducing or blocking sediment supply to the coast, causing vertical rising of saltwater intrusion, while intervening movement of littoral drift causes a series of erosion on the downdrift side (Mohanty et al., 2011; Kaliraj et al., 2014).

The Coastal Vulnerability Index (CVI) model is used to estimate potential vulnerability at pixel scale using multiple variables like shoreline changes, geomorphic units, sea-level rise rate, waves, and current pattern in the coastal area. The US Geological Survey (USGS) has implemented the CVI algorithm for the assessment of coastal vulnerability at a site-specific scale; particularly considered are six variables like geomorphology, shoreline change rates, coastal slope, relative sea-level rate, mean significant wave height, mean tidal range, which in general provided information regarding a vulnerable zone (Thieler and Hammer-Klose, 2001). Integrated remote sensing and GIS techniques offer the most viable technological solutions for generating databases on the coastal landform and its dynamic processes to date (Kaliraj and Chandrasekar, 2012). Satellite data provides baseline information on seashore line variation, geomorphology, and land use. GIS environment facilitates to collect, store, retrieve, and analyze spatial and nonspatial information from various resources, with more reliability (Rao et al., 1985). Richards and Jia (2006) have used a numerical algorithm to identify and devise the imagery into segments and classify them into the land and water pixels to extract land and water objects, and to trace the vector boundaries between land and water objects. The CVI model allows the users to identify a potential vulnerability regarding erosion, flooding, and habitats, and depend on user-defined input (Kaliraj and Chandrasekar, 2012). Many researchers have used the GIS-based CVI model to determine coastal vulnerability to sea-level rise, storm surge, erosion, and inundation worldwide (Gornitz et al., 1990; Pendleton et al., 2005; Abuodha and Woodroffe 2007; Srinivasa Kumar et al., 2010; Palmer et al., 2011; Kaliraj and Chandrasekar, 2012; Mohd Fauzi et al., 2014; Loinenak et al., 2015; Kantamaneni et al., 2017; Vousdoukas et al., 2018). Thieler and Hammar-Klose (1999) have developed the CVI model to demarcate coastal vulnerability for the US Atlantic coast based on the analysis of six types of physical variables that contribute to coastal change. The GIS modeling of the modified CVI algorithm offers an effective platform for any users to incorporate any number of influencing parameters or variables for the assessment of coastal vulnerability to physical landform damage due to sea-level rise (Kaliraj et al., 2014). The study of coastal vulnerability using geospatial technology is very important for sustainable coastal zone management in the coastal area at a local and regional scale. The GIS-based CVI model is an effective platform for mapping potential vulnerable zones to physical damage due to sea-level rise, and it is helpful for scientists and decision-makers to prepare long-term planning for sustainable coastal zone management.

2. Study area

The south Indian coastline comprises many urban and rural settlements with higher population density along with valuable environmental, aesthetic, and recreational resources, whereas the various hotspots have been noted as under vulnerability to physical and environmental damages (Mujabar and Chandrasekar, 2011; Kaliraj et al., 2013). The modified CVI model is applied to the coastline with the length of 58 km from the

Kanyakumari to Thengapattinam along the southern part of Tamil Nadu, India that covers the geographical coordinates of 77°9′49.23″E to 77° 34′14.99″E longitude and 8° 6′32.63″N to 8°14′15.30″N latitude. Fig. 18.1 shows the geographical location of the South Indian coastal stretch. The coastal configuration is gently sloping southward with respect to landward margin of the coastline that comprises beach landforms like sandy beaches, dunes, estuaries, and other fluvio-marine landforms (Srinivas and Sajan, 2010; Kaliraj et al., 2015). The coastal upland is mainly associated with rocky shore and offshore outcrops that act as a natural barrier to wave actions and storm surges. The sandy beaches and beach ridges are found in the Sanguthurai, Chothavilai, Pillaithoppu, Ganapathipuram, Rajakkamangalam, Colachel, and Simonkudiyiruppu coastal areas (Hentry et al., 2012). Onshore margin of the coast is composed of thick Late Quaternary deposits and sandy materials generally sloping toward the sea that cover settlements, coconut plantations, and shallow water bodies like backwater and creeks (Loveson et al., 1990). The rocky boulders and sea cliffs are found in the Muttam, Kanyakumari, and Cape Comorin coasts, and the sandstone found along the study area is made up of igneous rock and silt clay materials. The alluvial mixed clay is embraced at the mouth of

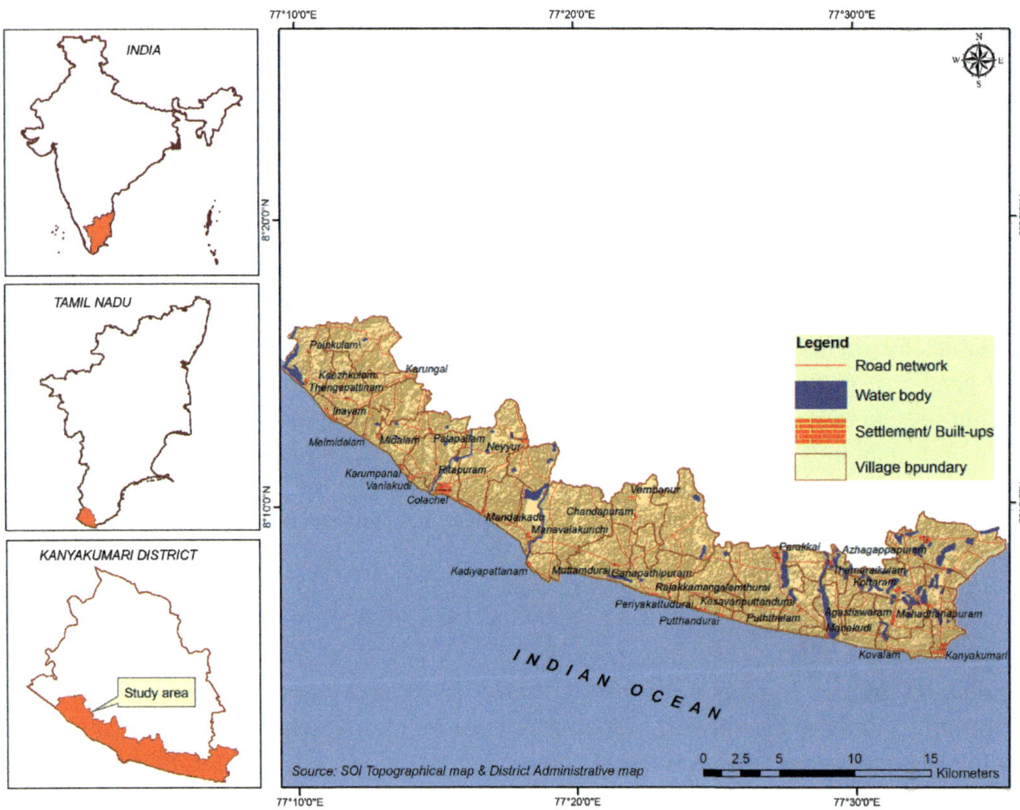

FIGURE 18.1 The geographical location of the South Indian coastal stretch.

the Thamirabarani estuary in the Thengapattinam and Pazhayar estuary in Manakudi. The three major drainage networks such as Pazhayar, Valliyar, and Thamirabarani and their tributaries are flowing in a southerly direction from the Western Ghats, and they are the primary sources for sediment discharge to maintain equilibrium status of coastal landforms during both southwest and northeast monsoons (Baba et al., 1987; Kaliraj et al., 2017a). The subtropical climatic condition prevails in the coastal region with the rainfall range of 826–1456 mm and temperature range of 23.78–33.95°C. Along the coastline, the human settlements have clustered in the nearshore and backshore, and this increased from 17.54 km^2 in 2000 to 39.22 km^2 in 2018. In the coastline, altering the hydrodynamic forces due to natural and anthropogenic forces produces adverse impact on coastal landforms and habitats for long-term scale. The coastal livelihoods and resources experience vulnerability to physical damage, hence it is very necessary for assessment and mapping the vulnerability using remote sensing and GIS for sustainable coastal management.

3. Materials and methods

GIS modeling of the modified CVI model is developed using the empirical algorithm for the assessment of potential vulnerability to physical damage to landforms due to sea-level rise, erosion, storm surge, and inundation. Thieler and Hammar-Klose (2001) have proposed the CVI model with six variables for the assessment of vulnerability to sea-level rise along the US coastline. The present chapter formulated a modified CVI model to make it suitable for the southern coast of India. The parameters are composed of nine types of geo-environmental variables include geomorphic landforms, shoreline erosion/accretion rate (m/yr), slope (or) topography (%), relative sea-level change rate (mm/yr), mean significant wave height (m), mean tidal range (m), storm wave run-up (m), regional elevation (m), and land use/land cover change (km^2). The multiple parameters (variables) driving site-specific coastal vulnerability is derived from various types of primary and secondary data sources using systematic GIS tools and functionalities. Landsat ETM+ and OLI (30 m) images (30 m) and IKONOS (2.5 m) are collected from the USGS online portal like Global Land Cover Facility, Earth Explorer, Digital Globe, and GeoEye for extracting land use/land cover and detecting shoreline change. The SRTM and ASTER DEM data sets (1 arc second or 30 m) are systematically incorporated with Landsat and IKONOS images for the assessment of relief range and geomorphic landforms change between 2000 and 2018. The in situ survey using Garmin GPS ETREX 30 is used to measure shoreline shift and areal beach erosion, locate benchmarks, and demarcate the high-tide line. Secondary data sources such as SOI topographical maps, Geological Survey of India (GSI) geology and geomorphology maps, NHO bathymetry chart and tide gauge data, wave atlas, buoys data, climate data, and census data have been collected from authenticated departments or agencies of the state or central government bodies. The Survey of India (SOI) topographical maps (Toposheet number 58 H/4, 58 H/8, 58 H/12 with the scale of 1:50,000) are used to extract physiographic details of the study area such as contour, drainage, water bodies and other natural, and cultural features at a specific scale. The GSI published maps of geology and geomorphology (scale 1:250,000) that are used for preparing geomorphic landforms. Relative sea-level change rate (mm/yr) in the offshore of the South Indian coastline is collected from the long-term observed data source of the USACE-Wave Information Study Atlas and Permanent Service for Mean Sea Level, UK. Meanwhile, the mean significant wave height (m) is estimated by incorporating the different data sources like water height, wind speed, wave runup, storm surge, and other ocean climatic information observed by buoys, and are collected from INCOIS, NIOT, NIO, and IMD, government of India. The GIS-based DSAS tool is used to evaluate shoreline changes. The per-year rate of shoreline change is estimated to determine the zone of erosion and accretion rate using multitemporal Landsat ETM+ and OLI images during 2000–2018 based on DSAS–net shoreline movement and end point rate statistical algorithms. The combined data sets of DEM and Landsat ETM+ and IKONOS images provide vital clues for mapping coastal geomorphic units such as coastal flood plains, beaches, swales, water bodies, swamps or marshlands, tidal flats, backwater creeks, and estuaries (Kaliraj et al., 2017a). The landsat ETM+ and OLI images are used for extracting the land use and land cover (LULC) feature based on the Level II category of the USGS-LULC Classification System using the maximum likelihood classifier (MLC) classifier of supervised classification technique. The data sources have been digitally converted and geographically referenced to UTM-WGS 84 projection and coordinate system by applying the ground control points obtained from the GPS field surveys using systematic functionalities of GIS software. The CVI parameters derived using various types of data sources are systematically processed using GIS tools and image processing functions to produce CVI parameters for mapping and assessment of coastal vulnerability.

3.1 GIS modeling of modified CVI algorithm

The GIS modeling of the modified CVI algorithm is formulated for analyzing the nine geo-environmental variables adopted from the standard USGS-CVI model developed by Thieler and Hammar-Klose (2000) for coastal vulnerability assessment and mapping. The CVI calculation is to formulate the key variables (parameters). The second step deals with assigning weights of parameters based on their potential magnitude to vulnerability. The CVI algorithm refers to the square root of the product of the weighted variables divided by the total number of variables (Thieler and Hammar-Klose, 2001; Hegde and Reju, 2007) and is expressed as

$$[\sqrt{a\ x\ b\ x\ c\ x\ d\ x\ e\ x\ f\ x\ g\ x\ h\ x\ i}]/9 \qquad (18.1)$$

where a refers geomorphology; b is shoreline erosion/accretion rate (m/yr); c is coastal slope (%); d is relative sea-level rise rate (mm/yr); e is mean significant wave height (m); f is mean tide range (m); g is storm wave run-up (m); h is regional elevation (m); i refers the landuse/land cover change (m^2). The mean wave height (e) is calculated using the wave energy equation proposed by Dyer (1986) based on input parameters like wave height (H), seawater density (ρ), and gravity (g), and it is expressed as

$$\left(1/8\,\mathrm{g}H^2\right) \tag{18.2}$$

This estimates quantitative relations of these parameters relative proportional to mean wave height. According to USGS (Thieler and Hammar-Klose, 2001) guidelines and expert suggestions, each CVI parameter and its corresponding feature classes are segmented to assign weighted values from 1 to 5 based on potential contribution to vulnerability with the range of very low to a very high level. The weighted value of 1 indicates a low contribution of a particular parameter to vulnerability, while the weighted value of 5 reflected a higher contribution. The weighted key variables are aggregated mathematically using the modified CVI algorithm. The modified CVI variables and their corresponding assigned weighted values are given in Table 18.1. The CVI value at each grid (pixel) of the raster is computed by calculating the square root of the weighted parameters by dividing the total number of parameters using the raster calculator tool in ESRI-ArcGIS software. The result of the modified CVI model is classified into five classes to represent site-specific vulnerability rate in the coastal region, and each class indicates its vulnerability to physical damages resulting from sea-level rise and coastal erosion impacts.

4. Results and discussion

Understanding the coastal vulnerability to physical damage to landforms is important for sustainable physical, environmental, and socioeconomic management. The CVI parameters and their potential impacts on vulnerability is described as follows.

4.1 Coastal geomorphic landforms

Coastal geomorphic landforms reflect the nature of coastal dynamic processes and have a certain degree of resistance to vulnerability. Each feature class is assigned to respective weights based on their susceptibility to physical damages (Thieler and Hammer-Klose, 2001). In general, the landforms along the coastal stretch experience dynamic changes due to erosion and accretion processes resulting from natural hydrodynamics and anthropogenic activities (Kaliraj et al., 2014). Fig. 18.2 shows the geomorphological landforms derived from Landsat and Ikonos images. Sandy beaches are showing a high rate of change in sediment loads due to back washing of high-energy waves and formed as cusps and scarps at the rate of 5.04–6.21 m/km^2 and 2.67–6.12 m/km^2, respectively, and hence its very high vulnerable weighted value is noted at 5. The cliffs are eroded at the rate of 3.28–6.52 m/km^2 along the rocky coast of Kanyakumari-Kovalam, Muttam, and Colachel coastal areas due to high-energy waves attacking the rocky shore resulting in the formation of wave-cut notches (4.68 m/km^2) and wave-cut platforms (6.15 m/km^2); the assigned weighted value of these features is 4. The alluvial plains are noticing

Table 18.1 The modified CVI variables and their corresponding assigned weighted values.

		Range and characteristics of CVI parameter for vulnerability rating				
		Very Low	**Low**	**Moderate**	**High**	**Very High**
Sl.No.	**Variables**	**1**	**2**	**3**	**4**	**5**
1	Geomorphic landforms	Rockyshore cliffs, coastal upland, offshore barrier coast	Terisand dunes, barrier dunes, indented coastal plains, deltaic plains	Alluvial plains, sand dunes, beach ridges; beach terraces, salt flats	Beach ridges, berms, swale, wave-cut notches, estuaries	Sandy beaches, cusps, scarps, sand spits, sand bars, shoal, salt flats, marshy/swamp, mudflats, tidal flats, wave-cut platforms, notches
2	Shoreline erosion/ accretion rate (m/yr)	> 10.0 (accretion)	1.0 and 10.0 (accretion)	−1.0 and 1.0 (seasonal accretion and erosion)	−10.0 and −1.0 (erosion)	< −10.0 (erosion)
3	Coastal slope (%)	< 0.5	0.5 − 1.5	1.5 − 3.0	3.0 − 7.5	> 7.5
4	Relative sea-level change rate (mm/yr)	< 1.5	1.5 − 3.0	3.0 − 4.5	4.5 − 6.0	> 6.0
5	Mean significant wave height (m)	< 0.2	0.2 − 0.6	0.6 − 1.0	1.0 − 1.4	> 1.4
6	Mean tidal range (m)	< 1.0	1.0 − 2.0	2.0 − 4.0	4.0 − 6.0	> 6.0
7	Storm wave run-up (m)	< 1.5	1.5 − 3.0	3.0 − 4.5	4.5 − 6.0	> 6.0
8	Coastal regional elevation (m)	> 40	30 − 40	20 − 30	10 − 20	< 10
9	Landuse/land cover change (km^2)	> 2 (increase in area)	1.0 − 2.0 (increase in area)	−1.0 and 1.0 (seasonal change)	−2.0 and −1.0 (decrease in area)	> −2.0 (decrease in area)

Negative value (−) refers to erosion and positive value (+) refers to accretion.

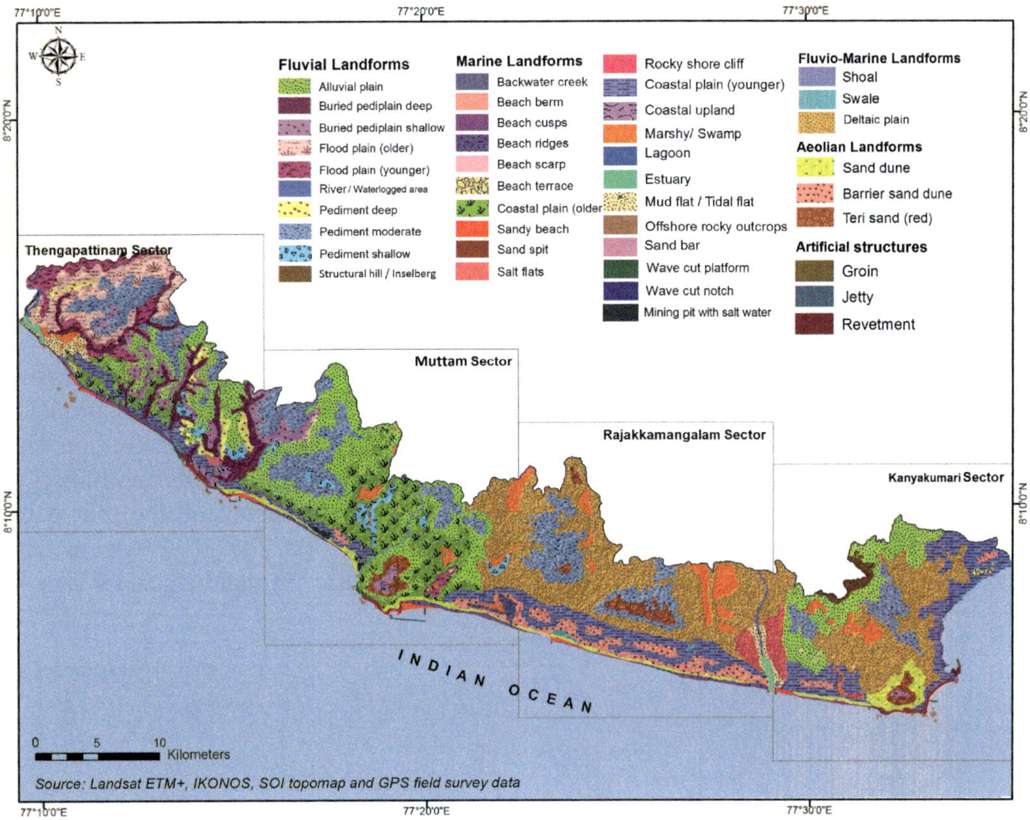

FIGURE 18.2 The geomorphological landforms derived from Landsat and Ikonos images.

significant volumetric changes in the rate of land lost 7.92 to 5.99 m/km^2 due to an erosion of sand and silt materials by the works of rivers and streams (Mujabar and Chandrasekar, 2011; Rejith et al., 2018). The flood plains and sand dunes is modified in their premorphological structures by removal of larger volume at the rate of 6.25–4.41 m/km^2 and 8.33–7.63 m/km^2, respectively, from the shallow land of river course due to flooding and surfacing runoff during the monsoons (Chandrasekar et al., 1996; Kaliraj et al., 2017a). The assigned weighted value of these features is 3. Likewise, the barrier dunes and terisand deposits are significantly decreased in their pre-morphological sediment volume at the rate of 6.68–6.36 m/km^2 and 9.42–5.52 m/km^2, respectively, due to wind action that frequently alters sediment quantity during both southwest and northeast monsoons, and the denoted weighted value is 2. The uplands and rocky shore in the Kovalam, Muttam, and Colachel coastal zones are protected by the natural coastal configuration structure, and offshore barrier rocks lead a formation of the young sandy beaches along the down the seaward slope (Kaliraj and Chandrasekar, 2012) and hence it is assigned with a weighted value of 1. Based on this view, the geomorphic features are assigned to vulnerability rank based on the degree of magnitude to landform deformation due to sea-level rise in the coastal area.

4.2 Shoreline erosion/accretion rate

Shoreline change in long-term periods indicates the erosion and accretion processes of the coast and is helpful to a certain extent in understanding the coastal response to future sea-level rise (Chandramohan et al., 1993; Bankoff et al., 2003; Mohanty et al., 2011). In this study, the site-specific shoreline change rate is estimated using the historical shoreline derived from the temporal Landsat ETM + images for the period 2000–2018. Fig. 18.3 shows the zone of long-term erosion and accretion in the South Indian coastal stretch. The shoreline change rate during these periods has seen severe erosion in various parts of the coastal zones interacting with high-energy waves divergent of littoral currents. The coastal zones, namely the headlands of Kanyakumari, Kovalam Pallam, and Inayamputhenthurai, are noticing the high rate of erosion (more than 10 m per year) with a weighted value of 5. The Manavalakurichi-Mandaikadu coastal tract is seriously depleted by high rates of erosion with an estimated value ranging from 3.73 to 4.51 m, respectively, with a weighted value of 4. This anomaly of the erosion process is mainly due to unregulated mechanized coastal mining that causes frequent changes in short-term periods (Narayana and Jayappa, 2009; Hentry et al., 2012; Baskar et al., 2015). The shoreline along the Puthenthurai, Midalam coastal zones is

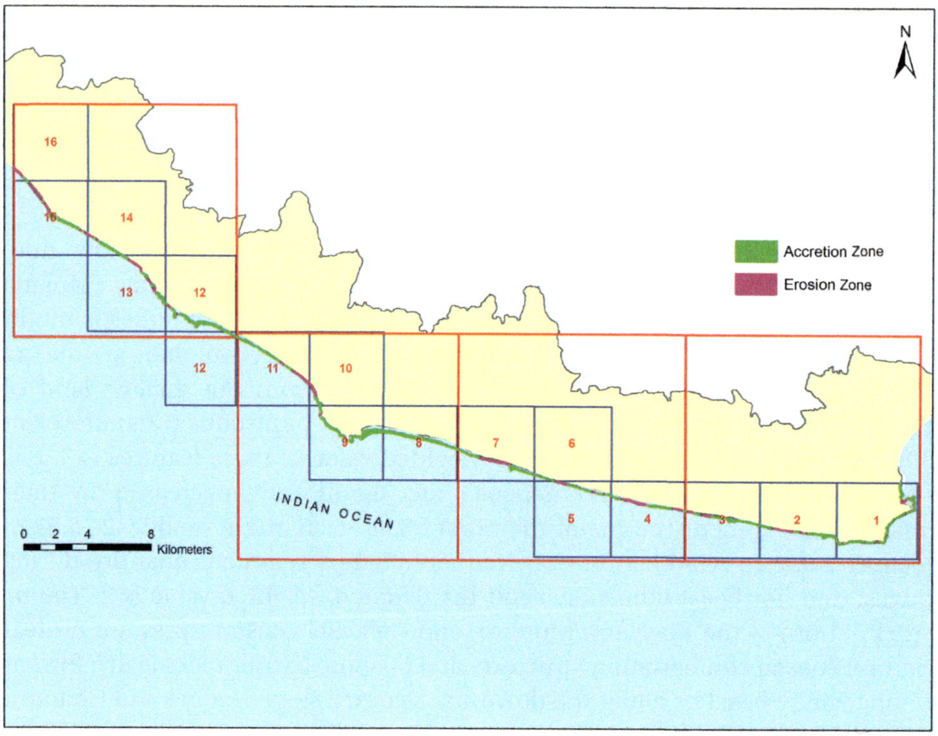

FIGURE 18.3 The zone of long-term erosion and accretion in the South Indian coastal stretch.

highly fluctuated due to erosion and accretion processes at the rate of −1.0–1.0 m with assigned weighted value of 3. The estuary landforms like sandbars, sand spits, and mudflats near Thengapattinam and Manakudi coastal areas are morphologically extended in shape and size due to deposition of large amounts of river discharge during the monsoons (Baba et al., 1991; Loveson and Rajamanickam, 1998; Jayappa and Narayana, 2009; Kaliraj et al., 2017a). Therefore these feature classes are assigned a weighted value of 2. The sand beaches, beach ridges, sand dunes, and other depositional features are accreted throughout the year in the Sanguthurai, Chothavilai, Ganapathipuram, Colachel, and Simonkudiyiruppu coastal zones at a rate of more than 10 m due to the swashing of sediments by low-energy wave actions and are assigned a weighted value of 1. The vulnerability rate is assigned to each segment with the respective rate of erosion that reflects the physical damages of landforms in the area.

4.3 Coastal slope (or) topography

The coastal slope is one of the major triggering factors for erosion followed by beach subsidence along the high-energetic wave zones. The slope represents the steepness, incline, gradient, or grade of a straight line along the coastline and the range of steepness is linked to the susceptibility of coast to physical vulnerability to sea-level rise (Srinivasa Kumar et al., 2010). In the study area, the slope map is prepared using the contours (1 m interval) from the ASTER DEM (30 m) acquired in 2011 with the reference of SOI topographic maps (1:50,000 scale) using the 3D Analyst module in ArcGIS 10.6 software. Fig. 18.4 shows the slope gradient of the coastal stretch. The total range of slope gradient along the coast is estimated in the range of 0.2% and 9.3%, in which the areas with a steep slope (7.5%) are considered a deterministic factor influencing the erosion process, and a high potential for beach subsidence; it is noted with a weighted value of 5. Similarly, the coastal locations having a land slope of 3.0%–7.5% are noticing significant erosion rate, and assigned the weighted value of 4. The moderately sloped coastal zones (1.5%–3.0%) have decreased beach width during high-wave action, and hence the assigned weighted value is 3. Slope gradient less than 1.5% found in the beaches and gentle slope areas are noted as higher susceptibility to back washing of sediments from the beaches (Murty and Varadachari, 1980; Jayappa, 1996; Chandrasekar, 2005; Kaliraj and Chandrasekar, 2012; Jeyabal and Ramasamy, 2014; Kaliraj et al., 2017b; Tripathy et al., 2018) and hence these zones are assigned weighted values 1 and 2. The slope gradient determines beach subsidence or land loss while interacting waves and storm surges cause severe vulnerability to sea-level inundation in the inland region.

4.4 Relative sea-level change rate

The changes occurring at sea level at a regional or local scale become a consequence of climatic factors that directly affect shoreline and associated landforms (Pirazzoli, 1996; Li, and Li, 2011; Saxena et al., 2013). The relative sea-level change rate is derived from the

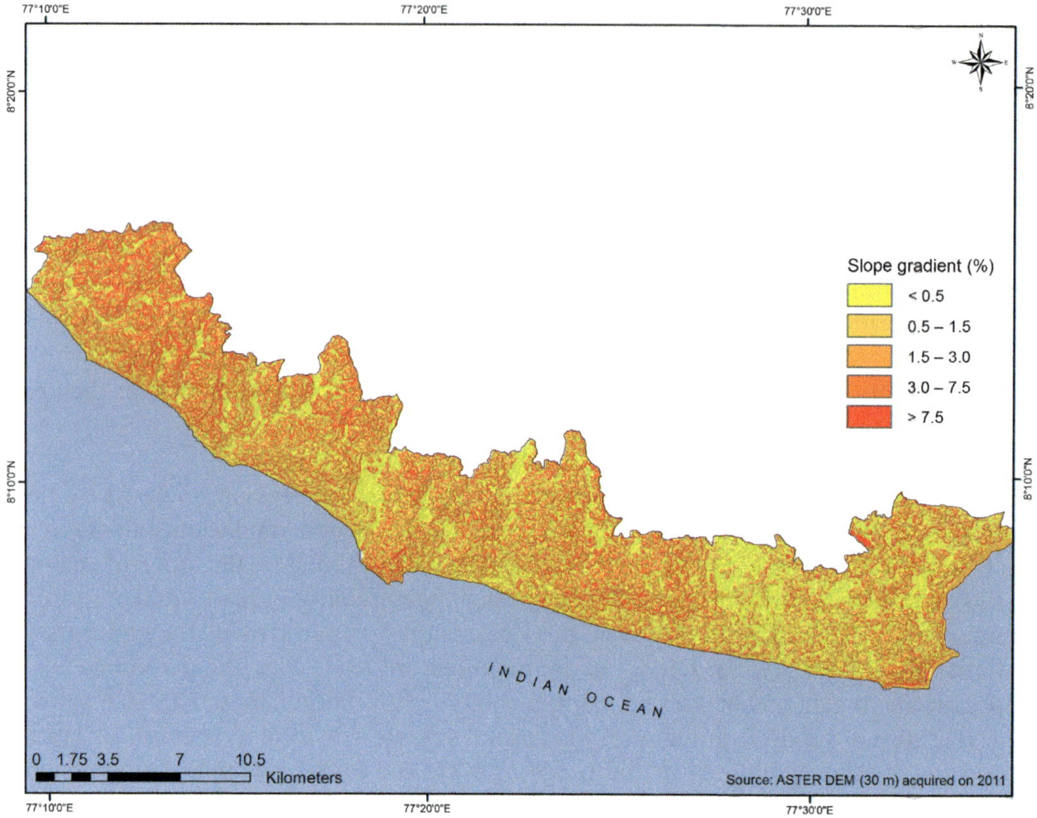

FIGURE 18.4 The slope gradient of the South Indian coastal stretch.

values of tide gauges and buoys placed along the southern coast of India during 1969–2011. Fig. 18.5 reveals the relative sea-level rise along the offshore of the South Indian coastline based on tide gauges and buoys data sets. The change in the rate of sea-level rise by gridding of offshore is estimated using a linear interpolation method in Arc GIS 10.6 software. The coastal zones seeing sea-level rise at the rate of more than 6.0 mm/yr is considered a potential vulnerability to physical damage due to waves and tides (Bankoff et al., 2003; Murthy et al., 2006; Mohanty et al., 2011; Kaliraj and Chandrasekar, 2012) and is assigned a weighted value of 5. Similarly, the coastal locations Manakudi, Kesavanputhenthurai, Pillaithoppu, and Midalam comprise sea-level change at the rate of 4.5–6.0 mm/yr with a weighted value of 4. The upland areas of Kanyakumari, Muttam, and Colachel coasts are falling within the very low vulnerability to sea-level change (less than 1.5 mm/yr) and they are assigned lower weighted values 1 and 2. The change in annual mean seawater elevation indicates a loss or gain of landforms due to transgression or regression processes in the coastal area.

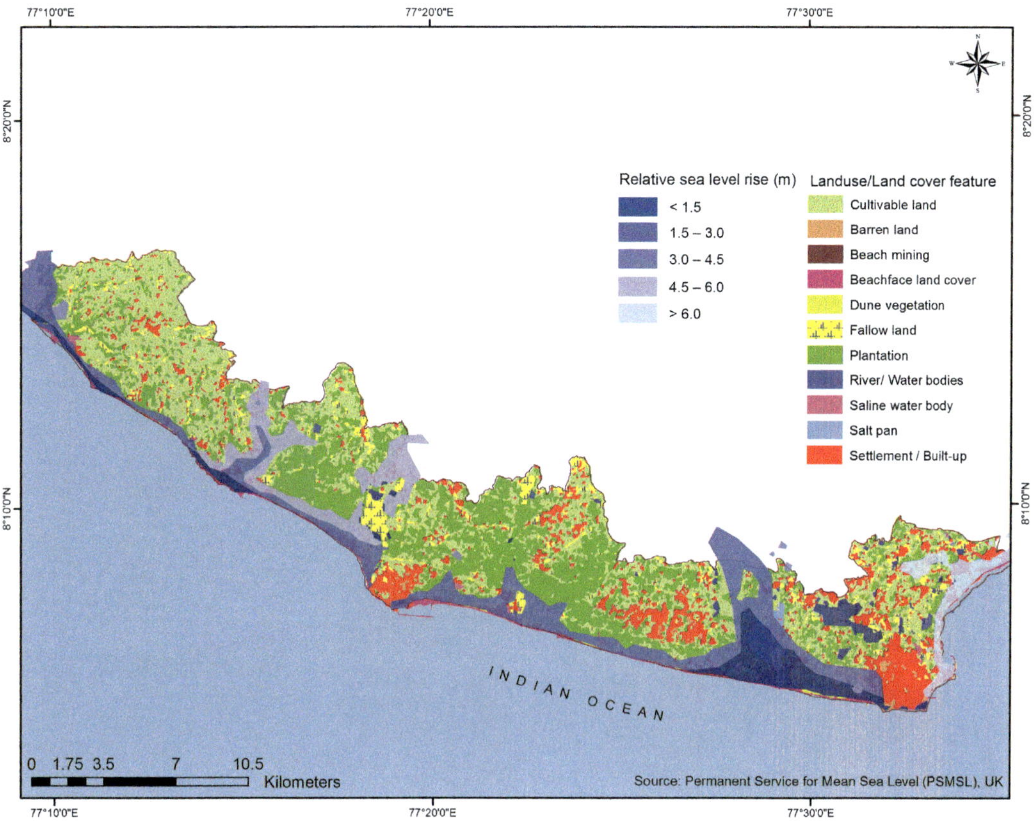

FIGURE 18.5 The relative sea-level rise along the offshore of the South Indian coastline based on tide gauges and buoys data sets.

4.5 Mean significant wave height

Wave height reflects the energy releasing to the coastline, which is one of the main factors in setting up a hazard management system along the coastal region (Chandramohan et al., 1991; Shahul Hameed et al., 2007; Ramana Murthy et al., 2009; Misra and Balaji, 2015). Mean significant wave height is calculated from the three physical parameters such as (1) wave height (H), (2) seawater density (ρ), and (3) gravity (g) using the wave energy equation ($E = 1/8\ \rho gH2$). Wave energy increases as the square of the wave height and it is able to mobilize and transport sediments; this results in loss of land area due to erosion and inundation along the coastline (Ashok Kumar et al., 2005; Kaliraj et al., 2014). Fig. 18.6 shows the mean significant wave height along the South India coastline. The coastal zones Inayamputhenthurai, Mandaikadu, Kovalam, and Cape Comorin show increased wave height of more than 1.4 m that releases higher wave-energy on the coast, resulting in severe erosion or subsidence of beaches along the

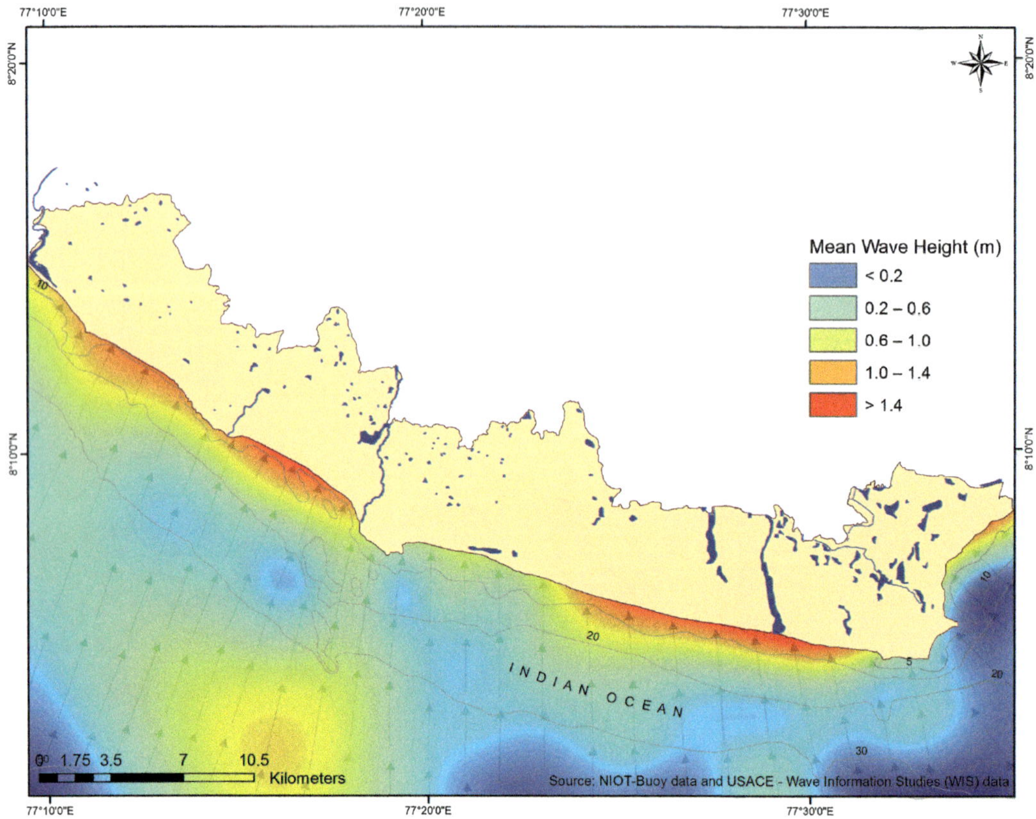

FIGURE 18.6 The mean significant wave height along offshore of the South India coastline.

shore (Hentry et al., 2012); hence a weighted value of 5 is assigned. The coastal zones with a wave height ranging from 1.0 to 1.4 m have frequently induced erosion, and the assigned weighted value is 4. Low-energy waves produce a lower height of waves (<1.0 m), which leads to accretion processes due to the swashing of sediments (Murty and Varadachari, 1980; Jena et al., 2001; Mishra et al., 2011). These areas, therefore, are assigned lower vulnerable weight values 1 and 2. It is noticed that the increase of wave height is causing physical vulnerability to the coastal landforms.

4.6 Mean tidal level

The coastal stretch has highly interfacing tidal waves, especially low-lying landforms that influence both permanent and episodic shoreline change (Baba and Nayak, 2002; Jeyabal and Ramasamy, 2014). The tidal range is the vertical difference between the highest high tide and the lowest low tide over time (Unnikrishnan et al., 2006; Mujabar and Chandrasekar, 2011; Sowmya and Jayappa, 2016). The mean tide range is estimated from the average of low tide and high tide values observed at eight tide gauge stations of hydrographic charts published by the National Hydrographic Office, India. The mean tidal range is more than 6.0 m (above the datum) in the low-lying zones, and these areas are considered highly vulnerable to the coastal landforms, and assigned weighted value 5, whereas the lower tidal range from 1.0 to 2.0 m in the Puthenthurai, Rajakkamangalam, Ganapathipuram, Pillaithoppu, and Mandaikadu are assigned lower vulnerable weighted values one and 2 (see Table 18.1). Subsequently, the vulnerability ranks are assigned to each segment based on the interpolated tidal range values of the nearest tide stations. The tidal range reflects the tendency of tidal waves to release suspended sediments to the coast over time.

4.7 Coastal relief (or) elevation range

The coastal regional elevation refers to the average elevation (topography) of a coastal zone above mean sea level (Tanh and Furukawa, 2007; Srinivasa Kumar et al., 2010; Jayappa et al., 2015). It is important to indicate the vertical movement of sediment loads of the landforms due to the removal or deposit of sediments over time and helps to identify the area threatened by physical vulnerability due to sea-level rise. The elevation of each grid zone is derived from ASTER DEM (30 m) acquired in 2011 and topographical maps (scale 1:50,000). Fig. 18.7 shows the coastal relief or elevation range based on the

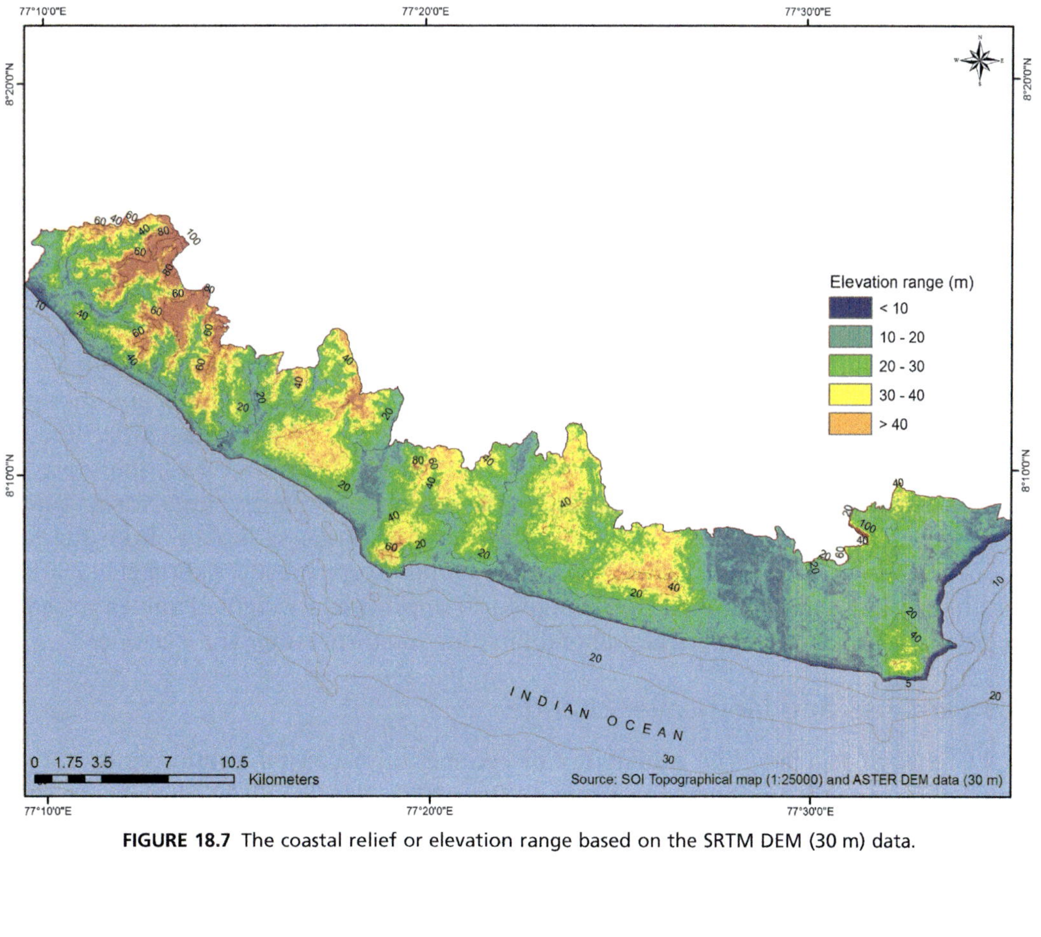

FIGURE 18.7 The coastal relief or elevation range based on the SRTM DEM (30 m) data.

SRTM DEM (30 m) data. The coastal landforms include estuaries, shallow marshy, backwater, creeks, salt pan, and human settlements, which are noticed at lower elevation (<10 m from MSL) with high potential for flooding or inundation of seawater due to sea-level rise (Bankoff et al., 2003; Jayappa and Kumar, 2006; Kaliraj et al., 2017a). Therefore, these areas are assigned the vulnerability weighted value 5. Similarly, the areas with the elevation ranging from 10 to 20 m are potentially vulnerable to inundation and sediment migration due to tsunami run-up and storm surge, and they are assigned vulnerability weighted value 4. The uplands and rocky shore areas with high elevation (more than 40 m above Mean Sea Level (MSL)) are considered less vulnerable areas because they provide more resistance for inundation against the rising sea level and are assigned lower weighted values of 1 and 2. The changing of the vertical movement rate of sediment load indicates the vulnerability characteristics of the beach landforms in the study area.

4.8 Storm surge (or) wave run-up

The wave run-up refers to the high-energy wave with increased elevation of sea level at the coast during storm surge (Murty and Flather, 1994; Ashok Kumar et al., 2005; IPCC, 2007; Li and Li, 2011). It is the maximum vertical extent of wave runup toward coastal landforms above the instantaneous water level or storm-tide level that causes severe land loss of short-term periods (Baba, 1985; Dolan, 1992; Chandramohan et al., 1993; Baskaran, 2004; Kaliraj et al., 2014; Jayappa et al., 2015). The combined action of storm-tide plus wave run-up damages the beaches and rises over dunes and seawalls in various parts of the study area. Wave run-up is calculated using the empirical equation proposed by Gibb (2001) based on the relationship between offshore significant wave height and maximum storm wave run-up, and it is expressed as ($Wave\ run\text{-}up\ (m) = 0.35\ \beta s(H_oL_o)^{1/2}$). Fig. 18.8 shows the mean wave height of storm surge in the South Indian coastal offshore. The coastal zones with wave run-up greater than 6.0 m are highly influencing the beach profile shape due to the removal of larger sediments during the storm surge and hence they are assigned a weighted value of 5. Wave run-up with a lower height (<3.0 m) in the uplands, rocky shore, shelves with a gentle slope, and mouth of the drainage (estuaries) has reduced impacts of storm surge and substantial flooding (Hegde and Reju, 2007; Huang and Hsieh, 2012). These feature classes, therefore, are assigned the weighted values of 1 and 2 (see Table 18.1). Storm-surge waves release higher energy at the breaking zones and backwashing larger sediments from the foreshore (Pradeep Kumar and Thakur, 2007; Kumar and Jayappa, 2009; Kaliraj et al., 2017a), and hence the remaining areas are assigned to appropriate weighted values based on degree of magnitude of physical vulnerability due to storm surge and cyclone.

4.9 Land use and land cover change

LULC change regulates the availability of sediments to coastal landforms due to the controlling erosion and accretion process (Rao, 2002; Bradley, 2009; Butt et al., 2015).

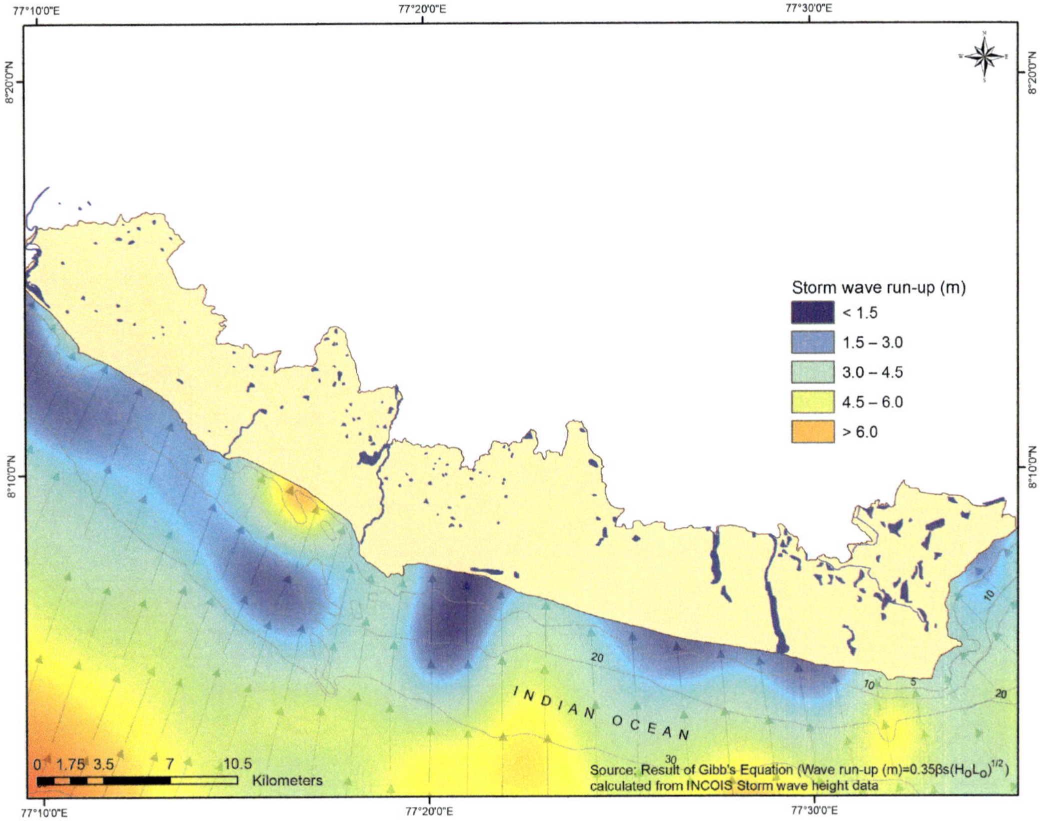

FIGURE 18.8 The mean wave height of storm surge in the South Indian coastal offshore.

The LULC features are extracted using Landsat ETM+ images (30 m) acquired in 2000 and 2018 using the MLC algorithm of supervised image classification technique in ERDAS Imagine 2014 software. Fig. 18.9 shows the spatial distribution of LULC in the South Indian coastal stretch. The beach landforms experience the decreasing rate more than 2.0 km^2 over time due to the impact of marine and coastal processes and other anthropogenic activities (Kaliraj and Chandrasekar 2012). Therefore, these areas have a high vulnerability weighted value of 5. Sand dunes have seen a significant amount of erosion due to the removal of dune vegetation and human encroachment activities (Huang and Hsieh, 2012), hence these areas are assigned weighted value 4. The cultivable lands show a significant decreasing trend in areas and hence their weighted value is 3. The sandy beaches and dunes increase in areal extent in the up-drift side of the coast and these accreted zones have weighted values 1 and 2. The LULC change is influencing the physical vulnerability to sediment loss or gain along the coastal area for long-term scale.

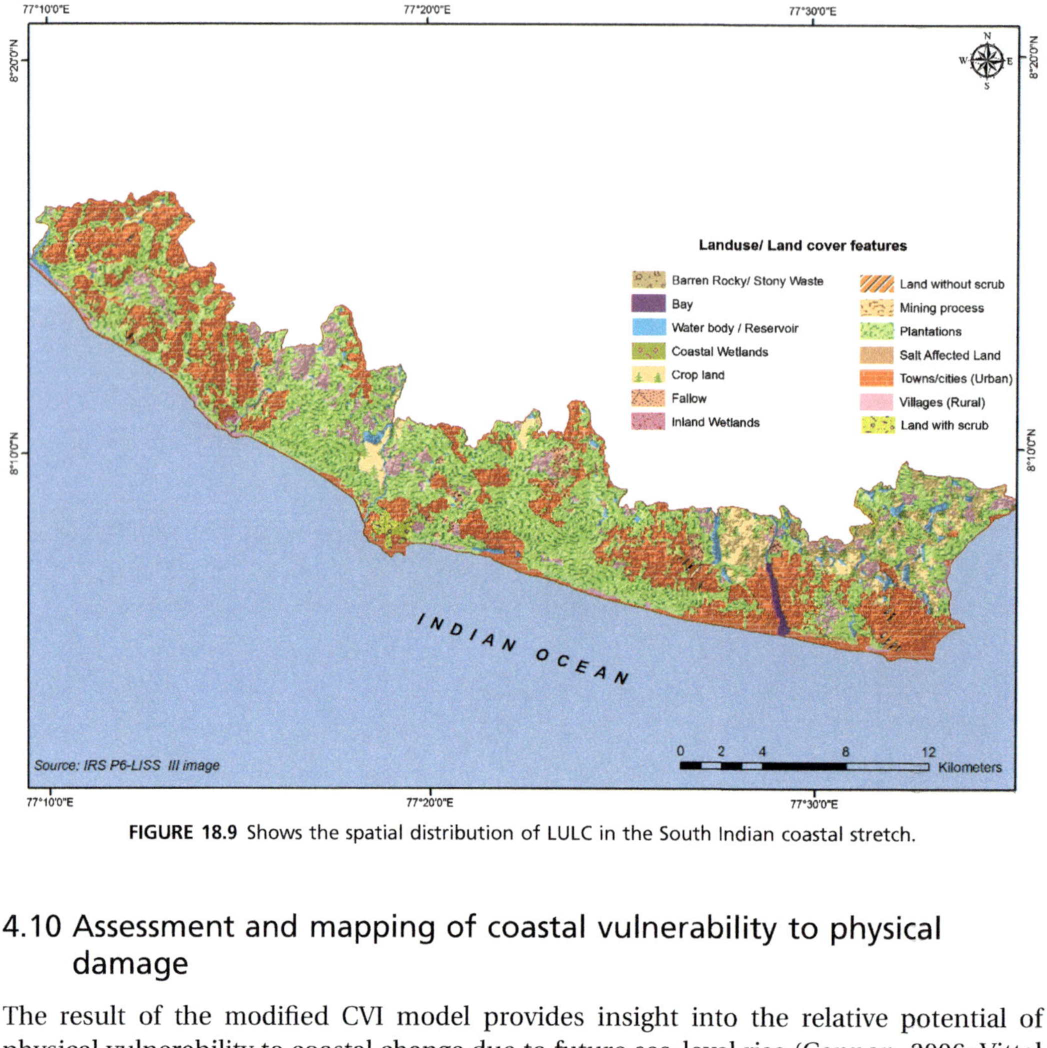

FIGURE 18.9 Shows the spatial distribution of LULC in the South Indian coastal stretch.

4.10 Assessment and mapping of coastal vulnerability to physical damage

The result of the modified CVI model provides insight into the relative potential of physical vulnerability to coastal change due to future sea-level rise (Cannon, 2006; Vittal et al., 2007; Srinivasa Kumar et al., 2010; Chandrasekar et al., 2013). The CVI model is performed using different geoprocessing tools for analyzing the nine types of weighted parameters for mapping the vulnerable zone. Fig. 18.10 shows the map of coastal vulnerable zones with its potential vulnerability characteristics at each grid along the South Indian coastal stretch. The result of the vulnerability index value ranges from 3.46 to 27.38, and the mean vulnerability value is 15.42. Based on the range of CVI values, the coastal zones are classified into five classes: very low, low, moderate, high, and very high vulnerability. Table 18.2 shows the CVI score and its potential vulnerability to physical damage of coastal landforms in the South Indian coastline. The vulnerability classification indicates an area of 52% in the Kanyakumari and 42% in the Thengapattinam

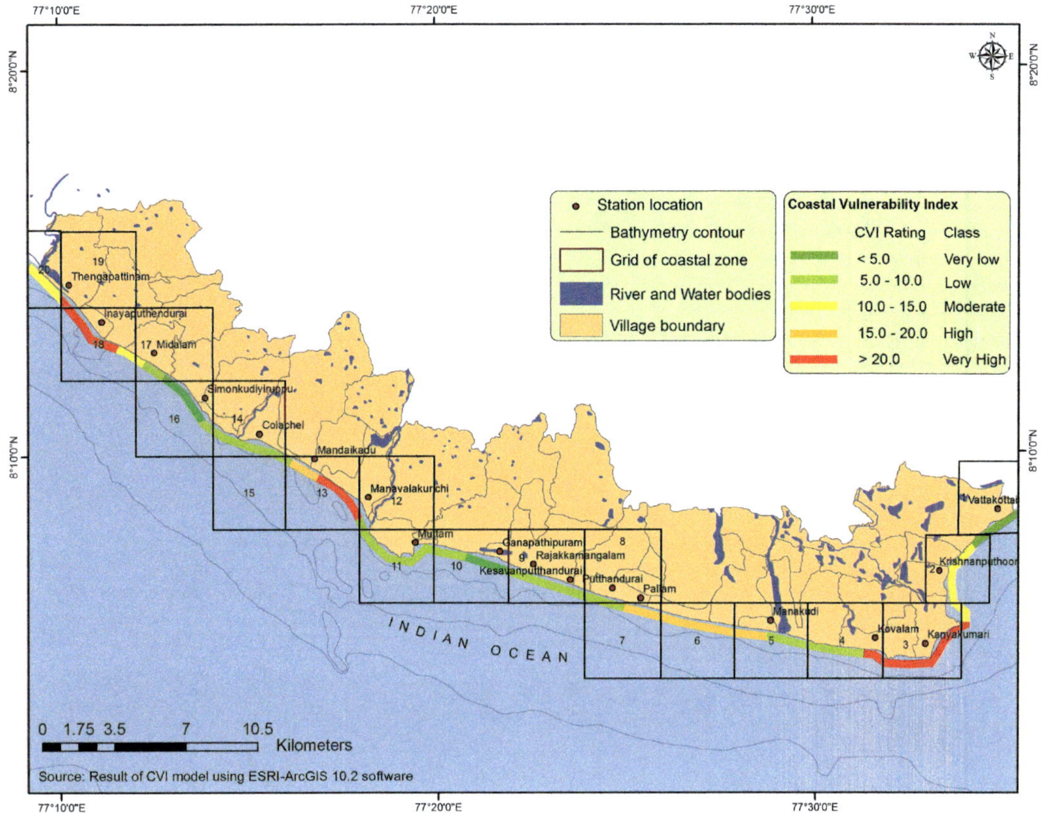

FIGURE 18.10 The map of coastal vulnerable zones with its potential vulnerability characteristics at each grid along the South Indian coastal stretch.

sectors with very high vulnerability conditions. The low-lying landforms such as estuaries, backwater creeks, mudflats, swamps, and salt pans are highly susceptible to physical damage due to sea-level rise (Kaliraj and Chandrasekar, 2012; Sowmya and Jayappa, 2016). However, more than 70% of the land area in the Rajakkamangalam and Colachel sectors have low vulnerability conditions. The remaining parts of the non-deltaic areas and Terisand dune-front sections come under moderate vulnerability conditions, while the rocky shore and uplands are have very low vulnerability to sea-level rise (Mujabar and Chandrasekar, 2011; Kaliraj et al., 2017a).

In the study area, a CVI value below 5 is considered a very low vulnerable zone, and this is seen in Kanyakumari northeast, Sanguthurai, Chothavilai, Ganapathipuram, and Simonkudiyiruppu coastal zones. These coastal zones have accretion activities due to natural barriers and offshore rocky outcrops leading to sandy beaches being extended in shape and size throughout the year. Similarly, the coastal uplands in the Colachel and Muttam coastal zones (elevation of above 6.0 m from MSL) are protected to physical

Table 18.2 The CVI score and its potential vulnerability to physical damage of coastal landforms in the South Indian coastline.

	Geomorphology	Shoreline Erosion /Accretion (m/yr)	Coastal Slope (%)	Relative Sea-Level Change (mm/yr)	Mean Wave Height (m)	Mean Tide Range (m)	Rank of Vulnerability	Remarks
Station 1 Mandakad	Barrier dunes, Fluvial-marine tidal flat	0.27	20	0.42	2.83	1.071	High	Shoreline accretion rate is decreasing and Mean Wave Height is increase
Station 2 Pillaitoppu	Barrier dunes, Fluvial-marine tidal flat	7.81	23	0.41	0.57	1.076	Low	Shoreline accretion rate is high and coastal slope is gentle
Station 3 Kesavanputtanturai	Beach with stacks and cliffs, Barrier dunes	−0.003	18	0.43	0.57	1.078	Moderate	Shoreline Erosion rate is increasing, Sea level change
Station 4 Mukkuranturai	Beach with stacks and cliffs	1.45	27	0.46	0.57	1.081	Low	Shoreline accretion and gentle slope and less impact of Wave Height
Station 5 Kovalam	Barrier dunes, Beach with stacks and cliffs	−3.17	3	0.52	2.83	1.081	Very High	Severe Shoreline Erosion, Very steep slope caused to damage of Coastal area, Wave Height is increasing, Raised marine terraces
Station 6 Cape Comorin	Marine-Barrier dunes predominated	−3.17	7	0.48	2.83	1.084	Very High	Severe Shoreline Erosion, Steep slope caused to dam.age of Coastal area, Raised marine terraces
Station 7 Muttam	Marine-Barrier dunes, Teri dune	1.77	11	0.47	0.57	1.084	Low	Shoreline accretion rate is increasing and less impact of Wave Height
Station 8 Lipuram	Barrier dunes, Fluvial tidal flats	−0.008	18	0.43	0.57	1.084	Low	Shoreline Erosion is beginning, however gentle slope with less Wave Height and minimum changes in Mean sea level

vulnerability due to sea-level rise. The CVI value of low vulnerability class ranges from 5 to 10. Thirty-two percent of the total area falls under this category. The coastal zones Manakudi, Putthandurai, Kesavanputhenthurai, Rajakkamangalamthurai, Colachel, and Midalam are classified as low vulnerability to physical landform damage due to sea-level rise, whereas the shoreline showed sandy beaches due to constructive wave actions (Kunte and Wagle, 2001; Lane et al., 2003; Marfai et al., 2008; Kaliraj et al., 2014). The coastal areas having CVI value ranges of 10–15 are classified as moderate vulnerability, which is noted in the Kanyakumari east, Mel Midalam, and Thengapattinam coastal zones. The coastal landforms are moderately vulnerable to physical damage due to seasonal erosion and accretion processes during monsoons (Ramana Murthy et al., 2009; Jayappa et al., 2015). Seasonal changes to wave height and littoral current directly affects coastal landforms by controlling sediment loads. High vulnerability class with CVI values of 15–20 is considered for Pallam, Mandaikadu, and South Thamaraikulam coastal zones. These coastal zones have prevailing high-energy wave actions that result in physical damage on coastal landforms due to flooding, seawater intrusion, and erosion activities during storm surges. Very high vulnerability (CVI is estimated as 20) is seen in the Kanyakumari, Kovalam, Manavalakurichi and Inayamputhenthurai and the low-lying estuary landforms in the Manakudi and Thengapattinam coastal zones. The low-lying wetlands and habitat areas are highly susceptible to flooding and seawater inundation vulnerability at sea-level rise at 0.5–1 m. In the middle-eastern parts of the coast, the rise of sea-level scenarios from 0.5 to 1.0 m leads to transgression of coastal landforms approximately 13 km^2 on both sides of estuaries in which lower-lying landforms include sandy beaches, salt pan, marshes, tidal flats, flood plain, creeks, habitats, and rural settlements under high vulnerability to sea-level rise (Kaliraj et al., 2013). Moreover, the wave run-up increases to above-average heights and produces physical damage of coastal landforms on the down-drift side of the groins during monsoons. Overall observation reveals that the low-lying areas prevailing high-energy wave actions may be the most vulnerable to the effects of future sea-level rise causing physical damage of coastal landforms and habitats in this region. The CVI model depicts vulnerability rates for each grid zone in the study area. It helps to identify the vulnerable areas of physical damage to coastal landforms due to sea-level rise for planning and management coastal resources.

5. Conclusion

The GIS modeling of the modified CVI model conclusively proves the potentially vulnerable zones at a site-specific scale based on remote sensing data and in situ observations. The advantage of the modified CVI model is formulated with nine types of parameters influencing coastal hydrodynamics of the southern coast of India and allows us to evaluate potential vulnerability site-specific scale. The result of the coastal vulnerability map serves as basic information sources for characterizing risks associated with coastal hazards for local people as well as stockholders. Systematic analysis of nine parameters using the CVI model indicates that the Kanyakumari and Thengapattinam sectors have very high vulnerability to physical damage of coastal landforms. Moreover, the low-lying landforms such as estuaries, backwater creeks, mudflats, swamps, and saltpans are highly susceptible to physical damage due to sea-level rise. In the coastal stretch, the uplands and some accreted beaches are noted with lower vulnerability due to the natural configuration of shoreline that acts like barriers from erosion and storm surge throughout the year. It is noted that man-made construction like seawalls, jetties, and groins cause extensive landform damage in the down-drift side, whereas the up-drift coastal tracts are accreted due to constructive wave actions. The GIS-based CVI model is useful to assess the vulnerability to physical and environmental hazards, and this model can be formulated according to user-defined input parameters for planning the coastal sustainable development and management.

Acknowledgments

The corresponding author is thankful to the Director, National Center for Earth Science Studies, Ministry of Earth Sciences, government of India, for providing support and encouragement to complete the research successfully. The authors would like to acknowledge SOI, Census of India, ICMAM-PD, INCOIS, USGS–Earth Explorer for providing necessary data sources.

References

Abuodha, P.A., Woodroffe, C.D., 2007. Assessing Vulnerability of Coasts to Climate Change: A Review of Approaches and Their Application to the Australian Coast. Australian National Centre for Ocean Resources and Security University of Wollongong, Wollongong, Australia, p. 458. http://citeseerx.ist.psu.edu/viewdoc/download?doi=10.1.1.461.2048&rep=rep1&type=pdf.

Ahnert, F., 1998. Introduction to Geomorphology. Arnold, London.

Antony, M.K., 1976. Wave refraction off Calangute Beach, Goa, with special reference to sediment transport and rip currents. Indian J. Mar. Sci. 5 (2), 1–8.

Ashok Kumar, K., Raju, N.S.N., Sanil Kumar, V., 2005. Wave Characteristics off Visakhapatnam Coast During a Cyclone. Annual Report. Goa, India: National Institute of Oceanography, Ocean Engineering Division.

Baba, M., 1985. Long term wave monitoring along the Kerala coast some results. Proc. Nat. Conf. on Dock & Harb. IIT Bombay 49–67.

Baba, M., Kurian, N.P., Shahul Hameed, T.S., Thomas, K.V., Harish, C.M., 1987. Temporal and spatial variation in wave climate off Kerala, Southwest coast of India. Indian Journal of Geo Marine science 16, 5–8.

Baba, M., Nayak, S., 2002. Muddy coasts of India. In: Healy, T., Wang, Y., Healy, J.A. (Eds.), Muddy Coasts of the World: Processes, Deposits and Function. Elsevier Science, pp. 375–390.

Baba, M., Shahul Hameed, T.S., Kurian, N.P., Thomas, K.V., Harish, C.M., Rameshkumar, M., Mohan, S., 1991. Longterm Wave and Beach Data of Kerala Coast Part I : 1980 '84. CESS, Thiruvananthapuram, pp. 132–141. Technical Report No. 57.

Bankoff, G., Frerks, G., Hilhorst, D., 2003. Mapping Vulnerability: Disasters, Development and People. Earthscan Publishers, UK, London, pp. 56–72 (Chapter 3).

Baskar, K., Sridhar, S.G.D., Sivakumar, T., Hussain, S.M., Maniyarasan, S., 2015. Temporal variation of physico chemical parameters and ostracoda population, off Rameswaram, Gulf of Mannar, southeast coast of Tamil Nadu, India. J. Geol. Soc. India 86 (6), 663–670.

Baskaran, R., 2004. Coastal erosion. Curr. Sci. 86, 25–30.

Blanchard, S.D., Rogan, J., Woodcock, D.W., 2010. Geomorphic change analysis using ASTER and SRTM digital elevation models in central Massachusetts, USA. GIScience Remote Sens. 47 (1), 1–24.

Bradley, B.A., 2009. Accuracy assessment of mixed land cover using a GIS-designed sampling scheme. Int. J. Rem. Sens. 30 (13), 3515–3529.

Butt, A., Shabbir, R., Ahmad, S.S., Aziz, N., Nawaz, M., Shah, M.T.A., 2015. Land cover classification and change detection analysis of Rawal watershed using remote sensing data. J. Biol. Environ. Sci. 6 (1), 236–248.

Cannon, T., 2006. Vulnerability analysis, livelihoods and disasters. In: Ammann, W., Dannemann, S., Vulliet, L. (Eds.), Coping with Risks Due to Natural Hazards in the 21st Century. Taylor & Frances/ Balkema, Leiden.

Chandramohan, P., Kumar, V.S., Nayak, B.U., 1993. Coastal processes along the shorefront of Chilka lake. Indian J. Mar. Sci. 22, 268–272.

Chandramohan, P., Nayak, B.U., Sanil Kumar, V.V., 1991. Sediment transport study using surf zone sediment traps. Technical Report, NIO., pp. 1–60. Goa, India.

Chandrasekar, N., 2005. Tsunami of 26th December 2004: observation on inundation, sedimentation and geomorphology of Kanyakumari coast, south India. In: Proc. 22nd IUGG Int. Tsunami Symp. Chania, Greece, pp. 49–56.

Chandrasekar, N., Joevivek, V., Saravanan, S., 2013. Coastal vulnerability and shoreline changes for southern tip of India Remote sensing and GIS approach. Earth Sci. & Clim. Change 4, 144–152.

Chandrasekar, N., Loveson, V.J., Rajamanickam, G.V., 1996. Longshore sediment transport along the southern coast Tamil Nadu. The IXth Convention on Geological Congress 60–63.

Dolan, D., 1992. An intensity scale for atlantic coast northeast storms. J. Coast Res. 8, 840–853.

Dyer, K.R., 1986. Coastal and Estuaries Sediment Dynamics. John Wiley and Sons, London, UK.

Gibb, J.G., 2001. Review of the 1995 Wainui Beach Coastal Hazard Zone. Coastal Management Consultancy Ltd Client Report Prepared for Gisborne District Council. C.R.2001/6.

Gornitz, V., White, T.W., Cushman, R.M., 1990. Vulnerability of the East coast, U.S.A. to future sea level rise. J. Coast Res. 9, 201–237.

Hegde, A.V., Reju, V.R., 2007. Development of coastal vulnerability index for Mangalore coast, India. J. Coast Res. 23, 1106–1111.

Hegde, V.,A., Reju, V.R., 2007. Development of coastal vulnerability index for Mangalore coast, India. J. Coast Res. 23 (5), 1106–1111.

Hentry, C., Chandrasekar, N., Saravanan, S., 2012. Beach dynamics of Colachel open coast, Kanyakumari District (SW India). Zeitschrift fur Geomorphologie. https://doi.org/10.1127/03728854/2012/0077.

Huang, S.W., Hsieh, H.I., 2012. The study of the land use change factors in coastal land subsidence area in Taiwan. In: Conference on Environment, Energy and Biotechnology (IPCBEE), vol. 3. IACSIT Press, Singapore, pp. 70–74.

IPCC, 2007. Summary for Policymakers. In: Soloman, S.D., Manning, Q.M., Chen, Z., Miller, H.L. (Eds.), Climate Change 2007: the Physical Science Basis. Contribution of Working Group I to the Fourth Assessment Report of the Intergovernmental Panel on Climate Change. Cambridge University Press, Cambridge, pp. 1–18.

IPCC, 2014. Climate change: impacts, adaptation, and vulnerability. In: Field, C.B., Barros, V.R., Dokken, D.J., Mach, K.J., Mastrandrea, M.D., Bilir, T.E., Chatterjee, M., Ebi, K.L., Estrada, Y.O., Genova, R.C., et al. (Eds.), Summaries, Frequently Asked Questions, and Cross-Chapter Boxes; A Contribution of Working Group II to the Fifth Assessment Report of the Intergovernmental Panel on Climate Change; World Meteorological Organization: Geneva, Switzerland, 2014, 190p. http://www.ipcc.ch/report/ar5/wg2/.

Jayappa, K.S., 1996. Longshore sediment transport along the Mangalore coast, west coast of India. Indian J. Mar. Sci. 25 (2), 157¬–159.

Jayappa, K.S., Kumar, G.T.V., 2006. Beach morphological studies in India a review. Geol. Soc. India 68 (5), 874–884.

Jayappa, K.S., Markose, V.J., Nagaraju, M., 2015. Quantitative Morphotectonic Analysis Using DEM and GIS – a Case Study from South West Coast of India. In: http://acrs2015.ccgeo.info/proceedings/TH4-8-3.pdf.

Jayappa, K.S., Narayana, A.C., 2009. Coastal Erosion and Management. Coastal Environments: Problems and Perspectives, I.K. International Publishing House Pvt. Ltd., New Delhi, pp. 183–¬189.

Jena, B.K., Chandramohan, P., Sanil Kumar, V., 2001. Longshore transport based on directional waves along north Tamilnadu Coast, India. J. Coast Res. 17, 322–327.

Jeyabal, G., Ramasamy, S., 2014. Down core variation in sediment characteristics and trace element geochemistry of a core sample in Pichavaram Mangrove area, Tamil Nadu, Southeast Coast of India. Environ. Geo Chimica Acta 1 (3), 206–214.

Kaliraj, S., Chandrasekar, N., 2012. Geo processing model on coastal vulnerability index to explore risk zone along the South West Coast of Tamilnadu, India. Int. J. Earth Sci. Eng. 5 (5), 1138–1147.

Kaliraj, S., Chandrasekar, N., Magesh, N.S., 2013. Evaluation of coastal erosion and accretion process along the South West coast of Kanyakumari, Tamil nadu using Geospatial techniques. Arab. J. Geosci. https://doi.org/10.1007/s1251701312167.

Kaliraj, S., Chandrasekar, N., Magesh, N.S., 2014. Impacts of wave energy and littoral currents on shoreline erosion/accretion along the south west coast of Kanyakumari, Tamil Nadu using DSAS and geospatial technology. Environ. Earth Sci. 71 (10), 4523–4542.

Kaliraj, S., Chandrasekar, N., Ramachandran, K.K., 2017a. Mapping of coastal landforms and volumetric change analysis in the south west coast of Kanyakumari, South India using remote sensing and GIS techniques. Egypt. J. Remote Sens. & Space Sci. 20 (2), 265–282. https://doi.org/10.1016/j.ejrs.2016.12.006.

Kaliraj, S., Chandrasekar, N., Ramachandran, K.K., Srinivas, Y., Saravanan, S., 2017b. Coastal landuse and land cover change and transformations of Kanyakumari coast, India using remote sensing and GIS. Egypt. J. Remote Sens. & Space Sci. https://doi.org/10.1016/j.ejrs.2017.04.003.

Kaliraj, S., Chandrasekar, N., Simon Peter, T., Selvakumar, S., Magesh, N.S., 2015. Mapping of coastal aquifer vulnerable zone in the south west coast of Kanyakumari, South India, using GIS based DRASTIC model. Environ. Monit. Assess. https://doi.org/10.1007/s1066101440732.

Kantamaneni, K., Du, X., Aher, S., Singh, R.M., 2017. Building Blocks: a quantitative approach for evaluating coastal vulnerability. Water 9, 905.

Krishnamoorthy, R., Bharathi, G.S., Periakali, P., Ramachandran, S., 2002. Coastal zone hazards in India: study based on remote sensing and GIS techniques. In: Observing Our Environment from Space: New Solutions for a New Millennium, Begni (Edi), 905809 2542.

Kumar, A., Jayappa, K.S., 2009. Long and short term shoreline changes along Mangalore coast, India. Int. J. Environ. Res. 3 (2), 177–188.

Kunte, P.D., Wagle, B.G., 2001. Littoral transport studies along west coast of India a review. Indian J. Mar. Sci. 30, 57–64.

Lane, S.N., Westaway, R.M., Hicks, D.M., 2003. Estimation of erosion and deposition volumes in a large, gravelbed, braided river using synoptic remote sensing. Earth Surf. Process. Landforms 28, 249¬–271.

Li, K., Li, G.S., 2011. Vulnerability assessment of storm surges in the coastal area of Guangdong Province. Nat. Hazards Earth Syst. Sci. 11, 2003–2010.

Loinenak, F., Hartoko, A., Muskananfola, M.R., 2015. Mapping of coastal vulnerability using the coastal vulnerability index and geographic information system. Int. J. Technol. 5, 819–827.

Loveson, V.J., Rajamanickam, G.V., 1998. Coastal geomorphology of southern Tamil Nadu. In: Bhanu, S. K., Jha, V.K. (Eds.), Proc. Of National Seminar on Remote Sensing in Land Transformation and Publ. NRSA. Hyderabad, pp. 38–46.

Loveson, V.J., Rajamanickam, G.V., Anbarasu, K., 1990. Remote sensing application in the study of sea level variation along the Tamilnadu coast, India. In: Rajamanickam, G.V. (Ed.), Sea Level Variation and its Impacts on Coastal Environments. Tamil University, Tanjavur, pp. 179–196.

Marfai, M.A., Almohammad, H., Dey, S., Susanto, B., King, L., 2008. Coastal dynamic and shoreline mapping: multi-sources spatial data analysis in Semarang Indonesia. Environ. Monit. Assess. 142, 297–308.

Mishra, P., Patra, S.K., Ramana Murthy, M.V., Mohanty, P.K., Panda, U.S., 2011. Interaction of monsoonal wave, current and tide near Gopalpur, east coast of India, and their impact on beach profile: a case study. Nat. Hazards 59, 1145–1159.

Misra, A., Balaji, R., 2015. Decadal changes in the land use/land cover and shoreline along the coastal districts of southern Gujarat, India. Environ. Monit. Assess. 187–461.

Mohanty, P.K., Patra, S.K., Bramha, S., Seth, B., Pradhan, U.K., Behera, B., Mishra, P., Panda, U.S., 2011. Impact of groins on beach morphology: a case study near Gopalpur Port, east coast of India. J. Coast Res. 28 (1), 132–142.

Mohd Fauzi, M., Lee Hin, L., Mohd Kamarul Huda, S., 2014. Coastal vulnerability assessment towards sustainable management of Peninsular Malaysia Coastline. Int. J. Environ. Sustain Dev. 5, 533–538.

Mujabar, P.S., Chandrasekar, N., 2011. A Shoreline change analysis along the coast between Kanyakumari and Tuticorin, India using remote sensing and GIS. Arab. J. Geosci. https://doi.org/10.1007/s1251701103944.

Murthy, K.S.R., Subrahmanyam, A.S., Murty, G.P.S., Sarma, K.V.L.N.S., Subrahmanyam, V., Rao, K.M., Rani, P.S., Anuradha, A., Adilakshmi, B., Sri Devi, T., 2006. Factors guiding tsunami surge at the Nagapattinam–Cuddalore shelf, Tamil Nadu, East Coast of India. Curr. Sci. 90 (11), 1535–1538.

Murty, T.S., Flather, R.A., 1994. Impact of storm surges in the Bay of Bengal. J. Coast Res. 2, 149–161.

Murty, C.S., Varadachari, V.V.R., 1980. Topographic changes of the beach at Valiyathura, Kerala. Indian J. Mar. Sci. 9, 31–34.

Narayana, A.C., Jayappa, K.S., 2009. Mineral Resources of Coastal Tracts of India. Coastal Environments: Problems and Perspectives. I.K. International Publishing House Pvt. Ltd., New Delhi, pp. 280–290.

Palmer, B.J., Van der Elst, R., Mackay, F., Mather, A.A., Smith, A.M., Bundy, S.C., Thackeray, Z., Leuci, R., Parak, O., 2011. Preliminary coastal vulnerability assessment for KwaZulu-Natal, South Africa. J. Coast Res. 64, 1390–1395.

Pendleton, E.A., Thieler, E.R., Williams, S.J., 2005. Coastal Vulnerability Assessment of War in the Pacific National Historical Park to Sea-Level Rise. U.S. Geological Survey Open-File Report 2005a-1056. https://pubs.usgs.gov/of/2005/1056/.

Pirazzoli, P.A., 1996. Sea level Changes: The Last 20,000 Years. Wiley, Chichester.

Pradeep Kumar, A., Thakur, N.K., 2007. Role of bathymetry in tsunami devastation along the east coast of India. Curr. Sci. 92 (4), 432–434.

Ramana Murthy, M.V., Arjun, S., Shamji, V.R., 2009. Wind waves and sediment transport regime off the south central Kerala coast, India. Nat. Hazards 49, 325–345.

Rao, D., 2002. Remote sensing application in geomorphology. Trop. Ecol. 43, 49–59.

Rao, P.P., Nair, M., Raju, D., 1985. Assessment of the role of remote sensing techniques in monitoring shoreline changes: a case study of the Kerala coast. Int. J. Remote Sens. 6, 549–558.

Rejith, R.G., Sundararajan, M., Kaliraj, S., 2018. A Geo-processing modeling of deltaic suspended sediment variability. J. Geovisualization & Spat. Anal. 2 (13). https://doi.org/10.1007/s41651-018-0020-z. Springer.

Richards, J.A., Jia, X., 2006. Remote Sensing Digital Image Analysis: An Introduction. Springer Heidelberg, New York, pp. 247–268.

Saxena, S., Geethalakshmi, V., Lakshmanan, A., 2013. Development of habitation vulnerability assessment framework for coastal hazards: Cuddalore coast in Tamil Nadu, India a case study. Weather & Clim. Extremes 2, 48–57.

Shahul Hameed, T.S., Kurian, N.P., Thomas, K.V., Rajith, K., Prakash, T.N., 2007. Wave and current regime off the southwest coast of India. J. Coast Res. 23 (5), 1167–1174.

Sowmya, K., Jayappa, K.S., 2016. Environmental sensitivity mapping of the coast of Karnataka, west coast of India. Ocean Coast Manag. 121, 70–87.

Srinivas, R., Sajan, K., 2010. Significance of textural analysis in the sediments of Kayamkulam lake, southwest coast of India. Indian J. Geo Mar. Sci. 39 (1), 92–99.

Srinivasa Kumar, T., Mahendra, R.S., Nayak, S., Radhakrishnan, K., Sahu, K.C., 2010. Coastal vulnerability assessment for Orissa state, East Coast of India. J. Coast Res. 26, 523–534.

Tanh, P.T.T., Furukawa, M., 2007. Impact of sea level rise on coastal zone of Vietnam. Bull. Fac. Sci. Univ. Ryukyus 84, 45–59.

Thieler, E.R., Hammer-Klose, E.S., 1999. National Assessment of Coastal Vulnerability to Sea-Level Rise, U.S. Atlantic Coast. US Geological Survey, Open-File Report 99–593. http://pubs.usgs.gov/of/1999/of99-593/index.html.

Thieler, E.R., Hammar-Klose, E.S., 2000. National Assessment of Coastal Vulnerability to Sea-level Rise: Preliminary Results for the US Gulf of Mexico Coast. Geological Survey Open-File Report; U.S. Geological Survey: Woods Hole, MA, USA.

Thieler, E.R., Hammer-Klose, E.S., 2001. National Assessment of Coastal Vulnerability to Sea-Level Rise. Preliminary Results for the US Pacific Coast; Open File Report 00–178. United States Geological Survey (USGS), Woods Hole, MA, USA, 2000; 1p. https://www2.usgs.gov/science/cite-view.php?cite=1442.

Tripathy, B.R., Kaliraj, S., Sajjad, H., Joshi, P.K., Chaudhary, B.S., Kumar, P., 2018. Monitoring of seasonal variability and movement of suspended sediment concentrations along the Thiruvananthapuram coast, southern India, using the Landsat OLI sensor. Ocean Sci. 14, 1085–1092. https://doi.org/10.5194/os-14-1085-2018.

UNEP (United Nations Environment Programme), 2007. Physical Alteration and Destruction of Habitats. https://www.unep.org.

Unnikrishnan, A.S., Rup Kumar, K., Fernandes, S.E., et al., 2006. Sea level changes along the Indian coast: observations and projections. Curr. Sci. 90, 362–368.

Vousdoukas, M.I., Mentaschi, L., Voukouvalas, E., Verlaan, M., Jevrejeva, S., Jackson, L.P., Feyen, L., 2018. Global probabilistic projections of extreme sea levels show intensification of coastal flood hazard. Nat. Commun. 1–12.

Further reading

López Royo, M., Ranasinghe, R., Jiménez, J.A., 2016. A rapid, low-cost approach to coastal vulnerability assessment at a national level. J. Coast Res. 32, 932–945.

19

An investigation of a credible strategy for coral reef bleaching and its management using a geospatial approach for the Gulf of Kutch

Meenu Rani[1], Ram Kumar Singh[2], Kalita Himangshu[3], B.S. Chaudhary[4], Pavan Kumar[5]

[1]DEPARTMENT OF GEOGRAPHY, KUMAUN UNIVERSITY, NAINITAL, UTTARAKHAND, INDIA; [2]DEPARTMENT OF NATURAL RESOURCES, TERI SCHOOL OF ADVANCED STUDIES, NEW DELHI, INDIA; [3]HARYANA SPACE APPLICATIONS CENTRE (HARSAC), (DEPARTMENT OF SCIENCE & TECHNOLOGY, HARYANA) CCS HAU CAMPUS, HISAR, HARYANA, INDIA; [4]DEPARTMENT OF GEOPHYSICS, KURUKSHETRA UNIVERSITY, KURUKSHETRA, HARYANA, INDIA; [5]COLLEGE OF HORTICULTURE AND FORESTRY, RANI LAKSHMI BAI CENTRAL AGRICULTURAL UNIVERSITY, JHANSI, UTTAR PRADESH, INDIA

Chapter outline

Remote Sensing of Ocean and Coastal Environments. https://doi.org/10.1016/B978-0-12-819604-5.00019-6

1. Introduction

In the variety of life in the marine ecosystem, the most splendid creation is the coral reef, formed primarily by calcium carbonate excreted by coral organisms. Colonies comprising tiny marine elements and sparse nutrients lead to the formation of coral reefs in the marine ecosystem (Elvidge et al., 2004). The most recent phenomenon threatening coral reefs worldwide is the loss of *Symbiodinium*. This is an intracellular endosymbiont, also known as zooxanthellae, the loss of which leads to coral bleaching (Smirnov et al., 2002). Wilkinson (2008) reported that 19% of coral reefs have been lost and 75% are threatened. Among all the coastal states of India, Gujarat is stated to have the largest coastline, i.e., 1600 km, along which the Gulf of Kutch has a length of 170 km covering an area of around 7300 km^2. The Gulf of Kutch is rich in biodiversity like few coastal zones of the world, presenting vast tidal mudflats with mangroves and coral making it as well a preferable zone for breeding of marine organisms. The combined effects of natural and anthropogenic pressures operating at both regional and global scales pose a major threat to coral reefs (Burke et al., 2011). Exploitation of coastal natural resources due to coastal development, tourism, coastal pollution, pisciculture, and runoff is responsible for the deterioration of the coral reef (Bruno et al., 2007). High-level atmospheric carbon dioxide and the breakout of global warming, one leading to the other, have contributed to the deterioration of ecosystems (Stow, 1999).

A huge number of coral reef monitoring programs by several agencies are ongoing, with field survey as well as data monitoring, and are available on respective sites. Some such coral monitoring events are Reef Check (www.reefcheck.org) and CoralWatch (www.coralwatch.org) (Hodgen, 2001). There are other programs, like Coral Reef Degradation in the Indian Ocean (CORDIO) (Obura, 2008) (www.cordioea.net), the Caribbean Coastal Marine Productivity Program (CARICOMP) (CARICOMP, 1999), and the Atlantic and Gulf Rapid Reef Assessment (AGRRA) program (http://www.agrra.org). The Global Coral Reef Monitoring Network provides biennial reports of the worldwide status of coral reefs (Wilkinson, 1998, 2000, 2002, 2004, 2008). Coral reef monitoring provides long-term trends in coral health and status, but at the same time metaanalysis using different data sources is a more tedious task and the use of these data sources is restricted due to higher surveying costs, map scale issues, and unsustainability of biological conditions if only various parameters related to coral reefs are used (Hasegawa et al., 1998; Hoegh-Gulberg, 1999; Edmunds et al., 1996; Hughes et al., 1999; Hochberg et al., 2003; Haltrin, 1999). Due to this, only scattered information has been collected by the coral reef monitoring system.

In addition to misinformation due to unavailability of identification data along with issues of inconsistency and in the data collection process, the further process of coral reef detection is being curtailed as global reach is defining an attenuated protocol for monitoring coral reefs (Ahmed et al., 1994; Bukata et al., 1995). The most effective and comprehensive surveys done by well-planned programs, such as the Australian Institute of Marine Science, struggle with the fact that field data surveys cover only a fraction of the entire work, which has to be done on a huge scale of kilometers (De'ath et al., 2012; Charpy et al., 1997). Some ongoing surveys such as autonomous camera surveys have provided details, for example, the Catlin Seaview Survey (Gonzalez-Rivero et al., 2014; Clark et al., 2000) combines with some computed automated image analysis for field survey to add detail to coral analysis (Beijbom et al., 2012). Some approaches are achieved after various standardized examinations, one of which is the XL Catlin Seaview Survey (Vermote et al., 1997). This is being operated (as of this writing) as a diver-based field survey, and researchers also use low-altitude unmanned vehicles for aerial surveys. Such kinds of techniques as those on the XL Catlin Seaview Survey are economically reasonable for humans performing survey analyses. Due to limited temporal repeatability and undeveloped physical conditions, monitoring such large distributions of coral reefs is still unreachable for human beings (Akatsuka, 2001).

Various techniques are being used to determine coral bleaching and are likely to detect it, with the caveats as previously mentioned (Mumby et al., 1998b). These techniques are spaceborne or airborne, large unscalable aerial survey, underwater survey, boat-based techniques, and large coverage by remote-sensing technologies. The sampling of physical areas achievable by field inspections is very cost effective (Jupiter et al., 2013) in contrast to habitat mapping and environmental pressure assessment by remote sensing, especially by spaceborne satellites. Mathematically, some large-scale patterns of complete coral coverage can be mapped, and the composition of such coral presented, but satellites inherently cannot accurately describe what is below the level of field survey or provide detailed underwater analysis (Andréfouët et al., 2002). But a huge algorithm based on statistical formulas and mapping through a coded form of live corals was previously used (Mumby et al., 2004). This was accompanied by a huge 6-day-long survey and acquisition of a solitary satellite. Actually, data acquired from an airborne survey are more valuable than a spaceborne data set. This is even potentially most cost effective, and a low acquisition cost is applicable. So collecting data in a valid form is more important when watching any coral reef events (Mumby et al., 1999).

Using satellite data on behalf of the environment and reef systems, some coral watch programs, like the US National Oceanic and Atmospheric Administration's (NOAA) Coral

Reef Watch program, observe systems such as reef temperature and sea surface temperature (SST), etc., which are some of the most targeting features in coral reef monitoring (Glynn, 1993). High-frequency observation of the spatial and temporal patterns of coral bleaching events is extensively assisted by remote-sensing data (Green et al., 1996). Several outstanding data sets, like characterization of a reef's biological condition and determining physical conditions, are obtained by combining remote-sensing data with some ecological factors acting on the specific area where the coral reef study is going on. An emphasizing situation for the base of coral reef monitoring activity on the surface of sea height over surface temperature change are being monitored and decadal change of coral bleaching may be impacted with it (Baker et al., 2008).

Biological responses are established to have some causative links and can be easily analyzed with stressors for potential conjunction in a coral reef watch (Hu et al., 2003). The present situation enables the easy assessment of conditions for bleaching by looking at some historical trends, allowing remote-sensing technologists a retrospective process and giving them real-time information on corals. The use of remote-sensing techniques allows one to initialize the data collection and helps scientists forecast conditions for bleaching; there is even a track-predictive model algorithm developed by a remote-sensing scientist for coral monitoring (Lirman and Biber, 2000). It is very easy now to evaluate the current state of the ecosystem with coral reef data synthesis from satellite imagery, and this is statistically dynamic for monitoring coral changes over time (Mumby et al., 2004a,b). Proper condition analysis with airborne system data collection and a sonar-based boat system is also potentially relevant for coral watch at present (Malthus et al., 2003). Using hyperspectral systems on airborne sensors can easily provide data that are cost effective over satellite data, and this may be more applicable in the future than satellite source use. Moreover, satellite data collection combined with a boat-based system with a sonar sensor is very good for this type of monitoring. Beyond basic maps of habitats or environmental variables, here we also consider concepts highly relevant to stakeholders, strategy architects, and community issues, such as biodiversity, environmental threats, and ecosystem services (Eakin et al., 2010). Between remote sensing, reef ecology, and reef preservation, this kind of work is the key to ongoing synergy (Lirman and Fong, 1997).

This chapter aims to identify coral reef bleaching and its impact on environmental parameters and to detect ecological diversity via remote-sensing techniques. This system of monitoring aims to identify the coral and provide a map of the coral reef in the study site. The progress and protection of the coral reef which helps in restoring the environment and how much generously be affecting the natural environment, and the effectiveness of this program are also discussed in this chapter.

2. Study area

The Gulf of Kutch is an indentation between the Saurashtra and the Kutch peninsulas. The Gulf of Kutch (22°15′ to 23°40′ N and 68°20′ to 70°40′ E) is an east–west-oriented indentation and bordered by the Kutch district on the north side. It is also headed by the Jamnagar district in south and a little portion is surrounded by the Rajkot district. The gulf is situated on Gujarat State in India. Declared by the government of Gujarat in 1982 as a marine national park, the Gulf of Kutch covers an area of about 457.92 km^2 as a marine sanctuary and 162.89 km^2 as the marine national park (Environment Agency, 1996, 1999, 2000). The variety of flora and fauna in the Gulf of Kutch is the main attraction for tourists in the marine national park. This is the first national marine park in India, and was established in 1982 under the provisions of the Wildlife (Protection) Act, 1972, of India. The geographical location and map of the study area are presented in Fig. 19.1.

2.1 Marine eco-biodiversity

The Gulf of Kutch Marine National Park, which is situated on the Jamnagar coast of Gujarat, has 42 islands and most of them are encircled by reefs. Considered an ecological

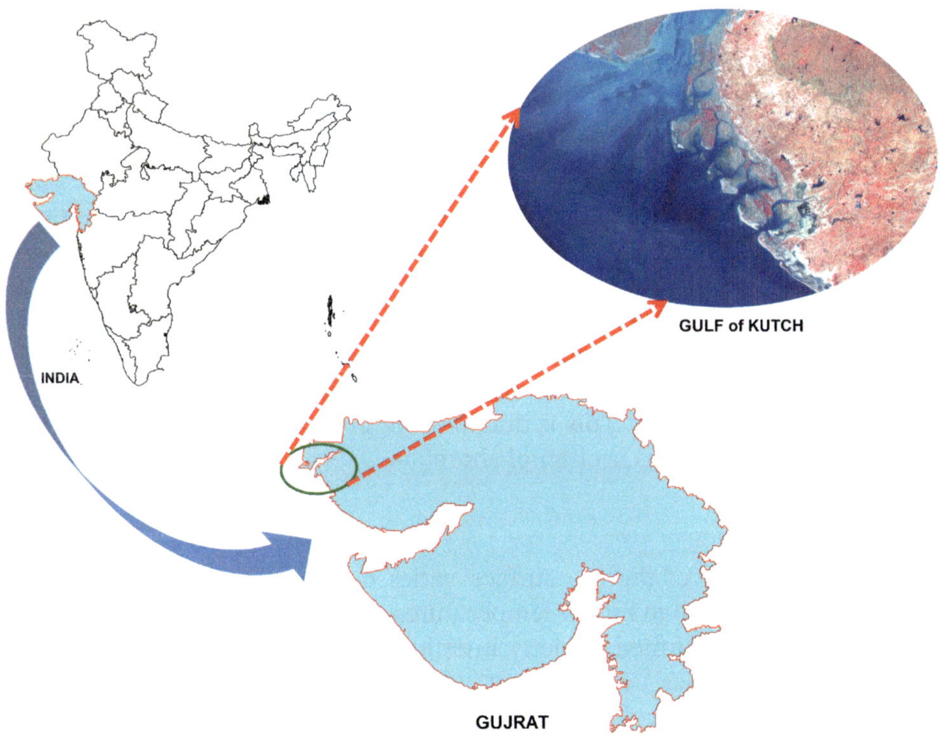

FIGURE 19.1 Geographic real-time location of the Gulf of Kutch.

hub of marine eco-biodiversity, the marine national park is surrounded by many major ecosystems, and this makes a unique piece for marine scientists to study in the Gulf of Kutch (Lubin et al., 2001). The most important ecosystem categories there are coral reefs, mudflats, mangroves, creeks, estuaries, seagrass and seaweeds, marshy areas, sandy strands to saline grasslands, and rocky shores (Mumbyet al., 2002).

2.2 Marine biology

A marine ecosystem is high in marine life and bird life. The Gulf of Kutch Marine National Park is such an area, reserved for the development of wildlife and an eco-sanctuary. In the park severe activity like forestry, hunting, poaching, grazing, cultivation, and development of human activities is not permitted. The Gulf of Kutch Marine National Park has a clearly marked boundary and well-developed circumscription.

2.3 Coastal activity

The tidal forms in the Gulf of Kutch are high and form a jetty condition with the salt water covering most of the region of the Gulf of Kutch Marine National Park. Small islands host a lot of mammalian (warm-blooded vertebrate animals) activity along with migratory birds and help to increase the fertility of these undersea angels. The Gulf of Kutch region forms the largest coastal habitat on the west coast of India. It has a generally shallow water level, with depths extending to 60 m at the head, and at the mouth to around 20 m, forming an average depth of only 30 m. So, using remote-sensing technology can be very helpful to monitor the reefs in the Gulf of Kutch.

2.4 Physiological conditions

The Gulf of Kutch Marine National Park experiences semidiurnal tidal conditions and large diurnal tidal conditions equally. These are the conditions mainly because of the complex undulation and rough topography of the sea bottom and a very multifaceted bathymetry. This develops the nonlinear tidal interaction with the shoreline and presents a smooth physiological condition for wildlife (Mumby et al., 2002). The tidal waves come from the western side and the amplitude of the tide increases with increase in distance upstream of Vadinar. This is due to some geographical conditions of the inner region and a narrowing cross section of shoreline.

2.5 Water quality

The water temperature of the sea surface varies between 20 and 30°C, even though localized environments go to higher temperatures, up to 35°C, which condition is due to isolated water puddles formed in petty intertidal depressions during small tides. This tidal-driven mixing causes no other vertical stratification in the ecological profile of salinity or temperature of the sea surface.

2.6 Sediment quality

Mainly divided into five parts, the central part of the Gulf of Kutch is rocky, with sediments being partial to the borders. A sediment quality of light gray silt, fines, and clay sand with patches of coarse sand is found in other sides. These sides are poorly arranged with erratic asymmetry of the sediments. The Indus River transports major sediments, which is the reason behind the increasing shore material in the Gulf of Kutch. Also, sediment condensation is very high due to low runoff in the sea bottom and reefs. Moreover, the discharging streams into the Gulf of Kutch are lower the lower the pitch.

3. Materials and methodology

3.1 IRS Linear Imaging Self-Scanning Sensor III data

With four multispectral cameras operating in four different bands, of which three are visible and NIR (near infrared) and the other is a short-wave infrared (SWIR) camera, the Linear Imaging Self-Scanning Sensor III (LISS-III) is an optical case for the IRS 1C/1D. All four bands have a resolution of 23.5 m. The newly featured LISS III camera consists of an SWIR band, which has a pixel stability of 1.55–1.7 μm and also provides data with spatial resolution, unlike the IRS 1C/1D. These data products are classified under standard and systematic accuracy level of study. The LISS III products are satellite images consisting of path and row–based, georeferenced, and quadrant products. These products come from the schematic design of each sensor from the LISS III camera. The along-track images from LISS III data are user-interested images that fall under two successive schematic scenes for the same path (Yamano et al., 2003). They are accessed via the along-track direction and data are supplied schematically for the LISS III satellite. LISS III full scenes are classified into four nominal and eight classes of derived quadrants. The scale for the LISS III camera is 1:125,000. All these products are truly north-oriented georeferenced. The data can be made available in the user site from Indian Space Research Organisation data set. The covered area in the imagery data for three specific bands is characterized by three different bands along the systematic path and row available. All LISS III products are path and row based with radiometrically and georeferenced enabled corrected image (Yu et al., 2003). The quadrant products cover an area of 70 × 70 km forming three or four bands.

3.2 Data processing

Using a geographic information system (GIS) environment to adopt the assessment of the coral reef in the Gulf of Kutch the data have been compiled with a space–time model of a spectral algorithm. The remotely sensed data of the IRS P-6 for 1989, 1999, and 2009 were compared. The primary advantage of using these IRS LISS III data is that they allow the local study at a residential level without any expensive requirements and consuming the least amount of time. This is used to mediate the historical data set and is reliable for accessing the data set for further study (Yamano et al., 2002). The IRS P-6 images and aerial level imagery represent a various but huge untapped data source to signify the entire environmental map. Also, the modeling application is usable with this kind of data set for monitoring time series processes. This three-decadal source of data can be characterized for GIS models to re-create contaminant expulsions in conjunction with remote-sensing techniques. The main aim of this study was to discover how remotely sensed data can be used via a GIS environment to study the connection between coral reef and sea surface eco-biodiversity. This is a comprehensive study using the satellite imagery archives and shows the strength of remote-sensing techniques to view the pros and cons of the coral reef in the Gulf of Kutch. This will develop into an oceanographic study focusing primarily on the chemistry of marine ecology and however within our reach of assessments (Singh et al., 2020).

3.3 Supervised classification

Supervised classification was performed on the three-decadal data, i.e., for the years 1989, 1999, and 2009, with maximum likelihood classifier to attempt to find the exact result-oriented bottom-level class (Myres et al., 1999; Yamano et al., 2000). The maximum likelihood algorithm assumes the training data set and their statistical regulation and fits it approximately with sample size and pixel counts. Here only the statistical characterization is approximated, followed by the conclusion of imparted pixel fitting sample size formation (Vanderstraete et al., 2004). Maximum likelihood classification technique is a practical allocation of bottom-type classification with the depth invariance index, and a training set area is used as the referenced data set (Lyzenga, 1981; Mumby et al., 1998). The following function is computed for each pixel while performing the maximum likelihood classifier operation and in this research we followed this predetermined equation for each pixel count:

$$\text{Pixel count} = gi(x) - \ln \rho(\omega_i) - \frac{1}{2}\ln\sum i - \frac{1}{2}(x - m_i)\sum_i^t(-1)(x - m_i) \tag{19.1}$$

where i is the class of pixel sorted, x is multidimensional data (probably n-dimensional), $\rho\ (\omega_i)$ is the probability class of the covariance matrix of data, $\sum i$ is the determinant of the covariance of data class, and m_i is the mean vector.

3.4 Sea surface temperature

NOAA has been measuring SST routinely since 1970. This involved many efficient systems up to the mid-1980s. This process has been used to analyze the SST, which is a fundamental environmental constraint for coral reefs. But since 1985, there has been a revolution in this field as daily SST measurement has been going on. Some post-processing techniques were performed for this data set by a student in the Advanced Very High Resolution Radiometer (AVHRR) Pathfinder SST project, which have provided consistent global SST data from 1985 to the present (Guest et al., 2012). NOAA has taken advantage of this data set, developing SST climatic parameters that work for the sea surface and show variations in SST. Initially it was thought that SST activity was not involved in coral bleaching nor that it affected the coral ecology. But natural hot spot anomalies and then the degree of heating showed SST differences attributable to coral reef (Yamano et al., 2002). A typical climatological SST irregularity does not establish the coral bleaching ecological defect. This is can be easily shown with the help of climatic factors in the seasonally hottest month compared with the coldest month. That SST climatology was affecting coral bleaching was derived by the Polar-orbiting Operational Environmental Satellite (POES) AVHRR. SST is not dynamic in time, but varies in space (Strong et al., 1997).

3.5 Radiance measurement

The radiance level was measured and attenuated with upwelling, as shown by this study when it was performed using satellite data processing (Yamano et al., 2001). The spectral scan recording algorithm was measured, while cross verification was underprocessed with an error matrix of supervised classification data. The spectral receptor of the 180-degree cosine vector was spectrally oriented in the 350 and 750 nm range and then a detailed analysis was performed of the coral reef formation in the Gulf of Kutch. The overview of the coral reefs was intensively analyzed and details of the survey is radiated form satellite imagery for the coral reef distribution with algae and sea grass detection. It was found that some specific reef areas had a high proportion of strong coral reef covering around 60% of the normal substrate (Environment Agency, 1996). Mass bleaching incidents (Wilkinson, 1998, 2000, 2002, 2004, 2008) in the Gulf of Kutch's coral reef have caused significant loss of healthy and strong corals (Kayanne et al., 2005). From this entire current study, a detailed map will be the final outcome based on the radiance scale.

3.6 Multispectral imagery

This location's position was accessed by a GPS satellite with a spectroradiometer with an accuracy of 10 m. Minimum surveyor's image processing practically investigated the coral coverage within an area of 50 m at each site. IRS P-6 imagery having a spectral resolution of 23.5 m was later compared with the survey data. Most interestingly, the signs of highest coral bleaching for the Gulf of Kutch were reported in the summer of 1989, prior to 1999, when roughly 60% of corals were reported to have been bleached (Thomes et al., 1997). This event in 1989 was previously unknown. A very high SST was shown to have occurred in 2009 and resulted in a high level of coral bleaching (Strong et al., 1997, 2002). But this event caused comparatively low bleaching and the amount of dead coral was lower at that time.

3.7 Pixel-based integration of live coral reef

The shallow essence of coral reefs at a depth of 1.2 m were used as an area of interest (AOI) for the designated event (Andréfouët et al., 2001). The dark sections of the coral show lighter patches indicating the bleaching event. The two main AOIs having a small center and heads to the north were roughly the same as the bleaching size. Located farther to the south, the third area was close to twice the size of the first two. For the IRS P-6 imagery, the demarcated areas for pixel integration could be easily seen as a set of black pixels within the AOI.

4. Results and discussion

4.1 Irradiation of vector data cloning

Previously, in a study by Yamano and Tamura (2002), it was shown that the reflectance of corals showed larger values and this was detected very easily with the likelihood classifier method in the satellite data of the Landsat Thematic Mapper (TM) or at higher resolutions (Holden and LeDrew, 1998). The differences in digital number (DN) values before and after the bleaching event in the years of study, i.e., 1989, 1999, and 2009, were collected, and the percentage of bleached corals in a TM pixel-based classification is shown. This difference in pixel count shows how healthy corals are degrading annually. As previously described, the coral bleaching is detectable with the pixel difference measurement in DN values and this minimizes the error count due to the use of 23.5 m IRS P-6 sensors. In this study, it was seen that initially, in the case of severe bleaching, an abundance of healthy corals was found, and maximum detection was enabled by using a combination of bands 1, 2, and 3. Actually, some initially covered areas were found to be 100% bleached in the Gulf of Kutch. The DN value differences in coral pixel count were compared, and after analyzing with different band consideration, it was found that for IRS LISS III (P-6) sensors, bands 1, 2, and 3 could detect at a difference rate of 9%, 15%, and 76% for the bleaching of coral reefs in the Gulf of Kutch.

4.2 Satellite data processing

For temporal changes, to find the difference between coral reefs in the interval of 1989–2009, we processed the IRS LISS III data in a normalization process. The differences between these values were slightly higher in some cases, and in nonbleached areas they were most stable for 1999 to 2009. Other satellite sensors like Landsat could not have detected this over more than 15% of the total bleach. But due to the maximum formation of bleaching between 1989 and 1999, the further IRS P-6 satellite was used. In this study, we conventionally stabilized the satellite data for preprocessing and post-processing to a minimum requirement for the correctness standard indices. The overall data set for stacking images contains some band-specific data, and this is highly observed for 1999 in the case of bands 1 and 2. The classification of ecological parameters of the Gulf of Kutch is tabulated follows. The present study indicates that the mass bleaching event in 1999 produced around 35%–45% bleached coral reef cover, and this is easily detectable by using IRS P-6 bands 1 and 2. After 1999, the band values for the data set of 2009 reflect a low progress of bleaching and coral cover detection. The facts are that the coral reef's reflected values showing the bleached portion progressively increased in 1989, then became huge in 1999 and were lower in 2009 (Fig. 19.2). For analysis of the satellite set, the data were categorized into band-wise results and the highest DN value interaction was found in bands 1 and 2.

4.3 Image correction and reformation

The Gulf of Kutch is a marine ecological ghat. This system has a sea bottom of various depths and this is suitable for the supervised method of classification. Satellite imagery of the Gulf of Kutch coral reef biosystem contains various noises. These are removed using a band-specific pixel count matrix before proceeding to supervised classification. The atmosphere and light-scattering noise is unveiled by the radiance removal method (Lyzenga's method) (Brodie et al., 1997). The algorithm has considerable correction methods for the noise specially based for IRS LISS III sensors. The radiance coming from sand in shallow water does not follow the presumption of radiance present in the NIR band that is totally absorbed by water (Kirk, 1994). Actually, the radiance emitted by sand to any NIR band is comparatively higher than any absorption from the water body.

4.4 Classification and statistical tabulation

In the present research area abundant forms of bottom types with shallow coral reefs are found along with sand, seagrass, rubble, and coral reefs (live and degrading). All these features most abundantly have spectral characteristics and they can be distinguished from one another with pixel classes showing a homogeneous spatial system (Kobayashi, 2004). These pixel classes will form a 5×5 m categorization. Moreover, some tough features that combine to a mixed bottom type of category define a nonseparable field category for the coral reef in the Gulf of Kutch. When performing supervised classification with maximum likelihood classifier, a training class of seven bottom surface point references is created. Based on adjustments in the invariance system of ecological category, using Lyzenga's method in the maximum likelihood classifier, pixel categorization was performed and three different maps of the Gulf of Kutch for the years 1989, 1999, and 2009 (Fig. 19.2) are shown (Lyzenga, 1978, 1981).

In Fig. 19.2, based on the combined invariance and depth of maximum likelihood system, these three decadal points of ecological characteristics are visually interpreted. The band combination of IRS P-6 for the band integral index value is normal here. This type of classification generally shows the ecological classes, with some misclassification for rubble and seagrass type, but categories are easily created for low turbid water, moderate turbid water, highly turbid water, dense mudflats, mudflats, dense mangrove, sparse mangrove, and salt-affected land. The combination of depth invariance value

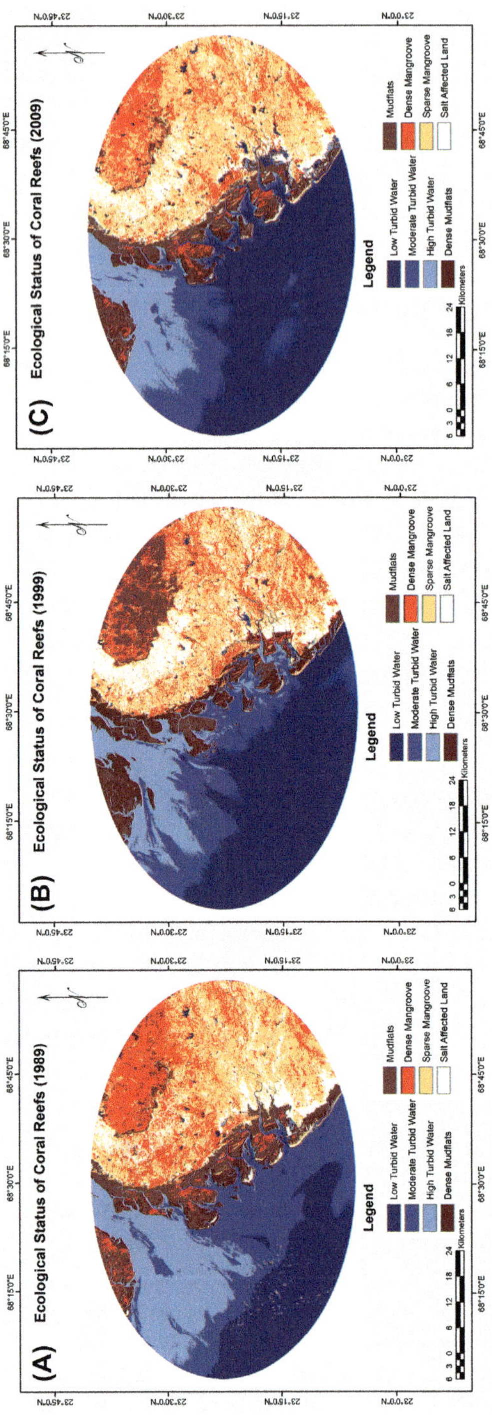

FIGURE 19.2 (A–C) Ecological parameters in the Gulf of Kutch during the coral watch (1989, 1999, and 2009).

Table19.1 Statistical measurement of ecological parameters in hundredth scale using Lyzenga's method.

Ecological Parameters	1989	1999	2009
Low Turbid Water	21%	18%	14%
Moderate Turbid Water	14%	21%	20%
High Turbid Water	10%	10%	16%
Dense Mudflats	9%	13%	16%
Mudflats	7%	10%	8%
Dense Mangrove	12%	8%	10%
Sparse Mangrove	11%	11%	8%
Salt Affected Land	6%	9%	8%

showing the index for three visible bands for IRS P-6 is separated for this categorization. Seagrass and rubble mudflats are standard systemic in mudflats. Table 19.1 shows the class categorization for the three years 1989, 1999, and 2009 with classes of ecological parameters showing overall affected area using Lyzenga's method (Lyzenga, 1978, 1981; Matsunaga et al., 2000).

4.5 Percentage of coral cover display

All three categorized images of data state that an equal trend of pixel loss is found for coral bleaching in each category of ecological parameters, while some coral-dominated pixels used in remote-sensing techniques provide coral cover in situ data and colony cover ex situ data. The coral covering the mudflats class is an esteemed class of coral reef cover found to produce AOIs, and this AOI declined from 28% in 1999 to 14% in 2009 as shown by the IRS LISS III images (Fig. 19.2). And this shows a total loss of 14%, half the previous coral reel cover in a 10-year period, and this is declining nearly 10%–15% per year (approximately 65.50 km^2). This is just a comparison of 10-year change, but in situ bottom level classes can be seen by using a heavy spectrometer survey and this will vary or increase. This coral change is expected to reach different levels in the data set in all areas. The main correlation between ecological parameters of the coral reef and SST was checked. The benthic change in a bathymetry study is always dependent on its eco-biodiversity, and the percentage coral versus year-wise data are shown (Fig. 19.3). Here it is confirmed that habitat loss is obvious here. The coral-covering mangrove and coral-dominated turbid water are distinctly classified with the band indices formation with bands 1, 2, and 3. The results are shown in Fig. 19.3 and their explanations are as follows:

1. The low turbid water ecosystem is the southernmost formation in the entire region.
2. The moderate turbid water pixels in the Gulf of Kutch remain concentric to everywhere.
3. The center zone is densely populated by high turbid water, mudflats, and sparse mangrove.

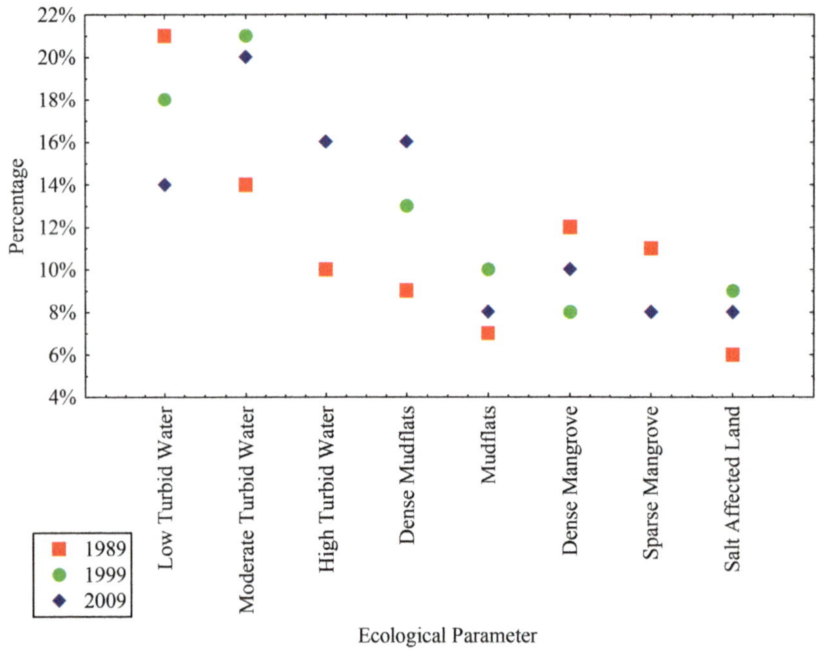

FIGURE 19.3 Statistical symbology showing coral percentage with time.

4. The northern area seems to have a huge coral reef found in the dense mangrove detected more highly in 1999 than in 2009 or 1989.
5. Overall loss of coral or coral bleach is found to vary at different times.

4.6 Crucial climate change and sea surface temperature

For any climatic change, temperature always plays a role in regulating the life habits. Here also, temperature is a crucial parameter in the regulation of coral reefs. Maximum SST for the working data from May to August is 28–31°C. The maximum sea surface variation is seen in the year 1989, followed by 2009 and then 1999. The Gulf of Kutch Marine National Park has two distinctive temperature classes, one is slightly shallow warm and the other is salty cool. The extensive classification of the resulting area for effective bleaching of coral reef has been surveyed on more than 50 training sets of data for SST measurement. The maximum depth that can be effective for coral reef due to SST is 10–20 m in the Gulf of Kutch Marine National Park. A huge amount of bleaching was recorded in the coral reef and foliose reef during 1999. The SST then lowered in 1999 and in 2009, and it seems with low bleaching. This temperature held from this year onward and the ecological system remained at normal level thereafter. Fig. 19.4 shows the graphical representation of SST in the Gulf of Kutch with the three sets of data. Due to this massive recovery, the observed marine ecology and coral behaviors remained steady from 1999 to 2009.

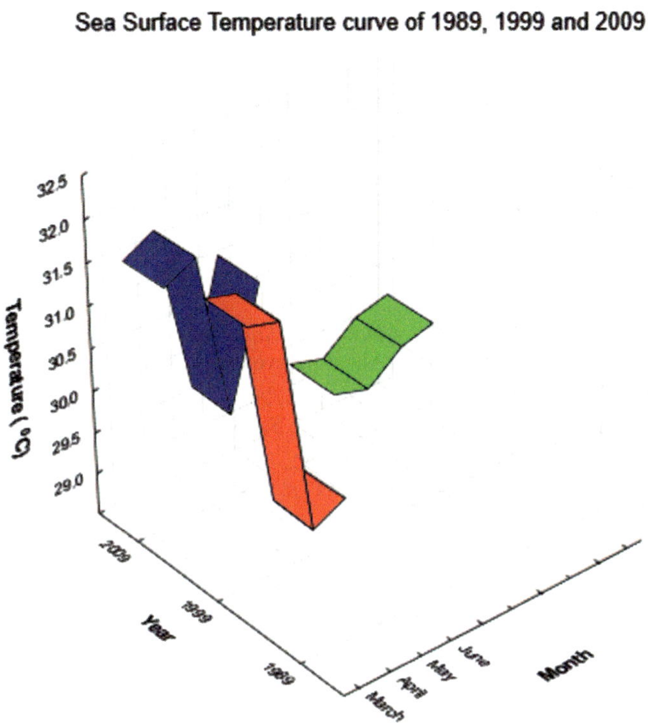

FIGURE 19.4 Kinetic relations among temperature, year and months.

References

Ahmad, W., Neil, D.T., 1994. An evaluation of Landsat Thematic Mapper (TM) digital data for discriminating coral reef zonation: heron Reef (GBR). Int. J. Rem. Sens. 15, 2583–2597.

Akatsuka, I., 2001. Seasonal and Yearly Changes of Seaweed Vegetation in Isigaki-Zima Island and Iriomote-Zima Island of Yaeyama Islands. WWF Japan-Funded Research Report 2000 (In Japanese).

Andréfouët, S., Muller-Karger, F.E., Hochberg, E.J., Hu, C., Carder, K.L., 2001. Change detection in shallow coral reef environments using Landsat 7 ETM+ data. Remote Sens. Environ. 78 (1–2), 150–162.

Andrefouet, S., Berkelmans, R., Odriozola, L., Done, T., Oliver, J., Müller-Karger, F., 2002. Choosing the appropriate spatial resolution for monitoring coral bleaching events using remote sensing. Coral Reefs 21 (2), 147–154.

Baker, A.C., Glynn, P.W., Riegl, B., 2008. Climate change and coral reef bleaching: an ecological assessment of long-term impacts, recovery trends and future outlook. Estuar. Coast Shelf Sci. 80, 435–471.

Beijbom, O., Edmunds, P.J., Kline, D.I., Mitchell, B.G., Kriegman, D., 2012. Automated annotation of coral reef survey images. In: Proceedings of the IEEE Conference on Computer Vision (CVPR), Providence, RI, USA, 16–21 June 2012.

Brodie, J.E., Furnas, M.J., Steven, A.D.L., Trott, L.A., Pantus, F., Wright, M., 1997. Monitoring chlorophyll in the great barrier reef lagoon: trends and variability. In: Proceedings of the 8th International Coral Reef Symposium, vol. 1, pp. 797–802.

Bruno, J.F., Selig, E.R., 2007. Regional decline of coral cover in the Indo-Pacific: timing, extent, and sub regional comparisons. PLoS One 2, e711.

Bukata, R.P., Jerome, J.H., Kondratyev, K.Y., Pozdnyakov, D.V., 1995. Optical Properties and Remote Sensing of Inland and Coastal Waters. CRC Press, Florida.

Burke, L., Reytar, K., Spalding, M., Perry, A., 2011. Reefs at Risk Revisited. World Resources Institute, Washington, DC, USA, p. 114.

CARICOMP, 1999. The Caribbean coastal marine productivity program (CARICOMP). Bull. Mar. Sci. 69, 819–829.

Charpy, L., Dufour, P., Garcia, N., 1997. Particulate organic matter in sixteen Tuamotu atoll lagoons (French Polynesia). Mar. Ecol. Prog. Ser. 151, 55–65.

Clark, C.D., Mumby, P.J., Chisholm, J.R.M., Jaubert, J., Andrefouet, S., 2000. Spectral discrimination of coral mortality states following a severe bleaching event. Int. J. Rem. Sens. 21, 2321–2327.

De'ath, G., Fabricius, K.E., Sweatman, H., Puotinen, M., 2012. The 27-year decline of coral cover on the Great Barrier Reef and its causes. Proc. Natl. Acad. Sci. U.S.A 109, 17995–17999.

Eakin, C.M., Nim, C.J., Brainard, R.E., Aubrecht, C., Elvidge, C., Gledhill, D.K., Muller-Karger, F., Mumby, P.J., Skirving, W.J., Strong, A.E., et al., 2010. Monitoring coral reefs from space. Oceanography 23, 118–133.

Edmunds, P.J., Bruno, J.F., 1996. The importance of sampling scale in ecology: kilometer-wide variation in coral reef communities. Mar. Ecol. Prog. Ser. 143, 165–171.

Elvidge, C.D., Dietz, R.J.B., Berklemans, R., Andréfouët, S., Skirving, W., Strong, A.E., Tuttle, B.T., 2004. Satellite observation of keppel islands (great barrier reef) 2002 coral bleaching using IKONOS data. Coral Reefs 23 (1), 123–132.

Environment Agency, 1996. Marine Biotic Survey (1989–1992) in the 4th National Survey on the Natural Environment: 1/10,000 Distribution Map of Coral Reefs.

Environment Agency, 1999.

Environment Agency, 2000. Report of Coral Reef Monitoring (Year of Heisei 11).

Glynn, P.W., 1993. Coral reef bleaching: ecological perspectives. Coral Reefs 12, 1–17.

Gonzalez-Rivero, M., Bongaerts, P., Beijbom, O., Pizarro, O., Friedman, A., Rodriguez-Ramirez, A., Upcroft, B., Laffoley, D., Kline, D., Vevers, R., et al., 2014. The Catlin Seaview Survey—kilometre-scale seascape assessment, and monitoring of coral reef ecosystems. Aquat. Conserv. Mar. Freshw. Ecosyst. 24, 184–198.

Green, E.P., Mumby, P.J., Edwards, A.J., Clark, C.D., 1996. A review of remote sensing for the assessment and management of tropical coastal resources. Coast. Manag. 24, 1–40.

Guest, J.R., Andrew, H., Baird, J.A., Maynard, E.M., Edwards, A.J., Campbell, S.J., Yewdall, K., Yang, A.A., Chou, L.M., 2012. Contrasting patterns of coral bleaching susceptibility in 2010 suggest an adaptive response to thermal stress. PLoS One 7 (3), e33353.

Haltrin, V.I., 1999. Chlorophyll-based model of seawater optical properties. Appl. Optic. 38, 6826–6832.

Hasegawa, H., Ichikawa, K., Kobayashi, M., Kobayashi, T., Hoshino, M., Mezaki, S., 1998. The mass-bleaching of coral reefs in the Ishigaki Lagoon, 1998. Galaxea, JCRS 1, 31–39 (in Japanese with English abstract).

Hochberg, E.J., Atkinson, M.J., Andréfouët, S., 2003. Spectral reflectance of coral reef bottom-types worldwide and implications for coral reef remote sensing. Rem. Sens. Environ. 85, 159–173.

Hodgson, G., 2001. Reef Check: the first step in community-based management. Bull. Mar. Sci. 69, 861–868.

Hoegh-Guldberg, O., 1999. Climate change, coral bleaching and the future of the world's coral reefs. Mar. Freshw. Res. 50, 839–866.

Holden, H., LeDrew, E., 1998. Spectral discrimination of healthy and non-healthy corals based on cluster analysis, principal components analysis, and derivative spectroscopy. Rem. Sens. Environ. 65 (2), 217–24.

Hu, C.M., Hackett, K.E., Callahan, M.K., Andrefouet, S., Wheaton, J.L.J., Porter, J.W., Muller-Karger, F.E., 2003. The 2002 ocean color anomaly in the Florida Bight: a cause of local coral reef decline? Geophys. Res. Lett. 30 (3), 1151. https://doi.org/10.1029/2002GL016479.

Hughes, T.P., Baird, A.H., Dinsdale, E.A., Moltschaniwskij, N.A., Pratchett, M.S., Tanner, J.E., Willis, B.L., 1999. Patterns of recruitment and abundance of corals along the great barrier reef. Nature 397, 59–63.

Jupiter, S., Roelfsema, C.M., Phinn, S.R., 2013. Science and management. In: Goodman, J.A., Phinn, S.R., Purkis, S. (Eds.), Coral Reef Remote Sensing. Springer, Berlin, Germany, pp. 403–427.

Kayanne, H., Hata, H., Kudo, S., HiroyaYamano, A., Watanabe, Y.I., Nozaki, K., Kato, K., Negishi, A., Saito, H., 2005. Seasonal and bleaching-induced changes in coral reef metabolism and CO2 flux. Global Biogeochem. Cycles 19 (3).

Kirk, J.T.O., 1994. Light and Photosynthesis in Aquatic Ecosystems. Cambridge University Press, New York.

Kobayashi, H., Matsunaga, T., Hoyano, A., Aoki, M., Komori, D., Boonyawat, S., 2004. Satellite estimation of photosynthetically active radiation in Southeast Asia: impacts of smoke and cloud cover. J. Geophys. Res. (in press).

Lirman, D., Biber, P., 2000. Seasonal dynamics of macroalgal communities of the northern Florida reef tract. Bot. Mar. 43, 305–314.

Lirman, D., Fong, P., 1997. Patterns of damage to the branching coral Acroporapalmata following Hurricane Andrew: damage and survivorship of hurricane-generated asexual recruits. J. Coast Res. 13, 67–72.

Lubin, D., Li, W., Dustan, P., Mazel, C.H., Stamnes, K., 2001. Spectral signatures of coral reefs: features from space. Rem. Sens. Environ. 75, 127–137.

Lyzenga, D.R., 1978. Passive remote sensing techniques for mapping water depth and bottom features. Appl. Optic. 17, 379–383.

Lyzenga, D.R., 1981. Remote sensing of bottom reflectance and water attenuation parameters in shallow water using aircraft and Landsat data. Int. J. Rem. Sens. 2, 71–82.

Malthus, T.J., Mumby, P.J., 2003. Remote sensing of the coastal zone: an overview and priorities for future research. Int. J. Rem. Sens. 24, 2805–2815.

Matsunaga, T., Hoyano, A., Mizukami, Y., 2000. Monitoring of coral reefs on Ishigaki Island in Japan using multitemporal remote sensing data. Proc. SPIE 4154, 212–222 (Sendai, Japan).

Mumby, P.J., Edwards, A.J., 2002. Mapping marine environments with IKONOS imagery: enhanced spatial resolution can deliver greater thematic accuracy. Rem. Sens. Environ. 82, 248–257.

Mumby, P.J., Clark, C.D., Green, E.P., Edwards, A.J., 1998a. Benefits of water column correction and contextual editing for mapping coral reefs. Int. J. Rem. Sens. 19, 203–210.

Mumby, P.J., Green, E.P., Clark, C.D., Edwards, A.J., 1998b. Digital analysis of multispectral airborne imagery of coral reefs. Coral Reefs 17, 59–69.

Mumby, P.J., Green, E.P., Edwards, A.J., Clark, C.D., 1999. The cost-effectiveness of remote sensing for tropical coastal resources assessment and management. J. Environ. Manag. 55, 157–166.

Mumby, P.J., Hedley, J.D., Chisholm, J.R.M., Clark, C.D., Ripley, H., Jaubert, J., 2004a. The cover of living and dead corals from airborne remote sensing. Coral Reefs 23, 171–183.

Mumby, P.J., Skirving, W., Strong, A.E., Hardy, J.T., LeDrew, E.F., Hochberg, E.J., Stumpf, R.P., David, L.T., 2004b. Remote sensing of coral reefs and their physical environment. Mar. Pollut. Bull. 48, 219–228.

Myers, M.R., Hardy, J.T., Mazel, C.H., Dustan, P., 1999. Optical spectra and pigmentation of Caribbean reef corals and macroalgae. Coral Reefs 18, 179–186.

Obura, D.O., Tamelander, J., Linden, O., 2008. Ten Years after Bleaching—Facing the Consequences of Climate Change in the Indian Ocean; CORDIO Status Report 2008; CORDIO (Coastal Oceans Research and Development. Indian Ocean)/Sida-SAREC: Mombasa, Kenya, p. 493.

Singh, R.K., Sinha, V.S.P., Joshi, P.K., Kumar, M., 2020. Modelling Agriculture, Forestry and Other Land Use (AFOLU) in response to climate change scenarios for the SAARC nations. Environ. Monit. Assess. 192, 236.

Smirnov, A., Holben, B.N., Kaufman, Y.J., Dubonik, O., Eck, T.F., Slutsker, I., Pietras, C., Halthore, R.N., 2002. Optical properties of atmospheric aerosol in maritime environments. J. Atmos. Sci. 59, 501–523.

Stow, D.A., 1999. Reducing the effects of mis-registration on pixel level change detection. Int. J. Rem. Sens. 20, 2477–2483.

Strong, A.E., Barrientos, C.S., Duba, C., Sapper, J., 1997. Improved satellite techniques for monitoring coral reef bleaching. In: Proceedings of the 8th International Coral Reef Symposium, vol. 2, pp. 1495–1498 (Balboa, Panama).

Strong, A.E., Liu, G., Kimura, T., Yamano, H., Tsuchiya, M., Kakuma, S., van Woesik, R., 2002. Detecting and monitoring 2001 coral reef bleaching events in Ryukyu Islands, Japan using satellite bleaching HotSpot remote sensing technique. In: Proceedings of International Geoscience and Remote Sensing Symposium 2002, vol. I, pp. 237–239. Toronto, Canada.

Thome, K., Markham, B., Barker, J., Slater, P., Biggar, S., 1997. Radiometric calibration of Landsat. Photogramm. Eng. Rem. Sens. 63, 853–858.

Vanderstraete, T., Goossens, R., Ghabour, T.K., 2004. Coral reef habitat mapping in the red sea (Hurghada, Egypt) based on remote sensing. EARSeL eProc. 3 (2), 191–207.

Vermote, E.F., Tanre', D., Deuze', J.L., Herman, M., Morcrette, J.-J., 1997. Second simulation of the satellite signal in the solar spectrum, 6S: an overview. IEEE Trans. Geosci. Rem. Sens. 35, 675–686.

Wilkinson, C., 1998. Status of Coral Reefs of the World. Australian Institute of Marine Science, Townsville, QLD, Australia, p. 184.

Wilkinson, C., 2000. Status of Coral Reefs of the World. Australian Institute of Marine Science, Townsville, QLD, Australia, p. 363.

Wilkinson, C., 2002. Status of Coral Reefs of the World. Australian Institute of Marine Science, Townsville, QLD, Australia, p. 378.

Wilkinson, C., 2004. Status of Coral Reefs of the World. Australian Institute of Marine Science, Townsville, QLD, Australia, p. 301.

Wilkinson, C., 2008. Status of Coral Reefs of the World. Global Coral Reef Monitoring Network and Reef and Rainforest Research Centre, Townsville, QLD, Australia, p. 296.

Yamano, H., Tamura, M., 2001. Use of Landsat TM data and radiative transfer models for monitoring coral reef bleaching. In: Proceedings of International Geoscience and Remote Sensing Symposium 2001, vol. V, pp. 2199–2201 (Sydney, Australia).

Yamano, H., Tamura, M., 2002. Can satellite sensors detect coral reef bleaching? A feasibility study using radiative transfer models in air and water. In: Proceedings of the 9th International Coral Reef Symposium, pp. 1025–1030 (Bali, Indonesia).

Yamano, H., Kayanne, H., Yonekura, N., Kudo, K., 2000. 21-year changes of backreef coral distribution: causes and significance. J. Coast Res. 16, 99–110.

Yamano, H., Tamura, M., Kunii, Y., Hidaka, M., 2002. Hyperspectral remote sensing and radiative transfer simulation as a tool for monitoring coral reef health. Mar. Technol. Soc. J. 36, 4–13.

Yamano, H., Tamura, M., Kunii, Y., Hidaka, M., 2003. Spectral reflectance as a potential tool for detecting stressed corals. Galaxea, JCRS 5, 1–10.

Yu, H., Dickinson, R.E., Chin, M., Kaufman, Y.J., Holben, B.N., Geogdzhayev, I.V., Mischenko, M.I., 2003. Annual cycle of global distributions of aerosol optical depth from integration of MODIS retrievals and GOCART model simulations. J. Geophys. Res. 108, 4128.

20

Assessment of shoreline vulnerability in parts of the coastline of Kasaragod district, Kerala, India

M. Vrinda[1], M.A. Mohammed-Aslam[2]

[1]DEPARTMENT OF GEOLOGY, GOVT. COLLEGE, KASARAGOD, KERALA, INDIA; [2]DEPARTMENT OF GEOLOGY, SCHOOL OF EARTH SCIENCES, CENTRAL UNIVERSITY OF KARNATAKA, GULBARGA, KARNATAKA, INDIA

Chapter outline

1. Introduction

The coastal zone represents the zone of conversion between the terrestrial and marine constituents on Earth's surface (Crossland et al., 2005) and the shoreline that overlaps with the physical margin of land and water resources (Dolan et al., 1980). It is one of the most dynamic areas in the world because the area is vulnerable to erosion, floods,

Remote Sensing of Ocean and Coastal Environments. https://doi.org/10.1016/B978-0-12-819604-5.00020-2

storms, and hurricanes (García-Rubio et al., 2012). Sustainability of the coastal environment is affected by pollution, eutrophication, industrialization, urbanization, land reclamation, agricultural production, overfishing, and exploitation. A major portion of the population in the world lives within 60 km of the coast, and it will increase to nearly three-quarters by the year 2020 (Prabaharan et al., 2010). The shoreline change study and the mapping of shoreline positions are very valuable in regard to climate change (Rajaiah and Ramanamurthy, 2016a,b). Temporal and spatial shoreline changes studies are very significant in order to establish a scientific background for disciplines such as environmental science, biology, and geography, and assist planners and policy makers in making relevant policies (Quang Tuan et al., 2017). So it needs to be continually monitored (Chand and Acharya, 2010).

The dynamic coastal zones produce diverse and productive ecosystems such as salt marshes, mangroves, wetlands, estuaries, and such, and those have been of great importance for human populations. In those, an estuary is a water body that is partially enclosed by saline water. More than one river or stream flows into the estuary and freely joins to the sea (Raj et al., 2019). Estuaries are the world's most productive ecosystem, and numerous animals are dependent on estuaries for food and shelter. These ecosystems produce more services relating to human well-being than most of the other systems and play a dynamic part for the regulation of climate and global ecosystems (Manonmani and Suganya, 2010). Still, coastal ecosystems are very sensitive to changes in the environment due to human activity like construction of structures, ports, harbors, jetties, groynes, and so on, and environmental challenges such as waves, wind, periodic storms, sea level changes, and so on.

The changes in the coasts are mainly depended upon the practices along the seashore (Mageswaran et al., 2015). The coastal erosion or coastal accretion makes changes in shoreline. Coastal erosion is a universal issue, which is disturbing almost every country with a coastline (Shetty et al., 2015). It is the process of long-term sediment subtraction and evacuation of the coastline landward by the action of waves, currents, and human interference (Anthony 2005). Natural influences of the shoreline erosion in the coastline of Kerala are heavy rainfall, loose sandy sea shore, sea level increases due to the melting of ice in the polar region, and devastation of mud deposits from the sea (Paul and Sumam, 2013). Coastline withdrawal has occurred over in 70% of the beaches in the world (Rajaiah and Ramanamurthy, 2016a,b), so it needs quick, precise and up-to-date information on every minor movement (Topouzelis et al., 2017). The accretion is the process of coastal sediment recurring on the visible region of a beach. Coastline accretion may generate extra land and it can be considered a positive change (Rahisha et al., 2016). Spatial changes differ from local to regional to worldwide scale (Sesli et al., 2009). Appropriate shoreline-change analysis entails a systematic understanding of the coastal processes, which causes the change as well as the coastal mapping methods. Therefore, consistent information about the shoreline is essential for scientists, coastal engineers, and coastal managers for developmental forecasting and threat zonation, analysis of erosion and accretion, and so on. Variations in the coastline and water level mainly reveal the changes in precipitation, evaporation, runoff, and so on (Li et al., 2016).

Moreover, the study on shoreline changes can be of vast benefit to the understanding of complex shoreline ecosystems because it is the bridge between aquatic and terrestrial life and usually a frail ecozone.

Long-term fluctuations in the coastline can be analyzed through field studies for the entire study time. This is a labor intensive, costly, and overwhelming process (Daniels 2012). Remote sensing and geographical information systems (GIS) provide fresh technologic benefits in coastal analysis studies (Maiti and Bhattacharya, 2009). Spatial data analysis and mapping, remote sensing, and GIS are widely applied for monitoring environmental and natural resources (Makota et al., 2004). Remote sensing data aids and substitutes the conventional study by its repetitiveness and lesser cost (Thieler et al., 2009). Changes can be obtained from comparing different dated satellite imagery, air photographs, and historical maps. By using satellite data and further data sources operated, manipulated, and combined into a GIS gives indispensable and valuable evidence based on coastline management. GIS is a computer-based tool that analyzes, stores, manipulates, and visualizes geographic data, generally in a map (https://gisgeography.com/what-gis-geographic-information-systems/). The computer-generated composite maps are very useful in defining geographica sections and quickly recognizing areas where major changes have occurred over time (Krishnamoorthy, 2013).

Satellite image resolutions are the major limit for mapping. The resolution manages the exactness and precisions, which depend on the nature of features, coverage, and the objectives of the analysis. Some studies with a small change rate in a huge area can fulfill the study with the medium-resolution satellite images. And those images acquired from Landsat and Sentinel can be very useful for global scale temporal change analysis mapping because images are accessible with high temporal resolution free of cost. Due to the dynamic nature of shoreline, high resolution images give accurate outcomes for the area. The accuracy of the shoreline change rate depends on more features such as the accurateness of the delineated shorelines, the shoreline's temporal resolution, the number of shoreline positions selected for change rate analysis, the period between the measurements of shoreline, and the technique used for the change rate calculation. Aerial photographs and light detection and ranging gives high-resolution images, but it is more expensive. The high-resolution satellite images from the Google Earth (GE) platform are freely accessible and it is very easy to access. Processes like shoreline change analysis and mapping of land use and land cover, which is not directly contingent on pixel values, can use GE images effectively. The major benefits of GE images are the obtainability of medium- to high-resolution images and the obtainability of every-time series data (Malarvizhi et al., 2016). The resolutions are available from 30 m (Landsat) to 0.31 m (Worldview-3/4) and data are available from 2002 to date from Google Earth Pro 2018. The margins of the land can be definitely delineated by digitizing by using GIS software. In this study, Google Earth Pro-2018 images were used mostly from other sources.

Features and subtleties of natural coastal ecosystems are environmentally connected to one another, so planning and implementation of sustainable coastal resource that are not implemented in a combined way would only harm the resources (Nandi, 2014). Estuaries also epitomize most iconic tourist terminuses and are areas of high financial worth for coastal real estate (Thrush et al., 2013). Shoreline changes draw more attention because they are the most significant environmental indicators, which unswervingly disturb financial growth and land organization (Natesan et al., 2015). So the aim of coastal zone management is to facilitate economic development and to reduce negative effects of all developmental activities and coastal hazards.

2. History

During past decades, to identify coastline changes researchers compared topographic maps and aerial photographs of different periods (Henry et al., 2013). All coastline maps had been created through field surveying between 1807 and 1927 (Alesheikh et al., 2007). Aerial photographs had come into use from 1927 to 1980. Black-and-white photographs have some limitations. From 1972, the Landsat imagery became the sole method of shoreline mapping because it provides digital images in infrared spectral bands. From these images the shoreline can define and extract perfectly and the info provided in a digital setup is effortlessly reachable to workers and policy makers for several applications and for decision-making in every field (Sheik and Chandrasekar, 2011). The world's coastal areas are extremely vulnerable for different hazards, so it is essential for well and effective procedures for coastal vulnerability (Vivek, 2016). Hence, it is confirmed that remote sensing data and image processing practices provide a conceivable result for creating and updating coastline maps (Alesheikh et al., 2007).

3. Study area

India has a 7500 km long sandy beach coastline, and it is composed of rocky outcrops and shores. In India, almost 25% of the population lives around 50 km of the coast (Kunte et al., 2014). The Indian coastline is very important because of high population pressure, growth of numerous industries, mistreatment of renewable and nonrenewable natural resources, release of sewages, and so on, and it faces serious threats from tsunami and tropical cyclones. The study was conducted along the two estuaries of Kasaragod district, Kerala state, Southwest India (i.e., Shiriya and Thrikaripur) (Fig. 20.1). The area is analyzed for shoreline changes to measure erosion and accretion for a period of 50 years (1967–2018). India's southwest coast spreading from Poovar (south) to Kasaragod (north) is one of the vastly fluctuating coastal areas of Indian (Sheela Nair et al., 2018). River Shiriya and a few minor rivers join together to form the estuary Shiriya. It is 10 km north of the Kasaragod town. Kariangode River joins the Arabian Sea from the Thrikaripur river mouth. Kariangode River is the second longest river in Kasaragod district. It flows through a total area about 64 km. The river mouth contains the residential zone and its beauty makes it a tourist attraction. The area contains small

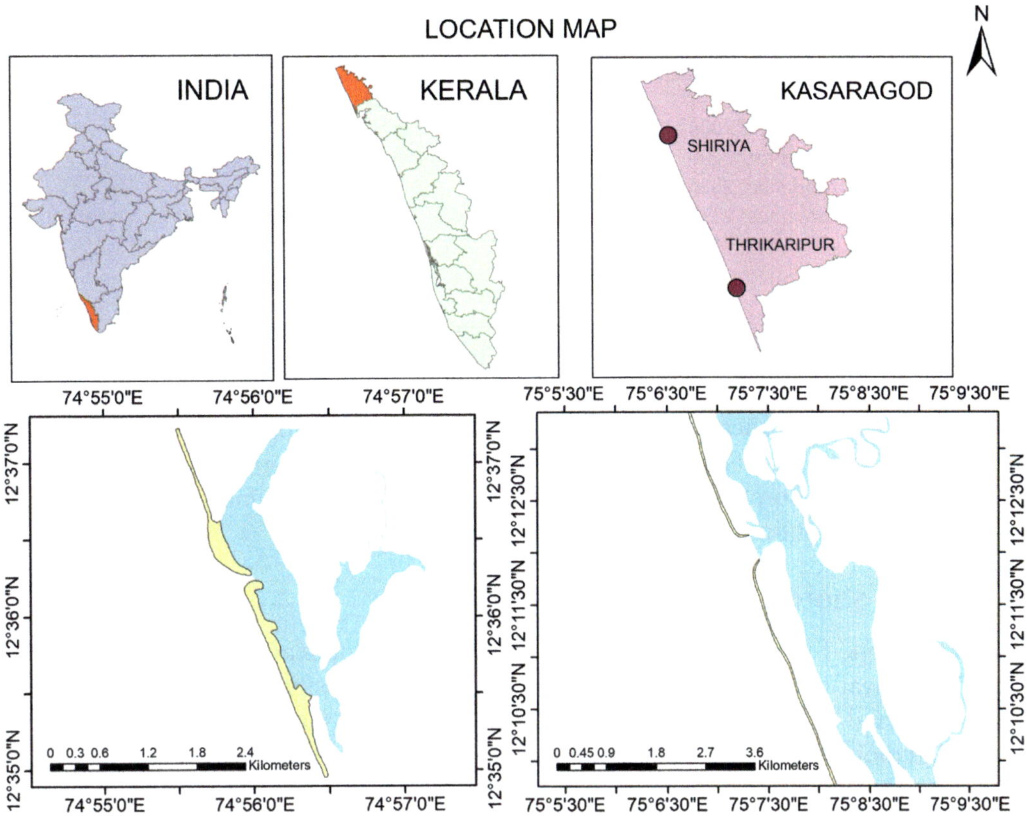

FIGURE 20.1 Location of study area.

industries and the area has every monsoon season and in the summer season, heat is moderate. The average annual temperature is 27.1°C and about 3825 mm of precipitation falls annually.

4. Materials and methods

The main aim of the procedure adopted for the present study is to increase the GE imagery uses to change studies in different areas. GE Engine is a very useful platform for geospatial analysis. It helps to study the variety of social issues such as deforestation, draught, disaster, disease, water management, analysis, and monitoring of climate and environmental protection (Gorelick et al., 2017). For the monitoring of the coastal region, extraction of coastline for different periods is a vital mission (Alesheikh et al., 2003). GE satellite images with high resolution, for the years of 2004, 2009, 2014, 2018 and toposheet of 1968, covering a period of 50 years, are the major data used for the completion of the objectives of the present study. Image acquisition date and image source are

Table 20.1 Information of data used for the study.

Acquisition month and year	Image source
1968	Toposheet
November 2004	Digital globe
November 2009	Digital globe
November 2014	Digital globe
November 2018	Digital globe

shown in Table 20.1. All the imageries were georeferenced with UTM projection with WGS84 datum. There are many shoreline indicators such as high water line, low water line, mean tide line, and so on. In these indicators the high-water line is easy to interpret, so it is the favored indicator for shoreline delineation (Mahapatra et al., 2013). Overlay analysis and measurement tool is used for the analysis of historical shoreline positions.

The digitized shoreline is applied in the digital shoreline analysis system (DSAS) software for computing the rate of shoreline change.

4.1 Digital shoreline analysis system

USGS developed the DSAS, which is an open source application that is active within ArcGIS software. DSAS has been established as an easily accessible extension to Environmental System Research Institute (ESRI)'s ArcGIS (Oyedotun, 2014). DSAS is an extension tool used for calculating and monitoring the statistics of shoreline change rate from numerous historical shoreline positions extracted from satellite images (Oyedotun, 2014). The software constructs transect from a baseline, which is parallel to the shoreline and along the transects. The transects evaluate the variation in shoreline and calculate the rate of accretion and erosion.

4.2 Analysis using DSAS in ArcGIS

In this study, calculation of the shoreline change rate was done by DSAS Version 5.0 along with the ARCGIS 10.6. The extracted shoreline should be in a projected coordinate system. A personal geodatabase is created to store the results. The shoreline of each year was coded with attribute fields for date, and the date on which the satellite data is taken is given as an attribute. All shoreline features were appended into a solo feature class in the geodatabase for calculating coastline change rates with DSAS. The OBJECTID, Shape, and Shape_Length are auto-generated by ArcGIS. The date appears as it is already given as attribute. The DATE_ and UNCERTAINTY fields are added to the attribute table. The DATE_ field data type should be text.

A baseline is created either offshore or onshore, which is the starting point for the transects. The baseline is created using the buffer tool, which is most landward and which will convert the polygon buffer to the polyline, and splitting and removal of the remaining segments will form a baseline.

Calculation of statistics are carried out using DSAS. In DSAS first the default parameters icon is clicked. In the set default parameters, the cast transect setting is defined and the baseline layer, its group, whether the transect should be cast offshore or onshore, and transect spacing and length are given as inputs. In shoreline calculation setting the shoreline layer used for analysis, the field containing the date of the shoreline, and the uncertainty field are given as inputs. The metadata set is used to input the details of the user and contact information. In the process, the personal geodatabase where the results are stored, the name of the transect layer that is to be generated, the casting method whether simple or smooth for straight coasts and curved coasts, respectively, are defined. The transects are generated and the attribute fields are also generated.

The transect layer is selected for the calculation and clicking on Calculate Change Statistics opens a dialog box in which the different methods of calculation are listed. The appropriate method is selected and once the tab clicked a table will be generated. If the name of the file is "transect," the file generated will be named transect_rates_yyymmdd.dbf. The file is added to the ArcGIS table of contents. Now the change rate statistics listed in the table generated are appended to the transect feature class shapefile using the common field.

4.2.1 Change rate calculation by EPR and LRR

There are several computational methods that are there for the rates of change of shoreline in DSAS such as shoreline change envelope, net shoreline movement, weighted linear regression, end point rate (EPR), and linear regression rate (LRR). In this present study, for finding the shoreline change per year, the EPR and LRR of shoreline changes are calculated in the period of 1968–2018. The positive EPR and LRR value represent the rate of accretion and a negative value represents erosion. EPR is intended by dividing the time intervened between the oldest and the youngest positions of shoreline. LRR find the shoreline change rate statistic by fitting a least squares regression to all shorelines at precise transects.

4.2.2 Ground truthing

Ground truthing on the two locations (Shiriya estuary and Thrikaripur estuary) was conducted for 2 days to assemble extra facts and evidence from the study area and from local people. The information collected by survey identified the condition of shoreline related with an earlier period and it agreed well with my findings.

5. Results and discussion

The coastline is the constantly fluctuating lined structures of the coastal region. The erosion and accretion pattern observed along the coast is influenced mainly by coastal processes and riverine inputs. The variation in river mouth morphology was quite significant and changes in rate were more all along the river mouth.

5.1 Geospatial analysis of coastline

The coastline change was determined by overlaying the digitized images. Fig. 20.2 shows the change that occurred in the land–water interface in the Shiriya estuary in the period of 1968–2018. From the figure, it is observed that in the study period, major shoreline erosion had taken place between 1968 and 2004. Around 0.16 km area had eroded in between 1968 and 2004 in the river mouth of Shiriya. From 2004 to 2014 the coastline exhibited minor changes in the width of the shoreline. From the multidated data, it was found that the mouth of the river has varied in its shape, pattern, and position during the study period. From 2014 to 2018, the river mouth had shifted to northern side around 0.46 km. In 2009, the spit in northern side of the estuary was facing erosion activity and the southern side was facing accretion, and it shifted to the northern side by the erosion and accretion changes stirring on the coast of the estuary. But in 2018, the river mouth shifting to the southern side due to erosion happens in the southern spit and accretion in the spit of northern side. Spit is a small point feature or thin ridge of land usually

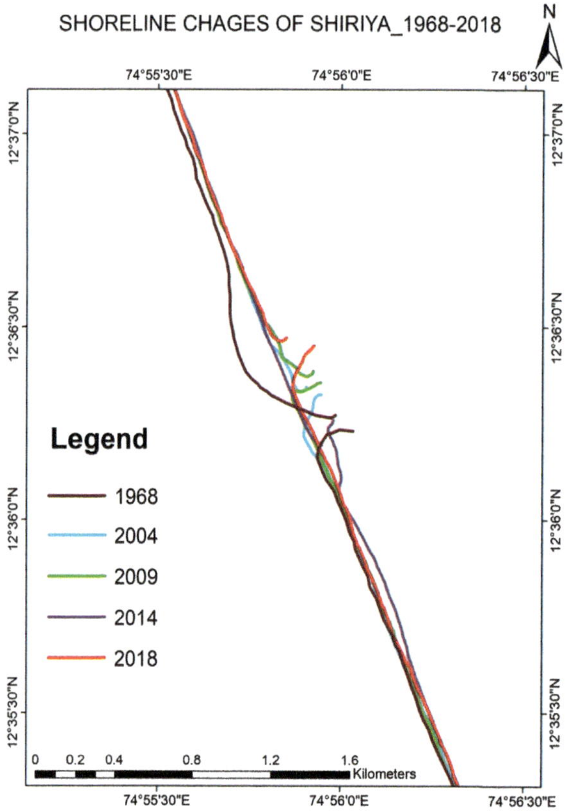

FIGURE 20.2 Shorelines of Shiriya river mouth 1968–2018.

composed of sand dumped due to long shore drift. Its one edge may connect to the coast and the next end terminates in water. Accretion was probably due to gradual increase in the area of land as a result of sedimentation.

The overlay analysis maps (Fig. 20.3) of Thrikaripur estuary say that major shoreline changes occurred during 1968–2014 and there has been more accretion than erosion after the period of constructing the sea wall. So not many changes were observed in the location of the river mouth. River mouth movement may influence coastal erosion and accretion. Construction of a seawall influenced the trend of accretion in the study area. It is clearly seen that after 2014 the river mouth did not have many changes. It was intended to avert the coastal erosion and other harm due to the action of waves and storm surges.

5.2 Shoreline change rate analysis

The DSAS software processed shoreline status for 50 years. DSAS creates transects, which are vertical to the baseline that the operator created. With orientation to the baseline, accretions are considered if the shifting of the shoreline is along transect, that is

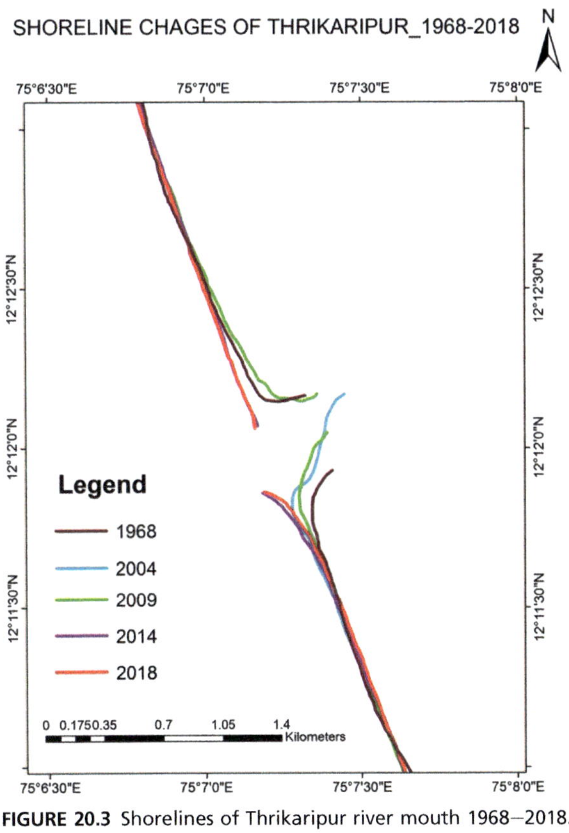

FIGURE 20.3 Shorelines of Thrikaripur river mouth 1968–2018.

to seaward, and the shifting to the landward is considered erosion. Shoreline change rate is calculated by the amount of shoreline shift along each transect by DSAS. The Shiriya river mouth has 77 transects and the Thrikaripur river mouth has 98 transects.

For classifying the shoreline change rate, we have adopted the classification criteria (Table 20.2) developed by the National Center for Sustainable Coastal Management, which comes under the Ministry of Environment and Forest and Climatic Change based in Chennai, which is a nodal agency to demarcate the high tide line along the Indian Coast. The shoreline change variations of Shiriya and Thrikaripur river mouths are shown in Figs. 20.4 and 20.5 respectively.

A shoreline change map of the Shiriya river estuary coastline represents the proper shoreline change that happened there for 50 years. The northern shoreline of the river mouth faced moderate erosion to very high erosion but the whole southern shoreline is facing moderate erosion. Accretion also takes place at different rates but it is relatively much less while comparing with erosion rate, so the cumulative effect is the erosion.

In the Thrikaripur river mouth, most of the shoreline is stable and a major part of the area of the remaining coast faces moderate accretion. The nearest shore of the river mouth faces high accretion. The accretions occur due to the deposition of sediment by the river. The shoreline change is mostly related to the geomorphic courses of erosion and accretion and human activities, and due to the actions of periodic storms, sea-level change, and so on (Salghuna and Bharathvaj, 2015).

The comparison of shoreline changes of Shiriya and the Thrikaripur river mouth represents relatively high erosion changes in Shiriya and accretion changes in the Thrikaripur river mouth. The Thrikaripur river mouth has the presence of rubble mound wall after 1968. That is the main cause of absence of erosion in the area. It is a coastal resistance technique constructed where the sea and other coastal actions disturb coastal landforms. The rubble mound sea wall is slanted and is made of loose material that guarantees maximum dissipation of wave energy and it is the lesser cost option (Kumar, 2016).

Table 20.2 Criteria for shoreline change analysis.

Sl. No	Rates (m)	Classification	Ranking	Symbols
1	<−5	Very high erosion	1	—
2	−5 to −2.5	High erosion	2	—
3	−2.5 to −0.5	Moderate erosion	3	—
4	−0.5 to 0.5	Stable coast	4	—
5	0.5 to 2.5	Moderate accretion	5	—
6	2.5 to 5	High accretion	6	—
7	>5	Very high accretion	7	—

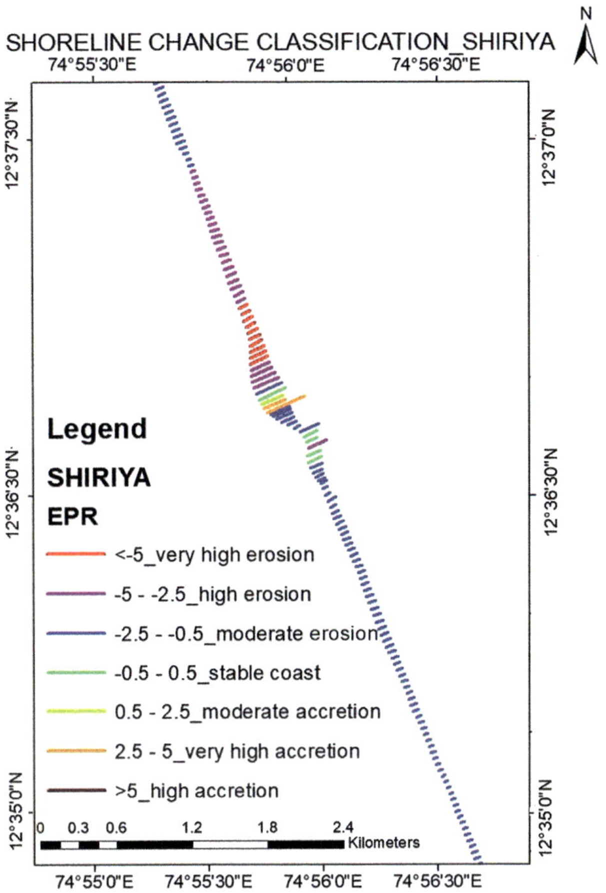

FIGURE 20.4 Shoreline classification of Shiriya river mouth.

Overall rates of changes range from 1968 to 2018 are represented in the graph, where negative values signify erosion and positive values signify accretion (Figs. 20.6 and 20.7).

In the Shiriya river mouth, 96% of the entire shoreline experienced erosion, and only 4% of the shoreline faces only accretion trend.

In the Thrikaripur river mouth, 61% of the entire shoreline experienced accretion while the remaining area has eroded.

6. Conclusion

The foremost purpose of the present study was to improve the method to analyze shoreline changes by satellite imageries, and identifying and quantifying the erosion and accretion rate and area of the shoreline of Northern Kerala. The multidated data indicated the growth and movement of spits, erosion, and accretion of coastline. The study

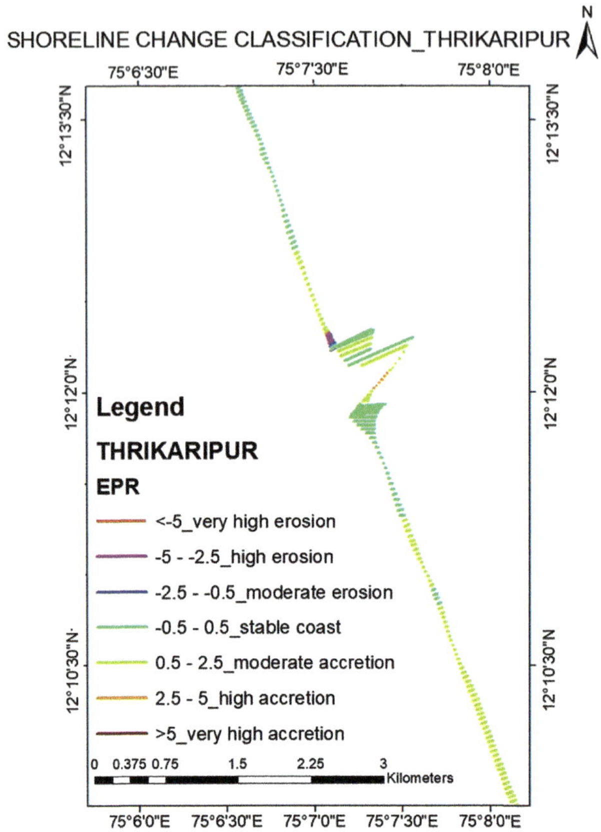

FIGURE 20.5 Shoreline classification of Thrikaripur river mouth.

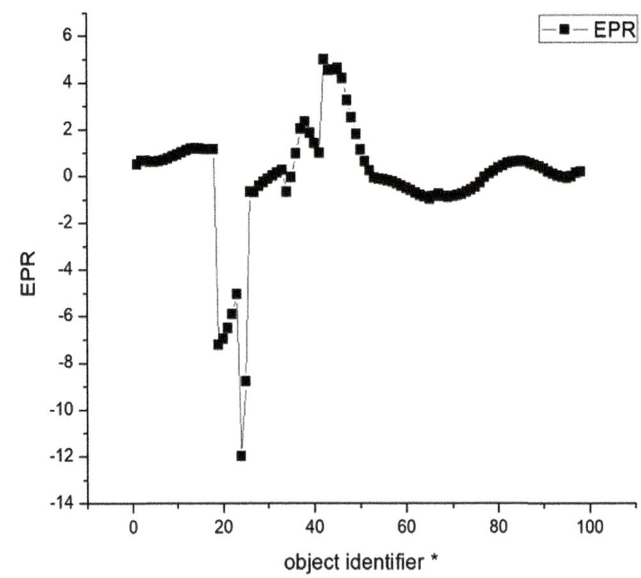

FIGURE 20.6 Shoreline change rate of Shiriya river mouth.

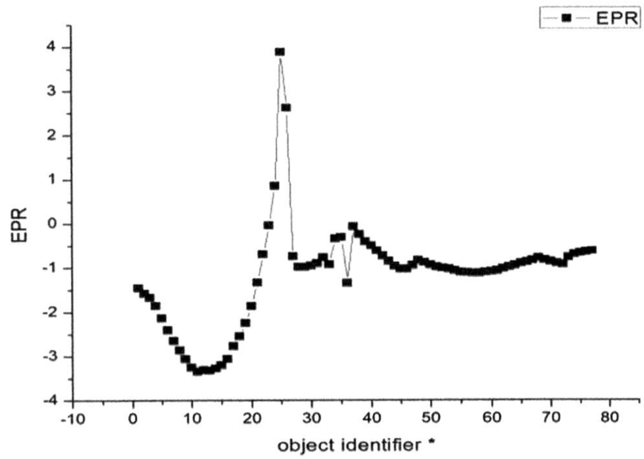

FIGURE 20.7 Shoreline change rate of Thrikaripur river mouth.

showed a high dynamic nature of the Shiriya river mouth because of the coastline erosion and accretion occurring frequently. In the case of the Thrikaripur river mouth, accretion is far more than erosion. The erosion and accretion change was perceived along the area of study and it was influenced mainly by riverine inputs and other coastal processes. The disparity in river mouth morphology was fairly significant. Both the river mouth's spit show remarkable changes. The general trend of erosion showed that the estuary had moved to a northern direction and southern direction due to erosion and accretion processes in a period of time. The influence of coastal management techniques such as rubble mound seawall helps to reduce the erosion in the Thrikaripur river mouth. Based on the time-based study of shoreline change and earlier analysis studies on the reasons of estuarine shoreline changes, it is confirmed that the fluctuating pattern of the shoreline was triggered by both natural developments in the estuaries and anthropological influences added by climate change. Coastal management techniques such as groynes, seawalls, revetments, and rock armor may be constructed to reduce or halt coastal erosion in this region. Most of India's coastline has faced positive or negative changes in the past few decades. So continuous shoreline monitoring is indispensable for coastal zone management. Studies in which connection with land use patterns, anthropological actions, and further happenings occur in the coastline are recommended for proper coastal zone management. The outcomes of the present study that correlate with other GIS data sets can monitor future shoreline change trends. Scientists can use GE in a better way, because of free cost, up-to-date maps, data availability, detailed information about the data, and so on. So it can conclude that GE and GIS is the best tool acquired to date and provides consistent shoreline information for shoreline management purposes.

References

Alesheikh, A.A., Ghorbanali, A., Nouri, N., 2007. Coastline change detection using remote sensing. Int. J. Environ. Sci. & Technol. 4 (1), 61–66, 1.

Alesheikh, A.A., Sadeghi Najafi, F., Talebzade, A., 2003. Improving classification accuracy using external knowledge. GIM Int. 17 (8).

Anthony, E.J., 2005. Beach Erosion. Encyclipedia of Coastal Science. Springer, pp. 140–144.

Chand, P., Acharya, P., 2010. Shoreline change and sea level rise along coast of Bhitarkanika wildlife sanctuary, Orissa: an analytical approach of remote sensing and statistical techniques. Int. J. Geomatics Geosci. 1 (3), 436–455.

Crossland, C.J., et al., 2005. Coastal Fluxes in the Anthropocene. https://doi.org/10.1007/3-540-27851-6.

Daniels, R.C., 2012. Using Arcmap to Extract Shoreline from Landsat TM and ETM+ Data Thirty-second ESRI international user conference proceedings, 2012, pp. 1–23.

Dolan, R., Hayden, B.P., May, P., May, S.K., 1980. The reliability of shoreline change measurements from aerial photographs. Shore and Beach 48 (4), 22–29.

García-Rubio, G., Huntley, D., Russell, P., 2012. Assessing shoreline change using satellite-derived shorelines in Progreso, Yucatán, México. Proc. Coast. Eng. Conf. (3918), 1–13.

Gorelick, N., Hancher, M., Dixon, M., Ilyushchenko, S., Thau, D., Moore, R., 2017. Google earth engine: planetary-scale geospatial analysis for everyone. Remote Sens. Environ. 202, 18–27. https://doi.org/10.1016/j.rse.2017.06.031.

Henry, U.U., Igbokwe, J., Chukwocha, A.C., 2013. Assessment of coastaline vulnerability for sustainable coastal development and planning in Nigeria using medium resolution satellite images and GIS. Int. J. Eng. 4 (4), 408–416.

https://gisgeography.com/what-gis-geographic-information-systems/.

Krishnamoorthy, G., 2013. Coastal Dynamics: Shoreline Change in the Le Morne Coast of Mauritius.

Kumar, S., August 2016. Protection of Shoreline at South Chennai.

Kunte, P.D., Jauhari, N., Mehrotra, U., Kotha, M., Hursthouse, A.S., Gagnon, A.S., 2014. Multi-hazards coastal vulnerability assessment of Goa, India, using geospatial techniques. Ocean Coast Manag. 95 (July), 264–281. https://doi.org/10.1016/j.ocecoaman.2014.04.024.

Li, X., Zhang, X.-Q., Zhao, Y., 2016. To cite this article: fatih Temiz and S. Sava Durduran IOP Conf. Ser. Earth Environ. Sci. 44. https://doi.org/10.1088/1755-1315/44/4/042033.

Mageswaran, T., Mohan, R.V., Selvan, C.S., Arumugam, T., Usha, T., Kankara, R.S., 2015. Assessment of shoreline changes along Nagapattinam coast using geospatial techniques. Int. J. Geomatics Geosci. 5 (4), 555–563. Retrieved from. www.landsat.org.

Mahapatra, M., Ratheesh, R., As, R., 2013. Shoreline change monitoring along the South Gujarat coast using remote sensing and Gis techniques. Int. J. Geol. Earth & Environ. Sci. 3 (2), 115–120.

Maiti, S., Bhattacharya, A., 2009. Shoreline change analysis and its application to prediction: a remote sensing and statistics based approach. Mar. Geol. 257 (1), 11–23.

Makota, V., Sallema, R., Mahika, C., 2004. wiojms3-1Makota.

Malarvizhi, K., Kumar, S.V., Porchelvan, P., 2016. Use of high resolution google earth satellite imagery in Landuse map preparation for urban related applications. Procedia Technol. 24, 1835–1842. https://doi.org/10.1016/j.protcy.2016.05.231.

Manonmani, R., Suganya, G.M.D., 2010. Remote sensing and GIS application in change detection study in Urban Zone using multi temporal satellite. Int. J. Geomatics Geosci. 1 (1), 60–65. https://doi.org/10.1186/1471-2148-11-104.

Nandi, 2014. Coastal conservation policies and integrated coastal zone management (ICZM) in Indonesia. Int. J. Conserv. Sci. 5 (3), 387–396.

Natesan, U., Parthasarathy, A., Vishnunath, R., Kumar, G.E.J., Ferrer, V.A., 2015. Monitoring longterm shoreline changes along Tamil Nadu, India using geospatial techniques. Aquat. Procedia 4, 325–332. https://doi.org/10.1016/j.aqpro.2015.02.044. Icwrcoe.

Oyedotun, T.D.T., 2014. Shoreline Geometry: DSAS as a tool for historical trend analysis (Online Edition). Geomorphol. Techn. 2, 1–12.

Paul, P., Sumam, K.S., 2013. Assessment of Coastal Vulnerability along Chavakkad , Kerala Coast in India, Using Geospatial Techniques, pp. 335–339.

Prabaharan, S., Srinivasa Raju, K., Lakshumanan, C., Ramalingam, M., 2010. Remote sensing and GIS applications on change detection study in coastal zone using multi temporal satellite data. Int. J. Geomatics Geosci. 1 (2), 159–166.

Quang Tuan, N., Cong Tin, H., Quang Doc, L., Anh Tuan, T., 2017. Historical monitoring of shoreline changes in the cua dai estuary, central Vietnam using multi-temporal remote sensing data. Geosciences 7 (3). https://doi.org/10.3390/geosciences7030072.

Rahisha, T., Goud, A.K.C., Sreedhara, V., September 2016. Change detection in Mangalore coastline using multi temporal remote sensing satellite data and GIS. Int. J. Adv. Mechan. & Civil Eng. 9–14.

Raj, N., Gurugnanam, B., Sudhakar, V., Glitson Francis, P., 2019. Estuarine shoreline change analysis along the Ennore river mouth, south east coast of India, using digital shoreline analysis system. Geodesy & Geodyn. 10 (3), 205–212. https://doi.org/10.1016/j.geog.2019.04.002.

Rajaiah, K., Ramanamurthy, M.V., 2016a. Shoreline change monitoring in Nellore coast at east coast Andhra Pradesh district using remote sensing and GIS. J. Fish. & Livest. Prod. 04 (01). https://doi.org/10.4172/2332-2608.1000161.

Rajaiah, K., Ramanamurthy, M.V., 2016b. Shoreline change monitoring in Nellore coast at east coast Andhra Pradesh district using remote sensing and GIS. J. Fish. & Livest. Prod. 04 (01), 316–324. https://doi.org/10.4172/2332-2608.1000161.

Salghuna, N.N., Bharathvaj, S.A., 2015. Shoreline change analysis for northern part of the Coromandel coast. Aquat. Procedia 4, 317–324. https://doi.org/10.1016/j.aqpro.2015.02.043. Icwrcoe.

Sesli, F.A., Karsli, F., Colkesen, I., Akyol, N., 2009. Monitoring the changing position of coastlines using aerial and satellite image data: an example from the eastern coast of Trabzon, Turkey. Environ. Monit. Assess. 153 (1–4), 391–403. https://doi.org/10.1007/s10661-008-0366-7.

Sheela Nair, L., Prasad, R., Rafeeque, M.K., Prakash, T.N., 2018. Coastal morphology and long-term shoreline changes along the Southwest coast of India. J. Geol. Soc. India 92 (5), 588–595. https://doi.org/10.1007/s12594-018-1072-x.

Sheik, M., Chandrasekar, 2011. A shoreline change analysis along the coast between Kanyakumari and Tuticorin, India, using digital shoreline analysis system. Geo Spat. Inf. Sci. 14 (4), 282–293. https://doi.org/10.1007/s11806-011-0551-7.

Shetty, A., Jayappa, K.S., Mitra, D., 2015. Shoreline change analysis of Mangalore coast and morphometric analysis of Netravathi-Gurupur and Mulky-pavanje spits. Aquat. Procedia 4, 182–189. https://doi.org/10.1016/j.aqpro.2015.02.025. Icwrcoe.

Thieler, E.R., Himmelstoss, E.A., Zichichi, J.L., Ergul, Ayhan., 2009. Digital shoreline analysis system (DSAS) version 4.0 — an arcGIS extension for calculating shoreline change. report 2008-1278. U.S. Geological Survey Open-File.

Thrush, S.F., Townsend, M., Hewitt, J.E., Davies, K., Lohrer, A.M., Lundquist, C., Cartner, K., 2013. The Many Uses and Values of Estuarine Ecosystems.

Topouzelis, K., Papakonstantinou, A., Doukari, M., December 2017. Coastline change detection using unmanned aerial vehicles and image processing techniques. Accept. Fresenius Environ. Bull. 7.

Vivek, G., March 2016. Shoreline Change Analysis for North East Coast of Andhra Pradesh, India. https://doi.org/10.13140/RG.2.2.19763.73767.

21

Insight to the spatial-temporal extent of mangrove forests in the northern coast of Kerala

K. Ibrahim-Bathis

DEPARTMENT OF ENGINEERING, POLITEKNIK NEGERI PONTIANAK, PONTIANAK, WEST KALIMANTAN, INDONESIA

Chapter outline

1. Introduction

Growing in the intertidal areas and estuaries at the land and sea, mangroves provide a typical habitat for a diverse aquatic and terrestrial flora and fauna (Tomlinson, 2016).The mangrove forest plays an imperative role in the coastal ecosystem and acts as a natural coastal guard to protect both the land and water resource (Sanford, 2009). It acts as a barrier for the soil and sand erosion in the coastal estuaries and river deltas, improves the water quality, enhances the groundwater table, controls the flood and river flow, inhabits diverse fauna and flora, accelerates the biodiversity and biomass, and economic

Remote Sensing of Ocean and Coastal Environments. https://doi.org/10.1016/B978-0-12-819604-5.00021-4

productivity, and promotes recreation and tourism (Tomlinson, 2016; Dahdouh-Guebas et al., 2005; Gilman et al., 2008). Its obstacles are the fertile soil and sand from erosion that trap sediment in their roots. Mangroves provide the chief economic mode for the local tribe as a means of timber for fuel, furniture, and construction; and are a source of charcoal, tannin, paper, dyes and chemicals, thatch, honey, and incense (Das and Crépin, 2013; Duke et al., 2007; Thirumalai and Radhakrishnan, 1999; Ellison et al., 1999). The tropical coastal regions are one of the hotspots for the economic and urban development in the modern century, and most of the megacity and industrial zone are situated in these regions. These types of development possess a threat to these unique tropical evergreen forests, which may be disappearing more quickly than inland tropical rainforests, and so far, with little public notice. Urbanization and increased pressure of the population of these unique natural habitats are destroyed and reoccupied by construction and man-made structures in the coastal zone (Das and Crépin, 2013; Duke et al., 2007; Thirumalai and Radhakrishnan, 1999; Ellison et al., 1999; Ibrahim-Bathis et al., 2012; Ward et al., 2016). Mangroves are generally distributed in deltas of large rivers and on the bays of tropical countries (Sanford, 2009; Ellison et al., 1999). The most luxuriant and densest marine tidal forests are found in the highest rainfall regions (http://mangroveworld.org/index.php). These plants adapted to anaerobic conditions of salt water, fresh water, and brackish environments on riverbanks and along coastal estuaries in the tropic and subtropical regions (Tomlinson, 2016; Ellison et al., 1999; Polidoro et al., 2010). Mangrove plants includes trees, shrubs, ferns, and palms that produce stilt roots, which project above the mud and water in order to absorb oxygen (Tomlinson, 2016). Sundarbans is the largest contiguous mangrove forest in the world situated on the deltas of Brahmaputra and Ganga River. This biomass-rich habitat of mangrove species diversity from 30 to 35 occupies the border part of India Bangladesh (Polidoro et al., 2010). Along the tidal sea edge, river mouth, or alluvial delta, coastal estuaries, and backwater of Tamil Nadu, Karnataka, Kerala, Goa, Maharashtra, and Gujrat are other regions where rich mangrove forests are recorded in India (Ibrahim-Bathis et al., 2012; Giri et al., 2015). The mangrove forest is very difficult to access because of the muddy soil characteristics. The mapping of the community or species level at the mangrove forest is not easy with traditional field-oriented cartographic or survey equipment and techniques (Jia et al., 2014). The satellite or airborne remote sensing coupled with a geographical information system (GIS) and global navigational and positioning system (GPS) can be used as an alternative to traditional field-oriented methods, especially for mangrove mapping. These geospatial technologies allow information and data gathering from the inaccessible environment of mangrove forest (Ibrahim-Bathis et al., 2012; Giri et al., 2015; Nayak and Bahuguna, 2001; Vo et al., 2013; Chen et al., 2017). It gives accurate, rapid, and cost-effective mapping techniques with different spectral, spatial, and temporal data for the effective study of the change detection, degradation causes, and well management of the mangrove forest.

1.1 Mangrove origin

According to scientists and the existing literature, the mangrove species originated in the Indo-Malayan region (Tomlinson, 2016).These regions are recorded with more mangrove species than anywhere else in the world, which may account for the theory of the origin of mangrove from this region (Tomlinson, 2016; Gilman et al., 2008; Ellison et al., 1999). These tidal inundated plants have a unique property of floating their propagules and seeds on the water for some time (Ellison et al., 1999). The research has a belief that certain of these early mangrove species spread across the world by ocean current and waves. Its propagation toward India and East Africa may have been carried by westward ocean currents and eastward ocean currents toward the continent of the Americas during the upper Cretaceous period and lower Miocene epoch, between 66 and 23 million years ago (Tomlinson, 2016). Mangroves may have spread during these periods toward the Caribbean Sea through an open seaway, and then ocean currents may have carried these mangrove seeds to the western coast of Africa and as far south as New Zealand. The scientist recognizes this open seaway might be where Panama lies today. These conclusions clarify why the mangroves of West Africa and the Americas have less species diversity, whereas Asia, India, and East Africa contain more mangrove species. There are about 39.3 million acres of mangrove forests in the warm coastlines of tropical oceans all over the world. More than 10.5 million acres or 27% of mangrove forests are found in Southeast Asia (Tomlinson, 2016; Ellison et al., 1999).

1.2 Ecology

Mangrove grows mostly in the saline coastal habitats of the tropic and subtropical latitudes between 25° N and 25° S (Das and Crépin, 2013; Giri et al., 2011; Heenkenda et al., 2014). They are found in the fine depositional sediments often with high organic and nutrient content, and they strive to protect themselves from high-energy wave action. Many species of trees and shrubs can tolerate saline conditions from brackish water, through pure sea water, to water concentrated by evaporation, to over twice the salinity of ocean sea water. The ecological gradients such as elevation, tidal inundation, water salinity, and soil pH are main governing factors in the occurrence of mangrove species (Tomlinson, 2016; Ibrahim-Bathis et al., 2012; Richards and Friess, 2016), and mangrove species are likely to grow within their own niches (Romañach et al., 2018; Makowski and Finkl, 2018). Plants in mangals are diverse, and they can create their habitat (the intertidal zone) by developing physiological adaptations to subdue the complications of anoxia, high salinity, and frequent tidal inundation. About 110 species belong to the mangal have their own solution of coping with the small environment variations within a mangal, exhibiting distinct mangrove tree species zonation in some shorelines (Anneboina and Kavi Kumar, 2017). Therefore, the tolerances of individual species to these physical conditions, including tidal inundation and salinity, and the influence other factors such as predation of plant seedlings by crabs, are partly determined by the mix of species in the mangal. The sea level and its fluctuations are a primary natural factor that affects the mangroves niche over the long term. The air temperature, salinity, ocean currents, storms, shore slope, and soil substrate are other short-term factors that affect the ecology of the mangrove (Tomlinson, 2016). Most mangroves are characterized by muddy soils, but they also can grow on sand, peat, and coral rock. The mangroves flourish far inland and along the upper reaches of coastal estuaries when tidal conditions are optimal for their adaptation. As discussed in the introduction, the dispersal and floating of certain mangrove propagules is another special adaptation that droops from the branches of mature trees. These propagules fall off and eventually take root in the surrounding soil of the parent tree or sometimes are carried to a far distant shoreline through the ocean current. Depending on the species and the prevailing climatic condition, these propagules may float for extended periods of a year and sometimes remain viable. Viviparity and the long-lived propagules allow these mangrove species to disperse over wide areas (Tomlinson, 2016).

1.3 Factors governing the mangrove ecosystem

A fragile complex and dynamic mangrove ecosystem is dependent on both biotic and abiotic factors. The most significant and important factors for natural growth, development, and succession of the mangroves are the intertidal climate condition (Tomlinson, 2016; Thirumalai and Radhakrishnan, 1999; Polidoro et al., 2010). The fluctuation between 20 and 35°C in the atmospheric temperature are favorable for the geographic distribution of the major species of mangroves in the world (Tomlinson, 2016; Field, 1999; Kristensen et al., 2008; Reef et al., 2010). The region experiencing the abundant and evenly distributed rainfall throughout the year with regular climatic conditions are developing the best mangrove forests in the world. Frequent flushing of rainwater washes out the surface salt and leach down the salt particles, which makes the land suitable for the growth of the mangrove (www.niobioinformatics.in/mangroves). The Southwest and Northeast are two monsoon seasons prevailing in the Indian coast. The coastal region of Kerala receives more than 3600 mm monsoon rainfall and decreases the amount further toward the Gujarat coast. The hasty wind flow effects mangrove trees, and several cyclones caused damage to the mangrove ecosystem (Nayak and Bahuguna, 2001; Romañach et al., 2018). The soil structure and salinity are the main agents that regulate the distribution of mangroves over the region. The northern coast of Kerala is represented by typical lateritic soil with decomposing plant material and dark gray soil. The tidal fluctuations have a major role in mangrove habitats and ecosystems. Mangrove growth is associated with the mean sea level (MSL) and low tidal region. In the northern coast of Kerala, including the Kasaragod region, the tidal amplitude is minimal; hence the mangrove is not affected with the high tidal inundation. But the mangrove in this region is affected by huge freshwater discharge from the inland river, particularly floods during southwest monsoon seasons (www.niobioinformatics.in/mangroves). There are 59 species in 41 genera and 29 families in the Indian mangrove diversities.

Thirty-four species belong to 25 genera, and 21 families are distributed along the west coast of India from the coast of Kerala to the coast of Gujrat. The species of mangroves like *Suaeda fruticosa, Sonneratia caseolaris,* and *Urochondrasetulosa* have been reported only from the west coast of India. In the Northern cost of Kerala, three to four species of mangrove are rarely found.

The planktonic and benthic animal communities also play a very significant role in the mangrove ecosystem as they are important as in the terrestrial animals (Tomlinson, 2016; Ellison et al., 1999; Romañach et al., 2018). Mudskippers, mullets, clupeids, serranids, hilsa, milkfish, carangids, seabass, and so on are the several fish species distributed in the west coast mangrove ecosystem. The wildlife of Indian mangrove forests is quite diverse and interesting including the Bengal tiger, but in the Kasaragod mangrove forest there are no wild animals that are commonly seen in the same habitat of the Sundarbans forest. Instead, the mangrove swamps of the west coast including in the Kasaragod coast are favored by a variety of birds, both migratory and resident. Microbial organisms like yeast, bacteria, and fungi play a very important and dominant role in the decomposition of mangrove foliage and regeneration of nutrients and mineralization (www.niobioinformatics.in/mangroves).

1.4 Mangroves: benefits

Mangrove forests minimize the damage prone to life and property in the coastal environment, by serving as a barrier against typhoons, cyclones, hurricanes, and tsunamis, and prevent coastal erosion, and flooding due to high tidal inundation (Sanford, 2009; Dahdouh-Guebas et al., 2005; Das and Crépin, 2013). Mangrove tree species that inhabit lower tidal zones can block high wave action at a measurable height of 30 m and several meters in circumference (Anneboina and Kavi Kumar, 2017). The mangrove trees shield the land from wind, trap sediment in their deep roots, and maintain a shallow slope on the seabed to absorb the energy of tidal surges. Mangrove scrubs are important for the purification of water quality from the huge amounts of organic matter, dissolved nutrients, pesticides, and other pollutants that are dumped by anthropogenic and agricultural activities (Ghajar et al., 2018). They also play an important role to prevent the contamination from excess nitrates and phosphates. These activities play a vital role in protecting seagrasses and coral reefs from eutrophication and siltation. Mangrove forests have a major rule in the global climate mitigation by sink for atmospheric carbon dioxide (Kristensen et al., 2008). They provide important nursery areas and habitats for commercially valuable shrimp, shellfish, and fish species. The healthy wetlands, such as mangroves, seagrasses, and coral reefs, are preserving nearly two-thirds of global fish resources. The mangrove deforestation and decline of its area are continuing; there will be a serious impact on ecosystems and the productivity of fisheries in the world (Tomlinson, 2016; Field, 1999). The mangrove forests proved to be the chief economic aspect for local people as wood for fuel, furniture, and construction; a source for charcoal, tannin, paper, dyes and chemicals, thatch, honey, and incense; and several more mangrove plants are used for traditional medicine. The mangrove scrubs are used to graze and feed livestock too. Dense mangrove forests reduce the devastating impact of high tide wave, tsunamis, and coastal storms by slowing downing the waves' energy. The dense mangroves areas in Pichavaram and Muthupet suffered fewer human casualties and less damage to the property as compared to areas without mangroves when the tsunami struck India's southern state of Tamil Nadu on December 26, 2004.

1.5 Geospatial application in mangrove mapping

Geospatial techniques will yield exclusive research study on mangrove forests, and will monitor drastic changes from accurate mapping on arterial extent, rapid, and cost-effective survey (Ghajar et al., 2018; Ibrahim-Bathis and Ahmed, 2019). It will agglomerate remote sensing, GIS and GPS. The use of remotely sensed data will offer many advantages including monitoring deforestation, aquaculture activity, environmental sensitivity analyses, seasonal growth study, and so on. Advance in the satellite and airborne remote sensing application made a viable tool, especially in the mangroves. GIS will act as the data rice resource inventory and exclusive spatial and temporal mapping purposes (Ibrahim-Bathis et al., 2012; Jia et al., 2014; Nayak and Bahuguna, 2001; Vo et al., 2013; Heenkenda et al., 2014; Kovacs et al., 2005; Pavithra et al., 2019). However, the accuracy of the derived map and information is highly influenced by the skill and ability of the cartographer and the mapping technology followed (Ibrahim-Bathis et al., 2012). The determining and evaluating change detection of a mangrove forest is a key application to monitor deforestation and distraction of these forests over a time schedule. The choice of an appropriate system and technique will depend on the objectives of the study and the size of the budget (Ibrahim-Bathis et al., 2012). Remote sensing images provide the synoptic view of mangroves and it is easy to map and locate the extent and distribution of mangrove trees. Along with the satellite images, Google Earth images are a very good tool for many applications, especially in mangrove mapping (Giri et al., 2011; Lefer et al., 2008; Mohammed-Aslam, 2015; Chen et al., 2017). Because of the inaccessibility to the mangrove ecosystem, the application of the remote sensing techniques is very important in the various aspects of the mangrove community such as mangrove diversities mapping, change detection, aerial extension, selecting sites for aquaculture, mangrove ecosystem study, mangrove management activities, and so on, mainly because the remote sensing images provide valuable information on the mangroves without the physical survey. The present project work mainly carried out regional mangrove mapping of the Kasaragod districts using thermal images in the Google Earth engine (Ibrahim-Bathis et al., 2012). The visual interpretation of the Google Earth images provides accurate timely information on the mangrove extension along the districts coast. The main aim of the present work is to map the aerial extent and assessment of mangroves.

2. Study area

Kasaragod is unique for its linguistic culture and northern-most district of Kerala state in India. Geographically it lies between 11°18′ and 12°48′ N latitudes and between 74°52′ and 75°26′ E longitudes. The name, Kasaragod, is said to be derived from the word Kusirakood, meaning Nuxvomica forests (Kanjirakuttom). The district is also known for its coir and handloom industries. Kasaragod district was organized as a separate district on May 24, 1985 (https://en.wikipedia.org/wiki/Kasaragod_district). The Western Ghats is an almost continuous mountain wall on the eastern side of the district and fringed coastal area with low cliffs in the western side of district, which made an attractive place for tourists (Balakrishnan, 2013). The district is bounded by the Canara district of Karnataka in the northern side and in the south by the Kannur district of Kerala (Fig. 21.1). The district has a rich coastal line of 80 kms from the Ullal Beach of Karnataka State to Payyannur Beach of Kannur districts in the south and north. Physiographically the district can be divided into three district units—the coastal plain land, the midlands or plateaus, and the eastern highland and hills in the Western Ghats regions (Ibrahim-Bathis et al., 2012; Balakrishnan, 2013). The coastal plains with an elevation of less than 10 m occur as a narrow belt of alluvial deposits parallel to the coast. To the east of the coastal belt is the midland region with altitude ranging from 10 to 300 m above MSL. The midland area is characterized by rugged topography formed by small hillocks separated by deep-cut valleys (Thirumalai and Radhakrishnan, 1999). The midland and hill ranges of the district present a rugged and rolling topography with hills and valleys. Along the midlands, the hills are mostly laterite and the valley is covered by valley-fill deposits. There are four major soil types encountered in the district (https://kasargod.gov.in/history/). They are lateritic soil, brown hydromorphic soil, alluvial soil, and forest loam. Lateritic soil is the most predominant soil type of the district; it occurs in the midland and hilly areas and is derived from laterites. Brown hydromorphic soil is confined to the valleys between undulating topography in the midlands and in the low-lying areas of the coastal strip. They have been formed as a result of transportation and sedimentation of materials from adjoining hill slopes. The alluvial soil is seen in the western coastal tract of the district. The coastal plain is characterized by secondary soils that are sandy and sterile with poor water holding capacity. The width of the zone increases toward the southern part of the district. Forest loamy soils are found in the eastern hilly areas of the district and are characterized by a surface layer rich in organic matter (Ibrahim-Bathis et al., 2012; Mohammed-Aslam, 2015; Balakrishnan, 2013).

2.1 Mangroves in Kasaragod

The patchy and thin mangrove formations are seen in almost the entire coast of the Kasaragod district (Ibrahim-Bathis et al., 2012; Ghajar et al., 2018; Mohammed-Aslam, 2015). But the large formations are seen only in several localities such as Kumbla backwaters, MogralPuthur River banks and Pallam Lake (Ibrahim-Bathis et al., 2012).

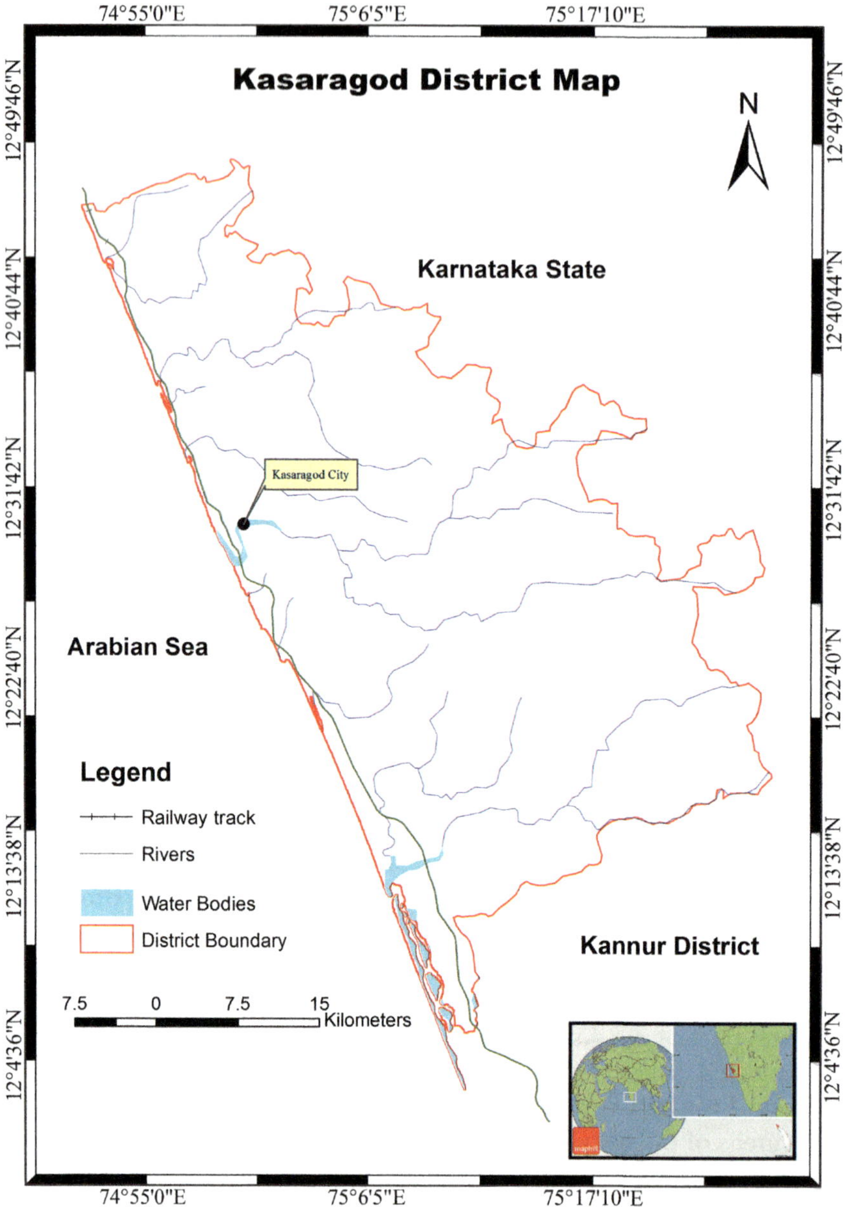

FIGURE 21.1 Location map of the Kasaragod district.

Uduma River, Chittari River, Azhitala, Orikkadavu, Vadekkekal, Kokkal, Edayilekkadu, and Madakkal are other mangrove growing places along the coastal line of the district. Of the entire district, 95% of total mangroves cover from Kumbla backwater, Mogralputhur River banks, and Pallam Lake or Chandragiri backwater (Fig. 21.2). From the available literature, about 0.45 km^2 of mangrove areas are estimated in the Kasaragod

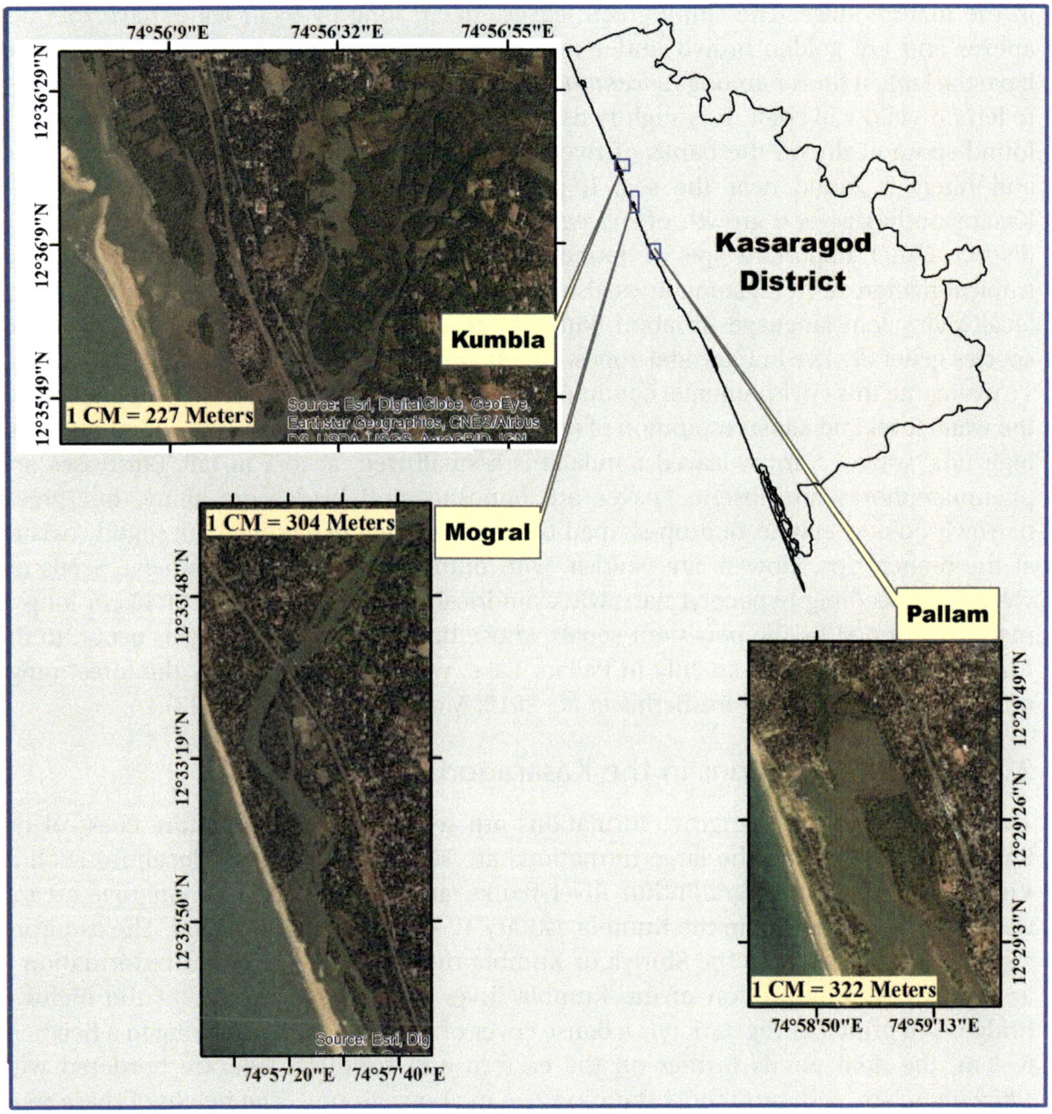

FIGURE 21.2 Location map of major mangrove sites along the Kasaragod coast. *Ibrahim-Bathis, K., Ahmed, S.A., Bhatt, S., Abdul-Karem, S.N., Nischitha, V., Kiran-Raj, S., 2012. Remote sensing and GIS based mapping and assessment of mangroves of Kasaragod district in Kerala, India. Front. Earth Sci. Res.*

(Ibrahim-Bathis et al., 2012; Mohammed-Aslam, 2015). The plant was represented by seven species and dominated by *Avicennia officinalis*, found periodically on the banks of rivers in Mogral, Kumbla, Chittari, and Pallam Lake, and rarely found near the seashore. *Avicennia officinalis* is a species of Indian mangrove named Uppatti in the Malayalam language. This tree forms a low, dense, bushy crown forming a columnar shape 15–30 m in the mature stage. The shiny green leaves, 10 cm long by 5 cm wide, have rounded apexes and are golden-brown under the leaves and grow in opposites. These species have the largest flower among *Avicennia* with a diameter of 6–10 mm. It is orange-yellow to lemon-yellow in color. It is slightly fissured and does not flake. *Avicennia officinalis* is found sporadically on the banks of rivers (Mogral, Kumbla, Chittari, and Pallam Lake) and rarely is found near the sea. It prefers clay soil and usually is found inland. Kasaragod has a good growth of *Avicennia officinalis*, which is a special feature of the district. Other important species found in Kasaragod is *Rhizophora* from a genus of tropical mangrove trees, sometimes also called true mangroves, or Cheru-kantalis in the local Malayalam language (Ibrahim-Bathis et al., 2012; Ghajar et al., 2018). *Rhizophora* species generally live in intertidal zones, which are frequently inundated by ocean waves. To overcome this environmental condition, they have stilt-roots that raise the plants above the water level and allow respiration of oxygen while their lower roots are submerged with high tidal waves. Narrow-leaved Kandalia is a small tree up to 7 m tall. Buttresses and pneumatophores are absent. Leaves are opposite, and blades are shiny, mid-green, narrowly oblong, elliptic, or drop-shaped-oblong. Stipules are flattened but slightly twisted at the branch tips. Flowers are whitish, with numerous protruding stamens. Seeds are viviparous, seedling hypocotyl narrowly cylindrical or club-shaped, up to 40 cm long at maturity, capped by the persistent sepals whose tips bend, where mangals occur, to the fruit stalk. They are seen mainly in Pallam Lake, which was planted by the forest range office in Kasaragod (Ibrahim-Bathis et al., 2012; Mohammed-Aslam, 2015).

2.2 Mangrove locations in the Kasaragod district

The patchy and thin mangrove formations are seen in almost the entire coast of the Kasaragod district, but the large formations are seen only in several localities such as Kumbla backwaters, MogralPuthur River banks, and Pallam Lake. A mangrove area of about 20 ha was located in the Kumbla estuary 10 km north of Kasaragod. The estuarine system is contributed by the Shiriya or Kumbla river basin. The mangrove formation is around the middle portion of the Kumbla River on the eastern side of the highway bridge. The principal vegetation is a dense cover of *Avicennia* trees growing to a height of 6–8 m, the tidal ponds further on the eastern and southern side are bordered with *Rhizophora* sp., with prop roots standing 2–4 m above ground. The height of these trees get progressively reduced where when joined by *Aegiceras corniculatum* and *Avicennia* shrubs. The mudflats get exposed during low tide, exposing the presence of oyster beds, gastropod shells, and an abundant population of fiddler crabs (Ghajar et al., 2018). There are a number of estuarine fishes in the creeks. MogralPuthur River is also one of the good mangrove formations in the district, located 7 km north of the Kasaragod. The mangrove formation is along the Mogral River on either side of the highway bridge. The principal vegetation is a dense cover of *Avicennia* trees growing to a height of 6–8 m, the tidal ponds further on the eastern and southern side are bordered with *Rhizophora* sp., with prop roots standing 2–4 m above ground. The eastern part of the river is characterized by an abundant population of fiddler crabs. There are a number of estuarine fishes in the creeks. Sand mining alone caused large-scale degradations of the natural mangrove species (Mohammed-Aslam, 2015). The Pallam Lake is situated near the Kasaragod coast on the Chandragiri backwaters. The dense formation of both *Avicennia* trees and *Rhizophora* sp. are seen in the lakes. This lake showed the best mangrove site for plantation of the *Rhizophora* sp. The Planted *Rhizophora* sp. trees have good growth in the lake mainly because of the lack of the anthropogenic activities such as sand mining and municipal waste disposal. The minor mangrove formations are distributed along the coastal line in the district from the extreme north to south. But the group of the mangrove is seen only in several parts of the district especially in the bank of the river, saline estuaries, and backwaters. Uduma River, Chittari River, Azhitala, Orikkadavu, Vadekkekal, Kokkal, Edayilekkadu, and Madakkal are other mangrove growing places along the coastal line of the district. Of the entire district, 95% of total mangroves cover from Kumbla backwater, Mogralputhur River banks, and Pallam Lake or Chandragiri backwater (Ibrahim-Bathis et al., 2012).

2.3 Threats of mangrove in Kasaragod locality

Mangroves are salt-tolerant trees that grow in coastal areas of the tropical and subtropical regions where rivers empty into the ocean. These trees have been able to successfully occupy the coastal environment where they have little or no competition from other species of plants. In order to do this, mangrove trees have had to cope with a number of problems including soft oxygen-poor soil, periodic flooding of their root zones, and a highly saline environment. Mangroves mainly occur in brackish water near the coast where they can pillar their root. But in case of the Kasaragod district, the majority of the mangroves are seen only on the backwaters and Kayals (Chandragiri backwater, Kumbla backwaters, Kavvai Lake) and along the banks of minor rivers (Mogral River, Uduma River, Shiriya River, and Chittari River). The mangroves are rare or minimal on the coast and along the major river. This is mainly because of the steep slope topography of the coastal beach and heavy flooding of major rivers in the monsoon seasons (Ghajar et al., 2018). The high flooding of the river in the monsoon season causes high sand erosion along the banks of the river. This is the major reason for the absence of mangroves along the major river banks. Such climate doesn't possess the mangrove seeds to pillar their root in the sediments. In the modern developing world, man acts as the destructing agent of the all-natural resources. From development in all fields, human use natural resources in improper ways. A natural scenario such as forest and trees are the main attraction toward development. Mangrove degradation mainly occurred in the developing coastal cities of the world. For the construction of the major infrastructures, industrial development and so on causes the degradation of the world of mangroves. But in the Kasaragod district, because of the lack of large-scale industries and lack of modern development, natural mangroves are not degraded too much. But compared to the earlier strength of natural mangroves, they are becoming degraded in the last few decades. The main reason for this is the uncontrolled and illegal sand mining in the river beds and digging of municipal waste along with the mangrove habitat (Ibrahim-Bathis et al., 2012). MogralPuthur and Shiriya Chandragiri are the main places where sand mining can be seen.

3. Methodology

The coastal evergreen mangrove forest has a very good appearance in the aerial images and their unique association along the estuaries and near the river delta provides a key for the interpretation. The visual interpretation of the Google Earth images provides accurate and timely information on the mangrove extension along the coast (Chen et al., 2017; Giri et al., 2011; Kovacs et al., 2005).The aerial extent and mapping of patchy coastal mangroves of Kasaragod is carried out using publicly available Google Earth images for the years 2004, 2009, and 2014. Field GPS points of true mangrove coordinates collected from the study area are uploaded to Google Earth to observe the mangroves feature on the image. According to the GPS point and with the visual interpretation keys, the entire mangrove feature (both natural and planted mangroves) was digitized from Google Earth. Some of the unique features are very important for the visual processing from the satellite images such as tone or the relative brightness or color of elements on a photograph, texture or the arrangement and frequency of tonal variation in the image pixel, associations or the relationship between other recognizable objects of element, for example, the mangrove always lies in connection with water and land in the coastal or estuary region. The digitized mangrove features are then converted to GIS shapefile format for the map composition and statistical area estimation in GIS software. The main aim of the present work is to map the aerial extent of mangroves as no perfect area estimation exists for the Kasaragod mangrove forest. Their aerial distribution over time is very important to monitor the change in the decade or any other time period because the mangrove is one of most rapidly deforesting entities in the world, especially in tropical and subtropical regions. The large-scale maps are prepared for localities such as Kumbala backwaters, MogralPuthur river bank, and Pallam Lake for the years 2004, 2009, and 2014, respectively, from the GIS environment.

4. Result and discussion

The Kerala coast has been observed to have very limited mangrove cover compared to other coastal states along the west and east coasts of India. However, species like *Hymenocallis littoralis* and *Excoecaria indica* have been reported only in Kerala along the west coast. The species like *A. officinalis* are found with enriched growth and are healthy on the coast of the Kasaragod. In general, the mangroves along the Kasaragod coast could be described as very sparse and thin. Despite intense anthropogenic pressures, particularly from urbanization and agriculture activities, several localities such as

Kumbla backwaters, MogralPuthur river banks, Pallam Lake, and Chittari River have sustained small to large patches of mangroves. The different temporal images such as those from 2004, 2009, and 2014 are used to digitize the mangrove features on Google Earth images. The regional distributions of mangroves are mapped for Kumbla backwater (Fig. 21.3), MogralPuthur river banks (Fig. 21.4), and Pallam Lake (Fig. 21.5). The result obtained from the Google Earth images showed the mangroves with a successive increment in spatial extensions. The area obtained from the year 2014 is more than the area in 2009 and 2004 (Table 21.1). This is mainly because of the lack of anthropogenic threat, lack of industrial treats, and successive achievement of the plantation of mangrove species by the Forest Range Office Kasaragod. Pallam Lake and Kumbla backwaters are characterized by the planted mangroves species of *Rhizophora kandalia*. These species are also seen in the MogralPuthur, Chittari, and several localities in the district. Though the aerial extent is increased in several localities, the natural mangrove bushes are still degrading from the last decade due to the cause of sand mining in the river and seashore and disposal of municipal waste in the mangrove habitat. In some areas, aerial extents are decreased from 2009 to 2014 because of the submergence of the mangrove bushes during monsoon flood. The regional map of Kumbla backwaters, Mogral River, and Pallam Lake are prepared for the years 2004, 2009, and 2014 showing the aerial extension. There is no massive industrial area in the Kasaragod coast, but the threat from the urbanization and increased population consequently put pressure on the mangrove forest. The earlier mangrove region is converted into the settlement and other built-up lands in order to sustain the demand of population in the coastal area of the district. Sand mining is one of the chief activities

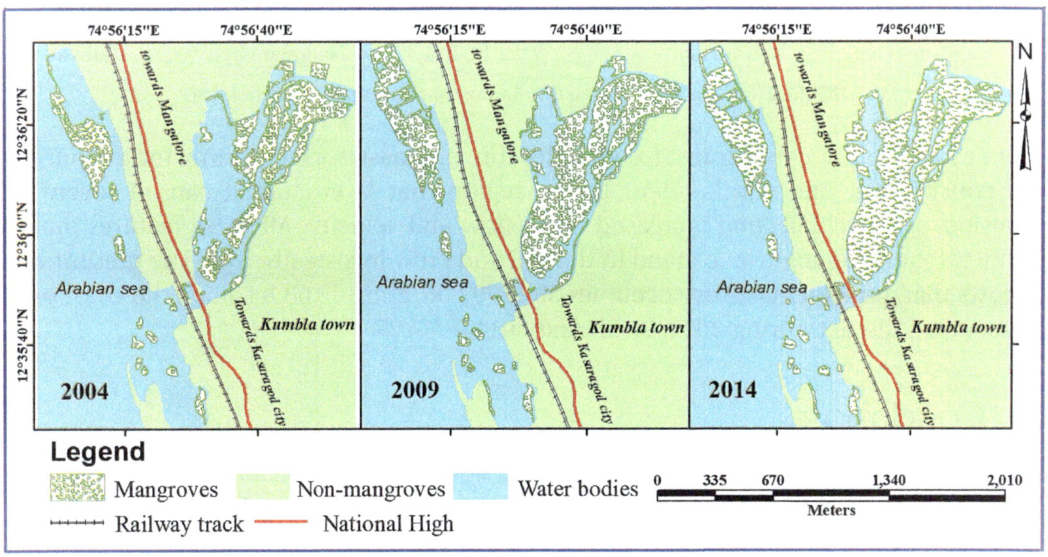

FIGURE 21.3 Map showing the distribution of mangroves in Kumbla.

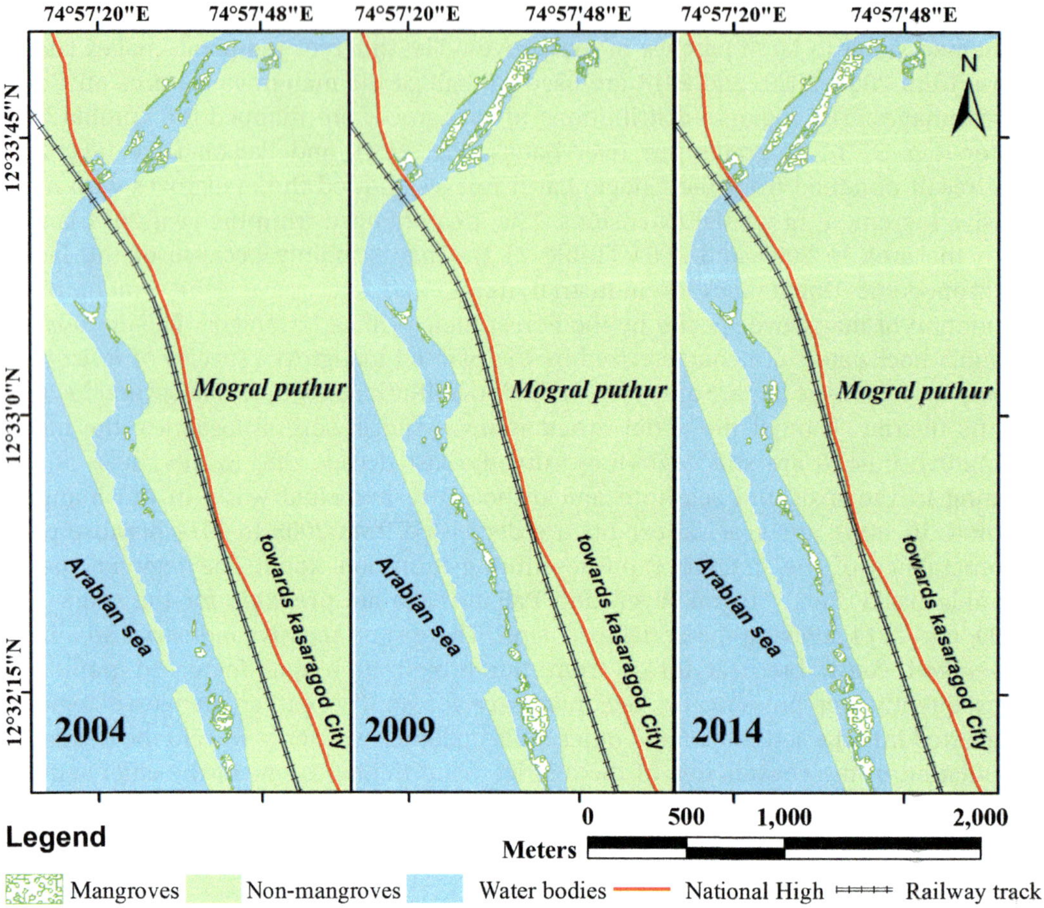

FIGURE 21.4 Map showing the distribution of mangroves in Mogral River.

that cause the mangrove forest to be under threat. Massive sand mining for the purpose of construction and development in the region that is evacuated caused severe soil erosion, sea-level rise, and increased water flow and velocity. All these features make it difficult for the mangrove to stand in the soil and grow in a steady way. The present work shows that the total mangrove occupies 59.58, 97.06, and 102.03 ha along the coast of the Kasaragod district during 2004, 2009, and 2014.

5. Conclusion

Mangrove forests are tropical to subtropical specialized ecosystems established along the intertidal zone of sea coasts, estuarine, and river deltas. Today less than 50% remain, and of this remaining forest, over 50% is degraded and not in good form. A rapid process of industrialization and urbanization in the tropical to subtropical regions of the world

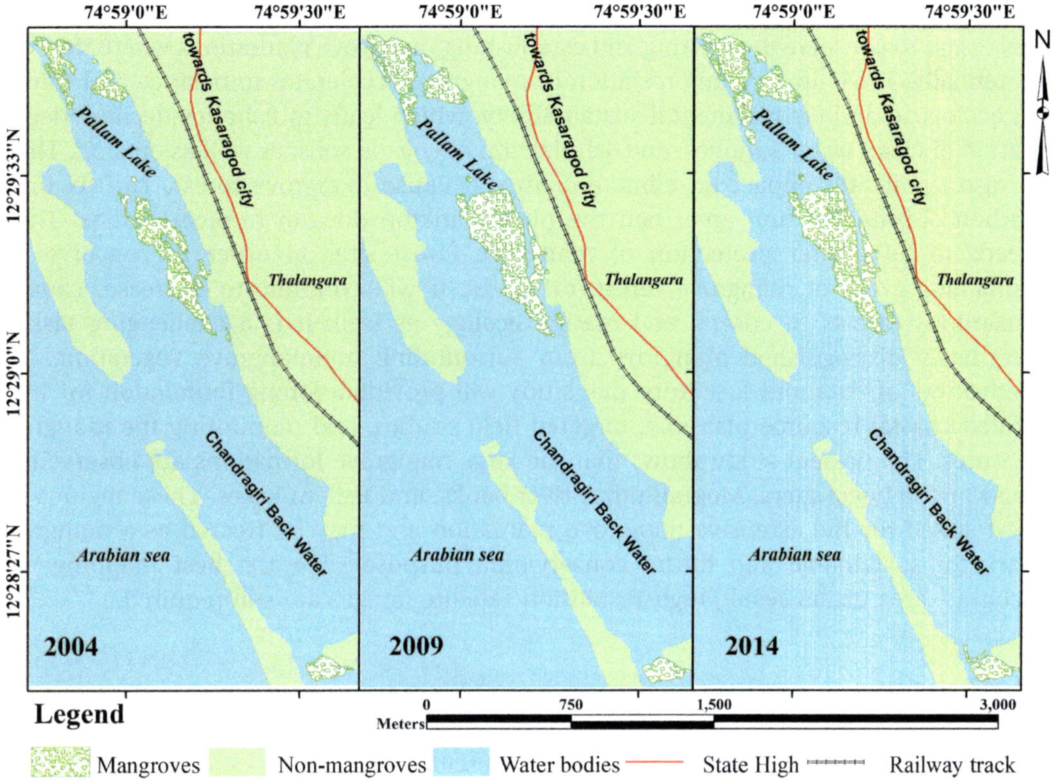

FIGURE 21.5 Map showing the distribution of mangroves in Pallam Lake and Chandragiri backwater.

Table 21.1 Statistical area summary of mangrove localities in the Kasaragod district.

		Area in hectare			Area in percentage		
Sl No	Places	2004[a]	2009[a]	2014	2004[a]	2009[a]	2014
1	Kumbla and Shiriya backwaters	34.87	53.9	55.7	58.53	55.53	54.59
2	MogralPuthur River bank	7.41	14.11	15.12	12.44	14.54	14.82
3	Chandragiri backwater and Pallam Lake	14.1	23.48	24.57	23.67	24.19	24.08
4	Part of Uduma River	0.17	0.3	0.3	0.29	0.31	0.29
5	Chittari River	0.81	0.91	0.99	1.36	0.94	0.97
6	Azhitala	0.1	0.1	0.15	0.17	0.10	0.15
7	Near Orikkadavu Bridge	0.02	0.1	1.2	0.03	0.10	1.18
8	Vadakkekad	0.1	0.15	0.15	0.17	0.15	0.15
9	Kokkal	1.1	1.48	1.52	1.85	1.52	1.49
10	Edayilekkadu	0.9	1.93	1.73	1.51	1.99	1.70
11	Madakkal	No data	0.6	0.6	No Data	0.62	0.59
Total		**59.58**	**97.06**	**102.03**	**100**	**100**	**100**

[a]Source: Ibrahim-Bathis et al. (2012).

have caused a major threat for the degradation of mangrove forests. Therefore, it is necessary to conserve the existing and establishing mangrove plantations where they can potentially thrive. In fact, deforestation of mangrove accelerates imbalance and threats in ecological and environmental sustainability, which leads to fisheries decline, degradation of clean water supplies, and salinization of coastal soils, as well as erosion. There is also significant impact on climate change because mangrove forests can fix more carbon dioxide per unit area than the phytoplankton does in tropical oceans. There needs to be greater protection of mangrove forest sites in order to conserve the remaining areas of mangrove forest, otherwise it will continue to decrease, causing hazardous threats to coastal and marine ecology as well. It is a challenging task to accurately distinguished mangrove from surrounding nonmangrove vegetation. The derivatives of data and fact from this study will provide a strong foundation for integrated coastal resource planning, targeted field studies, and monitoring the mangrove features. The present study shows that the large mangrove formations are observed on the Kumbla backwaters, MogralPuthur river banks, and Pallam Lakes. These regions are best suited for the extensive mangrove plantation and may be treated as a mangrove park for sustainable and future conservation purposes. For the best mapping and accurate area extent result, high-resolution satellite images are still required.

References

Anneboina, L.R., Kavi Kumar, K.S., 2017. Economic analysis of mangrove and marine fishery linkages in India. Ecosyst. Serv. https://doi.org/10.1016/j.ecoser.2017.02.004.

Balakrishnan, K., 2013. Ground Water Information Booklet of Kasargod District, Kerala State. Government of India Ministry of Water Resources Central Ground Water Board.

Chen, B., Xiao, X., Li, X., Pan, L., Doughty, R., Ma, J., et al., 2017. A mangrove forest map of China in 2015: analysis of time series Landsat 7/8 and Sentinel-1A imagery in Google Earth Engine cloud computing platform. ISPRS J. Photogramm. Remote Sens. 131, 104–120. https://doi.org/10.1016/j.isprsjprs.2017.07.011.

Dahdouh-Guebas, F., Jayatissa, L.P., Di Nitto, D., Bosire, J.O., Seen, D.L., Koedam, N., 2005. How effective were mangroves as a defence against the recent tsunami? Curr. Biol. https://doi.org/10.1016/j.cub.2005.06.008.

Das, S., Crépin, A.S., 2013. Mangroves can provide protection against wind damage during storms. Estuar. Coast Shelf Sci. https://doi.org/10.1016/j.ecss.2013.09.021.

Duke, N.C., Meynecke, J.O., Dittmann, S., Ellison, A.M., AngerK, Berger, U., Cannicci, S., Diele, K., Ewel, K.C., Field, C.D., Koedam, N., Lee, S.Y., Marchand, C., Nordhaus, I., Dahdouh-Guebas, F., 2007. A world without mangroves? Science 317 (5834), 41–42. https://doi.org/10.1126/science.317.5834.41b.

Ellison, A.M., Farnsworth, E.J., Merkt, R.E., 1999. Origins of mangrove ecosystems and the mangrove biodiversity anomaly. Glob. Ecol. Biogeogr. https://doi.org/10.1046/j.1466-822X.1999.00126.x.

Field, C.D., 1999. Rehabilitation of mangrove ecosystems: an overview. Mar. Pollut. Bull. https://doi.org/10.1016/S0025-326X(99)00106-X.

Ghajar, A.J., Korea, M., Muraleedharan, K.P., Swarupanadan, K., Anitha, V., Mini Mohandas Lekshmy, S. T.R., et al., 2018. The conservation of mangroves in Kerala: economic and ecological linkages. Kerala For. Res. Institu. https://doi.org/10.18520/cs/v114/i05/976-981.

Gilman, E.L., Ellison, J., Duke, N.C., Field, C., 2008. Threats to mangroves from climate change and adaptation options: a review. Aquat. Bot. https://doi.org/10.1016/j.aquabot.2007.12.009.

Giri, C., Ochieng, E., Tieszen, L.L., Zhu, Z., Singh, A., Loveland, T., et al., 2011. Status and distribution of mangrove forests of the world using earth observation satellite data. Glob. Ecol. Biogeogr. https://doi.org/10.1111/j.1466-8238.2010.00584.x.

Giri, C., Long, J., Abbas, S., Murali, R.M., Qamer, F.M., Pengra, B., et al., 2015. Distribution and dynamics of mangrove forests of South Asia. J. Environ. Manag. 148, 101–111. https://doi.org/10.1016/j.jenvman.2014.01.020.

Heenkenda, M.K., Joyce, K.E., Maier, S.W., Bartolo, R., 2014. Mangrove species identification: comparing WorldView-2 with aerial photographs. Rem. Sens. https://doi.org/10.3390/rs6076064.

Ibrahim-Bathis, K., Ahmed, S.A., Bhatt, S., Abdul-Karem, S.N., Nischitha, V., Kiran-Raj, S., 2012. Remote sensing and GIS based mapping and assessment of mangroves of Kasaragod district in Kerala, India. Front. Earth Sci. Res. 2012, 157–163.

Ibrahim-Bathis, K., Ahmed, S.A., 2019. Geospatial approach for evaluating LULC pattern in the Doddahalla Watershed, Chitradurga district, India. Remote Sens. Earth Syst. Sci. https://doi.org/10.1007/s41976-019-00017-7.

Jia, M., Zhang, Y., Wang, Z., Song, K., Ren, C., 2014. Mapping the distribution of mangrove species in the Core Zone of Mai Po Marshes Nature Reserve, Hong Kong, using hyperspectral data and high-resolution data. Int. J. Appl. Earth Obs. Geoinf. https://doi.org/10.1016/j.jag.2014.06.006.

Kovacs, J.M., Wang, J., Flores-Verdugo, F., 2005. Mapping mangrove leaf area index at the species level using IKONOS and LAI-2000 sensors for the Agua Brava Lagoon, Mexican Pacific. Estuar. Coast Shelf Sci. https://doi.org/10.1016/j.ecss.2004.09.027.

Kristensen, E., Bouillon, S., Dittmar, T., Marchand, C., 2008. Organic carbon dynamics in mangrove ecosystems: a review. Aquat. Bot. https://doi.org/10.1016/j.aquabot.2007.12.005.

Lefer, T.B., Anderson, M.R., Fornari, A., Lambert, A., Fletcher, J., Baquero, M., 2008. Using google earth as an innovative tool for community mapping. Publ. Health Rep. https://doi.org/10.1177/003335490812300408.

Makowski, C., Finkl, C.W., 2018. Erratum to: threats to mangrove forests. In: Makowski, C., Finkl, C. (Eds.), Threats to Mangrove Forests, Coastal Research Library, vol. 25. Springer, Cham. https://doi.org/10.1007/978-3-319-73016-5_32.

Mohammed-Aslam, M.A., 2015. Satellite Based Spectral Signatures of Coastal Wetlands and Mangrove Vegetation of Kasaragod. Biodivers. Conserv. - Challenges Futur., India. https://doi.org/10.2174/9781681080215115010024.

Nayak, S., Bahuguna, A., 2001. Application of remote sensing data to monitor mangroves and other coastal vegetation of India. Indian J. Mar. Sci. 30 (4), 195–213.

Pavithra, B., Kalaivani, K., Ulagapriya, K., 2019. Remote sensing techniques for mangrove mapping. Int. J. Eng. Adv. Technol. 8 (3S), 2249–8958.

Polidoro, B.A., Carpenter, K.E., Collins, L., Duke, N.C., Ellison, A.M., Ellison, J.C., et al., 2010. The loss of species: mangrove extinction risk and geographic areas of global concern. PloS One. https://doi.org/10.1371/journal.pone.0010095.

Reef, R., Feller, I.C., Lovelock, C.E., 2010. Nutrition of mangroves. Tree Physiol. https://doi.org/10.1093/treephys/tpq048.

Richards, D.R., Friess, D.A., 2016. Rates and drivers of mangrove deforestation in Southeast Asia, 2000-2012. Proc. Natl. Acad. Sci. U. S. A. https://doi.org/10.1073/pnas.1510272113.

Romañach, S.S., DeAngelis, D.L., Koh, H.L., Li, Y., Teh, S.Y., Raja Barizan, R.S., et al., 2018. Conservation and restoration of mangroves: global status, perspectives, and prognosis. Ocean Coast Manag. https://doi.org/10.1016/j.ocecoaman.2018.01.009.

Sanford, M.P., 2009. Valuating mangrove ecosystems as coastal protection in post-tsunami south Asia. Nat. Area J. https://doi.org/10.3375/043.029.0111.

Thirumalai, G., Radhakrishnan, C., 1999. Aquatic Hemiptera (INSECTA) of Kasaragod District, Kerala State. Rec ZoolSurv INDIA.

Tomlinson, P., 2016. The Botany of Mangroves. Cambridge University Press, Cambridge. https://doi.org/10.1017/CBO9781139946575.

Vo, Q.T., Oppelt, N., Leinenkugel, P., Kuenzer, C., 2013. Remote sensing in mapping mangrove ecosystems - an object-based approach. Rem. Sens. https://doi.org/10.3390/rs5010183.

Ward, R.D., Friess, D.A., Day, R.H., Mackenzie, R.A., 2016. Impacts of climate change on mangrove ecosystems: a region by region overview. Ecosys. Health Sustain. https://doi.org/10.1002/ehs2.1211.

Index

Note: 'Page numbers followed by "f" indicate figures and "t" indicate tables'.

C

T

U

9780128196045